11 GENERAL MATHEMATICS

PNG UPPER SECONDARY

Christine McRae
Esther Gawac Ehava
Mac Dandava
John Gesa Junior
Phyl Haydock
John Watson
Rory Barrett

OXFORD

OXFORD
UNIVERSITY PRESS

Oxford University Press is a department of the University of Oxford. It furthers the University's objective of excellence in research, scholarship, and education by publishing worldwide. Oxford is a registered trademark of Oxford University Press in the UK and in certain other countries.

Published in Australia by
Oxford University Press
Level 8, 737 Bourke Street, Docklands, Victoria 3008, Australia

First published 2012
Reprinted 2012, 2013, 2014 (twice), 2017 (twice), 2018, 2019, 2020, 2021, 2022, 2023, 2024, 2026

This book was originally published by ESA Publications, Auckland, New Zealand. This edition, specially adapted for the Grade 11 syllabus in Papua New Guinea, is published by arrangement with ESA Publications. Authors of the original work were Phyl Haydock, John Watson and Rory Barrett and this adaptation has been produced by Christine McRae with the assistance of Esther Gawac Ehava, Mac Dandava and John Gesa Junior.

ISBN 978 0 19 557869 0

Typeset by diacriTech
Printed in China by Golden Cup Printing Co. Ltd

Oxford University Press Australia & New Zealand is committed to sourcing paper responsibly.

Contents

Introduction

This book has been published to provide the information required by students in order to successfully complete the Upper Secondary course in General Mathematics at Grade 11 in Papua New Guinea.

The book is written in a manner that best develops an overall understanding of the mathematical concepts and processes set out in the PNG Grade 11 Syllabus. The order of Units in this book follows the order of Units presented in the General Mathematics syllabus for Grade 11:

Unit 11.1 Number and Application

Unit 11.2 Managing Money 1

Unit 11.3 Statistics

Unit 11.4 Geometry

Unit 11.5 Trigonometry

Within each Unit there are a number of Topics which follow the main sub-headings and bullet points set out within each Unit in the syllabus. The aim is to provide structure and content in a concise and compact format for students to use as an effective resource to support the classroom experience. It is acknowledged that Mathematics is more interesting and meaningful when students are exposed to a variety of resources and materials and we encourage students and teachers not to rely on this book as a sole source of information.

If you have any suggestions about how this book might be improved in future editions, please make contact with Oxford University Press:

Fax: 00 61 3 99349 100

Customer Service Email: exportsales.au@oup.com

We wish you every success in your studies.

The authors

Acknowledgments

There are many people to be thanked for their help and assistance in enabling the publication of this series to happen. First, we acknowledge the cooperation and generosity of Mark Sayes at ESA Publications in New Zealand, who responded with interest and support when the proposal to adapt his Study Guide series was put to him.

The main author of this book is Christine McRae, with input from Esther Gawac Ehava, Mac Dandava and John Gesa Junior, but with the approval of ESA Publications, she drew on material from *NCEA Level 1 Mathematics* by Phyl Haydock and John Watson, and *NCEA Level 2 Mathematics* by Rory Barrett. The objective has been to provide Grade 11 students in PNG with a book that they can use as a compact summary of content and skills related to the PNG Grade 11 General Mathematics syllabus.

We would also like to acknowledge many other individuals who have been happy to advise and assist in different ways: Ms Vagi Hanua Bino, Ms Betty Pulpulis, Ms Joy Sahumlal, Mr Greg Kapanombo, Mrs Anne Sangi and Mr Safak Deliismail.

Unit 11.1 Number and Application

Topic 1: Basic numeracy—the real number system

This Unit focuses on the mathematics we use in our daily lives. The Topic covers:

- The development of the real number system.
- Different types of numbers.
- Properties of whole numbers: closure, associative, commutative and distribututative laws.

Real numbers

The numbers and their symbols that we use in today's society have developed over many centuries. The original use of numbers in many cultures was to count objects, and, as communities became farmers and builders, the system and the way of writing numbers became more sophisticated. The decimal system of numbers that we use today was thought to originate with the Hindus.

Types of numbers

- **Natural numbers (Counting numbers):** The set of numbers 1, 2, 3, 4, 5, … is called the set of natural (or counting) numbers.
- **Whole numbers:** At some point in the development of numbers it was realised that there was a need for a number that represented 'none' and so the number zero was added to the natural numbers to form the set of whole numbers
- **Integers:** Later, the notion of 'owing' led to the development of negative numbers and so the set of integers was created : … –4, –3, –2, –1, 0, 1, 2, 3, 4, 5, …
- **Fractions and decimals:** The notion of 'part of a whole' led to the development of fractions and later their decimal representation.

Rational numbers

For many centuries whole numbers and fractions were thought to be the only types of numbers that we would ever have to deal with. They were collectively called the set of **rational numbers**.

Rational numbers:

- Have the special property that they can all be written as a **ratio**, ie as one integer divided by another.
- Can be positive or negative.
- Cannot have the number zero as the number in the denominator, eg $\frac{8}{0}$; these numbers are **undefined**.
- Can have the number zero in the numerator, eg $\frac{0}{8}$, as this number is equal to zero.
- In set notation, are described as $\frac{m}{n}, n \neq 0$ where $m, n \in J$.
- Were called **ratio**nal numbers because the first five letters spell the word '**ratio**'.

Irrational numbers

It became clear about 2 500 years ago that there existed another type of number that could not be written as a ratio.

This was in the time of the Greek philosopher and mathematician Pythagoras (580–500 BC) who belonged to a powerful brotherhood whose motto was 'All is number'. They believed that all things in the universe could be explained in terms of rational numbers. However, using a right-angled triangle and Pythagoras' rule they discovered and **proved** that there existed another type of number that could not be written as a ratio. These were numbers such as $\sqrt{2}$ $\sqrt{7.69}$, and π which appear as **infinite, non-recurring decimals**. Today these numbers are called **irrational numbers**. Pythagoras's brotherhood kept the existence of these irrational numbers secret for about 400 years as it ruined their 'All is number' theory.

Today, the set of **rational numbers** together with the set of **irrational numbers** make up the set of **real numbers**.

The set of real numbers is displayed on the following diagram. The capital letter next to the number headings represents commonly used letters for that set of numbers.

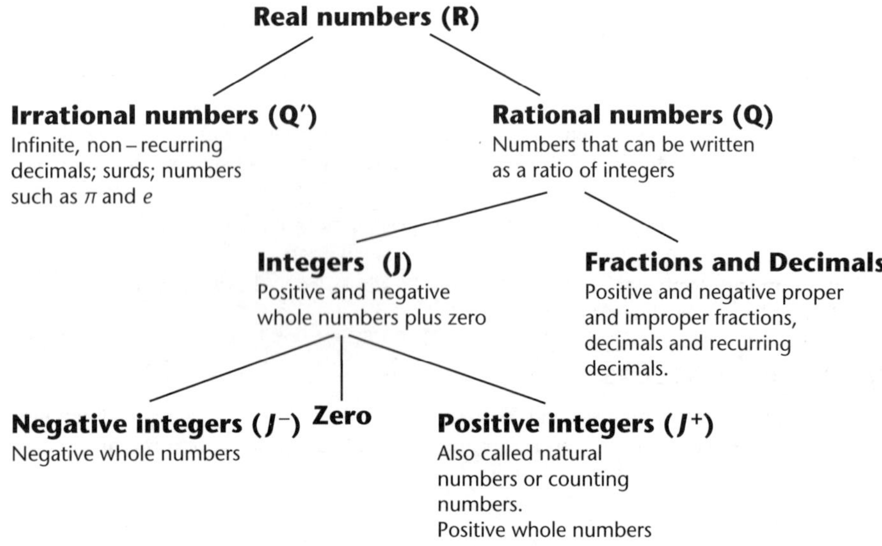

Showing that numbers are rational by changing them into a ratio of integers

i. **Whole numbers** and **integers** can be expressed as a ratio by using the whole number in the numerator and 1 in the denominator.

The number 3 can be written as $\frac{3}{1}$ which is a ratio of the whole numbers 3 and 1.

The integer –7 can be written as the ratio $\frac{-7}{1}$.

ii. **Proper fractions** and **improper fractions** are already given as a ratio; it is usual to give these in their simplest form.

The proper fraction $\frac{5}{12}$ is the ratio of 5 and 12.

The improper $\frac{24}{9}$ fraction is a ratio but should be written as $\frac{8}{3}$.

iii. Mixed numbers need to be changed to improper fractions.

For example: $5\frac{3}{4}$ can be written as $\frac{23}{4}$; the numerator of the improper fraction is found by multiplying the whole part (5) by the denominator (4) and adding the numerator (3) to give 23.

iv. Decimals: In the denominator, place a 1 below the decimal point and add as many zeros as needed to cover the decimal parts. Remove the decimal point from the numerator.

For example:
a. 0.83 can be written as $\frac{0\ 83}{100} = \frac{83}{100}$.
b. 25.6679 is written as $\frac{25\ 6679}{10000} = \frac{256\ 679}{10\ 000}$.
c. 0.0013 is written as $\frac{0\ 0013}{10000} = \frac{13}{10\ 000}$.

v. Recurring decimals can be written in several ways:

$2.\dot{6} = 2.666666...$ the 6 is the only repeated number.
$4.\underline{5027} = 4\ 502750275027...$ a dot above the beginning and the end of the repeated digits.
$0.5\overline{237} = 0.5237237237......$; a bar above the digits that are repeated.

Example A

Q. Express each of the following recurring decimals as a ratio of integers:

1. $0.\dot{6}$

2. $2.\overline{105}$

A. **1.** $0.\dot{6} = 0.666666...$

Let $x = 0.666666...$ (1)

so $10x = 6.6666...$ (2)

Subtract (1) from (2)

$10x - x = 6.6666... - 0.6666...$

$9x = 6$: Divide both sides by 9

$x = \frac{6}{9} = \frac{2}{3}$

So $0.\dot{6} = \frac{2}{3}$

2. $2.\overline{105} = 2.105105105......$

Let $x = 2.105105105...$ (1)

$1\ 000x = 2\ 105.105105...$ (2)

Subtract (1) from (2)

$1\ 000x - x = 2\ 105.105105... - 2.105105...$

$999\ x = 2103$: divide both sides by 999

$x = \frac{2103}{999}$

So $2.105 = \frac{2103}{999} = \frac{701}{333}$ (check this on your calculator by dividing 701 by 333)

Properties of rational numbers

The properties of whole numbers called axioms specifically apply to addition and multiplication only. An axiom does not need to be proved as it is assumed to be true.

Axiom 1: The closure law of addition

The closure law of addition states that when adding two whole numbers the sum is another whole number. It can be shown algebraically as, $a + b = c$ where a, b and c are all whole numbers. Example: 2 + 3 = 5

Axiom 2: Commutative law of addition

The law states that the whole numbers can be added in any order but their sum will be the same. That is shown algebraically as $a + b = b + a$. Example: $4 + 5 = 5 + 4 = 9$

Axiom 3: Associative law of addition

The order of association of whole numbers does not change their sum.
$(a + b) + c = a + (b + c)$ Example: $(3 + 4) + 5 = 3 + (4 + 5)$

Axiom 4: Identity law of addition

When adding zero to any whole number the sum is that number. The number did not lose its identity therefore zero is known as the identity element. That is algebraically represented as $a + 0 = 0 + a = a$. Example: $6 + 0 = 0 + 6 = 6$

Axiom 5: Closure law of multiplication

The law states that when multiplying any two whole numbers, the product is another whole number. That is shown algebraically as $a \times b = c$, where *a*, *b and c* are whole numbers. Example: $2 \times 3 = 6$

Axiom 6: Commutative law of multiplication

Whole numbers can be multiplied in any order but their product is unchanged. That is shown algebraically as $a \times b = b \times a$. Example: $3 \times 5 = 5 \times 3 = 15$

Axiom 7: Associative law of multiplication

The order in which numbers are associated in multiplication does not alter the product. That is shown algebraically as $(a \times b) \times c = a \times (b \times c)$. Example: $(2 \times 3) \times 4 = 2 \times (3 \times 4) = 24$

Axiom 8: Identity law of multiplication

One is the identity element for multiplication. When it is multiplied by any whole number, the product is the same as that number. That is, $a \times 1 = 1 \times a = a$. Example: $2 \times 1 = 1 \times 2 = 2$

Unit 11.1 Activity 1A: The real number system

1. By writing them in the form $\frac{m}{n} (n \neq 0)$, show that the following numbers are rational.

a. 15 **b.** -4 **c.** 0 **d.** 12.5 **e.** 0.045
f. $2\frac{3}{8}$ **g.** 23.224 **h.** $-6\frac{8}{9}$ **i.** -0.00005 **j.** $3\frac{23}{25}$

2. Show that the following recurring decimals are rational by converting them to the form $\frac{m}{n} (n \neq 0)$. Give your answers in simplest fractional form.:

a. 0.333333... **b.** $-0.\dot{7}$ **c.** 0.565656... **d.** $0.\overline{67}$
e. 2.555555... **f.** $12.\overline{45}$ **g.** $62.\overline{822}$

3. Explain the identity law of addition, giving an example. What is the identity element of addition?

4. Explain the identity law of multiplication, giving an example. What is the identity element of multiplication?

5. We know that $5 \times (3 \times 2) = 5 \times 6 = 30$ and $(5 \times 3) \times 2 = 15 \times 2 = 30$. Which of the axioms of rational numbers is this illustrating?

Unit 11.1 Number and Application

Topic 2: Basic numeracy—working with real numbers

The material in this Topic is about working with real numbers and solving straightforward number problems in context, and covers:

- Integers.
- Making sensible estimates and checking the reasonableness of results.
- Solving number problems in context, involving manipulation, several steps or reversing processes.

Introduction

Solving problems in context

The General Mathematics syllabus emphasises the importance of solving problems *in context* so that problems arising from practical situations can be correctly solved and the answers meaningfully interpreted.

Many straightforward calculations can be carried out mentally, or using pen and paper. It is most important to *understand* the processes involved in operations with numbers, and not to simply rely on a calculator. Unless these processes are understood thoroughly, it is difficult to succeed with algebra problems involving these skills (when a calculator can no longer be used).

Sets of numbers

In this Achievement Standard, various types of numbers are used.

Natural numbers	1, 2, 3, 4, 5, 6, 7, 8, 9, 10, …
Whole numbers	0, 1, 2, 3, 4, 5, 6, 7, 8, 9, 10, …
Integers	… –4, –3, –2, –1, 0, 1, 2, 3, 4, …
Rational numbers	Numbers which can be written in the form $\frac{\text{integer}}{\text{integer}}$. Rational numbers include fractions and finite or recurring decimals as well as integers.
Irrational numbers	Numbers which are not rational, such as $\sqrt{2}$ and π.
Real numbers	The combined set of all rational and irrational numbers.

Note: The Natural numbers are also called **Counting numbers**.

To solve number problems, numbers can be combined using **operations** such as addition, subtraction, multiplication and division.

Order of operations

In expressions involving more than one operation, there is an accepted order in which operations are performed. The *mnemonic* **BEMA** is useful for remembering the correct order:

B	Brackets first
E	Exponents (powers) next
M	Multiplication/division next
A	Addition/subtraction last

Note: **1.** Multiplication and division are inverses and are therefore the same 'type' of operation, eg division by 4 is the same as multiplication by $\frac{1}{4}$. Since neither operation 'outranks' the other, multiplication and division are carried out in order from left to right. (Remember 'of' is also used to mean multiplication.)

2. Similarly addition and subtraction are inverses, and are carried out in order from left to right.

Example A

Q. Calculate: **1.** $3 \times 2 - 16 \div 4$ **2.** $15 + 3 \times 2 - 4^2$ **3.** $5 + 3(15 - 3 \times 2)$

A. **1.** In $3 \times 2 - 16 \div 4$, there are no brackets or exponents so multiplication and division are done first. 3×2 is calculated before $16 \div 4$ because the multiplication appears before the division.

$3 \times 2 - 16 \div 4 = 6 - 16 \div 4$ [multiplication first]

$= 6 - 4$ [division next]

$= 2$ [subtraction next]

2. $15 + 3 \times 2 - 4^2 = 15 + 3 \times 2 - 16$ [the exponent 4^2 is done first]

$= 15 + 6 - 16$ [then multiplication]

$= 21 - 16$ [15 + 6 is done, not 6 – 16]

$= 5$

3. Here the bracket is done first. Within the bracket, the multiplication is done before the subtraction.

$5 + 3(15 - 3 \times 2) = 5 + 3(15 - 6)$ [multiplication in brackets first]

$= 5 + 3 \times 9$ [subtraction in brackets next]

$= 5 + 27$

$= 32$

Calculators

Scientific calculators are inexpensive and efficient tools for longer, more complex calculations. With scientific calculators such as the *Casio fx*-82, brackets may be entered, the correct order of operations is automatically followed and the **memory** is useful for storing parts of the answer as you go.

Example B

Q. Use a calculator to evaluate the expressions in Example A.

A. **1.** Press [3] [×] [2] [–] [1] [6] [÷] [4] [=]

to get the answer [2].

2. Press [1] [5] [+] [3] [×] [2] [–] [4] [x^2] [=]

to get the answer [5].

3. Press [5] [+] [3] [(] [1] [5] [–] [3] [×] [2] [)] [=]

to get the answer [32].

Note: In expressions where brackets are 'understood' they may not be shown.

For example, the fraction $\frac{47 + 53}{12 - 7}$ is understood to mean $\frac{(47 + 53)}{(12 - 7)}$, so press

(4 7 + 5 3) ÷ (1 2 – 7) =

to get answer 20.

Calculators vary so it is important that students become familiar with their own brand of calculator (ie read the instruction manual thoroughly) so that calculations are handled correctly.

Integers

The set of integers is the set of whole numbers and their 'opposites'.

The integers are ..., –3, –2, –1, 0, 1, 2, 3, ...

Adding and subtracting integers

A number line is useful for illustrating addition and subtraction of integers.

Example C

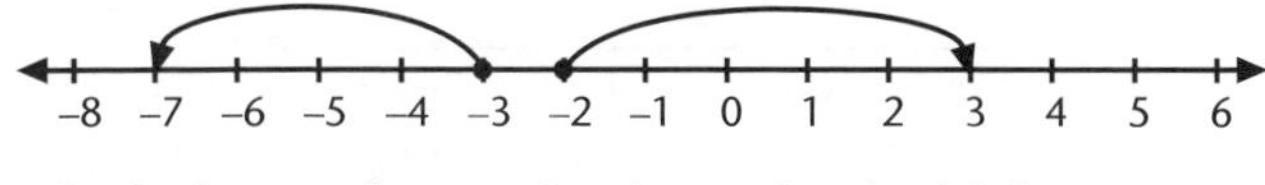

–2 + 5 = 3 [start at –2 and move 5 to the right]

–3 – 4 = –7 [start at –3 and move 4 to the left]

Two consecutive signs can be simplified to a single sign according to the following rule:

If two consecutive signs are the same change to '+'.
If two consecutive signs are different change to '–'.

Example D

1. 4 – –3 = 4 + 3 = 7 [subtracting a negative means adding]
2. 5 + –8 = 5 – 8 = –3 [adding a negative means subtracting]

Using a number line works well when adding or subtracting small integers. To add or subtract larger integers, the **size** and **sign** of the integers should be taken into account (for example –45 has a *size* of 45 and a *sign* that is negative).

When adding two integers:

- If the signs are the *same*, then *add* the sizes. The sign of the answer is the same as the signs of the two numbers, eg 40 + 60 = 100 and –40 + –60 = –100, ie –40 – 60 = –100.
- If the signs are *opposite* then *subtract* the sizes. The sign of the answer matches the sign of the number whose size is larger, eg –40 + 60 = 20 [60 is larger in size so answer is positive] and –80 + 30 = –50 [–80 is larger in size so answer is negative].

Problems expressed in words may be set out as numerical expressions and simplified.

Example E

Q. Vagi's bank balance is overdrawn by K153. She deposits K47. What is the current state of her bank balance?

A. Vagi still owes money, but not as much as before. Her account is now overdrawn by K(153 – 47) = K106.

Writing this in terms of integers:

$-153 + 47 = -106$ [negative since –153 is larger in size]

Example F

Q. The temperature at the top of Mt Wilhelm was –5°C at midnight, but by noon had risen by 11°C. Find the temperature at noon.

A. $-5 + 11 = 6$ [positive integer has larger size, so answer positive]

The temperature is 6°C at noon.

Multiplying and dividing integers

The rules for multiplying and dividing integers are more straightforward. When multiplying or dividing a *pair* of integers, 'like' signs give a positive answer, 'unlike' signs a negative answer. This is shown in the table:

+	x	+	gives	+	+	÷	+	gives	+
–	x	+	gives	–	–	÷	+	gives	–
+	x	–	gives	–	+	÷	–	gives	–
–	x	–	gives	+	–	÷	–	gives	+

Example G

1. $-40 \times 5 = -200$ [since signs are unlike answer is negative]
2. $-6 \times -3 = 18$ [since signs are like answer is positive]
3. $\frac{120}{-40} = -3$ [since signs are unlike answer is negative]
4. $\frac{-45}{-9} = 5$ [since signs are like answer is positive]

The rules for multiplying and dividing integers can be generalised as follows:

When an *even* number of negative numbers are multiplied together or divided, the result is a *positive* number.

When an *odd* number of negative numbers are multiplied together or divided, the result is a *negative* number.

Example H

Q. Calculate each of the following:

1. $-2 \times 4 \times -5$ **2.** $(-2)^3$ **3.** $\dfrac{-4 \times 15}{-3 \times -2}$ **4.** $\dfrac{-5 \times -6}{-2 \times -3}$

A. **1.** $-2 \times 4 \times -5 = 40$ [the answer is positive because there are two negative numbers in the calculation (two is even)]

2. $(-2)^3 = -2 \times -2 \times -2$
$= -8$ [the answer is negative because there are three negative numbers in the calculation (three is odd)]

3. $\dfrac{-4 \times 15}{-3 \times -2} = -\dfrac{60}{6}$
$= -10$ [the three negative numbers in the calculation give a negative answer]

4. $\dfrac{-5 \times -6}{-2 \times -3} = \dfrac{30}{6}$
$= 5$ [the four negative numbers in the calculation give a positive answer]

Note: Division by zero is **undefined**. This means that it is impossible to divide by zero (attempting to do so on a calculator will result in an error message). Thus, expressions such as $\frac{3}{0}$, or $\frac{-4}{0}$ cannot be simplified in any way to a finite value.

Integers and the calculator

Use the negative key  to enter negative numbers into the calculator; eg to enter –5 press .

Note: On most calculators the subtraction key [–] may be used instead of the negative key to enter negative numbers. Some calculators may have a [+/–] key which can be used to change the sign of a number. For example [2] [+/–] gives –2

Example I

Q. Calculate –45 × 78 × –23.

A. Enter the calculation by pressing

to get the answer 80 730 [positive since two negative integers in the calculation]

Unit 11.1 Activity 2A: Order of operations and integers

1. Use the correct order of operations to evaluate (where possible) each of the following *without* using a calculator:

a. $4 + 3 \times 2$ **b.** $4 \times 3 + 2$ **c.** $(5 + 3) \times 4$ **d.** $2\,(6 + 3)$

e. $12 - 6 \div 3$ **f.** $18 \div (6 + 3)$ **g.** $8 \times 5 - 3 \times 4$ **h.** $3^2 \times 2^2$

i. $5^2 - 4^2$ **j.** $8 - 5 + 3 \times 2$ **k.** $2 + 3^2 - 9 + 2$ **l.** $3 + 4^2 - 6 + 2$

m. $10 - (12 - 4 \times 2)$ **n.** $\dfrac{6 + 9}{7 - 4}$ **o.** $\dfrac{15 - 3}{3 \times 2}$ **p.** $\dfrac{5 \times 2 - 3 \times 2}{5 - 3}$

q. $\dfrac{3 + 12 \times 5}{9 - 2 \times 3}$ **r.** $\dfrac{4 - 4}{4 + 4}$ **s.** $\dfrac{4 + 4}{4 - 4}$ **t.** $\dfrac{6 \times 4 - 8 \times 3}{12 \times 2 - 8 \times 3}$

2. Peter purchases a phone for K157. He pays K35 and returns the next day to complete the payment on his phone. He wishes to buy an K18 phone case at the same time and to cash in a voucher for 'K10 off any purchase over K50'. How much does he need to pay the shop to settle his account?

3. Jane goes on a diet. She starts with a weight of 61 kg, and loses 2 kg every 3 weeks for 12 weeks. She abandons her diet at *the start* of a 2-week holiday, and gains back 1 kg per week. How much does she weigh at *the end* of her holiday?

4. Evaluate each of the following:

a. $8 - 12$ **b.** $5 - -2$ **c.** $9 - (7 - 10)$ **d.** $\frac{-12}{4}$

e. 5×-4 **f.** -3×-4 **g.** $-9 - 15$ **h.** $3 \times (-2)^2$

i. $-6 - -6$ **j.** $11 - (3 + 5)$ **k.** $\frac{35}{-5}$ **l.** $\frac{-3 \times -2 \times 4 \times -7}{8}$

m. $\frac{-2 \times 2 \times 4 \times -5}{16}$ **n.** $14 \div 7 \div -2$ **o.** $8 - 2(4 - 6)$ **p.** $\frac{2 - 3 - 4}{-2 - 3}$

q. $(6 - 11)^2$ **r.** $9 - (8 - 4)^2$ **s.** $\frac{(-4)^2 - 1}{-4 - 1}$ **t.** $((-7 + 4)^2 - 10)^2$

5. The 7 am temperature reading at Mt Wilhelm is –8°C. The noon reading is 10°C higher than this, but by midnight the temperature has fallen to 2°C below the 7 am reading.

a. What is the temperature at noon?

b. What is the temperature at midnight?

c. By how many degrees Celsius has the temperature fallen between noon and midnight?

6. Temperatures in Enga on three consecutive days are:

Day	Monday	Tuesday	Wednesday
Noon	2°C	–1°C	–3°C
Midnight	–5°C	–8°C	–9°C

a. By how many degrees did the temperature fall between noon and midnight on:
i. Monday? **ii.** Tuesday?

b. What was the change in temperature between Tuesday midnight and Wednesday noon?

c. The temperature at noon on the following Thursday was 7°C above the temperature at midnight on Wednesday. What was the Thursday noon temperature?

7. Pius had K32.50 in his cheque account. He wrote cheques for K25.60 and K45.80. He deposited K24. What was the balance after these transactions?

8. A bank offers an 'overdrawn' service in which customers are permitted a negative balance, usually with some maximum limit to the amount the customer may be overdrawn.

Paul has an account with this service, with a maximum 'overdrawn' limit of –K2 000.

Paul has his monthly salary of K1 980 credited automatically on the first of each month. On the first day of each month he automatically pays bills of K250 and withdraws K675 spending money as well as his monthly rent of K712. A friend who bought a car from him is repaying the debt at K425 per month by automatically crediting Paul's account on the first of each month.

At the end of May, Paul has just bought a new car and is overdrawn by K1 875.

a. If Paul spends as described above, how long will it be before Paul has at least K1 200 balance? Show all working clearly.

b. Paul wants to have a balance of at least K1 200 by the end of August. To do this he will reduce his monthly spending money. What is the maximum amount he should spend per month? Set out your working clearly.

Squares and square roots

The **square** of a number is found by multiplying that number by itself, eg the square of 5 is $5 \times 5 = 25$.

The square of a number, a, is $a \times a$, and is written a^2

It is useful to be familiar with the squares of the first dozen or so counting numbers: $1^2 = 1$, $2^2 = 4$, $3^2 = 9$, $4^2 = 16$, $5^2 = 25$, $6^2 = 36$, $7^2 = 49$, $8^2 = 64$, $9^2 = 81$, $10^2 = 100$, $11^2 = 121$, $12^2 = 144$, ... (These numbers are often called **perfect squares**.)

Scientific calculators have a squaring key .

Example J

1. $13^2 = 13 \times 13$
   $= 169$

2. $(-4)^2 = -4 \times -4$
   $= 16$

3. $-4^2 = -1 \times 4^2$
   $= -1 \times 16$
   $= -16$

Note: There is an important difference in meaning between the expressions $(-4)^2$ and -4^2 in part **2** and part **3** of the example. Use brackets carefully so that correct values are calculated.

The inverse of squaring a number is finding the **square root** of a number. For example, the square of 7 is 49 so the square root of 49 is 7 (ie, 7 is the number which squares to make 49). Similarly the square root of 100 is 10 since $10^2 = 100$.

The positive square root of a number a is written $\sqrt{a}$

For example, $\sqrt{49} = 7$ and $\sqrt{100} = 10$. Thus, even though $-4 \times -4 = 16$, the value of $\sqrt{16}$ is +4 (not −4).

Scientific calculators have a square root key $\sqrt{\ }$ for evaluating square roots.

Since squaring and taking the square root are inverses it follows that, for non-negative real numbers, a,

$$\left(\sqrt{a}\right)^2 = \sqrt{a} \times \sqrt{a} = a$$

For example $\sqrt{49} \times \sqrt{49} = 49$ and $\sqrt{3} \times \sqrt{3} = 3$.

Note: Since squares of numbers are always positive, the square root of a negative number is undefined (no real number multiplies by itself to make a negative number). For example, $\sqrt{-4}$ is undefined.

Example K

Q. 1. Without using a calculator,

a. Find $\sqrt{64}$. b. Find $\sqrt{81}$. c. Estimate $\sqrt{70}$.

2. Use a calculator to evaluate $\sqrt{70}$ to 3 dp.

3. Find $-\sqrt{144}$.

A. 1. a. $\sqrt{64} = 8$ [since $8 \times 8 = 64$]

b. $\sqrt{81} = 9$ [since $9 \times 9 = 81$]

c. 70 lies between the perfect squares 64 and 81 so $\sqrt{70}$ lies between $\sqrt{64}$ and $\sqrt{81}$, ie between 8 and 9. A reasonable approximation to $\sqrt{70}$ would be 8.5.

2. Using the square root key, $\sqrt{70} = 8.367$ (3 dp).

3. $-\sqrt{144} = -1 \times \sqrt{144} = -1 \times 12 = -12$

Order of operations with squares and square roots

Some expressions involving squares and square roots involve other operations as well. In the correct order of operations (mnemonic BEMA), brackets are done before exponents (powers). Thus, squaring and its inverse operation, taking the square root, are done *after* any expressions in brackets are evaluated. Note that expressions under the square root symbol are assumed to be in brackets (ie the brackets are implicit).

Example L

Q. Find the value of 1. $(5 + 6)^2$ 2. $\sqrt{100 + 44}$ 3. $\sqrt{5^2 - 3^2}$

A. 1. $(5 + 6)^2 = 11^2 = 121$ [adding inside bracket first]

2. $\sqrt{100 + 44}$ means $\sqrt{(100 + 44)}$ [brackets under square root symbol are implicit]

$\sqrt{100 + 44} = \sqrt{144} = 12$ [add before taking square root]

3. $\sqrt{5^2 - 3^2} = \sqrt{(25 - 9)} = \sqrt{16} = 4$

[implicit brackets so subtract before taking square root]

Higher powers and roots of numbers

In a similar way to squaring, numbers can be raised to higher powers. For example, to **cube** a number (raise it to the power of three), the number is multiplied by itself then by itself again, so that it appears *three* times in the product. Thus, $5^3 = 5 \times 5 \times 5 = 125$.

In general, to raise a number to any whole number power, n, the following definition holds:

$$x^n = \underbrace{x.x.x. \dots .x}_{n \text{ factors}}$$

For example, $2^6 = \underbrace{2 \times 2 \times 2 \times 2 \times 2 \times 2}_{6 \text{ factors}} = 64$ and $3^5 = \underbrace{3 \times 3 \times 3 \times 3 \times 3}_{5 \text{ factors}} = 243$.

Using calculators to evaluate higher powers and roots

Calculators are very useful tools for evaluating powers of numbers. Most scientific calculators have a cube key [x^3] for finding the cubes of numbers. To evaluate higher powers, the **power key** is used. On your calculator this may look like [x^y] or [^]. For example, to calculate 9^4, press [9][^][4][=] (or [9][x^y][4][=]) to get 6 561.

The inverse of *raising a number to the power of n is finding the nth root of a number*. For example, the cube of 5 is 125, so the **cube root** of 125 is 5 (ie, 5 is the number which cubes to make 125). Many scientific calculators have a cube root key [$\sqrt[3]{\ }$] for evaluating cube roots.

In general the **nth root** of a number *a* is written $\sqrt[n]{a}$, eg the 7th root of 2 187 is written $\sqrt[7]{2\,187}$.

To evaluate this number, the *n*th root key [$\sqrt[x]{\ }$] is used on the calculator (press SHIFT [x^y]). For example, to calculate $\sqrt[7]{2\,187}$, press [7][$\sqrt[x]{\ }$][2][1][8][7][=] to get the answer 3.

Note: Since numbers to even powers are positive, the *n*th root of an odd number is undefined when *n* is even. For example $\sqrt[4]{-16}$ is undefined since the fourth power of any number is always positive. Similarly $\sqrt[6]{-1}$, or $\sqrt[8]{-256}$ are undefined.

Example M

Q. Find
1. The volume of a cube whose side length is 16 cm.
2. The side length of a cube whose volume is 343 cm^3.

A. 1. Since volume = $(\text{side length})^3$ [formula for volume of cube]

Volume of cube of side length 16 = $16^3 = 4\,096$ cm^3

2. Since side length = $\sqrt[3]{\text{volume}}$ [cube root is inverse of cubing]

Side length of a cube whose volume is 343 = $\sqrt[3]{343} = 7$ cm

Unit 11.1 Activity 2B: Powers and roots of numbers

1. Evaluate each of the following to a maximum of 3 dp:

a. 5^2 **b.** 9^2 **c.** 25^2 **d.** 43^2 **e.** $\sqrt{36}$

f. $\sqrt{169}$ **g.** $\sqrt{3\,136}$ **h.** $\sqrt{9\,604}$ **i.** $\sqrt{346}$ **j.** $\sqrt{1\,234}$

k. $\sqrt{850}$ **l.** $\sqrt{60}$ **m.** $(\sqrt{3})^2$ **n.** $\sqrt{4^2}$ **o.** $(\sqrt{5})^2$

p. $\sqrt{9^2}$ **q.** $-\sqrt{121}$ **r.** $-\sqrt{400}$ **s.** $-\sqrt{600}$ **t.** $\sqrt{(-5)^2}$

2. Calculate each of the following using a calculator:

a. $8^2 - 4 \times 5^2$ **b.** $(2 - 3)^2$ **c.** $(-2 - 4)^2$

d. $5^2 - 5 \times 5^2 + 5^2$ **e.** $\sqrt{25^2 - 24^2}$ **f.** $8 - \sqrt{59 + 5}$

3. Use your knowledge of perfect squares to give two consecutive numbers these square roots lie between. Confirm using a calculator.

a. $\sqrt{140}$ **b.** $\sqrt{87}$ **c.** $\sqrt{27}$ **d.** $\sqrt{410}$

4. A square has an area of 1 369 cm^2.

a. What is its side length?

b. How much longer would the side of the square need to be to create a square with double the original area?

5. Two square gardens are planted. The first has a side length of 12 m. The second garden has a side length of 15 m.

a. How many square metres greater is the area of the second garden than the first?

b. A third square garden has an area equal to the difference between the areas of the first and second gardens. What is the side length of this garden?

6. Evaluate where possible to a maximum of 3 dp.

a. 8^3 **b.** 9^4 **c.** 3^5 **d.** 4^6 **e.** $\sqrt[3]{27}$

f. $\sqrt[4]{14\,641}$ **g.** $\sqrt[5]{1\,024}$ **h.** $\sqrt[7]{128}$ **i.** $\sqrt[6]{6}$ **j.** $\sqrt[4]{1}$

k. $\sqrt[10]{50}$ **l.** $\sqrt[3]{510}$ **m.** $(\sqrt[3]{2})^3$ **n.** $\sqrt[4]{5^4}$ **o.** $\sqrt[3]{7^3}$

p. $\sqrt[4]{2^8}$ **q.** $-\sqrt[3]{216}$ **r.** $\sqrt[3]{-216}$ **s.** $\sqrt[4]{-64}$ **t.** $\sqrt[7]{-9}$

7. Evaluate (where possible) to a maximum of 3 dp.

a. $\sqrt[4]{416-200}$ **b.** $(1+2)^4$ **c.** $\sqrt[4]{2^4-1^4}$ **d.** $\dfrac{\sqrt[5]{32}}{\sqrt[6]{64}}$

e. $\sqrt[3]{(8-6)^3}$ **f.** $\dfrac{1+\sqrt[3]{27}}{\sqrt[3]{-64}}$

8. A cube has side length 5 cm.

a. What is its volume?

A second cube has twice the volume of the first cube.

b. What is its side length?

9. A lump of clay has volume 1 000 cm^3. Four cubes are made from it. The first three cubes have equal volumes, but the third cube is double the volume of each of the other cubes.

a. What is the side length of the small cubes?

b. What is the side length of the larger cube?

c. It is suggested that the side length of the larger cube is equal to the product of the side length of the small cube and $\sqrt[3]{k}$ for some k. Find the value of k.

10. A block of chocolate has volume 2^{10} cubic centimetres. It is melted and poured into four identical moulds in the shape of cubes. There is 160 cm^3 quantity of chocolate left. What is the side length of each of these cube-shaped moulds?

Unit 11.1 Number and Application

Topic 3: Basic numeracy—revision of fractions

This Topic presents a revision of fractions and looks at problem solving in context. It covers:

- Fractions and mixed numbers.
- Solving fraction problems in context.
- Sharing quantities in a given ratio.
- Solve number problems in context, involving manipulation, several steps or reversing processes.

Introduction

The aim of this Topic is to solve practical problems involving fractions. Many operations with fractions should be familiar by now, and scientific calculators with a fraction key can be used to carry out fraction calculations rapidly. However, the processes involved must also be understood (without the use of a calculator) so that algebraic problems involving fractions can be solved.

Fractions

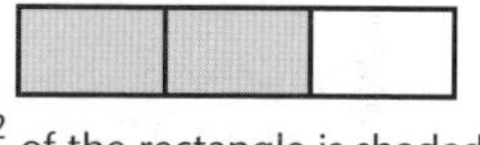

$\frac{2}{3}$ of the rectangle is shaded

A **fraction** is part of a whole. Fractions can be illustrated with diagrams which show the part compared to the whole. A fraction is commonly expressed as a **numerator** divided by a **denominator**. The numerator is the number on top of the fraction line and the denominator is the number below the fraction line. (The fraction line should be horizontal, not sloping.)

Equivalent fractions

Fractions representing the same quantity are called **equivalent fractions**. For example, $\frac{3}{4}, \frac{6}{8}, \frac{12}{16}$ are equivalent fractions, as the following diagrams show:

$\frac{3}{4}$

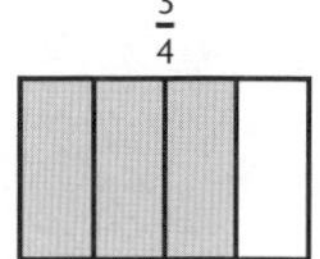

$\frac{6}{8}$

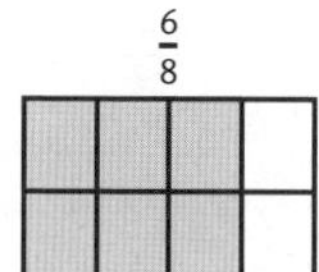

$\frac{12}{16}$

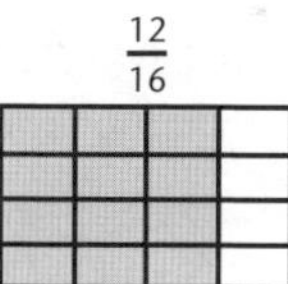

Forming equivalent fractions

Equivalent fractions can be formed by *multiplying* the numerator *and* denominator of a fraction by the *same* number.

Thus $\frac{3}{4}$ and $\frac{6}{8}$ are equivalent fractions because $\frac{3}{4} = \frac{6}{8}$

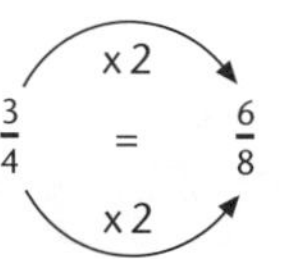

Equivalent fractions may also be formed by *dividing* the numerator *and* denominator of a fraction by the *same* number.

For example, $\frac{12}{16}$ and $\frac{3}{4}$ are equivalent fractions because $\frac{12}{16} = \frac{3}{4}$

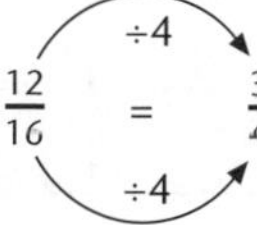

Simplifying fractions

Simplifying a fraction means writing it as an equivalent fraction with the smallest possible denominator. Simplifying involves '**cancelling**', which means *dividing* the numerator *and* denominator ('top and bottom') by the *same* number. Cancelling continues until the fraction is in its **simplest form** (the numerator and denominator have no more common factors).

Example A

Q. Express in simplest form: **1.** $\frac{6}{8}$ **2.** $\frac{16}{24}$

1. $\frac{6}{8}$ can be simplified to $\frac{3}{4}$ by dividing both 6 and 8 by 2.

This is written $\frac{6}{8} = \frac{6 \div 2}{8 \div 2} = \frac{3}{4}$

2. $\frac{16}{24} = \frac{4}{6}$ [dividing top and bottom by 4]

$= \frac{2}{3}$ [dividing top and bottom by 2]

Note: In part **2**, $\frac{4}{6}$ is *not* the answer because it can be simplified further.

An alternative procedure for part **2** involves dividing the top and bottom of the fraction by 8, the highest common factor of 16 and 24.

Thus $\frac{16}{24} = \frac{16 \div 8}{24 \div 8} = \frac{2}{3}$

Equivalent fractions are useful for **comparing** fractions (putting fractions in order of size). Express each fraction with the same denominator, and then compare numerators.

Example B

Q. Put in order of size from smallest to largest: $\frac{1}{3}, \frac{2}{5}, \frac{9}{25}$.

A. The lowest common multiple of 3, 5 and 25 is 75.

Expressing each fraction with denominator 75:

$\frac{1 \times 25}{3 \times 25} = \frac{25}{75}$ $\frac{2 \times 15}{5 \times 15} = \frac{30}{75}$ $\frac{9 \times 3}{25 \times 3} = \frac{27}{75}$

gives $\frac{1}{3}, \frac{9}{25}, \frac{2}{5}$ as the required order. [since $25 < 27 < 30$]

Types of fractions

Fractions can be classified as shown below:

Proper fraction	Improper fraction	Mixed number
The size of the numerator is *smaller* than the size of the denominator. Examples: $\frac{1}{2}, \frac{-3}{4}, \frac{15}{16}$	The size of the numerator is *bigger* than the size of the denominator. Examples: $\frac{3}{2}, \frac{4}{3}, \frac{-17}{9}$	An integer and a proper fraction are combined and written next to each other. Examples: $2\frac{3}{4}, 3\frac{1}{2}, -4\frac{1}{5}$

Mixed numbers and improper fractions

To change an **improper fraction** to a **mixed number**, divide the numerator by the denominator.

- The *quotient* is the integer part of the mixed number.
- The *remainder* is written over the divisor (which is the bottom part of the improper fraction).

Example C

The improper fraction $\frac{75}{8}$ is changed to a mixed fraction by dividing 75 by 8:

$$8\overline{)75}\quad 9,\quad 72,\quad 3$$

$\therefore\ \frac{75}{8} = 9\frac{3}{8}$ [the quotient 9 is the integer part, the remainder 3 is written over the divisor 8]

Note: The mixed number $9\frac{3}{8}$ means $9 + \frac{3}{8}$.

To change a mixed number to an improper fraction, multiply the *integer* part of the mixed number by the denominator of the fraction and add the result to the numerator of the fraction.

- The result of this calculation gives the numerator of the improper fraction.
- The denominator of the fraction remains unchanged.

Example D

Q. 1. Show that $3\frac{4}{5}$ is equivalent to the improper fraction $\frac{19}{5}$.

2. Illustrate the relationship between $3\frac{4}{5}$ and $\frac{19}{5}$ diagrammatically.

3. Change $7\frac{3}{8}$ to an improper fraction.

A. 1. The mixed number $3\frac{4}{5}$ becomes $\frac{3 \times 5 + 4}{5} = \frac{19}{5}$.

[multiply integer 3 by denominator 5 and add the numerator 4]

2. $3 + \frac{4}{5} = \frac{19}{5}$

3. $7\frac{3}{8} = \frac{7 \times 8 + 3}{8} = \frac{59}{8}$

Unit 11.1 Activity 3A: Equivalent fractions and mixed numbers

1. Express the shaded parts as a fraction of the whole:

a.

b. 

c. 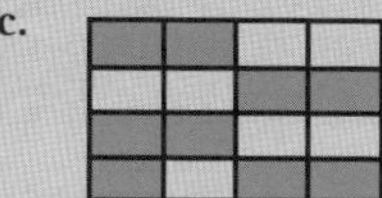

2. Jeanne has a bag containing 8 red, 4 orange and 5 green sweets.

a. What fraction of the sweets are red?

b. What fraction of the sweets are green?

c. What fraction of the sweets are *not* green?

3. Eri purchased a K25 hammer, K10 worth of nails and a K15 metal ruler. What fraction of the total cost was

   a. The hammer? b. The nails? c. The ruler?

   Simplify your answers.

4. Write each of the following fractions in their simplest form:

   a. $\frac{4}{8}$ b. $\frac{6}{24}$ c. $\frac{28}{60}$ d. $\frac{125}{1\,000}$ e. $\frac{72}{132}$ f. $\frac{5}{10}$

   g. $\frac{18}{24}$ h. $\frac{75}{125}$ i. $\frac{175}{250}$ j. $\frac{625}{1\,000}$

5. Change each of the following to mixed numbers:

   a. $\frac{19}{2}$ b. $\frac{48}{5}$ c. $\frac{79}{6}$ d. $\frac{87}{10}$ e. $\frac{100}{12}$ f. $\frac{11}{2}$

   g. $\frac{17}{3}$ h. $\frac{28}{5}$ i. $\frac{47}{6}$ j. $\frac{95}{10}$

6. Change each of the following to improper fractions:

   a. $2\frac{1}{2}$ b. $4\frac{2}{3}$ c. $5\frac{3}{4}$ d. $5\frac{7}{8}$ e. $12\frac{7}{20}$ f. $2\frac{1}{3}$

   g. $3\frac{4}{5}$ h. $2\frac{11}{20}$ i. $11\frac{5}{8}$ j. $12\frac{11}{15}$

7. Put in order of size (from smallest to largest):

   a. $\frac{3}{8}, \frac{2}{7}, \frac{1}{3}$ b. $\frac{1}{6}, \frac{3}{20}, \frac{2}{15}$ c. $\frac{7}{8}, 1\frac{1}{5}, \frac{4}{3}$ d. $2\frac{3}{4}, \frac{12}{5}, \frac{8}{3}$

8. Shift workers in a factory were offered part shifts as overtime. Anne worked $\frac{1}{3}$ of a shift, Bob worked $\frac{3}{8}$ of a shift and Carol worked $\frac{5}{14}$ of a shift as overtime. Who did

   a. The most overtime? b. The least overtime?

9. Three out of every five employees in a clinic are male. If there are 36 males, how many workers are there in the clinic?

10. Takis is a triathlete. Each day he runs, swims and cycles. Takis runs 12 km each day.

    a. After he has run 8 km what fraction of his run remains?

    Takis swims 3 km each day.

    b. What fraction of his total running and swimming distance is the run?

    Takis cycles 25 km each day.

    c. What fraction of the total training distance is the cycling?

    Give all fractions in simplest form.

Addition and subtraction of fractions

When adding (or subtracting) fractions with the *same denominators*:

- The numerators are added together (or subtracted).
- The denominator of the answer is the same as the original denominators.

Example E

Q. Calculate **1.** $\frac{1}{5} + \frac{2}{5}$ **2.** $\frac{3}{5} - \frac{1}{5}$

A. 1. $\frac{1}{5} + \frac{2}{5} = \frac{1 + 2}{5}$ [add numerators (1 and 2), keep the denominator the same]

$= \frac{3}{5}$

This can be shown in the diagram shown alongside.

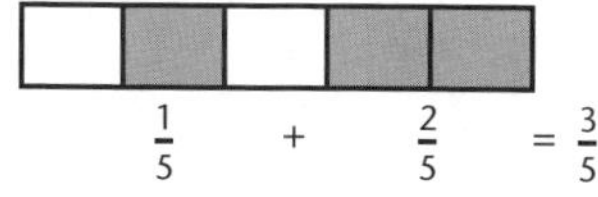

2. $\frac{3}{5} - \frac{1}{5} = \frac{3 - 1}{5}$ [subtracting numerators, keeping denominator the same]

$= \frac{2}{5}$

If the *denominators are different*, the fractions must first be changed into equivalent fractions with the same denominators. Then proceed as before.

Example F

Q. Calculate: **1.** $\frac{2}{3} + \frac{1}{4}$ **2.** $\frac{7}{8} - \frac{2}{5}$ **3.** $1 - \left(\frac{2}{3} + \frac{1}{4}\right)$

A. 1. Since denominators are different, change into equivalent fractions with a common denominator of 12. [12 is the LCM of 3 and 4]

$$\frac{2}{3} + \frac{1}{4} = \frac{2 \times 4}{3 \times 4} + \frac{1 \times 3}{4 \times 3}$$
$$= \frac{8}{12} + \frac{3}{12}$$
$$= \frac{11}{12}$$

2.
$$\frac{7}{8} - \frac{2}{5} = \frac{7 \times 5}{8 \times 5} - \frac{2 \times 8}{5 \times 8}$$
$$= \frac{35}{40} - \frac{16}{40}$$
$$= \frac{19}{40}$$

Note: An alternative setting out is:

$$\frac{7}{8} - \frac{2}{5} = \frac{7 \times 5 - 2 \times 8}{40}$$
$$= \frac{35 - 16}{40}$$
$$= \frac{19}{40}$$

[changing $\frac{7}{8}$ and $\frac{2}{5}$ to fractions with a denominator of 40, since 40 is the LCM of 5 and 8]

3. $1 - \left(\frac{2}{3} + \frac{1}{4}\right) = 1 - \frac{11}{12}$ [from part **1**]

$= \frac{12}{12} - \frac{11}{12}$ [since $1 = \frac{12}{12}$]

$= \frac{1}{12}$

Mixed numbers can be added or subtracted by changing them to improper fractions and continuing as before. The final answer is expressed in mixed number form, where appropriate.

Example G

Q. Calculate: $8\frac{2}{3} - 2\frac{5}{12}$

A. $8\frac{2}{3} - 2\frac{5}{12} = \frac{26}{3} - \frac{29}{12}$ [changing to improper fractions]

$= \frac{104}{12} - \frac{29}{12}$ [change $\frac{26}{3}$ to a fraction with a denominator of 12]

$= \frac{75}{12}$

$= 6\frac{3}{12}$ [change to mixed number]

$= 6\frac{1}{4}$ [simplifying]

Note: Many calculators have a fraction key [$a\frac{b}{c}$] so that addition, subtraction, multiplication and division of fractions may be carried out with the answer given as a fraction.

For example, to find the answer to Example G, press

[8] [$a\frac{b}{c}$] [2] [$a\frac{b}{c}$] [3] [−] [2] [$a\frac{b}{c}$] [5] [$a\frac{b}{c}$] [1] [2] [=] to get the answer

[6 ⅃ 1 ⅃ 4] which is $6\frac{1}{4}$.

Multiplying fractions

To multiply proper or improper fractions:

- Multiply numerators together.
- Multiply denominators together.

Note that the word 'of' is used to mean multiplication eg $\frac{1}{2}$ of $\frac{5}{12}$ means $\frac{1}{2} \times \frac{5}{12} = \frac{5}{24}$.

Example H

Q. One fifth of a garden is planted in crops. Two thirds of the crops are taro. What fraction of the garden is planted with taro?

A. As the diagram illustrates, $\frac{2}{3}$ of $\frac{1}{5}$ of the garden is in taro.

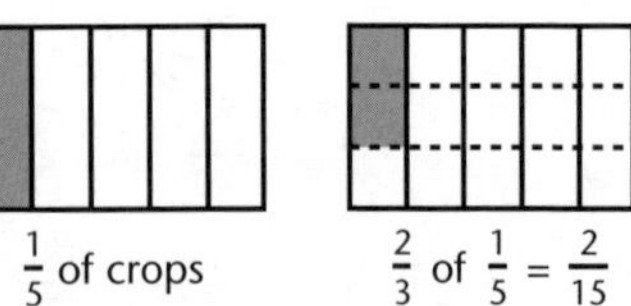
$\frac{1}{5}$ of crops $\frac{2}{3}$ of $\frac{1}{5} = \frac{2}{15}$

$\frac{2}{3} \times \frac{1}{5} = \frac{2 \times 1}{3 \times 5}$ ['of' means multiply]

$= \frac{2}{15}$

So $\frac{2}{15}$ of the garden is planted with taro.

Often the multiplication process can be simplified by cancelling first.

Example I

Q. Twenty-eight boxes of fruit are delivered of which five boxes contain guavas. $\frac{4}{15}$ of the guavas are Indian variety. What fraction of the delivery is Indian?

A. $\frac{5}{28}$ of the delivery are guavas. $\frac{4}{15}$ of $\frac{5}{28}$ are Indian.

$$\frac{4}{15} \times \frac{5}{28} = \frac{{}^{1}\not{4}}{\not{15}_{3}} \times \frac{\not{5}^{1}}{\not{28}_{7}}$$

[dividing left numerator and right denominator by 4; dividing left denominator and right numerator by 5]

$$= \frac{1 \times 1}{3 \times 7}$$

$$= \frac{1}{21}$$

To multiply mixed numbers, change them to improper fractions and then proceed as in the previous example.

Example J

$$2\frac{2}{3} \times 5\frac{3}{4} = \frac{{}^{2}\not{8}}{3} \times \frac{23}{\not{4}_{1}}$$

[changing to improper fractions and cancelling]

$$= \frac{2 \times 23}{3 \times 1}$$

$$= \frac{46}{3}$$

$$= 15\frac{1}{3}$$

Fractions may be raised to powers, using the rules for multiplication of fractions. Again, change mixed numbers to improper fractions first.

Example K

Q. Calculate **1.** $\left(\frac{3}{4}\right)^3$ **2.** $\left(2\frac{2}{3}\right)^2$

A. **1.** $\left(\frac{3}{4}\right)^3 = \frac{3 \times 3 \times 3}{4 \times 4 \times 4}$

$$= \frac{27}{64}$$

2. $\left(2\frac{2}{3}\right)^2 = \left(\frac{8}{3}\right)^2$ [changing to improper fraction]

$$= \frac{8 \times 8}{3 \times 3}$$

$$= \frac{64}{9} \text{ or } 7\frac{1}{9}$$

Division of fractions

The inverse of multiplying by a fraction $\frac{a}{b}$ is either

- Dividing by $\frac{a}{b}$ (since multiplication and division are inverses), or
- Multiplying by $\frac{b}{a}$, (since $\frac{a}{b} \times \frac{b}{a} = \frac{ab}{ba} = 1$).

Thus, dividing by a fraction $\frac{a}{b}$ is the same as multiplying by the inverted fraction $\frac{b}{a}$.

> To divide by a fraction $\frac{a}{b}$, multiply by $\frac{b}{a}$

For example, division by $\frac{2}{3}$ is changed to multiplication by $\frac{3}{2}$.

Example L

$\frac{2}{3} \div \frac{1}{5} = \frac{2}{3} \times \frac{5}{1}$ [division by $\frac{1}{5}$ becomes multiplication by $\frac{5}{1}$]

$= \frac{2 \times 5}{3 \times 1}$ [multiplying numerators and multiplying denominators]

$= \frac{10}{3}$ or $3\frac{1}{3}$

To invert a mixed number, express it as an improper fraction first. For example to invert $2\frac{1}{2}$, change it to $\frac{5}{2}$ then invert it to get $\frac{2}{5}$.

Example M

$2\frac{3}{4} \div 3\frac{2}{3} = \frac{11}{4} \div \frac{11}{3}$ [change mixed numbers to improper fractions]

$= \frac{{}^{1}\cancel{11}}{4} \times \frac{3}{\cancel{11}_{1}}$ [changing '$\div \frac{11}{3}$' to '$\times \frac{3}{11}$' and cancelling]

$= \frac{3}{4}$ [multiplying numerators and multiplying denominators]

Unit 11.1 Activity 3B: Operations with fractions

1. Calculate each of the following, giving answers as fractions or mixed numbers in simplest form.

a. $\frac{3}{5} + \frac{1}{5}$ **b.** $\frac{3}{10} + \frac{4}{10}$ **c.** $\frac{5}{8} + \frac{1}{8}$ **d.** $\frac{1}{2} + \frac{1}{3}$ **e.** $\frac{2}{3} + \frac{1}{5}$

f. $\frac{5}{8} + \frac{2}{5}$ **g.** $\frac{1}{4} + \frac{1}{5}$ **h.** $\frac{2}{5} + \frac{1}{6}$ **i.** $\frac{4}{5} + \frac{3}{4}$ **j.** $1\frac{1}{5} + \frac{1}{5}$

k. $2\frac{1}{2} + 1\frac{1}{4}$ **l.** $1\frac{3}{4} + 2\frac{5}{6}$

2. Calculate each of the following, giving answers as fractions or mixed numbers in simplest form.

a. $\frac{5}{8}-\frac{2}{8}$ **b.** $\frac{7}{10}-\frac{4}{10}$ **c.** $\frac{11}{12}-\frac{4}{12}$ **d.** $\frac{1}{2}-\frac{1}{3}$ **e.** $\frac{2}{3}-\frac{1}{4}$

f. $\frac{5}{9}-\frac{3}{8}$ **g.** $\frac{1}{4}-\frac{1}{5}$ **h.** $\frac{7}{8}-\frac{5}{6}$ **i.** $\frac{5}{12}-\frac{3}{8}$ **j.** $1\frac{1}{3}-\frac{2}{3}$

k. $2\frac{4}{5}-1\frac{2}{3}$ **l.** $3\frac{1}{4}-1\frac{3}{8}$

3. Calculate each of the following, giving answers as fractions or mixed numbers in simplest form.

a. $\frac{1}{2}\times\frac{1}{3}$ **b.** $\frac{5}{8}\times\frac{4}{15}$ **c.** $\frac{7}{12}\times\frac{2}{5}$ **d.** $\frac{1}{4}\times\frac{1}{5}$ **e.** $\frac{7}{8}\times\frac{16}{21}$

f. $\frac{8}{15}\times\frac{25}{36}$ **g.** $2\frac{2}{3}\times 4$ **h.** $6\frac{1}{2}\times 3\frac{4}{5}$ **i.** $4\frac{3}{8}\times 2\frac{4}{5}$ **j.** $3\frac{2}{5}\times 5$

k. $5\frac{1}{3}\times 1\frac{1}{4}$ **l.** $10\frac{2}{3}\times 5\frac{3}{8}$

4. Calculate each of the following, giving answers as fractions or mixed numbers in simplest form.

a. $\frac{1}{2}\div\frac{1}{3}$ **b.** $\frac{3}{4}\div\frac{1}{2}$ **c.** $\frac{5}{8}\div\frac{2}{3}$ **d.** $\frac{1}{4}\div\frac{1}{5}$ **e.** $\frac{2}{5}\div\frac{6}{11}$

f. $\frac{7}{15}\div\frac{6}{35}$ **g.** $1\frac{1}{2}\div\frac{3}{4}$ **h.** $\frac{4}{5}\div 4$ **i.** $4\frac{1}{6}\div 3\frac{1}{3}$ **j.** $\frac{2}{3}\div 3$

k. $2\frac{3}{5}\div 3\frac{1}{2}$ **l.** $4\frac{2}{3}\div 3\frac{4}{7}$

5. Evaluate each of the following as fractions.

a. $\left(\frac{2}{3}\right)^2$ **b.** $\left(\frac{5}{9}\right)^3$ **c.** $\frac{2}{3}$ of 24 **d.** $\frac{3}{4}$ of 30 **e.** $\left(1\frac{1}{2}\right)^2$

f. $\left(-2\frac{1}{3}\right)^2$ **g.** $\left(3\frac{4}{5}\right)^3$ **h.** $\left(-1\frac{2}{3}\right)^4$

6. Give your answers as mixed numbers in simplest form.

a. A square has side length $3\frac{3}{4}$ m. What is its area?

b. A cube has side length $2\frac{1}{4}$ cm. What is its volume?

7. An area of land is divided into blocks of each area $\frac{5}{8}$ hectare. If the total area is 25 hectares, how many blocks are there?

8. At the Port Moresby swimming pool the children's pool is divided into six lanes, five for swimming lessons and one for free play. In the free play area a slide takes up $\frac{2}{5}$ of the lane. What fraction of the pool is taken up by the slide?

9. Sai did $\frac{3}{5}$ of her project on Saturday and another $\frac{1}{4}$ on Monday. What fraction of her project is now complete?

10. Timi and Pala order a pizza. Together they eat $\frac{7}{8}$ of the pizza. If Timi ate $\frac{1}{2}$ of the pizza, what fraction of the pizza did Pala eat?

11. Carol sat a multichoice test of 60 questions. She got $\frac{11}{15}$ of them correct. How many did she get wrong?

12. Susan used $\frac{2}{3}$ cup of plain flour and $\frac{1}{2}$ cup of wholemeal flour in a recipe. How much flour did she use?

Fractions and order of operations

Many fraction calculations involve more than one operation. The expression must be evaluated following the correct order of operations (BEMA).

Example N

Q. Calculate $3\frac{3}{4} \times (\frac{2}{3} - \frac{1}{2})$

A. $3\frac{3}{4} \times (\frac{2}{3} - \frac{1}{2}) = 3\frac{3}{4} \times (\frac{4}{6} - \frac{3}{6})$ [subtracting inside the brackets first]

$= 3\frac{3}{4} \times \frac{1}{6}$ [subtracting]

$= \frac{15}{4} \times \frac{1}{6}$ [changing to a mixed number]

$= \frac{15}{24}$ [multiplying]

$= \frac{3}{8}$ [simplifying]

Alternatively, enter the calculation into a scientific calculator (using fraction key and brackets) and the correct order of operations will be followed automatically.

Solving word problems

When solving problems involving fractions, be sure to identify clearly the operation(s) involved. Remember that all the fractions of a quantity total 1 (the whole quantity).

Example O

Q. Max spends $\frac{2}{5}$ of his money on rent, and $\frac{1}{8}$ on petrol. What fraction of his money remains?

A. Fraction spent $= \frac{2}{5} + \frac{1}{8}$

$= \frac{16 + 5}{40}$ [40 is the LCM of 5, 8]

$= \frac{21}{40}$

Fraction remaining $= 1 - \frac{21}{40}$

$= \frac{40}{40} - \frac{21}{40}$ [creating equivalent fractions with the same denominator]

$= \frac{19}{40}$

Note: A statement about what is being calculated should be included with each separate calculation.

Alternatively the calculation is worked out to be $1 - (\frac{2}{5} + \frac{1}{8})$ or $1 - \frac{2}{5} - \frac{1}{8}$ and this is entered into a scientific calculator to arrive at the answer.

Unit 11.1 Activity 3C: Problem solving involving fractions

1. Evaluate the following. Give answers as fractions or mixed numbers.

a. $\frac{3}{7} + \frac{1}{2} \times \frac{3}{7}$ **b.** $2\frac{4}{5} - 6\frac{1}{2} \times 3$ **c.** $3\frac{1}{2} \times 2\frac{3}{4} - 4\frac{1}{4} \times 5\frac{1}{5}$

d. $\left(\frac{1}{7} \times \frac{2}{3}\right) \times \frac{3}{2}$ **e.** $\frac{4}{5} \times \left(\frac{5}{4} \times \frac{3}{25}\right)$ **f.** $5 - \left(\frac{7}{8} + \frac{3}{4}\right)$

g. $\left(1\frac{1}{2} + \frac{3}{4}\right)^2$ **h.** $\dfrac{\frac{7}{10} - \frac{2}{5}}{\frac{1}{3} + \frac{2}{5}}$ **i.** $\frac{3}{4} + 1\frac{1}{2} \times 2\frac{2}{3}$

j. $\left(5\frac{3}{4} - 4\frac{2}{5}\right)^2$ **k.** $8\frac{5}{8} - 5\frac{1}{4} \div \frac{3}{4}$ **l.** $\sqrt[3]{\frac{3}{16} + \frac{5}{64} + \frac{5}{32}}$

2. Kila brings $1\frac{1}{4}$ kg of sausages and $2\frac{1}{2}$ kg of steak to a function. If $\frac{4}{5}$ of the meat is cooked and eaten, what weight of meat is that?

3. David competed in a triathlon which involved swimming, cycling, and running. If the length of the swim was $\frac{1}{12}$ of the total distance of the triathlon and the cycling was $\frac{2}{3}$ of the total distance, what fraction of the total distance was the run?

4. In her will, Mrs Diwai leaves $\frac{1}{8}$ of her money to her son, twice as much to her daughter, and half as much as her son receives to her sister. The remainder is left to her husband. What fraction does her husband get?

5. Rawa ate $\frac{3}{5}$ of a pizza. The next day he ate a third of the remaining pizza. What fraction of the pizza has Rawa eaten altogether?

6. Sharon spends $\frac{2}{3}$ of her money at a circus, then half of what is left on afternoon tea. What fraction of her money remains?

7. Mrs Karam brings a dozen fruit pies to a function. At the end $1\frac{1}{2}$ pies remain. What fraction of the total number of pies was eaten?

8. Two friends go shares in a car. Brian pays $\frac{1}{4}$ and Gabriel pays $\frac{3}{8}$ of the total price of the car. The balance of K300 is loaned to them by Gabriel's mother. How much does the car cost?

9. A water tank is $\frac{4}{5}$ full at the beginning of a holiday. During the holiday 600 L of water are used, leaving the tank $\frac{1}{2}$ full. How much water does the tank hold when full?

10. A survey found that two out of every five school students listen to music while doing their homework. If there were 18 students in the survey who listened to music while doing their homework, how many students were there in the survey?

Unit 11.1 Activity 3D: Multiple choice – fractions

1. Which one of the following is **not** equal to $6\frac{1}{4}$?

A. $6+\frac{1}{4}$ **B.** 6.25 **C.** $\frac{75}{12}$ **D.** $\frac{61}{4}$

2. $\left(2\frac{2}{3}\right)^2$ is equal to:

A. $4\frac{4}{9}$ **B.** $7\frac{1}{9}$ **C.** $\frac{64}{3}$ **D.** $\frac{16}{9}$

3. $2\frac{1}{3}-\frac{1}{3}\times\frac{4}{5}=$

A. $\frac{8}{5}$ **B.** $\frac{31}{15}$ **C.** $\frac{26}{15}$ **D.** $\frac{6}{5}$

4. $4\frac{8}{9}\div 2\frac{2}{3}=$

A. $1\frac{5}{6}$ **B.** 2 **C.** $\frac{41}{8}$ **D.** $2\frac{4}{3}$

5. $3\frac{3}{4}+2\frac{1}{3}=$

A. $5\frac{3}{7}$ **B.** $\frac{22}{7}$ **C.** $6\frac{1}{4}$ **D.** $6\frac{1}{12}$

6. Manus weeded $\frac{3}{5}$ of his garden in the morning and another $\frac{1}{3}$ after school. The remainder of his garden that is not weeded is:

A. $\frac{1}{5}$ **B.** $\frac{1}{2}$ **C.** $\frac{1}{15}$ **D.** $\frac{1}{8}$

7. John and Amos buy 16 mandarins. If John eats 5 mandarins and Amos eats a quarter of the total, what fraction of the total is remaining?

A. $\frac{7}{16}$ **B.** $\frac{9}{16}$ **C.** $\frac{2}{9}$ **D.** $\frac{7}{9}$

8. Josie spent two-fifths of her money on food and then one-third of the remainder on clothing. If she had K36.00 to start with, how much does she have left?

A. K9.60 **B.** K9.00 **C.** K14.40 **D.** K12.00

9. At the beginning of the week $\frac{4}{9}$ of a garden is planted with sweet potato and $\frac{1}{3}$ with taro. At the end of the week half of the sweet potato and $\frac{3}{7}$ of the taro is harvested. What fraction of the garden now contains no plants?

A. $\frac{23}{63}$ **B.** $\frac{37}{63}$ **C.** $\frac{40}{63}$ **D.** $\frac{107}{252}$

Unit 11.1 Number and Application
Topic 4: Basic numeracy—decimals and approximations

This topic is also about solving number problems in context. It covers:
- Decimals.
- Making sensible estimates and checking the reasonableness of results.
- Using numbers expressed in standard form.
- Solve number problems in context, involving manipulation, several steps or reversing processes.

Introduction

In a decimal number such as 16.345, the **decimal point** separates the *whole part* of the number, 16, from the *fractional part* of the number, 345 thousandths. Numbers with a decimal point are frequently referred to as **decimals**. In certain calculations it is useful to treat a whole number as a decimal, so a decimal point and extra zeros are added, eg 25 may be written as 25.0 or 25.00, etc.

Changing fractions to decimals

To change a fraction to a decimal the **numerator** of the fraction is divided by the **denominator** of the fraction.

Example A

Q. Change $\frac{1}{4}$ to a decimal.

A. The division $1 \div 4$ is done, as the following sequence shows:

$$\frac{1}{4} \longrightarrow 4\overline{)1} \longrightarrow \begin{array}{r}0.25\\4\overline{)1.00}\end{array}$$ [inserting decimal point and extra zeros]

$\therefore \frac{1}{4} = 0.25$

This can be carried out using the division key on the calculator:

press [1] [÷] [4] [=] to get = 0.25.

Notes: 1. Once a fraction is entered into a scientific calculator and [=] is pressed, pressing the fraction key [$a^{b/c}$] again will automatically convert the fraction to a decimal.

2. Changing fractions to decimals is a useful alternative method for comparing the size of fractions, eg $\frac{3}{8} = 0.375$, $\frac{2}{5} = 0.4$. So $\frac{2}{5}$ is bigger than $\frac{3}{8}$.

3. It is useful to be familiar with the decimal form of simple fractions such as $\frac{1}{2}, \frac{1}{4}, \frac{1}{5}, \frac{3}{4}$ etc; and vice versa. [$\frac{1}{2} = 0.5$, $\frac{1}{4} = 0.25$, $\frac{1}{5} = 0.2$, $\frac{3}{4} = 0.75$]

To convert a mixed number to a decimal, convert the fraction to a decimal first, then add the whole number, eg $15\frac{1}{4} = 15.25$.

Changing decimals to fractions

To change a decimal number to a fraction:

- The numerator of the fraction is equal to the whole number formed by removing the decimal point from the decimal.
- The denominator is a power of ten equal to the number of **digits** after the decimal point in the original decimal.

The resulting fraction should be simplified if possible.

Example B

Q. Change to fractions: **1.** 0.3 **2.** 0.47 **3.** 0.0125

A. **1.** $0.3 = \frac{3}{10}$ [numerator is 03 = 3, denominator = 10^1]

2. $0.47 = \frac{47}{100}$ [numerator is 047 = 47, denominator = 10^2]

3. $0.0125 = \frac{125}{10\,000}$ [numerator is 00125 = 125, denominator = 10^4]

$= \frac{1}{80}$

Note: **1.** The denominators in the fractions above have as many zeros as there are digits after the decimal point: 0.3 has one digit after the decimal place and therefore one zero in the denominator, 0.47 has two digits, so two zeros, 0.0125 has four digits, so four zeros, and so on.

2. By pressing the fraction key [$a^{b/c}$] after a decimal is entered and [=] is pressed, many scientific calculators will automatically convert a decimal to a fraction in its simplest form. For example, pressing [•] [2] [=] enters the decimal 0.2 into the calculator. Pressing [$a^{b/c}$] after this automatically converts the calculator display to [1⌟5] which is the fraction $\frac{1}{5}$.

Recurring decimals

When using division to change a fraction to a decimal, the same digit or group of digits after the decimal point may repeat. This type of decimal is called a **recurring** or **repeating decimal**. To indicate a recurring part, the repeating part is written once with a *dot above the digits at the beginning and end of it*.

Example C

1. 0.123123 ... is written $0.\dot{1}2\dot{3}$

2. $\frac{1}{3} = 0.333\ldots$ and is written $0.\dot{3}$

3. $\frac{5}{6} = 0.8333\ldots$ and is written $0.8\dot{3}$

4. $15\frac{4\,508}{9\,990} = 15.4512512\ldots$ and is written $15.4\dot{5}1\dot{2}$

Decimals and the basic operations

A calculator makes short work of decimal calculations involving the basic operations of addition, subtraction, multiplication and division.

Multiplying by powers of 10

It is important to understand the rule for multiplying decimals by powers of ten. To multiply a decimal number by a power of 10, the decimal point is moved the same *number* of places as there are zeros in the power of 10.

- The decimal point is moved to the *right* for positive powers.
- The decimal point is moved to the *left* for negative powers.

Note: Powers of ten include: $10^0 = 1$, $10^1 = 10$, $10^2 = 100$, $10^3 = 1\ 000$, $10^4 = 10\ 000$ etc

$10^{-1} = \frac{1}{10} = 0.1$, $10^{-2} = \frac{1}{10^2} = \frac{1}{100} = 0.01$, $10^{-3} = \frac{1}{10^3} = \frac{1}{1\ 000} = 0.001$, etc

Further explanation on powers can be found under *Indices* on pages 5–8 in Chapter 1.

Example D

1. $16.23 \times 10 = 162.3$ [decimal point is moved one place to the right]
2. $2.46 \times 1\ 000 = 2\ 460$ [decimal point is moved three places to the right]
3. $0.2437 \times 10^2 = 24.37$ [decimal point is moved two places to the right]
4. $1.5 \times 10^{-2} = 0.015$ [decimal point is moved two places to the left]
5. $34.86 \times 10^{-4} = 0.003486$ [decimal point is moved four places to the left]

Note: The '0' in 2 460 is required as a **place holder** in **2**. (Zeros are also used in **4** and **5**.)

Unit 11.1 Activity 4A: Calculations involving decimals

1. Change the following fractions to decimals:

a. $\frac{4}{5}$ **b.** $3\frac{19}{25}$ **c.** $\frac{5}{6}$ **d.** $18\frac{7}{12}$ **e.** $\frac{13}{99}$

Change the following decimals to fractions (or mixed numbers) in their simplest form:

f. 0.546 **g.** 3.02 **h.** $4.\dot{3}$ **i.** $0.\dot{4}$ **j.** −11.001

2. Calculate each of the following:

a. $1.2 \times 10 + 3.2$ **b.** $3.2 \times 10^{-1} + 1.5$ **c.** $\frac{0.3 + 4.6}{13 \times 0.2}$ **d.** $\frac{0.4 \times 0.2^2}{0.8}$

e. $\frac{5.8 + 4.6}{13 \times 0.2}$ **f.** $\frac{4.9 \times 2.5}{1.5 - 0.25}$ **g.** Subtract 9.4 from 6.3^2.

h. Multiply (5.2×10^3) by (1.2×10^{-4}).

3. Evaluate each of the following (without a calculator):

a. 10×5.4 **b.** 0.7×100 **c.** 4.2×10^2 **d.** 23.8×10^{-2}

e. 5.763×10^4 **f.** 2.958×10^{-3} **g.** 13.581×10^3

4. Four towns lie along a straight road:

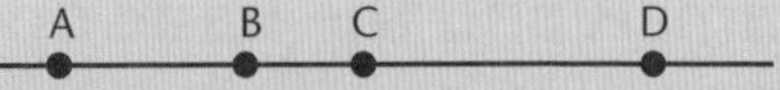

The distance between town A and town C is 18.5 km and the distance between town B and town D is twice that distance. If the distance between town A and town D is 45 km, find the distance between town B and town C.

5. I have K40 to spend during Monday, Tuesday and Wednesday. On Monday and Tuesday I spend a total of K26.45. On Tuesday and Wednesday I spend a total of K15.30. If I have K5.20 left at the end of Wednesday, how much did I spend on Tuesday?

6. Using his calculator, Silas inputs a number, multiplies it by 4.8, presses [=] and gets 16.896. What number was input?

7. In Sam's short answer test he got 13 marks out of 20. In the long answer section he got 49 out of 80.

   In which part of the test did Sam do better?

8. Janet is 1.5 m tall and her brother is 1.2 times taller. Their father is half their combined height, which is 0.1 m more than their mother's height.

   a. What is the combined height of Janet's mother and father?

   b. Is Janet more or less than half of her parents' combined height?

9. To calculate Peter's mark, the average of 2.4, 5.76 and 3.09 is added to the (positive) difference between 9.76 and 5.81. The answer is then multiplied by 10. What is Peter's mark?

10. Beth's telephone bill is made up of a charge for line rental and a charge for calls. Beth budgets K70 per month for her phone bill. In July she made K27.45 worth of calls and her bill was K12.72 above her budgeted amount. How much is Beth charged for line rental?

Significant figures

- Every non-zero digit in a number is a **significant figure** (**sf**).
- Any zero that is not at the end of a whole number or at the beginning of a decimal fraction is also a significant figure.

To count the number of significant figures in a number, start counting (from the left) at the first non-zero digit and continue to the end of the number (excluding zeros on the end of a whole number). All zeros within the number and at the end of a decimal are counted.

Example E

Q. State the number of significant figures in each of the following numbers:

1. 641 **2.** 0.0036 **3.** 3 425 000 **4.** 5.009 **5.** 89.0

A.

1. 641	3 significant figures	[the first is 6, the second is 4 and the third is 1]
2. 0.0036	2 significant figures	[count from 3 (leading zeros are not significant)]
3. 3 425 000	4 significant figures	[zeros at the end of a whole number are not significant – see note]
4. 5.009	4 significant figures	[zeros within a number are significant]
5. 89.0	3 significant figures	[zeros at the end of a decimal are significant (the final zero indicates that the number is accurate to the nearest tenth)]

Note: In part **3**, the number 3 425 000 may be accurate to 4, 5, 6 or 7 significant figures, depending on how many of the final zeros are actual (counted) values, instead of placeholder zeros resulting from rounding. For example, there may be exactly 3 425 000 people on a mailing list (the figure is accurate to 7 sf). Alternatively, a country's population may be given to the nearest hundred as 3 425 000 in which case the figure is accurate to 5 sf. Consequently, all zeros at the end of a whole number are assumed to be not significant unless otherwise stated.

Rounding

Rounding a number means writing the number with fewer significant figures. The rounded number will be approximately equal in size to the original number.

To **round** a number to a particular number of significant figures:

- Start from the first significant figure in the number and count to the right the number of significant figures required.
- If the next digit is *5 or greater*, add a 1 to the last significant figure and use zeros where necessary as place holders.
- If the next digit is *less than* 5, leave the last significant figure unchanged and use zeros where necessary as place holders.

Example F

1. 3 486 = 3 500 (2 sf) [The first two significant figures are 3 and 4. The next digit is 8, which is bigger than 5, so 1 is added to the second significant figure, 4. Two zeros are used as place holders.]
2. 0.0642 = 0.064 (2 sf) [The first two significant figures are 6 and 4. The next digit is 2, which is less than 5, so the 4 is left unchanged. Two zeros are used as place holders.]
3. 20.850 = 20.9 (3 sf) [The first three significant figures are 2, 0 and 8. The next digit is 5, so 1 is added to the third significant figure 8. No zeros are needed as place-holders.]

Note: 1. A common error is to forget place holder zeros when rounding whole numbers. For example in part **1** the rounded number 3 500 is similar in size to 3 486. A rounded value of 35 would be absurd.

2. Do not write the answer in part **3** as 20.900 since this would imply that the answer is accurate to the nearest one thousandth (ie 20.900 has 5 sig figs).]

When rounding a number to a particular number of **decimal places**, the digits *after* the decimal point are counted.

Example G

1. 18.346 = 18.35 (2 dp) [The first two digits after the decimal point are 3 and 4. The next is 6, which is bigger than 5, so 1 is added to the 4 in the second decimal place.]
2. 4.0523 = 4.052 (3 dp) [The first three digits after the decimal point are 0, 5, and 2. The next is 3, which is less than 5, so .052 is left unchanged.]
3. 1.98 = 2.0 (1 dp) [The first digit after the decimal point is 9 and the next Is 8, which is greater than 5, so 1 must be added to 9.]

Note: **1.** The answer to part **1** is 18.35, *not* 18.350, because the zero in 18.350 shows three digits after the decimal point, not two.

2. In part **3**, the 0 in the 2.0 is essential because it shows the rounding to 1 dp has been done. In other words, the 0 is a significant figure.

Rounding answers to calculations

Calculator answers are often given to too many decimal places when considering the accuracy of the figures started with. For example, in the calculation $16.2 \div 1.72 = 9.418604651$, numbers with 3 sf give rise to an answer with 10 sf.

Answers to calculations should be rounded off to a *sensible* number of significant figures or decimal places. (A problem may specify exactly what accuracy is required eg round off to 2 decimal places.)

Whenever rounding is done, the accuracy of the rounding should be stated: 'to 3 sf', 'to 2 dp', etc.

Generally, rounding is done so that *the answer has the same number of significant figures as the number in the calculation with the least number of significant figures*.

This rule is flexible.

Example H

In the calculation $16.2 \div 1.72$, the answer should be rounded to 3 sf.

Thus $16.2 \div 1.72 = 9.42$ (3 sf)

Example I

Q. Calculate $\frac{19.35 \times 4.2}{6.81}$, rounding your answer sensibly.

A. $\frac{19.35 \times 4.2}{6.81} = 12$ (2 sf)

[Using a calculator, the answer is 11.9339207. Noting that 19.35 has 4 sf, 4.2 has 2 sf and 6.81 has 3 sf, the answer should have 2 sf. However, rounding to 2 sf is not a 'sensible' answer in some situations because it may suggest that the answer is *exactly* 12. The number may therefore be rounded to 3 sf to give 11.9.]

Estimates

An exact answer to a calculation is not always required. In many practical situations, an **estimate** (an **approximate answer**) may be sufficient.

An estimate is made by rounding the numbers in a calculation to one (sometimes two) significant figures. When a calculation involves division, the divisor is rounded to 1 sf, and the numerator is rounded to a multiple of the divisor (which may be to 2 sf if necessary). An estimate can be used to check the accuracy of a calculator answer.

Example J

Q. Find estimates for the following calculations:

1. 28.3×846.7 2. $\frac{47.56}{6.34}$ 3. $\frac{734 \times 43.7}{8.87}$

A. 1. $28.3 \times 846.7 \approx 30 \times 800$ [rounding to 1 sf]

$= 24\,000$

2. $\frac{47.56}{6.34} \approx \frac{48}{6}$ [6.34 is rounded to 6 (1 sf) and 47.56 is rounded to 48, which is the closest multiple of 6 to 47.56]

$= 8$

3. $\frac{734 \times 43.7}{8.87} \approx \frac{700 \times 40}{9}$ [rounding all to 1 sf]

$= \frac{28\,000}{9}$ [simplifying]

$\approx \frac{27\,000}{9}$ [rounding numerator to a multiple of 9]

$= 3\,000$

Note: These estimates are not unique. They vary according to how the numbers involved are rounded.

Unit 11.1 Activity 4B: Significant figures and rounding

1. Write each of the following numbers correct to the number of significant figures shown in brackets:

a. 115 (2) **b.** 5 914 (3) **c.** 31 786 (3)

d. 14.713 (3) **e.** 89.863 (3) **f.** 0.0487 (2)

2. Write each of the following numbers correct to the number of decimal places indicated in brackets:

a. 1.537 (2) **b.** 11.705 (1) **c.** 0.4137 (3)

d. 89.8632 (2) **e.** 0.00413 (4) **f.** 9.97 (1)

3. Estimate answers to each of the following (without a calculator) giving answers to 1 or 2 significant figures.

a. 86×149 **b.** 43.4×847 **c.** 348×0.843

d. $\frac{855}{28}$ **e.** $\frac{98.7}{5.37}$ **f.** $\frac{3\,584}{84.7}$

g. $\frac{15.7 \times 2.8}{0.34}$ **h.** $\frac{437 \times 75.4}{29.1}$ **i.** $\frac{431 \times 682.5}{74}$

4. The volume of a cylinder is given by the formula $V = \pi r^2 h$, where r is the radius of the circular base, and h is the height. If $r = 1.89$, $h = 59.28$ and $\pi = 3.14159$ (6 sf):

a. Use appropriate whole number approximations to estimate the value of V.

b. Evaluate V correct to 2 decimal places, using the values given for r and h and your calculator value for π.

5. Sila used her calculator and the formula $A = \pi r^2$ to find the area of a circle with a radius of 6.2 cm. She used her calculator value for π. The answer her calculator gave was 120.76282.

a. Explain why she should not record her answer as 120.76282 cm^2.

b. Write down a sensible answer that Sila might use.

6. Christine was laying lino tiles on her kitchen floor. The area of the floor is 11.48 m^2 and the area of each tile is 0.0625 m^2. Tiles come in packets of 20. Assuming no tiles need to be cut, how many packets should Christine buy? Explain your answer.

7. The volume of a sphere is $\frac{4}{3}\pi r^3$. Use this formula:

a. To estimate the volume of a sphere whose radius is 2.75 cm. Give your answer to 1 sf.

b. To calculate the volume of a sphere whose radius is 2.75 cm, rounding your answer sensibly.

8. A rectangular garden is 4.75 m long and 3.9 m wide.

a. Estimate its area to 1 sf.

b. Calculate its area, rounding sensibly.

Standard form

A number is said to be in **standard form** if it is expressed in the form

(number between 1 and 10) × (power of ten)

For example the numbers 1.4×10^4 and 3.6743×10^{-8} are in standard form.

Writing numbers in standard form has a number of advantages:

- Large numbers, particularly those with many zeros, occupy considerably less space when written in standard form.
- It is often easier to round a number if it is written in standard form and to determine the number of significant figures it has.
- The sizes of very large, or of very small numbers are more readily compared when expressed in standard form.

To express a number in standard form, move the decimal point to directly after the first significant digit of the number, then multiply by the appropriate power of ten (so that the value of the original number is kept the same).

Example K

Q. Write each of the following in standard form:

1. 340 **2.** 0.0045 **3.** 16.3×10^2

A. **1.** $340 = 3.4 \times 10^2$ [Place the decimal point after the first significant figure (3). The decimal point moved 2 places so 3.4 needs to be multiplied by 10^2 in order to equal the original number, 340.]

2. $0.0045 = 4.5 \times 10^{-3}$ [The decimal point is placed after the first significant figure (4). The point moved three places, so 4.5 needs to be multiplied by 10^{-3}, in order to equal the original number 0.0045.]

3. $16.3 \times 10^2 = 1.63 \times 10 \times 10^2$ [expressing 16.3 in standard form]]

$= 1.63 \times 10^3$ [since $10 \times 10^2 = 10^3$]

Alternatively, $16.3 \times 10^2 = 1\ 630 = 1.63 \times 10^3$

Note: A good rule to remember is that when numbers are expressed in standard form, large numbers (greater than 10) have a positive power of ten, whereas small numbers (less than 1) have a negative power of ten.

To change a number written in standard form back to an 'ordinary' number, move the decimal point in the usual manner for multiplying by powers of ten. Often zeros must be inserted as place-holders.

Example L

Q. Write each of the following as ordinary numbers:

1. 1.34×10^2 **2.** 2.4×10^4 **3.** 3.6×10^{-3}

A. **1.** $1.34 \times 10^2 = 134$ [The 2 in 10^2 shows that the decimal point needs to move two places to the right.]

2. $2.4 \times 10^4 = 24\ 000$ [The 4 in 10^4 shows that the decimal point is moved four places to the right (insert three zeros as place-holders.]

3. $3.6 \times 10^{-3} = 0.0036$ [The –3 in 10^{-3} indicates the decimal point needs to move three places to the left (insert two zeros as place-holders).]

Standard form and calculators

Scientific calculators automatically express very large or very small answers to calculations in standard form. The answer is displayed as a number between one and ten, followed by a gap, then the power of ten required. Since a calculator has a fixed number of digits in its display, this enables a much larger range of answers to be displayed.

Numbers can be entered into the calculator using the exponent [EXP] key. The calculator will leave the number in standard form if it is a very large or very small number. Otherwise it will convert it to an 'ordinary' number.

Example M

1. To enter 2.598×10^{-6}, press [2] [·] [5] [9] [8] [EXP] [(–)] [6].

Pressing [=] will give [2.598 –06] or similar.

2. To enter 3.54×10^4 press [3] [·] [5] [4] [EXP] [4] [=] to get a display [35400].

The calculator has converted the number since it is in the acceptable range of values for display in the 'ordinary' form.

Note: On some calculators [+/–] is pressed instead of [(–)].

Scientific calculators and the [EXP] key make fast work of calculations done with numbers in standard form.

Example N

Q. Calculate the following, leaving your answer in standard form.

1. $2.6 \times 10^{-4} \times 4 \times 10^{8}$ **2.** $\frac{2.6 \times 10^{-4}}{4 \times 10^{8}}$

A. 1. By calculator, press [2] [•] [6] [EXP] [(–)] [4] [×] [4] [EXP] [8] [=]

to get a display [104000].

This can be converted to standard form in the usual way to 1.04×10^{5}.

2. By calculator, press [2] [•] [6] [EXP] [–] [4] [÷] [4] [EXP] [8] [=]

to get 6.5×10^{-13}.

Note: The calculator treats a number such as 2.6 EXP – 4 as a single number so no brackets are needed around this expression when dividing on a calculator.

Rounding numbers in standard form

Numbers in standard form are rounded by rounding the 'number between 1 and 10' in the usual way (the power of ten is unchanged).

Example O

Q. Calculate $(765\ 258)^2$, rounding your answer appropriately.

A. Enter [765258], then press [x^2] and [=].

Depending upon the type of calculator used, the calculator displays

[$5.856198066 \times 10^{11}$] *or* [5.856198066 11] *or* [5.856198066 E+11]

which is the number 5.85620×10^{11} (6 sf) [rounding]

Note: The number of significant figures is clear from the number 5.85620 whose final zero would not be included unless it was significant. (6 sf is chosen since 765 258 has 6 sf.)

Comparing numbers in standard form

To compare numbers in standard form, compare their powers of ten first. The larger the number, the higher the power of ten it has. If powers of ten are the same, compare their initial numbers (between one and ten).

Example P

Q. Put in order of size from smallest to largest:

3.56×10^{4}, 1.9×10^{-2}, 3.8×10, 4.6×10^{-2}, 9.019×10^{5}

A. The lowest power of 10 is –2, then 1, 4, 5.

Comparing the two numbers with the same power of 10, note that
$1.9 \times 10^{-2} < 4.6 \times 10^{-2}$ [since $1.9 < 4.6$]

The required order is, therefore:
1.9×10^{-2}, 4.6×10^{-2}, 3.8×10, 3.56×10^{4}, 9.019×10^{5}

Unit 11.1 Activity 4C: Standard form

1. Write each of the following in standard form:

a. 487 **b.** 9 120 **c.** 763 000 **d.** 0.0743

e. 0.00518 **f.** 0.00000451 **g.** 69 500 000 **h.** 495.8

i. 16.401 **j.** 126.5×10^2 **k.** 0.0032×10^4 **l.** 48.59×10^{-3}

2. Write each of the following as ordinary numbers:

a. 2.51×10^2 **b.** 6.39×10^3 **c.** 8.412×10^4 **d.** 3.57×10^{-1}

e. 1.7×10^{-3} **f.** 9.75×10^{-4} **g.** 3.48×10 **h.** 4.6×10^4

i. 7.83×10^{-2} **j.** 18.45×10^3 **k.** 126.4×10^{-2} **l.** 0.57×10^{-3}

3. Put in order of size, from smallest to largest, each of the following sets of numbers:

a. 2.4×10^3, 3.65×10^2, 1.901×10^4, 3.7×10^2

b. 4.1×10^{-2}, 0.0014, 2.6×10^{-1}, 0.002

c. 12.6×10^3, 1.3×10^4, 12 965, 9.9×10^3

4. Write each of the following correct to the number of significant figures shown in brackets.

a. 4.754×10^3 (2) **b.** 5.6648×10^{-3} (4) **c.** 8.98×10^4 (2)

d. 7.841×10^{-2} (2) **e.** 3.76×10^5 (1) **f.** 9.95×10^{-3} (2)

5. Find approximate answers to each of the following, then verify by calculation, leaving your answer in standard form, correct to three significant figures.

a. $8.5 \times 10^2 \times 2.6 \times 10^3$ **b.** $(7.48 \times 10^5) - (4.7 \times 10^4)$

c. $(6.92 \times 10^4)^2$ **d.** $\dfrac{8.74 \times 10^3}{6.5 \times 10^2} \times \dfrac{43 \times 10^{-2}}{7.3 \times 10^{-3}}$

6. A website gave the estimated population of New Zealand on 12 October 2004 as 4 071 829. The population of Auckland, the largest city in New Zealand, was given as 1.05×10^6. Using these data:

a. Express the population of New Zealand in standard form to 3 significant figures.

b. Approximately how many New Zealanders do not live in Auckland?

c. Approximately 8 out of every 37 people in New Zealand live in the South Island. Estimate the population of the South Island on 12 October 2004. Give your answer in standard form to 3 significant figures.

7. The mass of an electron at rest is 9.109×10^{-31} kg. The mass of a proton at rest is 1.673×10^{-27} kg, both figures correct to 4 significant figures. How many times greater is the mass of the proton, when compared with the mass of the electron?

8. The planet Pluto is 5 913 000 000 km from the sun. Neptune, another planet, is 4.498×10^9 km from the sun. How much further from the sun is Pluto than Neptune? Give your answer in standard form.

9. In 2003 the population of China was approximately 1.29×10^9 and the population of New Zealand was 3.95×10^6.

a. How much greater was the population of China than the population of New Zealand. Give your answer in standard form to 4 significant figures. Explain why an answer to 3 significant figures would be misleading.

b. How many times greater was the population of China than the population of New Zealand? Give your answer correct to 3 significant figures.

10. A confectionery manufacturer produces 1.76×10^6 individual sweets and 2.41×10^5 individual chocolates per year. He has been producing chocolates for 15 years and sweets for 10 years.

a. Which type of confectionery has the manufacturer produced more of since he has been in business, sweets or chocolates? How many more? (Give your answer in standard form to 3 significant figures.)

b. $\frac{4}{5}$ of the sweets produced by the manufacturer are unwrapped. How many wrapped sweets are produced per year? Give answer in standard form.

c. The manufacturer produces twice as many soft-centred chocolates as hard-centred ones. How many soft-centred chocolates are produced annually? Give answer in standard form to 3 significant figures.

Unit 11.1 Activity 4D: Multiple choice – decimals

1. The number 8476 written correct to 2 significant figures and 3 significant figures, respectively, is:

A. 8500 and 8470 **B.** 85 and 848

C. 8500 and 8480 **D.** 8400 and 8480

2. For the number 654386, which one of the following is **not** a correct rounding?

A. 654000 to 3 significant figures **B.** 654300 to 4 significant figures

C. 700000 to 1 significant figure **D.** 654390 to 5 significant figures

3. 963.44×10^5 written in standard form is

A. 963.44×10^5 **B.** 9.6344×10^3

C. 9.6344×10^7 **D.** 96344000

4. $8 \times 10^2 \times 12 \times 10^{-3}$ is equal to:

A. 96×10^{-6} **B.** 9.6×10^{-1}

C. 96×10^1 **D.** 9.6

5. $\frac{2.4 \times 10^3}{0.00008}$ written in standard form is:

A. 3×10^7 **B.** 0.3×10^8

C. 3×10^8 **D.** 3×10^{-3}

Unit 11.1 Number and Application
Topic 5: Basic numeracy—irrational numbers and surds

This Topic covers the content as set out on p. 13 of the syllabus: apply the properties of surds, simplify and rationalise surds. It includes:

- The number π.
- The number e.
- Simplifying surds.
- Addition, subtraction and multiplication of surds.
- Conjugation of surds.
- Rationalising the denominator of fractional surds.

The number π

The ratio of the circumference to the diameter of any circle is a constant number called **pi**, usually represented by the Greek letter π.

Ancient civilisations knew that there was a ratio between the circumference and the diameter of a circle and that it was approximately equal to 3. Pi proved to be a number that fascinated mathematicians and the approximation to pi was improved over the centuries until the mathematician Lambert, in1761, proved that pi was an **irrational number**; it could *not* be written as a ratio of integers.

- $\pi = 3.141\ 593\ 654\ 589\ 793\ 238\ 46$; correct to 20 decimal places. It is an infinite non-recurring decimal.
- Archimedes calculated pi as a number between $\frac{223}{71}$ and $\frac{22}{7}$ in 250 BC.
- The Chinese used the approximation $\frac{355}{113}$ in AD 450.
- Many famous mathematicians have developed formulae to calculate pi (these can be found if you research pi on the Internet).
- Computers have been put to the task of calculating pi and it can now be written correct to millions of decimal places.

The number e

It was noticed, when calculating compound interest, that the formula $\left(1+\frac{1}{n}\right)^n$ converged to a constant number between 2 and 3 as n became larger. This number, denoted by e, is commonly called Euler's number; it is named after the mathematician Leonhard Euler, who first referred to it in 1727. The number e also fascinated many mathematicians who developed formulae for its calculation.

- ***e*** is an irrational number; an infinite non-recurring decimal.
- $e = 2.718\ 281\ 828\ 459\ 045\ 235\ 36$; correct to 20 decimal places.
- You can find the number e on your calculator by calculating e^1.

Unit 11.1 Activity 5A: The numbers π and e

1. Use your calculator to determine each of the following approximations to pi. State the accuracy correct to the relevant number of decimal places.

   a. $\frac{223}{71}$ **b.** $\frac{22}{7}$ **c.** $\sqrt{2}+\sqrt{3}$ **d.** $\frac{355}{113}$

   e. $4\left(\frac{1}{1}-\frac{1}{3}+\frac{1}{5}-\frac{1}{7}\ \ldots\ldots -\frac{1}{15}\right)$ Try extending this pattern further.

   f. The square root of the following number: $6\left(\frac{1}{1^2}+\frac{1}{2^2}+\frac{1}{3^2}\ \ldots\ldots\ldots\ \frac{1}{10^2}\right)$. Try extending this pattern further.

2. Use your calculator to determine each of the following approximations to *e*. State the accuracy correct to the relevant number of decimal places.

   a. $\left(1+\frac{1}{10}\right)^{10}$ **b.** $\left(1+\frac{1}{100}\right)^{100}$

   c. $\left(1+\frac{1}{1000}\right)^{1000}$ **d.** $1+\frac{1}{1!}+\frac{1}{2!}+\frac{1}{3!}+\frac{1}{4!}+\frac{1}{5!}=1+\frac{1}{1}+\frac{1}{2}+\frac{1}{6}+\frac{1}{24}+\frac{1}{120}$

   (Note: $4! = 1 \times 2 \times 3 \times 4 = 24$ etc.)

   e. Extend the formula from part **d** until you get an approximation that is correct to 4 decimal places. Write down the fractions that you need to add to your answer from part **d**.

Surds

Surds are **irrational** numbers. Surds cannot be written as a ratio of integers.

Surds like $\sqrt{2}$ or $\sqrt[3]{7}$ are **infinite, non-recurring decimals**. This means that the decimal parts go on forever, not repeating the pattern of decimals.

The **exact value of a surd** is written as $\sqrt{2}, \sqrt[3]{5}, \sqrt{17}, \sqrt{6.85}, \sqrt{0.0258}$ … etc although approximations can be found for all of them.

There are some square roots that are not surds, ie they are rational numbers.

- $\sqrt{4} \times \sqrt{4} = 4$ but we know that $2 \times 2 = 4$ so $\sqrt{4} = 2$, which is a rational number.
- $13 \times 13 = 169$ so $\sqrt{169} = 13$; a rational number.
- $1.1 \times 1.1 = 1.21$ so $\sqrt{1.21} = 1.1 = \frac{11}{10}$; a rational number.
- The numbers 4, 9, 16, 25, 36, 49, 1.21 etc are called **perfect squares** because they have **rational** square roots.

Note: It is not possible to find the square root of a negative number, ie it is impossible to find a number n such that $n \times n$ = a negative number.

Some numbers are perfect cubes and so their cube root is not a surd.

- $2^3 = 8$ so $\sqrt[3]{8} = 2$; a rational number.
- $\sqrt[3]{-27} = -3$; a rational number, because $-3 \times -3 \times -3 = -27$.

Common perfect cubes are 8, 27, 64, 125, 216, 343 …

Simplifying surds

$\sqrt{x}$ is called a **simple surd** if none of the factors (excluding 1) of x are perfect squares; eg $\sqrt{6}$ is a simple surd because it has factors (1), 2, 3 and 6 and none of these are perfect squares.

Surds that have one or more factors that are perfect squares can be **simplified**.

$\sqrt{12}$ can be simplified because it has a factor, 4, that is a perfect square

$\sqrt{12} = \sqrt{4 \times 3} = \sqrt{4} \times \sqrt{3} = 2 \times \sqrt{3} = 2\sqrt{3}$.

$\sqrt{12}$ is called a **single surd** and $2\sqrt{3}$ is the **simplified** version of $\sqrt{12}$.

Unit 11.1 Activity 5B: Surds

1. Which of the following numbers are surds?

$\sqrt{101}$, 81, $\sqrt{36}$, $\sqrt{17}$, 1.732, $\sqrt[3]{0.125}$, $\sqrt{8}$, $\sqrt{83.4}$, $\sqrt[3]{216}$, $\sqrt{1.44}$, $\sqrt{4900}$

2. a. Use your calculator to find a decimal value for the following numbers, rounding to two decimal places where necessary.

$\sqrt{64}, \sqrt{0.25}, \sqrt[3]{81}, \sqrt{400}, \sqrt{4000}, \sqrt[3]{-0.216}, \sqrt{-0.36}, \sqrt{0.0016}, \sqrt[3]{-0.8}, \sqrt{250000}$

b. Which of the numbers in part **a**. are perfect squares?

3. Identify the following numbers as irrational (Q') or rational (Q):

a. $\sqrt{0.49}$ **b.** $-\sqrt{0.4}$ **c.** $\sqrt{0}$ **d.** $1 + \sqrt{1.6}$

e. $5\sqrt[3]{8}$ **f.** $4 - 5\sqrt[3]{0.08}$ **g.** $\sqrt[5]{-32}$ **h.** $7 - 3\sqrt[3]{64}$

4. Simplify the following surds:

a. $\sqrt{50}$ **b.** $\sqrt{48}$ **c.** $\sqrt{98}$ **d.** $\sqrt{320}$

e. $\sqrt{675}$ **f.** $\sqrt{108}$ **g.** $\sqrt{324}$ **h.** $\sqrt{8}$

5. Express as a single surd:

a. $5\sqrt{3}$ **b.** $-2\sqrt{7}$ **c.** $10\sqrt{10}$ **d.** $5\sqrt{6}$

e. $-3\sqrt{5}$ **f.** $4\sqrt{2}$ **g.** $2\sqrt{11}$ **h.** $0.2\sqrt{14}$

Addition and subtraction of surds

Only 'like' surds can be added or subtracted. Like surds have the same number under the root sign. For example $\sqrt{3}, -2\sqrt{3}$, and $5\sqrt{3}$ are like surds.

Example A

Q. Simplify:

1. $\sqrt{3} - 2\sqrt{3} + 5\sqrt{3}$
2. $6\sqrt{5} + 5\sqrt{3} - 4\sqrt{5}$
3. $\sqrt{12} + \sqrt{8} - \sqrt{2}$

A. 1. $\sqrt{3} - 2\sqrt{3} + 5\sqrt{3} = (1 - 2 + 5)\sqrt{3} = 4\sqrt{3}$

2. $6\sqrt{5} + 5\sqrt{3} - 4\sqrt{5} = 2\sqrt{5} + 5\sqrt{3}$

3. $\sqrt{12} + \sqrt{8} - \sqrt{2} = \sqrt{4 \times 3} + \sqrt{4 \times 2} - \sqrt{2}$

$= 2\sqrt{3} + 2\sqrt{2} - \sqrt{2}$

$= 2\sqrt{3} + \sqrt{2}$

Unit 11.1 Activity 5C: Simplifying surds

1. Simplify where possible.

a. $5\sqrt{7} + 7\sqrt{3} - 2\sqrt{7}$ **b.** $3\sqrt{3} - \sqrt{2} + 4\sqrt{2} - \sqrt{3}$

c. $\sqrt{6} + 2\sqrt{3} - 6\sqrt{3} + 3$ **d.** $5\sqrt{2} + 3\sqrt{5} - 2\sqrt{10}$

2. Simplify:

a. $\sqrt{24} - \sqrt{6}$ **b.** $\sqrt{45} + \sqrt{80}$

c. $\sqrt{75} - \sqrt{147}$ **d.** $2\sqrt{3} + \sqrt{300}$

e. $\sqrt{75} - 2\sqrt{12}$ **f.** $\sqrt{125} - \sqrt{5}$

g. $\sqrt{242} + \sqrt{99}$ **h.** $\sqrt{32} + \sqrt{50} - 3\sqrt{2}$

3. Simplify the following surd expressions:

a. $2 + \sqrt{12} - 3 + \sqrt{8}$ **b.** $3\sqrt{5} - 2\sqrt{18} + 3\sqrt{8} - \sqrt{75}$

c. $4 - 3\sqrt{24} - \sqrt{144} + 4\sqrt{48}$ **d.** $\sqrt{6} + 2\sqrt{3} - 5\sqrt{12} - \sqrt{24} + 2\sqrt{72}$

e. $\sqrt{2} + \sqrt{4} + \sqrt{8} + \sqrt{16} + \sqrt{32} + \sqrt{64} + \sqrt{128}$

f. $\sqrt{3} - \sqrt{9} + \sqrt{27} - \sqrt{81} + \sqrt{243} - \sqrt{729}$

Multiplication of surds

Surd terms can be multiplied together:

$a\sqrt{m} \times b\sqrt{n} = ab\sqrt{mn}$, with the surd $\sqrt{mn}$ being simplified if possible.

Example B

Q. Simplify:

1. $\sqrt{5} \times \sqrt{3}$

2. $3\sqrt{6} \times 5\sqrt{3}$

3. $\sqrt{3} \times 2\sqrt{3} \times 5\sqrt{3}$

A. 1. $\sqrt{5} \times \sqrt{3} = \sqrt{5 \times 3} = \sqrt{15}$

2. $3\sqrt{6} \times 5\sqrt{3} = 3 \times 5\sqrt{6 \times 3} = 15\sqrt{18} = 15 \times 3\sqrt{2} = 45\sqrt{2}$

3. $\sqrt{3} \times 2\sqrt{3} \times 5\sqrt{3} = 2 \times 5\sqrt{3 \times 3 \times 3} = 10\sqrt{27} = 10 \times 3\sqrt{3} = 30\sqrt{3}$

Expanding brackets containing surds

The distributive law can be used to expand surd expressions:

$a(b + c) = ab + ac$

$(a + b)(c + d) = ac + ad + bc + bd$

Example C

Q. Expand and simplify:

1. $3\sqrt{2}(2\sqrt{6}+5)$
2. $\left(2\sqrt{5}-\sqrt{2}\right)\left(\sqrt{15}+7\right)$
3. $\left(\sqrt{7}+\sqrt{5}\right)\left(\sqrt{7}-\sqrt{5}\right)$

A. 1. $3\sqrt{2}\left(2\sqrt{6}+5\right)=3\times2\sqrt{2\times6}+3\times5\sqrt{2}$

$=6\sqrt{12}+15\sqrt{2}$

$=6\times2\sqrt{3}+15\sqrt{2}$

$=12\sqrt{3}+15\sqrt{2}$

2. $\left(2\sqrt{5}-\sqrt{2}\right)\left(\sqrt{15}+7\right)=2\sqrt{5\times15}+2\times7\sqrt{5}-\sqrt{2\times15}-7\sqrt{2}$

$=2\sqrt{75}+14\sqrt{5}-\sqrt{30}-7\sqrt{2}$

$=2\times5\sqrt{3}+14\sqrt{5}-\sqrt{30}-7\sqrt{2}$

$=10\sqrt{3}+14\sqrt{5}-\sqrt{30}-7\sqrt{2}$

3. $\left(\sqrt{7}+\sqrt{5}\right)\left(\sqrt{7}-\sqrt{5}\right)=\sqrt{7\times7}-\sqrt{7\times5}+\sqrt{5\times7}-\sqrt{5\times5}$

$=\sqrt{49}-\sqrt{35}+\sqrt{35}-\sqrt{25}$

$=7-5$

$=2$

Conjugate surds

The surd expressions $a\sqrt{m}+b\sqrt{n}$ and $a\sqrt{m}-b\sqrt{n}$ are called **conjugate pairs**. The terms are the same in each expression but the connecting sign is different.

When conjugate pairs are multiplied the result is a **rational number**.

$$(a\sqrt{m}+b\sqrt{n})(a\sqrt{m}-b\sqrt{n})=a\times a\sqrt{m\times m}-ab\sqrt{mn}+ab\sqrt{mn}-b\times b\sqrt{nn}$$

$$=a^2m-b^2n;\text{ a rational number}$$

$$=(\text{first term})^2-(\text{second term})^2$$

Example D

Q. Simplify:

1. $\left(\sqrt{3}+\sqrt{2}\right)\left(\sqrt{3}-\sqrt{2}\right)$
2. $\left(2\sqrt{7}+\sqrt{3}\right)\left(2\sqrt{7}-\sqrt{3}\right)$
3. $\left(3\sqrt{11}-5\right)\left(3\sqrt{11}+5\right)$

A. 1. Checking that the terms in each bracket are the same but that the connecting sign is different we can see that the expressions to be multiplied are a conjugate pair.

Hence $\left(\sqrt{3}+\sqrt{2}\right)\left(\sqrt{3}-\sqrt{2}\right)=$ *(first term)*$^2-$ *(second term)*^{2}

$=\left(\sqrt{3}\right)^2-\left(\sqrt{2}\right)^2$

$=3-2$

$=1$

2. $\left(2\sqrt{7}+\sqrt{3}\right)\left(2\sqrt{7}-\sqrt{3}\right)$: terms are conjugate pairs.

$$\begin{aligned}\left(2\sqrt{7}+\sqrt{3}\right)\left(2\sqrt{7}-\sqrt{3}\right) &= \left(2\sqrt{7}\right)^2-\left(\sqrt{3}\right)^2 \\ &= 2\times 2\times\left(\sqrt{7}\right)^2-3 \\ &= 4\times 7-3 \\ &= 25\end{aligned}$$

3. $\left(3\sqrt{11}-5\right)\left(3\sqrt{11}+5\right)$: terms are conjugate pairs.

$$\begin{aligned}\left(3\sqrt{11}-5\right)\left(3\sqrt{11}+5\right) &= 3\times 3\times\left(\sqrt{11}\right)^2-5\times 5 \\ &= 9\times 11-25 \\ &= 99-25 \\ &= 74\end{aligned}$$

Unit 11.1 Activity 5D: Simplifying and expanding surds

1. Simplify:

a. $\sqrt{5}\times\sqrt{5}$ **b.** $\sqrt{15}\times\sqrt{5}$ **c.** $\sqrt{18}\times\sqrt{3}$

d. $\sqrt{45}\times\sqrt{10}$ **e.** $2\sqrt{3}\times 3\sqrt{2}$ **f.** $5\sqrt{12}\times\sqrt{6}$

2. Expand and simplify where possible:

a. $\left(2+\sqrt{5}\right)^2$ **b.** $\left(3\sqrt{2}-2\sqrt{3}\right)^2$ **c.** $\left(3\sqrt{5}-\sqrt{2}\right)^2$

d. $\left(4\sqrt{2}+3\right)^2$ **e.** $\left(12-5\sqrt{3}\right)^2$

3. Expand and simplify where possible (Hint: Simplify surds first):

a. $\sqrt{3}\left(2\sqrt{3}-1\right)$ **b.** $\left(\sqrt{7}+\sqrt{3}\right)\left(\sqrt{7}+\sqrt{3}\right)$

c. $\left(2\sqrt{3}+3\sqrt{2}\right)\left(\sqrt{3}-\sqrt{2}\right)$ **d.** $\left(\sqrt{12}-\sqrt{8}\right)\left(\sqrt{3}+1\right)$

e. $\left(\sqrt{50}-\sqrt{8}\right)\left(\sqrt{2}-\sqrt{75}\right)$ **f.** $\left(\sqrt{81}-\sqrt{27}\right)\left(1+2\sqrt{3}\right)$

4. Write, in simplest surd form, the conjugate of:

a. $2\sqrt{3}-4$ **b.** $2+\sqrt{24}$

c. $5-\sqrt{45}$ **d.** $\sqrt{125}+\sqrt{64}$

5. Simplify each of the following, writing your answer as a rational number:

a. $\left(2+\sqrt{5}\right)\left(2-\sqrt{5}\right)$ **b.** $\left(\sqrt{7}+\sqrt{4}\right)\left(\sqrt{7}-\sqrt{4}\right)$

c. $\left(\sqrt{11}-\sqrt{7}\right)\left(\sqrt{11}+\sqrt{4}\right)$ **d.** $\left(3\sqrt{2}+\sqrt{5}\right)\left(3\sqrt{2}-\sqrt{5}\right)$

e. $\left(4\sqrt{2}-3\right)\left(4\sqrt{2}+3\right)$ **f.** $\left(3\sqrt{7}+5\sqrt{2}\right)\left(3\sqrt{7}-5\sqrt{2}\right)$

g. $\left(12-5\sqrt{2}\right)\left(12+5\sqrt{2}\right)$

Rationalising the denominator of fractional surds

Fractional surds, for example $\frac{\sqrt{3}}{2\sqrt{5}+\sqrt{7}}$, are much easier to work with if they do not have surds in their denominators. The process of changing a fractional surd, with surds in the denominator, to one with a rational number in the denominator is called **rationalisation**.

- To rationalise the denominator of a fraction with a simple surd in the denominator, for example $\frac{a}{\sqrt{b}}$, we multiply that fraction by 1 in the form of $\frac{\sqrt{b}}{\sqrt{b}}$. This does not change the value of the fraction because we are multiplying by 1.

  $\frac{a}{\sqrt{b}} \times \frac{\sqrt{b}}{\sqrt{b}} = \frac{a\sqrt{b}}{b}$; the resulting fraction has a rational denominator.

- To rationalise the denominator of a fraction such as $\frac{\sqrt{3}}{2\sqrt{5}+\sqrt{7}}$ we multiply the fraction by 1 in the form of $\frac{\textit{conjugate of the denominator}}{\textit{conjugate of the denominator}}$. Multiplying by 1 does not change the fraction.

  $$\frac{\sqrt{3}}{2\sqrt{5}+\sqrt{7}} = \frac{\sqrt{3}}{2\sqrt{5}+\sqrt{7}} \times \frac{2\sqrt{5}-\sqrt{7}}{2\sqrt{5}-\sqrt{7}} = \frac{\sqrt{3}\left(2\sqrt{5}-\sqrt{7}\right)}{\left(2\sqrt{5}\right)^2-\left(\sqrt{7}\right)^2} = \frac{\sqrt{3}\left(2\sqrt{5}-\sqrt{7}\right)}{20-7} = \frac{\sqrt{3}\left(2\sqrt{5}-\sqrt{7}\right)}{13}$$

  The denominator is rationalised.

Example E

Q. Write the following fractional surds with a rational denominator.

1. $\frac{\sqrt{18}}{\sqrt{8}}$ **2.** $\frac{5}{\sqrt{3}}$ **3.** $\frac{7\sqrt{3}}{2\sqrt{5}}$ **4.** $\frac{3}{\sqrt{5}+\sqrt{2}}$ **5.** $\frac{\sqrt{5}}{3\sqrt{5}-\sqrt{7}}$

A. **1.** $\frac{\sqrt{18}}{\sqrt{8}} = \frac{\sqrt{9\times 2}}{\sqrt{4\times 2}} = \frac{\sqrt{9}\times\sqrt{2}}{\sqrt{4}\times\sqrt{2}} = \frac{3\sqrt{2}}{2\sqrt{2}} = \frac{3}{2}$

2. $\frac{5}{\sqrt{3}} = \frac{5}{\sqrt{3}} \times \frac{\sqrt{3}}{\sqrt{3}} = \frac{5\times\sqrt{3}}{\sqrt{3}\times\sqrt{3}} = \frac{5\sqrt{3}}{3}$: the denominator is rational

3. $\frac{7\sqrt{3}}{2\sqrt{5}} = \frac{7\sqrt{3}}{2\sqrt{5}} \times \frac{\sqrt{5}}{\sqrt{5}} = \frac{7\sqrt{3}\times\sqrt{5}}{2\sqrt{5}\times\sqrt{5}} = \frac{7\sqrt{15}}{2\times 5} = \frac{7\sqrt{15}}{10}$

4.
$$\frac{3}{\sqrt{5}+\sqrt{2}} \times \frac{\textit{conjugate of the denominator}}{\textit{conjugate of the denominator}} = \frac{3}{\sqrt{5}+\sqrt{2}} \times \frac{\sqrt{5}-\sqrt{2}}{\sqrt{5}-\sqrt{2}}$$
$$= \frac{3\left(\sqrt{5}-\sqrt{2}\right)}{\left(\sqrt{5}+\sqrt{2}\right)\left(\sqrt{5}-\sqrt{2}\right)}$$
$$= \frac{3\left(\sqrt{5}-\sqrt{2}\right)}{5-2}$$
$$= \frac{3\left(\sqrt{5}-\sqrt{2}\right)}{3}$$

5. $$\frac{\sqrt{5}}{3\sqrt{5}-\sqrt{7}} \times \frac{\textit{conjugate of the denominator}}{\textit{conjugate of the denominator}} = \frac{\sqrt{5}}{3\sqrt{5}-\sqrt{7}} \times \frac{3\sqrt{5}+\sqrt{7}}{3\sqrt{5}+\sqrt{7}}$$
$$= \frac{\sqrt{5}\left(3\sqrt{5}+\sqrt{7}\right)}{\left(3\sqrt{5}-\sqrt{7}\right)\left(3\sqrt{5}+\sqrt{7}\right)}$$
$$= \frac{3\times\sqrt{5}\times\sqrt{5}+\sqrt{5}\times\sqrt{7}}{45-7}$$
$$= \frac{15+\sqrt{35}}{38}$$

Unit 11.1 Activity 5E: Simplest form

1. Write the following surd fractions in simplest form, with a rational denominator.

a. $\frac{\sqrt{2}}{\sqrt{50}}$ **b.** $\frac{\sqrt{24}}{\sqrt{54}}$ **c.** $\frac{3\sqrt{40}}{\sqrt{250}}$ **d.** $\frac{\sqrt{14}}{\sqrt{2}}$ **e.** $\frac{\sqrt{30}}{2\sqrt{5}}$

2. Rationalise the denominators of the following fractions, expressing the answer in simplest form.

a. $\frac{1}{\sqrt{5}}$ **b.** $\frac{1}{2\sqrt{2}}$ **c.** $\frac{3}{\sqrt{3}}$ **d.** $\frac{1}{\sqrt{24}}$ **e.** $\frac{5+\sqrt{8}}{\sqrt{2}}$

3. Express the following surd fractions with rational denominators.

a. $\frac{1}{\sqrt{3}-\sqrt{2}}$ **b.** $\frac{\sqrt{5}}{\sqrt{5}+2}$ **c.** $\frac{2\sqrt{3}}{\sqrt{3}-1}$

d. $\frac{\sqrt{7}-2}{\sqrt{7}+2}$ **e.** $\frac{2\sqrt{5}}{2\sqrt{5}-1}$ **f.** $\frac{3\sqrt{6}-2}{3\sqrt{6}+2}$

4. Express the following fractions with a rational denominator, in simplest surd form.

a. $\frac{3+2\sqrt{2}}{3-2\sqrt{2}}$ **b.** $\frac{2\sqrt{3}-\sqrt{6}}{2\sqrt{3}-1}$ **c.** $\frac{5-\sqrt{12}}{2+\sqrt{18}}$ **d.** $\frac{2\sqrt{20}+1}{\sqrt{45}-1}$

e. $\frac{3+\sqrt{8}}{2+\sqrt{2}}$ **f.** $\frac{2\sqrt{10}+1}{\sqrt{40}-1}$ **g.** $\frac{\sqrt{3}}{\sqrt{5}+\sqrt{3}}$ **h.** $\frac{4\sqrt{7}}{2\sqrt{3}+\sqrt{7}}$

i. $\frac{2+\sqrt{6}}{\sqrt{11}-\sqrt{6}}$ **j.** $\frac{4\sqrt{5}-\sqrt{3}}{3\sqrt{5}-\sqrt{3}}$

Unit 11.1 Number and Application
Topic 6: Basic numeracy—indices

This Topic presents an introduction to indices in line with the General Mathematics syllabus, p. 14. It covers:

- Use of fractional and negative indices.
- Simplifying algebraic expressions with indices.
- Using calculators to simplify calculations.
- Solution of equations involving indices.

Introduction

Indices involve expressions of the form x^n, where x is the base and n is the power or index or exponent. The index notation x^n is a short form of expressing a series of multiplication of the same number. That is, $\underbrace{x \times x \times x \times \ldots \times x}_{\text{n times}}$ is written in short form as x^n.

Indices involve expressions of the form x^n, where x is the **base** and n is the **power** or **index** or **exponent**.

Some important properties of indices already learned in NCEA Level 1 Mathematics are:

$x^m \times x^n = x^{m+n}$	$\frac{x^m}{x^n} = x^{m-n}$ $[x \neq 0]$
$(xy)^m = x^m y^m$	or $\frac{1}{x^{n-m}}$
$(x^m)^n = x^{mn}$	$\left(\frac{x}{y}\right)^m = \frac{x^m}{y}$
$x^0 = 1$ $[x \neq 0]$	

Negative indices

Using the division law for powers, $\frac{x^2}{x^5} = x^{2-5} = x^{-3}$. Simplifying $\frac{x^2}{x^5}$ by writing in full and cancelling gives $\frac{\cancel{x} \times \cancel{x}}{\cancel{x} \times \cancel{x} \times x \times x \times x} = \frac{1}{x^3}$. This example illustrates the rule for negative indices:

$$x^{-n} = \frac{1}{x^n} \text{ where } x \neq 0$$

Zero indices

The same law can also be used to illustrate the rule for zero indices. That is,

$$\frac{x^2}{x^2} = x^{2-2} = x^0$$

By simplifying through cancellation:

$$\frac{x^2}{x^2} = \frac{\cancel{x} \times \cancel{x}}{\cancel{x} \times \cancel{x}} = 1$$

By directly comparing the two results:

$x^0 = 1$ for all a except $a = 0$

Example A

Q. Use the properties of indices to calculate the following, giving the answers as fractions or whole numbers.

1. 2^{-3} **2.** $\frac{1}{3^{-2}}$ **3.** $\left(\frac{2}{5}\right)^3$ **4.** $\left(\frac{3}{7}\right)^{-2}$ **5.** $\frac{2^{-3}}{3^{-2}}$ **6.** $\frac{2^{-1}}{7}$

7. $2^5 \times 2^3$ **8.** $\frac{5^3}{5^2}$ **9.** 7^0 **10.** $2 + 5^0$ **11.** $2 + 3^0 + 2^{-1}$

A. **1.** $2^{-3} = \frac{1}{2^3}$ [since $x^{-m} = \frac{1}{x^m}$]

$= \frac{1}{8}$ [$2^3 = 8$]

2. $\frac{1}{3^{-2}} = 3^2$ [since $\frac{1}{x^m} = x^{-m}$]

$= 9$

3. $\left(\frac{2}{5}\right)^3 = \frac{2^3}{5^3}$ [since $\left(\frac{x}{y}\right)^m = \frac{x^m}{y^m}$]

$= \frac{8}{125}$

4. $\left(\frac{3}{7}\right)^{-2} = \left(\frac{7}{3}\right)^2$ [since $\left(\frac{x}{y}\right)^{-m} = \left(\frac{y}{x}\right)^m$]

$= \frac{7^2}{3^2}$ [since $\left(\frac{y}{x}\right)^m = \frac{y^m}{x^m}$]

$= \frac{49}{9}$

5. $\frac{2^{-3}}{3^{-2}} = \frac{3^2}{2^3}$ [since $\frac{x^{-n}}{y^{-m}} = \frac{y^m}{x^n}$]

$= \frac{9}{8}$

6. $\frac{2^{-1}}{7} = \frac{1}{7} \times 2^{-1}$ [as dividing by 7 is the same as multiplying by $\frac{1}{7}$]

$= \frac{1}{7} \times \frac{1}{2}$ [as $2^{-1} = \frac{1}{2^1} = \frac{1}{2}$]

$= \frac{1}{14}$ [multiplying fractions]

7. $2^5 \times 2^3 = 2^{5+3}$

$= 2^8$

$= 256$

8. $\frac{5^3}{5^2} = 5^{3-2}$

$= 5^1$

$= 5$

9. $7^0 = 1$

10. $2 + 5^0 = 2 + 1$

$= 3$

11. $2 + 3^0 + 2^{-1} = 2 + 1 + \frac{1}{2}$

$= 3\frac{1}{2}$

Unit 11.1 Activity 6A: Negative indices

1. Calculate the following, giving the answers to no more than 4 decimal places:

a. 2^{-4} **b.** 1.2^{-5} **c.** 0.37^{-2} **d.** $0.37^{-1} + 2.3$

e. $(2.3)^{-2} + 4$ **f.** $\frac{1}{3^{-2}}$ **g.** $\frac{2}{(2.4)^{-3}}$ **h.** $(-3.4)^{-2}$

i. $\frac{1}{0.56^{-1}} + \frac{2}{0.5^{-2}}$ **j.** $(3.4^{-1} - (0.9)^{-2})^{-3}$ **k.** $(2.3)^{-3} + (-2.3)^{-2}$

2. Use the properties of indices to evaluate the following, giving the answers as fractions or whole numbers.

a. 3^{-2} **b.** 4^{-3} **c.** 2^{-4} **d.** $\left(\frac{3}{4}\right)^{3}$

e. $\frac{2^3}{5^2}$ **f.** $\frac{1}{3^{-2}}$ **g.** $\frac{1}{4^{-2}}$ **h.** $\left(\frac{3}{5}\right)^{-2}$

i. $\left(\frac{4}{5}\right)^{-3}$ **j.** $\frac{2^{-2}}{3^{-3}}$ **k.** $\frac{3^{-3}}{5^{-2}}$ **l.** $\frac{3^{-1}}{8}$

m. $\frac{5^{-1}}{2}$ **n.** $7(5^{-1})$ **o.** 8×3^{-1} **p.** $2^{-1} \times 3^{-1}$

q. $3^{-1}\,5^{-1}$ **r.** $3^{-1}\,2^{-2}$ **s.** $\frac{5}{2^{-1}}$ **t.** $\frac{3}{2^{-2}}$

u. $\frac{3^{-2}}{4}$ **v.** $\frac{5^{-2}}{7}$ **w.** $3^{-1}\,6 + 3^{0}$ **x.** $3^{-2}\,6^{2} + 6^{0}$

y. $2^{-1} \div 7$ **z.** $3^{-1} \div 4^{-1}$

3. Calculate the following, giving the answers as fractions:

a. $2^{-1} \times 3$ **b.** $3^{-2} \times 4^{-1}$ **c.** $(2^{-2} \times 3^{-1})^{-2}$ **d.** $(3^{-1}2^{-3})^{2}$

e. $2^{-1} + 3^{-1}$ **f.** $3^{-2} + 6^{-1}$ **g.** $2^{-1}\,3 + 5^{-1}\,3^{2}$ **h.** $2^{-2} - 8^{-1}$

i. $4^{-2} \div 8^{-1}$ **j.** $\frac{3^{-1}}{2^{-1}} + \frac{4^{-1}}{3^{-1}}$ **k.** $\frac{3^{-2}}{2^{-2}} - \frac{2^{-1}}{3^{-2}}$ **l.** $(3^{1}\,2^{-1} - 2^{1}\,3^{-1})^{-1}$

m. $(2^{2}\,3)^{-1} + 2^{-1}\,3^{-2}$ **n.** $5(2^{3})^{-1} + 4^{-2}$ **o.** $7(2^{-1}) - 3(2^{-2})$ **p.** $5^{0} - 5^{-1}$

4. Write the following as fractions or integers with no powers:

a. 2×4^{-1} **b.** $2^{-1} \times 3^{-1}$ **c.** 4×2^{-1} **d.** $3^{-2} \times 5$

e. $\left(\frac{2}{3}\right)^{-1}$ **f.** $\left(1\frac{1}{2}\right)^{-1}$ **g.** $3\left(\frac{1}{3}\right)^{-1}$ **h.** $\frac{2}{3}\left(\frac{2}{3}\right)^{-2}$

i. $\frac{2^{-1}}{5} \times \left(\frac{1}{4}\right)^{-2}$

5. Write each of the following as the product of **primes** raised to powers:
eg, $28 = 2^{2} \times 7^{1}$, $\frac{3}{8} = 3 \times 2^{-3}$.

a. 12 **b.** 200 **c.** 45 **d.** $\frac{25}{16}$ **e.** $\frac{2}{3}$

f. $\frac{1}{27}$ **g.** $\frac{16}{81}$ **h.** $\frac{64}{81}$ **i.** $2\frac{1}{4}$

6. Write the following as powers of 2, 3 and 5: eg $\frac{1}{25} = \frac{1}{5^2} = 5^{-2}$

a. 16 **b.** 125 **c.** 81 **d.** $\frac{1}{16}$

e. 64 **f.** $\frac{1}{625}$ **g.** $\frac{1}{9}$ **h.** $\frac{1}{243}$

7. Evaluate each of the following:

a. $(-5)^2$ **b.** -5^2 **c.** $(-5)^{-2}$ **d.** $3^{-1} + 2^{-1}$

e. $3^{-2} \times 2^{-2}$ **f.** $7^{-1} - 7^0$ **g.** $(-1)^{19}$

8. Write the following as powers of 2 and 3: eg $2^n \times 8 = 2^n \times 2^8 = 2^{n+2}$, $\frac{4}{2^n} = \frac{2^2}{2^n} = 2^{2-n}$ and $27 \times 9 = 3^3 \times 3^3 = 3^{3-2} = 3^5$.

a. $\frac{2^c}{4}$ **b.** $\frac{2^m}{2^{-m}}$ **c.** 8×2^t **d.** $\left(2^{x+1}\right)^2$

e. $\left(2^{1-n}\right)^{-1}$ **f.** $\frac{2^{x+1}}{2^x}$ **g.** $\frac{4}{2^{1-x}}$ **h.** 9×3^x

i. $\frac{3^y}{27}$ **j.** 243^a **k.** $\frac{9^{n+1}}{3^{2n-1}}$ **l.** $\frac{9^a}{3^{1-a}}$

Simplifying algebraic expressions with indices

Algebraic expressions may need to be simplified so that indices have positive powers only.

Example B

Q. Write with positive indices: **1.** x^{-5} **2.** $8a^{-2}$ **3.** $\frac{2^{-1}a^{-3}}{5}$

A. **1.** $x^{-5} = \frac{1}{x^5}$ [since $x^{-r} = \frac{1}{x^r}$]

2. $8a^{-2} = 8 \times a^{-2}$

$= 8 \times \frac{1}{a^2}$ [since $a^{-2} = \frac{1}{a^2}$]

$= \frac{8}{a^2}$

3. $\frac{2^{-1}a^{-3}}{5} = 2^{-1} \times a^{-3} \times \frac{1}{5}$ [dividing by 5 is the same as multiplying by $\frac{1}{5}$]

$= \frac{1}{2} \times \frac{1}{a^3} \times \frac{1}{5}$ [since $2^{-1} = \frac{1}{2}$ and $a^{-3} = \frac{1}{a^3}$]

$= \frac{1}{10a^3}$

Example C

Q. Express $\frac{15x^{-3}y^2}{(3x^{-2}y^4)^2}$ without negative indices.

A. $\frac{15x^{-3}y^2}{(3x^{-2}y^4)^2} = \frac{15x^{-3}y^2}{3^2(x^{-2})^2(y^4)^2}$ [since $(xy)^n = x^ny^n$]

$= \frac{15x^{-3}y^2}{9x^{-4}y^8}$ [since $(x^n)^m = x^{nm}$]

$= \frac{5x^1y^{-6}}{3}$ $[\frac{x^m}{x^n} = x^{m-n}$ and $\frac{15}{9} = \frac{5}{3}]$

$= \frac{5x}{3} \times y^{-6}$

$= \frac{5x}{3} \times \frac{1}{y^6}$ [since $y^{-n} = \frac{1}{y^n}$]

$= \frac{5x}{3y^6}$

Unit 11.1 Activity 6B: Simplifying expressions with indices

1. Write each of the following without negative indices and as simply as possible:

eg, $3x^{-1} = \frac{3}{x}$

a. x^{-3}	**b.** y^{-2}	**c.** a^{-5}	**d.** b^{-4}	**e.** z^{-7}
f. y^{-6}	**g.** z^{-11}	**h.** $(xy)^{-1}$	**i.** $(abc)^{-1}$	**j.** $(xy)^{-2}$
k. $(2x)^{-1}$	**l.** $(2x)^{-2}$	**m.** $(3a)^{-2}$	**n.** $(2x)^{-3}$	**o.** $3x^{-1}$
p. ab^{-3}	**q.** $a^{-3}b^2$	**r.** $a^{-3}a^2$	**s.** $\frac{x^{-1}}{2}$	**t.** $\frac{x^{-3}}{4}$
u. $\frac{3x^{-2}}{5}$	**v.** $\frac{2a^{-3}}{7}$	**w.** $\frac{4a^{-2}}{9}$	**x.** $\frac{3x^{-2}}{5x^2}$	**y.** $\frac{4x^{-3}}{5x^2}$

2. Write the following using negative indices for the variable: eg, $\frac{3}{x^2} = 3x^{-2}$

a. $\frac{1}{x^6}$	**b.** $\frac{1}{y^5}$	**c.** $\frac{1}{z^2}$	**d.** $\frac{4}{x}$	**e.** $\frac{5}{y^2}$
f. $\frac{7}{z^5}$	**g.** $\frac{1}{2x^2}$	**h.** $\frac{1}{3a^4}$	**i.** $\frac{1}{5y^4}$	**j.** $\frac{2}{5x^2}$
k. $\frac{3}{4y^3}$	**l.** $\frac{3}{(2x)^2}$	**m.** $\frac{4}{(5y)^2}$	**n.** $\frac{x^2}{(2x)^3}$	**o.** $\frac{y^2}{(2x)^3}$

3. Write the following as simply as possible without using negative indices and as a single fraction:

a. $\frac{4x^{-2}}{8y^{-3}}$	**b.** $\frac{6a^{-3}}{9b^{-4}}$	**c.** $\frac{5x^{-2}}{25x^{-3}}$	**d.** $\frac{5^{-1}x^{-2}}{10^{-1}y^{-4}}$	**e.** $\frac{(2x)^2}{4x^{-3}y}$
f. $\frac{(4x^{-1})^2}{8x^{-1}y^{-3}}$	**g.** $\frac{(2x^{-2})^3y^{-2}}{x}$	**h.** $\frac{5^{-1}x^{-2}}{5x}$	**i.** $\frac{2^{-1}x^{-1}}{3y}$	**j.** $\frac{3^{-1}x^{-2}}{(3x)^2}$
k. $2^{-1}y^{-2}x$	**l.** $3^{-2}y^{-3}x^2$	**m.** $2(3^{-1}x^{-2})$	**n.** $4(5^{-1}x^{-3})y^2$	
o. $x^{-1} + y^{-1}$	**p.** $x^{-1} + x^{-2}$	**q.** $x(2^{-1}y + y^{-1}x^{-1})$		

Rational powers (fractional indices)

Fractional indices refer to powers that appear as fractions and are used to express roots of numbers. For example, the square root of a number $\sqrt{x}$ can be expressed as $x^{\frac{1}{2}}$ and the cube root of a number $\sqrt[3]{x}$ can be expressed as $x^{\frac{1}{3}}$.

Generally, the nth root of a number can be written as:

$$\boxed{x^{\frac{1}{n}} = \sqrt[n]{x}}$$

Proof: The following steps set out the proofs for $n = 2$ and $n = 3$ and can be applied to any other value of n.

1. The square root of a number can be symbolically expressed as $\sqrt[2]{x}$ or $\sqrt{x}$ and the index representing the square root of a number is obtained from:

$$x^{\frac{1}{2}} \times x^{\frac{1}{2}} = \left(x^{\frac{1}{2}}\right)^2 \quad [\text{ since } (x^m)^n = x^{mn}]$$

$$= x^1$$

$$= x$$

$$\therefore x^{\frac{1}{2}} = \sqrt[2]{x} \text{ or } \sqrt{x}$$

2. The index representing the cube root of a number is obtained from:

$$x^{\frac{1}{3}} \times x^{\frac{1}{3}} \times x^{\frac{1}{3}} = \left(x^{\frac{1}{3}}\right)^3$$

$$= x^1$$

$$= x$$

$$\therefore x^{\frac{1}{3}} = \sqrt[3]{x}$$

Generalising further:

$$\boxed{x^{\frac{m}{n}} = \sqrt[n]{x^m} = \left(\sqrt[n]{x}\right)^m}$$

Example D

$$(-27)^{\frac{5}{3}} = \left[(-27)^{\frac{1}{3}}\right]^5 \qquad [\text{since } (x)^{\frac{m}{n}} = \left[(x)^{\frac{1}{n}}\right]^m]$$

$$= \left(\sqrt[3]{-27}\right)^5 \qquad [\text{since } (x)^{\frac{1}{n}} = \sqrt[n]{x}]$$

$$= (-3)^5$$

$$= -243$$

Note: When you raise a negative number to an even power the answer is positive, but if you raise a negative number to an odd power the answer is negative. Therefore, if you take an even root of a negative number, the result will be undefined.

When simplifying algebraic expressions with fractional powers, the answer is often expressed in the form ax^n.

Example E

Q. Write the following in simplest index form:

1. $\sqrt{x}$ **2.** $3\sqrt[5]{x^2}$ **3.** $\dfrac{2}{\sqrt[3]{x^2}}$

A. 1. $\sqrt{x} = \sqrt[2]{x}$ [from the definition of $\sqrt{\ }$]

$= x^{\frac{1}{2}}$ [since $\sqrt[y]{x} = x^{\frac{1}{y}}$]

2. $3\sqrt[5]{x^2} = 3 \times \sqrt[5]{x^2}$

$= 3x^{\frac{2}{5}}$ [since $\sqrt[n]{x^m} = x^{\frac{m}{n}}$]

3. $\dfrac{2}{\sqrt[3]{x^2}} = 2 \times \dfrac{1}{\sqrt[3]{x^2}}$

$= 2 \times \dfrac{1}{x^{\frac{2}{3}}}$ [since $\sqrt[n]{x^m} = x^{\frac{m}{n}}$]

$= 2 \times x^{-\frac{2}{3}}$ [since $\dfrac{1}{x^n} = x^{-n}$]

$= 2x^{\frac{-2}{3}}$

Unit 11.1 Activity 6C: Fractional indices

1. Express the following in index form as simply as possible. eg, $3\sqrt[4]{x^5} = 3x^{\frac{5}{4}}$

a. $\sqrt{a}$ **b.** $\sqrt{c}$ **c.** $\sqrt[2]{x}$ **d.** $\sqrt[5]{a}$ **e.** $\sqrt[6]{u}$

f. $\sqrt[5]{x^2}$ **g.** $\sqrt[7]{x^3}$ **h.** $\sqrt[8]{a^3}$ **i.** $\dfrac{1}{\sqrt{x}}$ **j.** $\dfrac{1}{\sqrt{y}}$

k. $\dfrac{1}{\sqrt[3]{a}}$ **l.** $\dfrac{1}{\sqrt[5]{x}}$ **m.** $\dfrac{1}{\sqrt[7]{x^2}}$ **n.** $\dfrac{1}{\sqrt[9]{x^5}}$ **o.** $\dfrac{1}{\sqrt[7]{a^3}}$

p. $\dfrac{1}{\sqrt[9]{a^8}}$ **q.** $4\sqrt[3]{a}$ **r.** $\dfrac{6}{\sqrt[3]{y}}$ **s.** $\dfrac{8}{\sqrt[3]{x^7}}$ **t.** $\dfrac{1}{7\sqrt[3]{x^2}}$

u. $\dfrac{2}{5\sqrt[4]{x^3}}$ **v.** $(\sqrt{x})^{-2}$ **w.** $\dfrac{\sqrt{x}}{\sqrt[3]{x}}$ **x.** $\dfrac{1}{\sqrt[3]{a}\,\sqrt[3]{a^4}}$ **y.** $\dfrac{\sqrt{x}}{\sqrt[3]{x}\,\sqrt[4]{x}}$

2. Express the following using radical (or root) signs: eg, $x^{\frac{2}{3}} = \sqrt[3]{x^2}$

a. $a^{\frac{3}{7}}$ **b.** $b^{\frac{3}{4}}$ **c.** $x^{\frac{2}{5}}$ **d.** $x^{\frac{-2}{3}}$ **e.** $y^{\frac{-3}{4}}$

3. Express the following in index form and as simply as possible:

a. $\sqrt{xy}$ **b.** $\sqrt{xy^2}$ **c.** $\sqrt{\dfrac{x^2}{y}}$

d. $\sqrt[3]{\dfrac{x^6}{y}}$ **e.** $\sqrt[5]{\dfrac{32x^5}{y^3}}$ **f.** $\sqrt[4]{x^{-5}y^4}$

g. $\sqrt{x} \times \sqrt[3]{a}$ **h.** $\dfrac{1}{\sqrt{a}} \times \sqrt{c}$ **i.** $\dfrac{1}{\sqrt[4]{b^{-2}}} \times \sqrt{b}$

4. Simplify fully, writing with no negative powers or radical signs:

a. $(16x^4)^{\frac{-3}{4}}$ **b.** $(27x^9)^{\frac{-2}{3}}$ **c.** $(32y^{10})^{-\frac{2}{5}}$ **d.** $\left(\sqrt[5]{243A^5}\right)^3$

e. $\left(\sqrt[6]{64B^{12}}\right)^{-5}$ **f.** $\dfrac{1}{\left(\sqrt[3]{8x^6}\right)^2}$ **g.** $\dfrac{1}{\left(\sqrt[4]{16A^4B^4}\right)^{-2}}$ **h.** $\left(\sqrt[3]{125A^3B^6}\right)^{-2}$

i. $\dfrac{3}{\left(\sqrt{x}\right)^{-3}}$ **j.** $\dfrac{8}{\left(\sqrt{64x^{12}}\right)^{-\frac{5}{6}}}$

Using calculators to evaluate $(-x)^{\frac{m}{n}}$

Calculators which have exponent keys such as x^y or ^ can simplify calculations involving indices. However, many calculators do not accept a negative base (such as –27.5) raised to a fractional power (such as $\frac{2}{3}$) and if the calculation $(-27.5)^{\frac{2}{3}}$ is entered, an error message results. In such cases the following procedure is used.

Calculation of $(-x)^{\frac{m}{n}}$ where *m* and *n* are integers and *x* is positive

The index $\frac{m}{n}$ should be simplified as much as possible. Then, regardless of whether $\frac{m}{n}$ is negative or positive:

- If n is even, $(-x)^{\frac{m}{n}}$ cannot be calculated. [cannot take square root, 4th root, etc, of a negative number]
- If n is odd and m is even, $(-x)^{\frac{m}{n}} = (x)^{\frac{m}{n}}$ and the answer is positive.
- If n and m are odd, $(-x)^{\frac{m}{n}} = -\left[(x)^{\frac{m}{n}}\right]$ and the answer is negative.

Example F

$(-27.5)^{\frac{2}{3}} = (27.5)^{\frac{2}{3}}$ [since 2 is even and 3 is odd]

$= 9.11$ [using a calculator, to 2 decimal places]

Note: Be sure to put brackets around fractional exponents. On a calculator $(-27.5)^{\frac{2}{3}}$ would be calculated by pressing:

2 7 · 5 ^ (2 $a\frac{b}{c}$ 3) =

Unit 11.1 Activity 6D: Calculating with indices

Use a calculator to work out each of the following:

1. 2^4

2. 3^{-2}

3. 0.23^{-3}

4. $(-2)^5$

5. $\left(\frac{2}{3}\right)^{-2}$

6. $\left(1\frac{3}{4}\right)^{\frac{1}{2}}$

7. $(-3.47)^{\frac{3}{5}}$

8. $(-2.49)^{\frac{-1}{2}}$

9. $\left(\frac{12.37}{2.69}\right)^{\frac{14}{17}}$

10. $(3.52^4 - 2.15^3)^{\frac{3}{2}}$

11. $(17.63)^{\frac{1}{2}} - (3.165)^{\frac{1}{3}}$

12. $(15.86)^{\frac{-3}{7}} + (-4.39)^{\frac{2}{7}}$

13. $\frac{(11.45)^{0.467}}{(2.78)^{-3.1}}$

14. $\frac{(-17.45)^{\frac{13}{27}}}{(15.35)^{\frac{3}{4}}}$

15. $(3.2)^{\frac{1}{3}} \times (2.5)^{\frac{1}{2}} - (1.36)^{\frac{1}{4}}$

16. $(-3.2)^{\frac{2}{3}}$

17. $(56)^{\frac{2}{3}}$

18. $(63)^{\frac{-3}{5}}$

19. $(-4.57)^{\frac{-3}{4}}$

20. $(-31.2)^{\frac{-4}{5}}$

Solution of equations involving indices

Equations with indices often simplify to equations of the form $x^n = C$. By raising the expression on each side of the equation to the **reciprocal of the power of *x***, the solution can be found (the reciprocal of n is $\frac{1}{n}$).

Example G

Q. Solve $8x^4 - 3 = 2$, correct to 4 significant figures.

A. $8x^4 - 3 = 2$

$\therefore\ 8x^4 = 5$ [adding 5]

$x^4 = \frac{5}{8}$ [dividing by 8]

$\therefore\ (x^4)^{\frac{1}{4}} = \left(\frac{5}{8}\right)^{\frac{1}{4}}$ [taking reciprocal powers]

$\therefore\ x = 0.8891$ (4 sf)

The same technique works for fractional indices (the reciprocal of $\frac{m}{n}$ is $\frac{n}{m}$).

Example H

Q. Solve $(x)^{\frac{15}{23}} = 26.4$, correct to 4 significant figures.

A. $(x)^{\frac{15}{23}} = 26.4$

$\therefore \left[(x)^{\frac{15}{23}}\right]^{\frac{23}{15}} = (26.4)^{\frac{23}{15}}$ [raising both sides to the **reciprocal power** to isolate the variable]

$\therefore x = 151.3$ [using a calculator]

Example I

Q. Solve $(2A + 3)^{\frac{-4}{7}} = 8$, to 3 significant figures.

A. $(2A + 3)^{\frac{-4}{7}} = 8$

$\therefore \left[(2A + 3)^{\frac{-4}{7}}\right]^{\frac{-7}{4}} = (8)^{\frac{-7}{4}}$ [raising to the reciprocal power]

$\therefore 2A + 3 = (8)^{\frac{-7}{4}}$

$\therefore 2A = (8)^{\frac{-7}{4}} - 3$ [subtracting 3]

$\therefore A = \dfrac{(8)^{\frac{-7}{4}} - 3}{2}$ [dividing by 2]

$\therefore A = -1.49$

Sometimes *the unknown occurs as part of the index*. Equations of this nature are called **exponential** or **indicial** equations. When solving such equations, create same bases on both sides of the equation then equate the indices to one another and solve for the unknown. $2^x = 8$, $3^x = 10$ and $\frac{1}{10^x} = 100$ are examples of indicial equations.

Example J

Q. Solve the following exponential equations:

a. $2^x = 4$ **b.** $9^x = \frac{1}{27}$ **c.** $2 \times \frac{1}{4^x} = 32$ **d.** $5^{x+2} = \frac{1}{\sqrt{5}}$

A.

a. $2^x = 4$

$\therefore 2^x = 2^2$

$\therefore x = 2$

b. $9^x = \frac{1}{27}$

$\therefore 3^{2(x)} = \frac{1}{3^3}$

$\therefore 3^{2x} = 3^{-3}$

$\therefore 2x = -3$

c. $2 \times \frac{1}{4^x} = 32$

$\therefore \frac{1}{4^x} = 16$

$\therefore \frac{1}{(2^2)^x} = 2^4$

$\therefore 2^{-2x} = 2^4$

d. $5^{x+2} = \frac{1}{\sqrt{5}}$

$\therefore 5^{x+2} = \frac{1}{5^{\frac{1}{2}}}$

$\therefore 5^{x+2} = 5^{-\frac{1}{2}}$

$\therefore x + 2 = -\frac{1}{2}$

$\therefore x = \frac{-3}{2}$ $\quad \therefore -2x = 4$ $\quad \therefore x = -2\frac{1}{2}$

$= -\frac{3}{2}$ $\quad \therefore x = -2$

Unit 11.1 Activity 6E: Equations with indices

1. Solve each of the following equations:

a. $x^3 = 2$ **b.** $x^5 = 243$ **c.** $x^{\frac{1}{4}} = 3$ **d.** $a^{\frac{1}{2}} = 5$

e. $a^{\frac{1}{5}} = \frac{2}{3}$ **f.** $y^{-2} = 25$ **g.** $x^{-4} = 256$ **h.** $y^{\frac{3}{4}} = 27$

i. $x^{\frac{2}{5}} = 4$ **j.** $\frac{y^{\frac{4}{3}}}{2} = 8$ **k.** $x^{\frac{-5}{3}} = 32$ **l.** $3y^{\frac{-7}{3}} = 384$

m. $2x^{\frac{1}{2}} - 1 = 3$ **n.** $4x^{\frac{1}{3}} + 1 = 9$ **o.** $2(x-1)^{\frac{1}{2}} = 4$ **p.** $\frac{x^{\frac{2}{3}}+1}{2} = 5$

q. $\frac{3}{x^{\frac{1}{3}}}$ **r.** $x^{\frac{-1}{2}} = 2$ **s.** $3y^{\frac{-3}{2}} - 1 = 80$ **t.** $\frac{z^{\frac{-2}{5}}}{3} - 1 = 2$

2. Solve the following equations:

a. $(x)^{\frac{2}{3}} = 16$ **b.** $(A)^{\frac{3}{7}} = 28$ **c.** $(A)^{\frac{-7}{3}} = 43$

d. $(p)^{1\frac{3}{4}} = 7$ **e.** $(p)^{\frac{4}{3}} = -2.3$ **f.** $(p)^{\frac{3}{5}} = -4$

g. $2(x)^{\frac{2}{3}} = 7$ **h.** $\left[(x)^{\frac{7}{9}} + 3\right]^{\frac{1}{2}} = 9$ **i.** $(2x-3)^{\frac{-3}{2}} = 5$

3. Solve the following indicial equations:

a. $2^x = 8$ **b.** $4^x = \frac{1}{2}$ **c.** $4^x = 32$ **d.** $2^x = 1$

e. $3^x = 0$ **f.** $3^{-x} = \frac{1}{9}$ **g.** $3^{2-x} = \frac{1}{27}$ **h.** $5^{x+1} = \frac{1}{25}$

i. $5^x = 125$ **j.** $10^x = 100\sqrt{10}$ **k.** $10^x = 0.001$ **l.** $5^{1-2x} = 1$

m. $49^x = \frac{1}{7}$ **n.** $4^x = \frac{1}{\sqrt{2}}$

4. Find the value for x for each of the following exponential equations:

a. $3 \times 2^x = 24$ **b.** $7 \times 2^x = 56$ **c.** $3 \times 2^{x+1} = 24$

d. $12 \times 3^x = \frac{4}{3}$ **e.** $9 \times 5^x = 225$ **f.** $5 \times \left(\frac{1}{2}\right)^x = 20$

5. The relationship connecting length, L, to cost, C, of a type of material is given by $C = 5L^{\frac{2}{3}} + 6$, where C is given in Kina and L in metres.
   a. What is the cost when 27 metres are purchased?
   b. What length can be purchased for K100?
6. The relationship between the volume, height and base radius of a shape is given by $V = KR^{\frac{5}{2}}H^{\frac{1}{2}}$, where R is base radius, H is height and K is a constant. When the height and base radius are both 2 the volume is 16:
   a. Find the volume when the radius is 16 and the height 49.
   b. Find the height when the volume is 128 and the radius is 4.
   c. Find the radius when the volume is 486 and the height is 3 125.
7. T and U are related by the formula $T\sqrt[5]{U^4} = 13.5$.
   a. Describe what happens to T as U gets larger.
   b. If the minimum value of T is 500, find the maximum value of U.

Unit 11.1 Number and Application

Topic 7: Basic numeracy—logarithms

Following the General Mathematics syllabus, p.14, this Topic covers the manipulation of algebraic expressions and solving equations:

- Use elementary properties of logarithms.
- Solve simple logarithmic equations.
- Solve problems in context.
- Choose algebraic techniques and strategies to solve problems.

Logarithmic functions

Any function of the type $y = a^x$ is called an **exponential function**. The inverse of this function is called the **logarithmic function to base *a*** and is written $\log_a x$.

Most scientific calculators have the logarithmic function to base 10 ($\log_{10} x$) and the exponential function, 10^x. On most calculators, $\log_{10}$ has a key labelled [log] and the exponential function, 10^x, is the inverse of the log key.

Example A

1. $\log_{10} 1\,000 = 3$ [since $1\,000 = 10^3$]
2. $\sqrt{\log_{10} 17} = \sqrt{1.2304489}$
   $= 1.11$ (2 dp)
3. $(10^{1.2} - 1)^{\frac{1}{3}} = (15.848932 - 1)^{\frac{1}{3}}$
   $= 2.46$ (2 dp)
4. $10^{\log_{10} 5} = 5$ [as '10^x' and 'log' are inverse functions]
5. $10^{2x} = 6$
   $\therefore\ 2x = \log_{10} 6$ [taking logs]
   $\therefore\ x = 0.39$ (2 dp)

Note: To evaluate powers of 10 press [shift] [10^x] or use the general power key 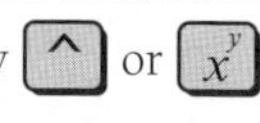.

[On some calculators [inv] replaces [shift]]

Unit 11.1 Activity 7A: Other logarithmic functions

Calculate to 2 decimal places using the [10^x] and [log] keys on your calculator:

1. $\log_{10} 125.7$
2. $10^{2.3}$
3. $10^{3.4} - \log_{10} 2347.5$
4. $\dfrac{10^{-2.3}}{\log_{10} 1.37}$
5. $10^{\log_{10} 3.678}$
6. $\log_{10} (10^{2.5})$

General properties of logarithmic functions

- If a is any positive number then $y = a^x$ and $y = \log_a x$ are **inverse functions**.

  Note: $\log_a x$ is the logarithm of x to the base a.

- The function $y = \log_a x$ is defined by:

  $y = \log_a x$ if and only if $x = a^y$

- $a^{\log_a x} = x$

  $a^{\log_a x} = x$ because a^x and $\log_a x$ are inverse functions of each other.

- $\log_a a^x = x$
- For all $a > 0$: $\log_a a = 1$ and $\log_a 1 = 0$
- $\log_a (xy) = \log_a x + \log_a y$

  Proof: Let $u = \log_a x$ and $v = \log_a y$

  $\therefore\ x = a^u$ and $y = a^v$ [from definition of log]

  $\therefore\ xy = a^u a^v$ [multiplying x and y]

  $\therefore\ xy = a^{u+v}$ [adding indices]

  $\therefore\ u + v = \log_a xy$ [by definition of log]

  $\therefore\ \log_a x + \log_a y = \log_a xy$, as required [since $u = \log_a x$, $v = \log_a y$]

- $\log_a\left(\dfrac{s}{t}\right) = \log_a s - \log_a t$

Proof: Let $x = \log_a s$ and $y = \log_a t$

$\therefore s = a^x$ and $t = a^y$

$\therefore \dfrac{s}{t} = \dfrac{a^x}{a^y}$ [divide by]

$\therefore \dfrac{s}{t} = a^{x-y}$ [subtract indices]

$\therefore x - y = \log_a\left(\dfrac{s}{t}\right)$

$\therefore \log_a s - \log_a t = \log_a\left(\dfrac{s}{t}\right)$ [since $x = \log_a s, v = \log_a t$]

- $\log_a x^n = n\log_a x$

Proof: $\log_a x^n = \log_a \underbrace{(x \times x \times x \times \ldots x)}_{n \text{ terms}}$

$= \underbrace{\log_a x + \log_a x + \ldots \log_a x}_{n \text{ terms}}$

$= n\log_a x$

Worked examples

Example B

Write the equation $128 = 2^7$ in logarithmic form.

Solution

$128 = 2^7$

$\therefore\ 7 = \log_2 128$ [since $y = \log_a x$ when $x = a^y$]

$\therefore$ The required equation is $\log_2 128 = 7$.

Example C

Solve $4^x = 8$.

Solution

$$4^x = 8$$

$$\log_{10}(4^x) = \log_{10} 8 \quad \text{[taking logs of both sides]}$$

$$x\log_{10} 4 = \log_{10} 8 \quad \text{[since } \log A^n = n\log A\text{]}$$

$$x = \frac{\log_{10} 8}{\log_{10} 4} \quad \text{[dividing by } \log_{10} 4\text{]}$$

$$x = \frac{\log_{10} 2^3}{\log_{10} 2^2}$$

$$= \frac{3\log_{10} 2}{2\log_{10} 2}$$

$$= \frac{3}{2}$$

Example D

Solve the equation $5^{3t-1} = 8$.

Solution

$$5^{3t-1} = 8$$

$$\therefore\ \log_{10} 5^{3t-1} = \log_{10} 8 \quad \text{[taking logs of both sides]}$$

$$\therefore\ (3t-1)\log_{10} 5 = \log_{10} 8 \quad \text{[since } \log A^n = n\log A\text{]}$$

$$\therefore\ 3t - 1 = \frac{\log_{10} 8}{\log_{10} 5} \quad \text{[dividing by } \log_{10} 5\text{]}$$

$$\therefore\ 3t = \frac{\log_{10} 8}{\log_{10} 5} + 1 \quad \text{[adding 1]}$$

$$\therefore\ t = \left(\frac{\log_{10} 8}{\log_{10} 5} + 1\right) \div 3$$

$$= 0.764 \ \ (3 \text{ dp})$$

Note: $\log_{10} 8$ and $\log_{10} 5$ are evaluated last to avoid rounding errors.

Example E

Find the value of $\log_{16} 15$.

Solution

Let $x = \log_{16} 15$

$\therefore 16^x = 15$ [since $x = a^y$ when $y = \log_a x$]

$\therefore \log_{10} 16^x = \log_{10} 15$ [taking logs of both sides]

$\therefore x\log_{10} 16 = \log_{10} 15$ [since $\log A^n = n \log A$]

$\therefore x = \dfrac{\log_{10} 15}{\log_{10} 16}$

$\therefore x = 0.977$ (3 dp)

Note: The **base changing rule** is: $\boxed{\log_a x = \dfrac{\log_b x}{\log_b a}}$

Here, $\log_{16} 15 = \dfrac{\log_{10} 15}{\log_{10} 16} = 0.977$.

Example F

Solve $\log_3 (2x + 1) = 5$.

Solution

$\log_3 (2x + 1) = 5$

$\therefore 2x + 1 = 3^5$ [$x = a^y$ when $y = \log_a x$]

$\therefore 2x + 1 = 243$ [since $3^5 = 243$]

$\therefore 2x = 242$ [subtracting 1]

$\therefore x = 121$ [dividing by 2]

Logarithm laws can be used to simplify expressions involving logarithms.

Example G

Express $2\log_a x - 3\log_a y + 4$ as a single logarithm.

Solution

$2\log_a x - 3\log_a y + 4 = \log_a x^2 - \log_a y^3 + 4$ [$n\log u = \log u^n$]

$= \log_a \dfrac{x^2}{y^3} + 4$ [$\log u - \log v = \log \dfrac{u}{v}$]

$= \log_a \dfrac{x^2}{y^3} + 4\log_a a$ [since $\log_a a = 1$]

$= \log_a \dfrac{x^2}{y^3} + \log_a a^4$ [$n\log u = \log u^n$]

$= \log_a \dfrac{x^2 a^4}{y^3}$ [$\log u + \log v = \log uv$]

Unit 11.1 Activity 7B: Using logarithmic properties

1. Write the equivalent logarithmic statement for each of the following:

a. $2^3 = 8$ **b.** $3^4 = 81$ **c.** $5^2 = 25$ **d.** $6^3 = 216$ **e.** $x^a = b$

2. Write the equivalent exponential statement for the following:

a. $\log_3 9 = 2$ **b.** $\log_9 3 = \frac{1}{2}$ **c.** $\log_{25} 125 = 1.5$

d. $\log_{1000} 10 = \frac{1}{3}$ **e.** $\log_3 \frac{1}{3} = -1$

3. Solve the following for x (give answers as fractions or integers):

a. $2^x = 64$ **b.** $2^x = \frac{1}{2}$ **c.** $2^{x-1} = 32$ **d.** $2^x = 128$

e. $3^x = \frac{1}{3}$ **f.** $3^x = \frac{1}{27}$ **g.** $(3)^{\frac{1}{x}} = 9$ **h.** $125 = 5^x$

i. $9^x = 3$ **j.** $16^x = 2$ **k.** $25^x = 5$ **l.** $25^x = 125$

m. $25^x = 625$ **n.** $64^x = 32$ **o.** $49^x = \frac{1}{7}$ **p.** $121^x = 11^2$

q. $2^{1-x} = 4$ **r.** $\frac{1}{2^x} = 8$ **s.** $27^x = 243$ **t.** $81^x = 27$

4. Solve the following equations to 3 decimal places:

a. $10^x = 9$ **b.** $10^{-x} = 131$ **c.** $10^x = 323.3$

d. $2.5^x = 14.06$ **e.** $10^{3x} = 822.36$ **f.** $3^x = 4$

g. $2^{3x-1} = 9$ **h.** $3^x = 2^{x+1}$ **i.** $4^{2x} = 5^{x+2}$

j. $3.6^{4x+5} = 2.9^{2x+13}$ **k.** $500 = 200 \times 1.03^{2x}$

5. Calculate the following:

a. $\log_4 4$ **b.** $\log_3 3$ **c.** $\log_2 8$ **d.** $\log_9 3$

e. $\log_{16} 2$ **f.** $\log_{16} 32$ **g.** $\log_{625} 125$ **h.** $\log_{196} 14$

i. $\log_{32} 16$ **j.** $\log_{0.1} 10$

6. Solve the following equations for x:

a. $\log_2 x = 6$
b. $\log_{3x} x = 4$
c. $\log_4 x = \frac{1}{2}$
d. $\log_6 x = 3$
e. $\log_x 3 = 2$
f. $\log_x 4 = \frac{1}{3}$
g. $\log_x 9 = \frac{1}{2}$
h. $\log_x 3 = 1$
i. $\log_x 16 = 4$
j. $\log_x 216 = \frac{3}{2}$
k. $\log_2 32 = x$
l. $\log_3 27 = x + 2$
m. $\log_{\frac{1}{4}} 64 = x$
n. $\log_5 25 = 3x + 1$
o. $\log_8 4 = x$

7. Express each of the following as a single logarithm:

a. $\log x + \log y + \log z$
b. $2\log x + \log y$
c. $\log x + \log y + \log y$
d. $\log x + \log y - \log z$
e. $\log a + 3\log b - 2\log c$
f. $\frac{1}{2}\log p - \frac{3}{2}\log q$
g. $\log xy + \log z$
h. $\log xyz - \log x - \log z$
i. $\log x^2y^3 - \log x + \log y$
j. $\log \left(\frac{xy}{z}\right) + \log \left(\frac{z}{x}\right)$
k. $2\log \left(\frac{x}{y}\right) - \log x^2$
l. $\log \left(\frac{1}{x}\right) + \log x + 2\log y$
m. $\frac{1}{2}\log x - \log \sqrt{x} + \log xy$
n. $\log \frac{x}{yz} + \log \frac{1}{z} - \log \frac{x}{y}$
o. $2\log xy - \log \frac{x}{y}$
p. $2\log_a x + 5$

8. Express the following as logarithms of single numbers:

a. $\log 12 + \log 5$
b. $\log 60 - \log 5$
c. $2\log 5 + \log 3$
d. $\frac{1}{2}\log 9 + \frac{2}{3}\log 8$
e. $\frac{3}{2}\log 4 + \frac{1}{4}\log 81$
f. $\frac{1}{3}\log 27 + \frac{1}{5}\log 32 - \log 7$
g. $\frac{2}{5}\log 32 - \log 4$
h. $3\log 5 - \log 15$
i. $7\log 2 - \frac{1}{2}\log 16 + \log 3$
j. $\log 15 + \log 20 - \log 25$

9. Simplify to a single number, given as a fraction or integer without using a calculator:

eg, $\frac{\log 7}{\log 49} = \frac{\log 7}{\log 7^2} = \frac{\log 7}{2\log 7} = \frac{1}{2}$

a. $\frac{\log 8}{\log 2}$ **b.** $\frac{\log 25}{\log 125}$ **c.** $\frac{\log 36}{\log 216}$ **d.** $\frac{\log 625}{\log 125}$

e. $\frac{\log 64}{\log 32}$ **f.** $\frac{\log 196}{\log 14}$ **g.** $\frac{\log 64 + \log 32}{\log 32}$ **h.** $\frac{\log 7 + \log 7}{\log 49}$

i. $\frac{\log 32 - \log 8}{\log 4}$ **j.** $\frac{\log 625 + \log 25}{\log 125}$

10. Solve the index equations

a. $2^x \times 3^x = 10$ **b.** $5^{2x} \times 25^x = 100$

c. $8^x \times 4^{x+1} = 48$ **d.** $2^x \times 3^{x+1} = 73$

Unit 11.1 Activity 7C: Further logarithmic problems

1. The population p of a species of creatures after t years is given by $\frac{6(3)^t - 2}{2(3)^t + 5}$ where p is measured in millions.

a. What is the initial population?

b. What is the population after 3 years?

c. After how many years is the population equal to 2 million?

d. What size does the population get close to as the years pass?

2. The number of cassowaries in a bush after t years is $N = 2\,000 - 1\,000(2)^{-t}$.

a. How many cassowaries are there in the bush when records are first collected?

b. How many cassowaries are there in the bush $3\frac{1}{2}$ years later?

c. What is the maximum number of cassowaries which can live in the bush?

d. At what time is the population of cassowaries equal to 1 500?

3. The value of a computer is given by the equation $V = 4\,500(0.3)^{0.25t}$, where:
V = value of computer in dollars, and
t = time in years since the computer was purchased.

a. How much was the computer worth initially?

b. How much was the computer worth after three years?

c. When will the computer be worth $1 500?

d. How long is it before the computer loses half its value?

4. The cost of a hectare of land is believed to be $A(2.7)^{kt}$ in Kina, where t is the number of years, and A and k are constants. After 3 years the value of the land is K112 486 and after 7 years it is K131 593.

a. Find the values of A and k.

b. Find the value of the land after 10 years.

c. When will the land have a value of K1 000 000?

5. The number of particles of ash deposited up a chimney is given by $\log_{10} N = A \log_{10} h + \log_{10} k$, where N is the number of particles, h is the height of the chimney, and A and k are constants. At a height of 1 metre, the number of particles deposited is 1 000. At a height of 5 metres the number of particles deposited is 25 000. Find the height of a chimney if 100 000 particles are deposited.

Unit 11.1 Number and Application

Topic 8: Units of measurement

Units of measurement is one sub-section within Unit 11.1 of the Grade 11 syllabus (see Syllabus p. 14). This topic deals with problem solving involving measurement of everyday objects and covers:

- Choosing and making measurements using appropriate equipment, scales and units.
- Conversion between units.
- Problems involving length, mass, capacity and time (including 24-hour clock time.
- Calculation and interpretation of rates.
- The precision (limits of accuracy) of a measurement.

Introduction

The **Imperial** system of units was defined by the British in 1824 and was adopted by countries in the British Empire. By the end of the 20th century most of these countries had converted to metric units.

Papua New Guinea declared the **International System of units (SI)** to be the legal units of measurement in 1980. It is illegal to use any units of measurement other than the SI units for trade purposes in PNG.

The International System of Units (SI) defines seven units of measure as a basic set from which all other SI units are derived. These SI base units and their physical quantities are:

- Metre for length.
- Kilogram for mass.
- Second for time.
- Ampere for electric current.
- Kelvin for temperature.
- Candela for luminous intensity.
- Mole for the amount of substance.

If the base unit of length is the metre then all the units for length, area and volume are defined in terms of this basic unit. Many other units, such as the litre, are not formally part of the SI, but are accepted for use with the SI.

The metric system

Appropriate units must be used when solving problems.

In the **metric system**, each quantity measured has a **basic unit** (gram, metre, litre), of which other units are multiples. These other units are identified by a **prefix** such as 'kilo', 'centi', and 'milli'. By using prefixes measurements can be expressed with suitably sized numbers, eg not 45 000 g but 45 kg.

The table shows commonly used metric prefixes.

Prefix	kilo, k	centi, c	milli, m
Means	10^3 1 000 one thousand	10^{-2} $\frac{1}{100}$ one hundredth	10^{-3} $\frac{1}{1\,000}$ one thousandth

Note: Less commonly used prefixes include hecto (h) meaning 100, deci (d) meaning $\frac{1}{10}$ and micro (μ) meaning $\frac{1}{1\ 000\ 000}$.

Example A

The kilometre, centimetre, and millimetre are all units *derived* from the basic unit, the metre.

1. The length and width of a classroom would probably be measured with the basic unit, the metre.
2. To measure a longer distance, such as the distance between Port Moresby and Brisbane, the kilometre is a more suitable unit to use.
3. For smaller lengths, such as the thickness of a telephone book, the unit millimetre might be preferred. Alternatively, the centimetre may be used.

Changing from one metric unit to another

To change from one unit to another, either **multipliers** or **divisors** are used. This **conversion** method is shown in the following diagram for the prefixes kilo, centi, and milli.

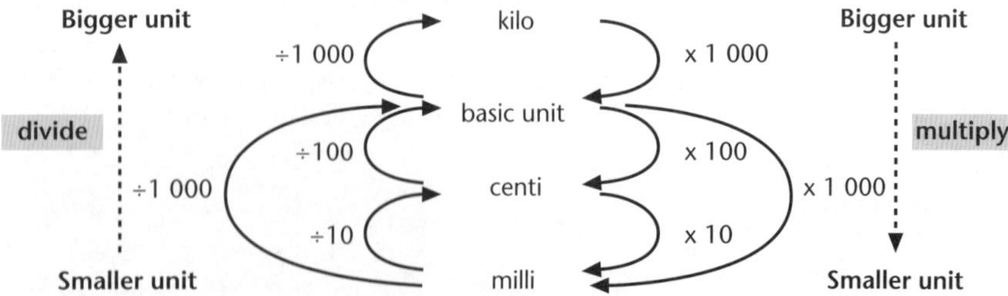

To convert from one unit to another follow the arrows, eg to change from the basic unit to kilo, divide by 1 000; to change from centi to milli, multiply by 10.

- To change from a bigger unit to a smaller unit (eg centi to milli), use multiplication, since there will be more of the smaller unit.
- To change from a smaller unit to a bigger unit, use division, since there will be fewer of the larger unit.

Example B

Q. Convert: **1.** 146 m to cm **2.** 13 400 mg to kg **3.** 1.75 km to cm

A. **1.** m to cm is from a bigger unit to a smaller unit, so multiplication is used:

146 m = 146 x 100 cm [changing m to cm]
= 14 600 cm

2. mg to kg is from a smaller unit to a bigger unit, so division is used:

13 400 mg = 13 400 ÷ 1 000 g [changing mg to g]
= 13.4 g
= 13.4 ÷ 1 000 kg [changing g to kg]
= 0.0134 kg

3. 1.75 km = 1.75 x 1 000 m [changing km to m by multiplying]
= 1 750 x 100 cm [changing m to cm]
= 175 000 cm

Some important metric quantities

Length

The basic unit of length in the metric system is the **metre** (symbol **m**). Other commonly used units for length are **kilometre** (**km**), **centimetre** (**cm**), and **millimetre** (**mm**).

1 km = 1 000 m	100 cm = 1 m	1 000 mm = 1 m

It is useful to know approximately the width of your own handspan, the length of your foot or your 'pace' for estimating lengths in practical situations. Find out, for example, how far apart your hands need to be to approximate a metre.

Mass

The common units of mass in the metric system are the **gram** (symbol **g**) and **kilogram** (**kg**). Other units commonly used for mass are **milligram** (**mg**), and **tonne** (**t**).

1 kg = 1 000 g	1 000 mg = 1 g	1 tonne = 1 000 kg

Note: Mass is commonly (but incorrectly) referred to as weight. In this book mass will, on occasions, be referred to as weight.

Capacity

The capacity of a container is the amount of liquid (or other material) it will hold. The **litre** (**L**) and **millilitre** (**mL**) are the commonly used units of capacity in the metric system.

1 kL = 1 000 L	1 000 mL = 1 L

It is useful to know that a teaspoon holds 5 mL, a tablespoon holds 15 mL and a standard cup holds 250 mL. Also, 1 mL of water weighs 1 g and 1 L of water weighs 1 kg.

Capacity is commonly referred to as volume (see page 110).

Money

The basic unit of money in Papua New Guinea is the **Kina** (symbol **K**). The other unit is the **toea** (symbol **t**): K1 = 100 toea.

Note: 48 toea can be written 48t or K0.48. Do *not* write 48 toea as 0.48t.

Temperature

The temperature scale is measured in degrees, using the **Celsius**, or **centigrade scale** (**°C**). This scale is based on the freezing and boiling points of water.

Freezing point of water is 0°C. Boiling point of water is 100°C.

Time

Time is the only commonly used measurement that is not in the metric system.

60 seconds = 1 minute	60 minutes = 1 hour	24 hours = 1 day	
7 days = 1 week	2 weeks = 1 fortnight	365 days = 1 year	366 days = 1 leap year
1 century = 100 years			

You should also be familiar with the number of days in each month of the year.

There are two systems used to record time: the **12-hour clock** and the **24-hour clock**. It is essential, when writing time using the 12-hour clock, that either am or pm (whichever is appropriate) is written as well.

In the 24-hour clock, the hours in a day are numbered from 0–24, with 0 being midnight. Four digits are always used to express the time on a 24-hour clock.

- The first two digits represent the hours after midnight.
- The second two digits represent the minutes past the hour.

Example C

1. 3.00 am (in 12-hour time) is 0300 hours (in 24-hour time).
2. 1200 is midday or 12 noon. [12 hours after midnight]
3. 5.30 pm is 1730 hours. [1200 + 0530]
4. 1408 hours is 2.08 pm. [1408 – 1200 → 2 hours and 8 minutes after midday]
5. 1 minute after 1559 hours is written 1600 hours. [1559 + 0001]

When using 24-hour time, remember that the final two digits are minutes (60 minutes = 1 hour) and do not treat them as you would decimal (base 10) numbers.

Example D

Q. Elijah's plane leaves at 1015 and arrived at its destination at 2105. How long was the journey?

A. Number of minutes from 1015 to 1100 = 45 minutes

Hours and minutes from 1100 until 2105 = 10 hours 5 mins

Total time = 10 hours 50 mins [adding]

Note: The usual technique of finding the difference by subtracting the times (as decimal numbers) cannot be used for 24-hour times (2105 – 1015 = 1090).

Unit 11.1 Activity 8A: Units of measurement

1. From the list **i.– xx.** below, choose the measurement that best fits the following:

a. The distance between Goroka and Kavieng.
b. The height of a man.
c. The time for a good swimmer to swim 100 m.
d. The length of a pen.
e. The capacity of the petrol tank of a car.
f. The thickness of a match.
g. The length of a playing field.
h. The weight of a brick.
i. The temperature inside a refrigerator.
j. The diameter of a 10t coin.
k. The time for a car to travel 90 km.
l. The capacity of a teaspoon.
m. The weight of an average woman.
n. The height of Mt Wilhelm.
o. The temperature of a recently made cup of tea.

Choose from:

i. 2 mm	**ii.** 1 hour	**iii.** 5 mL	**iv.** 4509 m	**v.** 15 cm
vi. 60 g	**vii.** 1.8 m	**viii.** 2.3 cm	**ix.** 5° C	**x.** 717 km

xi. 50 L	**xii.** 55 kg	**xiii.** 10 sec	**xiv.** 30° C	**xv.** 300 m
xvi. 80°C	**xvii.** 23 cm	**xviii.** 2 kg	**xix.** 1 min	**xx.** 100 L

2. Change each of the following quantities to the unit in the brackets:

a. 8 m (cm)	**b.** 4 cm (mm)	**c.** 4 000 g (kg)
d. 3.5 m (mm)	**e.** 0.3 km (cm)	**f.** 600 mL (L)
g. 4.8 km (m)	**h.** 8 300 kg (tonne)	**i.** 0.37 g (mg)
j. 0.4 m (cm)	**k.** 75 cL (mL)	**l.** 945 mg (g)
m. 140 000 mm (km)	**n.** 13 700 g (tonne)	**o.** 45 000 mm (km)
p. 4.7 tonne (kg)	**q.** 120 L (kL)	**r.** 640 m (km)
s. 0.004 km (mm)	**t.** 7.3 tonne (g)	**u.** 186 cm (m)
v. 63 000 000 mg (kg)	**w.** 3.8 kg (g)	**x.** 9.37 L (mL)

3. a. A rope is 20 metres long. A piece 150 cm long is cut off it. What length (in metres) remains?

b. Mary has a handspan of approximately 20 cm. She finds that a table top is 12 handspans long. What is the approximate length of the table?

c. A child's building blocks are 35 mm cubes. How tall (in centimetres) is a tower of 11 blocks?

4. a. An average apple weighs 155 g and an average orange weighs 175 g. What is the approximate weight in kg of a bag of 3 apples and 5 oranges?

b. A plastic container of identical buttons weighs 150 g. If the container weighs 30 g, and there are 30 buttons in it, how much does each button weigh?

c. If an egg weighs 53 g and a carton containing a dozen eggs weighs 776 g, how much does the empty carton weigh?

5. a. Ravi entered a race, which began at 9.45 am. She finished 3 hours and 37 minutes later. At what time did she finish?

b. Express 1412 hours using the 12-hour clock.

c. Express half-past-four in the morning, using the 24-hour clock.

d. On March 5, Pania calculates that there are 30 days until her birthday. When is her birthday?

e. A parade starts at 1130 hours and finishes at 1315 hours. How long did the parade last?

6. a. An aeroplane, due to arrive at 4.10 pm, is 35 minutes early. When does it arrive?

b. A traveller leaves Lae at 9.05 am and reaches Kainantu at 11.10 am. She stops for refreshments then starts again at 11.47 am. She reaches Goroka at 3.14 pm. How long did it take her to get from Lae to Goroka, excluding stopping time at Kainantu?

c. A traveller left Port Moresby on Tuesday at 1425 and arrived at her destination the following Wednesday at 0835. How much time did her journey take?

d. How many seconds are there in the month of March?

e. A flight leaves Auckland, New Zealand at 2130 for Port Moresby. The flight takes 4 hours and 55 minutes. In Port Moresby, time is 3 hours behind Auckland time. What time will it be in Port Moresby when the plane arrives (local PNG time)?

Scales and dials

Measurements are usually made by using various tools and instruments such as measuring tapes, stop watches or meters. On many such instruments taking a measurement means reading a **scale** or **dial**.

A scale, in this context, is shown by marks at set intervals, such as those on a ruler or protractor, the speedometer of a car or bathroom scales. Small divisions between markings on a scale indicate intermediate values that can be read with reasonable accuracy.

Example E

Q. The scale indicates the number of kilograms an object weighs. The pointer stops as shown. What weight is the object?

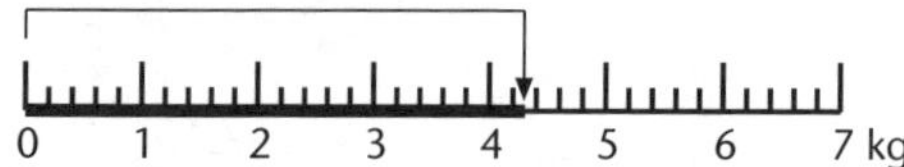

A. Between each kilogram marking are 5 equal divisions. Each division therefore represents $\frac{1}{5}$ kg = 0.2 kg.
The pointer has stopped halfway between the first and second division, which is $0.2 + \frac{1}{2} \times 0.2 = 0.3$ kg beyond 4 kg. The weight is 4.3 kg.

Circular dials, speedometers and scales on measuring jugs are other examples of linear scales, which are read in a similar way (be aware that some dials are read anticlockwise).

If a jug has sloping sides the scale may not be linear, ie the gaps between divisions may not be equal. Even if gaps between divisions on a scale are of unequal size, the divisions marked on the scale indicate measures of equal size. For example, in the scale shown below, the interval between 0 and 1 has been divided into five parts of unequal length; however, A is 0.2, B is 0.4, C is 0.6 and D is 0.8.

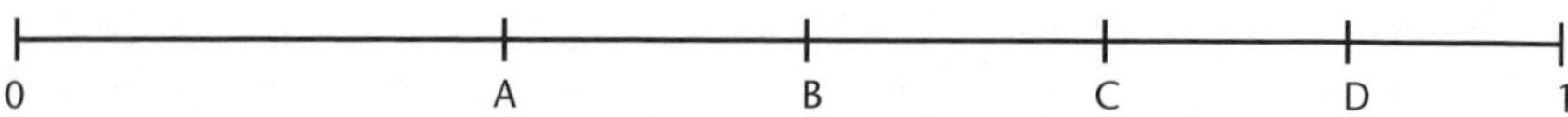

Example F

Q. A measuring cup has a capacity of 300 mL. A point labelled B is halfway between the base of the cup and the 100 mL mark. Which of the three positions A, B or C shown is the most likely to be the 50 mL mark?

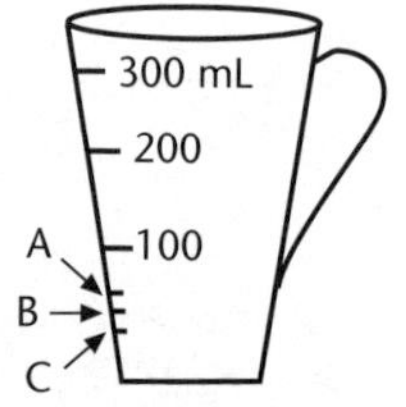

A. Position A is the most likely to be the 50 mL mark.

[the capacity up to level B is smaller than the capacity between level B and the 100 mL mark]

Unit 11.8 Activity 8B: Reading measuring instruments

1. Flour is weighed for a recipe (see diagram).

a. Give the weight in kilograms.

b. How many grams did the flour weigh?

c. Copy the diagram and draw an arrow to show a weight of 1.75 kg.

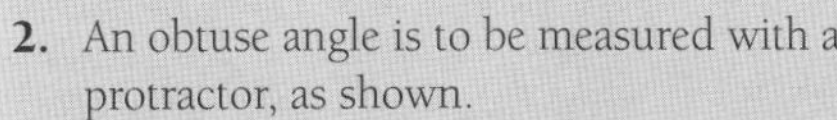

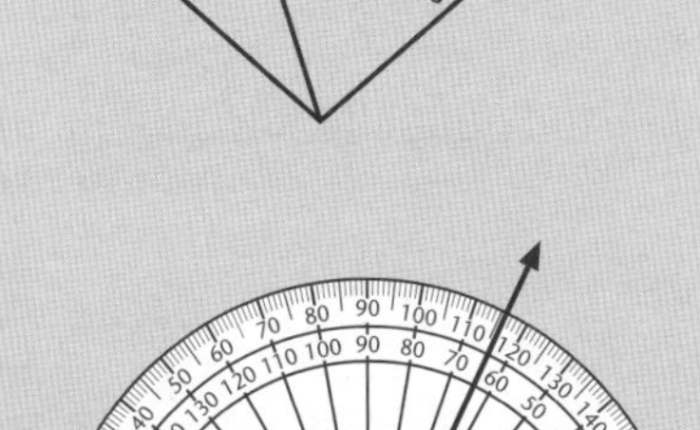

2. An obtuse angle is to be measured with a protractor, as shown.

a. How large is the angle to the nearest degree?

b. Two angles are supplementary if they add up to 180°. What size is the supplement of the angle shown?

3. For the following scales, give the reading or measurement shown by each arrow. Give units with each answer.

a.

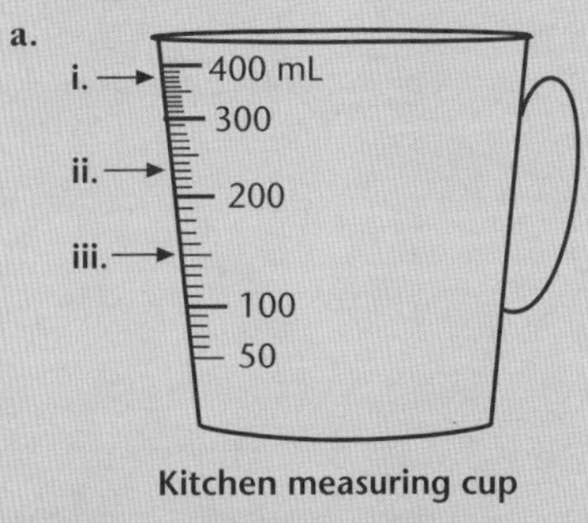

Kitchen measuring cup

b.

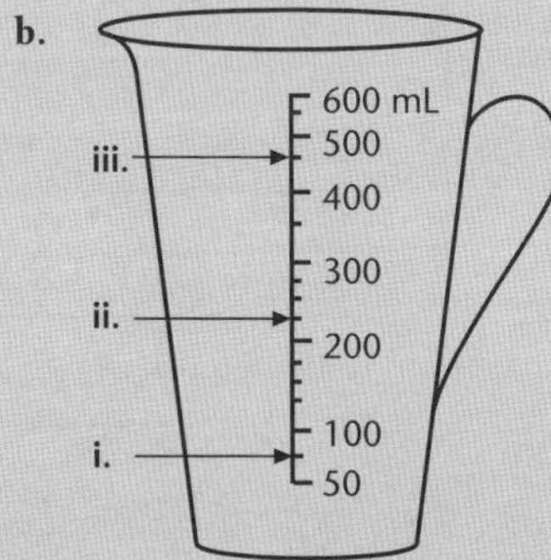

Kitchen measuring jug

c.

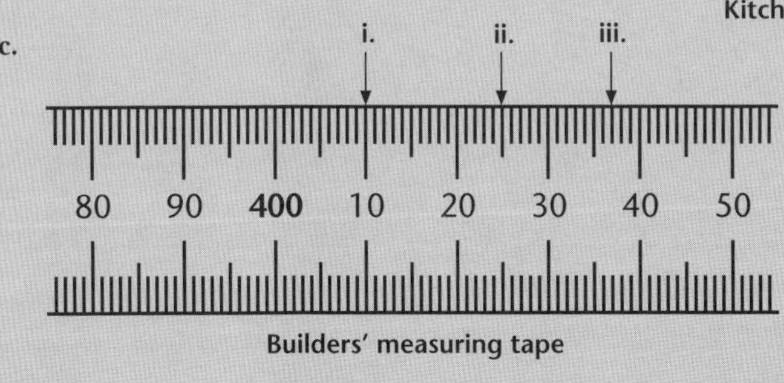

Builders' measuring tape

d.

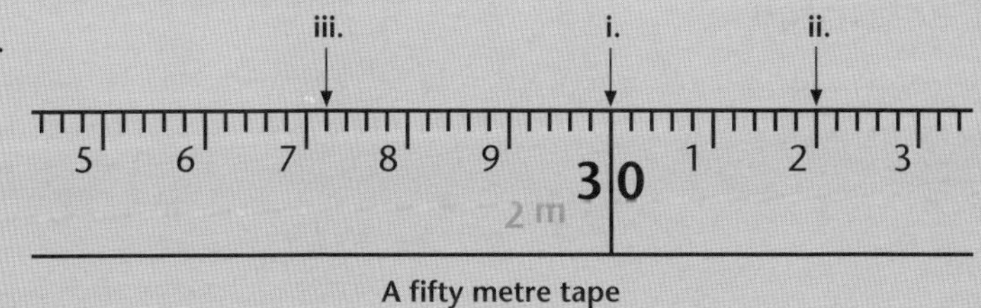

A fifty metre tape

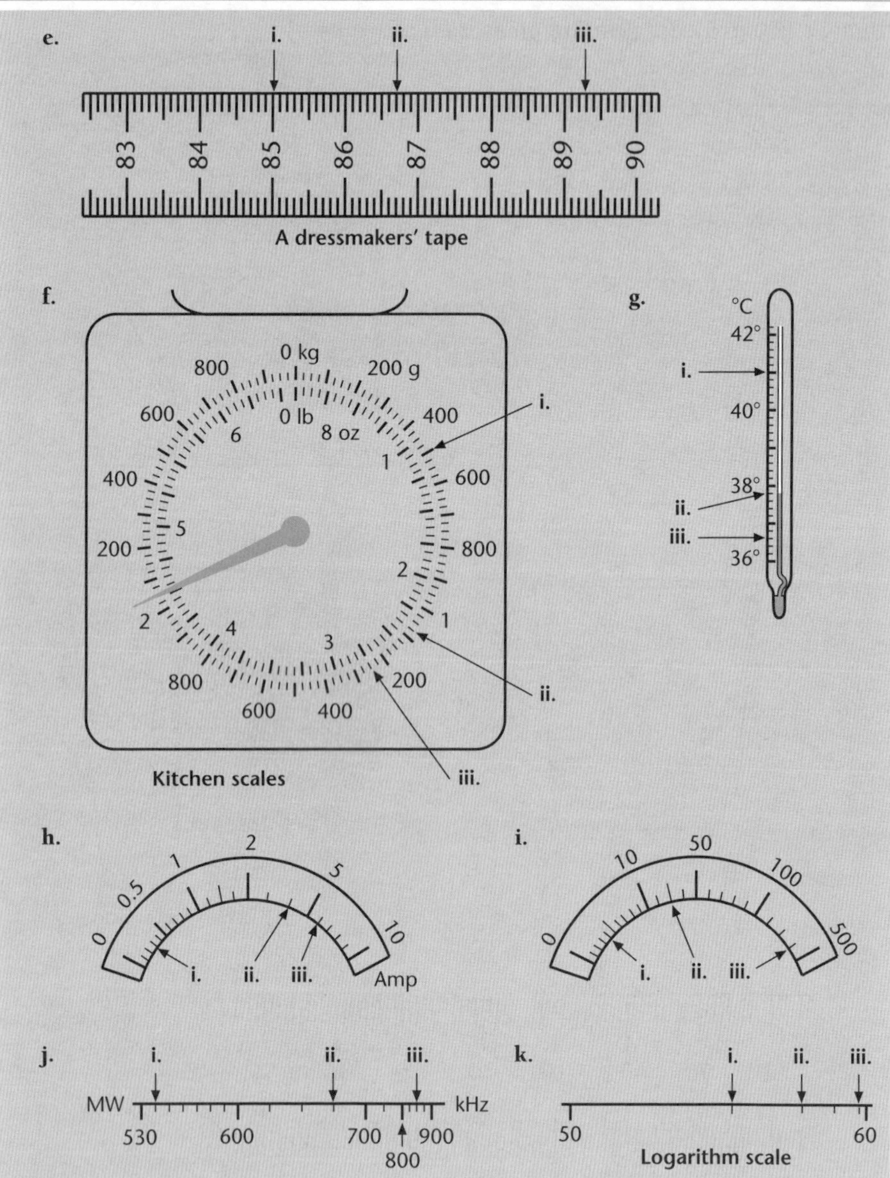

Limits of accuracy

When a quantity is measured, the measurement is never exact. It is only accurate to *half of the place value of the last* ***significant figure***.

Example G

1. If the length of a pen is recorded as being 14.4 cm, its actual length lies between 14.35 cm and 14.45 cm.
   [any measurement in this range of values rounds to 14.4 cm (3 sf)]

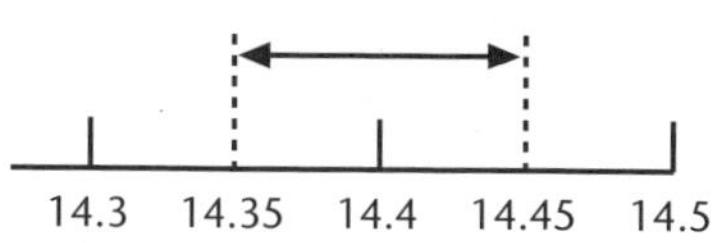

2. The time taken for a car to travel down a street was written down as 12 s. The actual time taken lies between 11.5 s and 12.5 s.

The limits of accuracy are often expressed as an **inequality**. In Example G, the actual length of the pen, x, recorded as being 14.4 cm, lies between 14.35 cm and 14.45 cm. This can be written as $14.35 \leq x < 14.45$ (note that the upper value 14.45 rounds to 14.5 cm, so is excluded from the set of values x can take).

- The number 14.35 is the **lower limit** for the measurement 14.4 cm.
- The number 14.45 cm is the **upper limit** for the measurement 14.4 cm.

To calculate the limits of accuracy for any measurement, proceed as follows:

Lower limit: reduce the last significant figure by 1 then 'add a 5' in the next place right (often the end of the number), adding a decimal point if required.

Thus 14.4 becomes 14.35, and 12 becomes 11.5. Note that 2.50 becomes 2.495.

Upper limit: 'add a 5' in the place to the right of the last significant figure (often the end of the number), adding a decimal point if required.

Thus 14.4 becomes 14.45, and 12 becomes 12.5. Note that 4 700 becomes 4 750.

Example H

Q. The side length of a square is given as 11.5 cm. Between what limits does its perimeter lie?

A. The lower limit of accuracy of the measurement is 11.45 cm. The upper limit of accuracy of the measurement is 11.55 cm. Therefore the perimeter lies between 4 x 11.45 = 45.8 cm and 4 x 11.55 = 46.2 cm, ie 45.8 cm ≤ Perimeter < 46.2 cm.

Unit 11.1 Activity 8C: Limits of accuracy

1. Write the limits of accuracy for each of the following measurements:

a. 6 m	**b.** 4.3 cm	**c.** 5.24 m	**d.** 1.8 cm	**e.** 24 mm
f. 758 km	**g.** 1.04 L	**h.** 0.6 mL	**i.** 7.0 mg	**j.** 9.00 m

2. A stopwatch shows a swimmer's time for a 25 m freestyle event to be 15.42 seconds. What are the limits of accuracy for this time?

3. The length of a laboratory bench is measured and found to be 3.40 m long. Give the limits of accuracy for this measurement.

4. The dimensions of a room are given as 3.2 m by 2.1 m. Find the limits of accuracy for the perimeter of the room.

5. The capacity of a pool is given as 48 000 L. Between what limits does the capacity lie?

6. The old record for a running race was given as 12.4 seconds. At a recent event, and using an electronic timer, a runner was timed as running the event in 12.41 seconds. Is the runner necessarily slower in this event than the record-holder?

Conversion of units

Many countries, although having officially converted to metric (SI) units, have had trouble convincing their population to work in these units. This is particularly evident in Britain and the United States of America where there are often references to the Imperial system of units.

Conversion tables

Length

In the Imperial system of units:

- Small lengths are measured in inches (in). For example, the length of my index finger is 3.5 inches.
- Medium lengths are measured in feet (ft) or yards (yd). For example, the length of my desk is two feet, the width across our classroom is 6 yards.
- Large distances are measured in miles. For example, the distance between Lae and Madang is 133 miles.

The following tables can be used for conversion from Imperial units to Metric units and vice versa.

Imperial		Metric
1 inch [in]		2.54 cm
1 foot [ft]	12 in	0.3048 m
1 yard [yd]	3 ft	0.9144 m
1 mile	1760 yd	1.6093 km
1 nautical mile	2025.4 yd	1.853 km

Metric		Imperial
1 millimetre [mm]		0.03937 in
1 centimetre [cm]	10 mm	0.3937 in
1 metre [m]	100 cm	1.0936 yd
1 kilometre [km]	1000 m	0.6214 mile

Area

Imperial units for area are square inches (in^2), square feet (ft^2), square yards (yd^2), square miles (mile^2) and acres where 1 acre = 4840 yd^2.

Metric		Imperial
1 sq cm [cm^2]	100 mm^2	0.1550 in^2
1 sq m [m^2]	10,000 cm^2	1.1960 yd^2

Metric		Imperial
1 hectare [ha]	10,000 m^2	2.4711 acres
1 sq km [km2]	100 ha	0.3861 square miles

Imperial		Metric
1 sq inch [in^2]		6.4516 cm^2
1 sq foot [sq ft]	144 in^2	0.0929 m^2
1 sq yd [yd^2]	9 sq ft	0.8361 m^2
1 acre	4840 yd^2	4046.9 m^2
1 sq mile [$mile^2$]	640 acres	640 acres

Volume

Imperial units for volume are cubic inches (in^3), cubic feet (ft^3), and cubic yards (yd^3). For liquids the units are the fluid ounce (fl oz), the pint, where 1 pint = 20 fl ounces, and the gallon, where 1 gallon = 8 pints.

Metric		Imperial
1 cubic cm [cm^2]		0.0610 in^3
1 cubic decimetre [dm^3]	1,000 cm^3	0.0353 ft^3
1 cubic metre [m^3]	1,000 dm^3	1.3080 yd^3
1 litre [l]	1 dm^3	1.76 pt

Imperial		Metric
1 cubic inch [in^3]		16.387 cm^3
1 cubic foot [ft^3]	1,728 in^3	0.0283 m^3
1 fluid ounce [fl oz]		28.413 ml
1 pint [pt]	20 fl oz	0.5683 l
1 gallon [gal]	8 pt	4.5461 l

Example I

Q. Convert

a. 5 metres to feet

b. 10 miles to kilometres

c. 3 hectares to acres

d. 15 square feet to square metres

e. 2 litres to pints

f. 20 gallons to litres

A. **a.** 5 (metres) × 1.0936 (yards) × 3 (feet) = 16.404 feet

b. 10 (miles) ×1.6093 (km) = 16.093 km

c. 3 (hectares) × 2.4711 (acres) = 7.4133 acres

d. 15 (sq ft) × 0.0929 (m^2) = 1.3935 m2

e. 2 (litres) × 1.76 (pints) = 3.52 pints

f. 20 (gallon) × 4.5461 (litres) = 90.922 litres

Unit 11.1 Activity 8D: Conversion of units

1. Convert the following lengths:

a. 3 metres to yards **b.** 200 miles to kilometres **c.** 40 inches to metres

d. 5 kilometres to yards **e.** 1 nautical mile to metres **f.** 10 centimetres to inches

2. Convert the following areas:

a. 1000 m^2 to square yards **b.** 2000 ha to square miles **c.** 1 km^2 to acres

d. 2000 square yards to hectares **e.** 3 square inches to mm^2 **f.** 0.35 acres to m^2

3. Convert the following volumes:

a. 4 m^3 to cubic feet **b.** 204 litres to cubic feet **c.** 60 fluid ounces to litres

d. 2 pints to cm^3 **e.** 350 cm^3 to pints **f.** 0.75 litres to cubic inches

Unit 11.1 Activity 8E: Multiple choice – measurement

1. 2.563 kilometres, in centimetres, is:

A. 256.3 **B.** 2563 **C.** 25630 **D.** 256300

2. One cubic centimetre is equivalent to:

A. 10 mm^3 **B.** 100 mm^3 **C.** 1000 mm^3 **D.** 1 000 000 mm^3

3. A rectangle has a length of 1.2 metres and a width of 80 centimetres. The area of this rectangle, in square metres, is:

A. 0.96 **B.** 9.6 **C.** 96 **D.** 9600

4. The limits of accuracy for a measurement of 2.54 m are:

A. 2.53 and 2.55 **B.** 2.545 and 2.546

C. 2.535 and 2.545 **D.** 2.50 and 2.60

5. The interval below is divided into 5 parts of unequal length but they indicate measures of equal size.

The measurement indicated by the arrow on the scale is closest to:

A. 83.25 **B.** 86.4 **C.** 87.0 **D.** 87.25

6. Jak leaves at 22:15 on a flight that takes 3 hours and 55 minutes. His destination is in a time zone that is 2 hours ahead of his departure time zone. The local time when Jak arrives at his destination is:

A. 1:10 **B.** 2:10 **C.** 3:10 **D.** 4:10

7. 0.9144 metres is equivalent to 36 inches in Imperial units. 15 inches, in centimetres, is equivalent to:

A. $\frac{15}{36} \times 0.9144 \times 100$ **B.** $\frac{36}{15} \times 0.9144 \times 100$

C. $\frac{15}{0.9144} \times 36 \times 100$ **D.** $\frac{15 \times 36}{0.9144 \times 100}$

Unit 11.1 Number and Application

Topic 9: Ratio and proportion

Ratio and proportion is a sub-section within Unit 11.1 of the Grade 11 syllabus (see Syllabus p. 14). This Topic introduces ratio and covers:

- Applying scales on a map with actual lengths on the ground.
- Solving problems on direct and inverse variation.

Ratios

A **ratio** is a *comparison* of two or more *like* quantities.

A ratio can be expressed in different ways. The different quantities can be compared using colons (:) between the quantities, eg 3 : 5 or 1 : 2 : 4. Two quantities can also be compared by writing their ratio as a fraction.

Example A

If a bag contains 30 black and 40 white marbles, then the ratio of black marbles to white marbles can be expressed as 30 : 40. This can also be expressed as the fraction $\frac{30}{40}$.

Simplifying ratios

Since a ratio can be expressed as a fraction, it can be **simplified** like a fraction by dividing by a common factor. Thus, in Example P, the ratio of the black marbles to white marbles can be written more simply as 3 : 4 or $\frac{3}{4}$. A ratio is in its **simplest form** when all terms are *whole numbers with no common factors.*

Example B

1. 80 : 40 = 2 : 1 [dividing both terms by 40]
2. $1\frac{1}{2} : 2\frac{1}{4}$ = 6 : 9 [multiplying by 4 to remove fractions]
   = 2 : 3 [dividing by 3 to simplify]
3. 0.5 : 1.5 : 2.25 = 50 : 150 : 225 [multiplying each term by 100 to remove decimal points]
   = 2 : 6 : 9 [dividing by 25]

Note:
1. A ratio should *not* include fractions or decimal points in any term.
2. A ratio does *not* have a unit.
3. Ratio pairs may be simplified using the fraction key on a calculator.

Quantities with different units must be expressed in the same unit before they can be compared as a ratio.

Example C

16 mm to 4 cm = 16 mm to 40 mm [changing 4 cm to 40 mm so that both terms have the same unit]
= 16 : 40
= 2 : 5 [dividing each term by 8]

Calculations with ratios

Calculations with ratios can be done by treating ratios in a similar way to equivalent fractions.

Example D

Q. A supermarket ordered lamb chops and chickens for Christmas sales in the ratio 3 : 5. If 60 chickens arrived in one order, how many lamb chops arrived in the same order?

A. Let the number of lamb chops in the order be x.

Then, lamb chops : chickens = 3 : 5 = x : 60

$$\therefore \frac{3}{5} = \frac{x}{60} \qquad \text{[expressing the ratios as fractions]}$$

$$\frac{3}{5} \times \frac{60}{1} = x \qquad \text{[multiplying by 60]}$$

$$x = 36$$

∴ there were 36 lamb chops in the order.

Sharing quantities in a given ratio

To share a quantity in a given ratio:

- *Find the total number of parts* by adding all numbers in the ratio.
- *Find the value of each part* by dividing the quantity by this total.
- Multiply each number in the ratio by the value of each part.

Example E

Q. Divide 16 oranges between two people in the ratio 3 : 5.

A. The ratio 3 : 5 gives 8 parts, since 8 = 3 + 5.

Therefore each part is $\frac{16}{8}$ = 2 oranges.

One share is 3 x 2 = 6 oranges and the other is 5 x 2 = 10 oranges.

So, 16 oranges divided in the ratio 3 : 5 gives 6 oranges to the first person and 10 oranges to the second person.

This process works no matter how many numbers are in the ratio.

Example F

Q. Divide K36 in the ratio 2 : 3 : 5.

A. Number of parts = 2 + 3 + 5 = 10   Value of each part = K36 ÷ 10 = K3.60

K36 in the ratio 2 : 3 : 5 divides into: 2 x 3.60 : 3 x 3.60 : 5 x 3.60

= 7.20 : 10.80 : 18.00   [simplifying]

The K36 divides into K7.20, K10.80, and K18.00.

Note: The final figures must always add up to the number or quantity being divided. In Example E, 6 oranges and 10 oranges totalled 16 oranges; and in Example F, K7.20 + K10.80 + K18.00 = K36.00.

Unit 11.1 Activity 9A: Ratios

1. Simplify each of the following ratios:

a. 8 : 16 **b.** 5 : 15 **c.** 20 : 12 **d.** 0.5 : 4

e. $\frac{1}{2} : \frac{1}{3}$ **f.** K2 to 50t **g.** 1 cm to 1 mm **h.** 20 min to 1 hr

i. 0.4 kg to 20 g **j.** 4 m to 1 km **k.** 10 mL to 5 L **l.** $4\frac{1}{5} : \frac{3}{10}$

2. a. Divide 40 in the ratio 3 : 5. **b.** Divide 56 in the ratio 5 : 2.

c. Divide 90 in the ratio 1 : 4 : 5. **d.** Divide K8.00 in the ratio 2 : 3.

e. Divide K32.40 in the ratio 5 : 7. **f.** Divide 180 in the ratio 2 : 3 : 5.

3. Mr Arua and Ms Tai bought a truck for K8 000. Mr Arua paid K5 000. They then hired the truck out for K300.

a. How much did Ms Tai pay for her share of the truck?

b. What should be Ms Tai's share of the $300?

4. Mr Bik invested K50 000 to buy a K70 000 company in partnership with Mr Liklik, who paid the balance.

a. How much did Mr Liklik pay?

b. What was the ratio of the investments, Mr Liklik's to Mr Bik's?

c. A profit of $175 000 was shared between the two in the same ratio as their investments. How much did each get?

5. A fruit concentrate is sold in 250 mL packs. The instructions are 'Add water to make one litre of ready-to-drink fruit juice'.

a. What is the ratio of concentrate to water in the made-up juice?

b. How many litres of water are required to dilute 2 L of concentrate?

c. Find the volume of water in 6 L of made-up juice.

d. If one litre of made-up juice fills 5 glasses, how many packs of concentrate are needed to make the juice to fill 80 glasses?

6. A nut mixture has peanuts, almonds and cashews in the ratio 5 : 3 : 1.

a. If Helen buys 2.7 kg of nuts, what weight of almonds will she have?

b. How much nut mixture should Helen buy is she wants to ensure she has 750 g of almonds in the mix?

Map scales

Ratio scales

A ratio scale is often found on a map, usually at the top or the bottom of the map.

For example, if the ratio scale is 1 : 1 000 000, then this means that 1 unit, measured on the map, represents 1 000 000 units in actual distance.

Simple scales

A simple scale is a diagram on the map showing a map measurement and stating the actual distance that this represents.

For example, if the simple scale is,

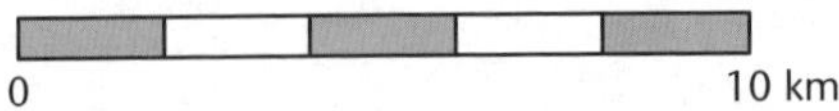

where each division on the scale is exactly 1 cm long, then 5 centimetres on the map represents 10 kilometres in actual distance (or 1 cm on the map represents 2 km in actual distance)

Example G

Q. Below is a map of the National Capital District

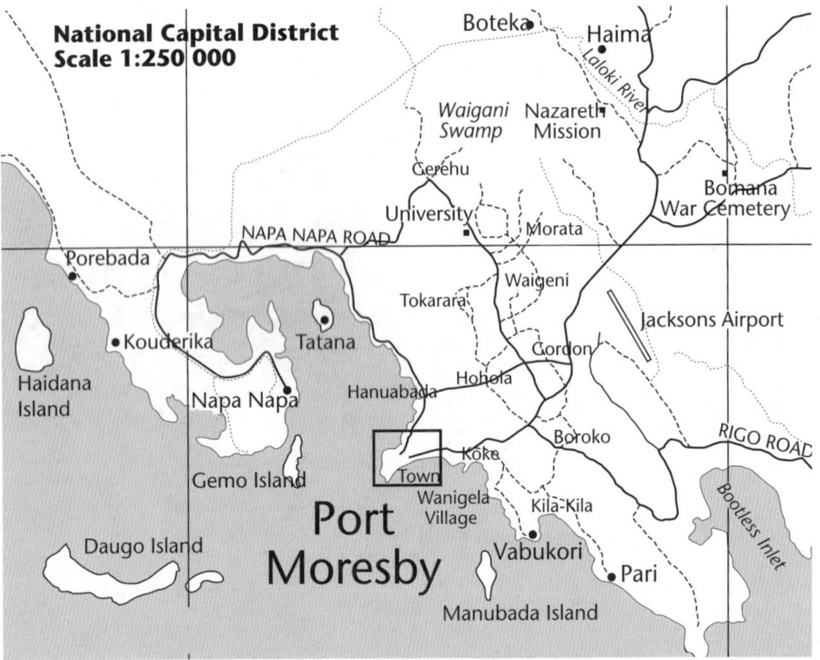

- **a.** Use the map and the ratio scale to find the actual straight line distance between Haima and Pari.
- **b.** Draw a simple scale for this map.

A. a. The straight line distance between Haima and Pari on the map is measured as 7 centimetres.

Substituting in the ratio:

Map distance : Actual distance = 1 : 250 000

7: Actual distance = 1 : 250 000 (centimetres)

$$\frac{\textit{Actual distance}}{7} = \frac{250\,000}{1}$$

Actual distance = 7 × 250 000
= 1750 000 centimetres
= 17 500 metres
= 17.5 kilometres

b. Map distance : Actual distance = 1 : 250 000

So 1 cm on the map is equivalent to 250 000 cm actual distance.

250 000 cm = 2500 metres

= 2.5 kilometres

1 centimetre on the map is equivalent to 2.5 kilometres in actual distance so 4 cm is equivalent to 10 km.

Simple scale:

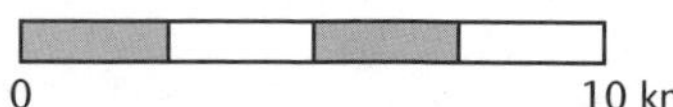

Example H

Q. Two towns that are known to be 20 kilometres apart measure 5 cm apart on a map. What is the ratio scale of the map?

A. Map distance : Actual distance = 5 cm : 20 km

The quantities have different units so we need to convert the larger unit, kilometres, to the smaller unit, centimetres:

Map distance : Actual distance = 5 cm : 20 km
= 5 cm : 20 000 metres
= 5 cm : 2 000 000 cm
= 5 : 2 000 000 (Units are now the same)
= 1 : 400 000 (Dividing by 5 for simplest form)

Multiply kilometres by 1000 to convert to metres

The ratio scale of the map is 1 : 400 000

Example I

Q. Convert the following simple scale to a ratio scale.

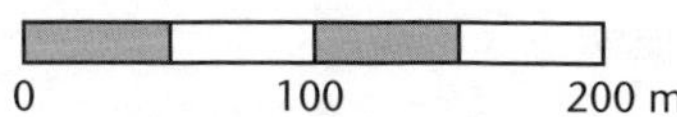

A. The marks on the scale are 1 cm apart.

This simple scale tells us that 4 centimetres on the map represent 200 metres in actual distance (or 1 centimetre represents 50 metres)

To convert this to a ratio scale:

Map distance : Actual distance = 4 cm : 200 metres

= 4 : 20 000 centimetres

= 1 : 5000 (Dividing by 4 for simplest form)

The ratio scale is 1 : 5000

Unit 11.1 Activity 9B: Map scales

1.

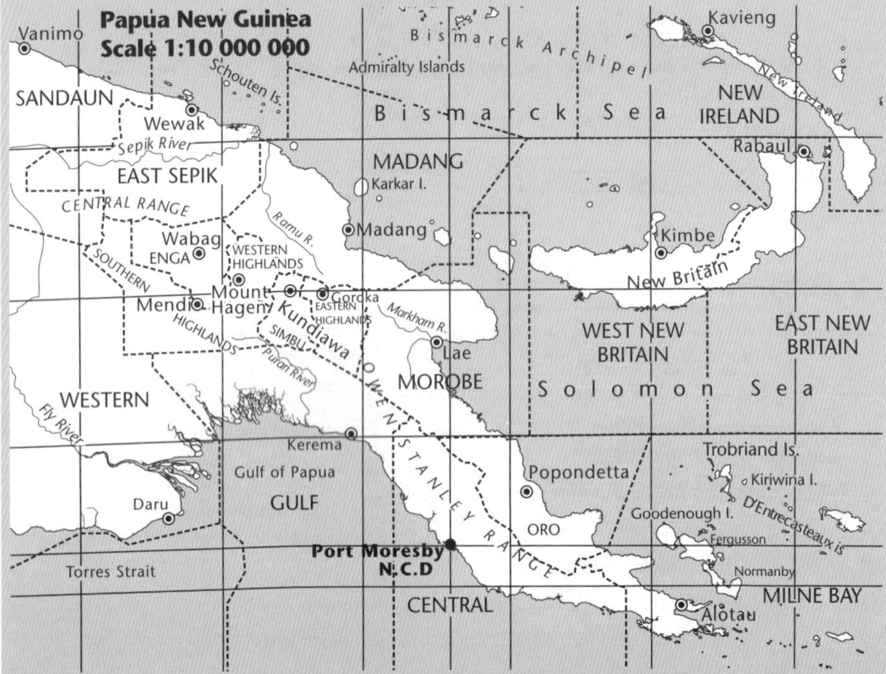

a. The ratio scale for the map above is 1 : 10 000 000. Convert this so that you can say 'One centimetre on the map represents .... kilometres'.

b. Construct a simple scale for this map.

c. Measure the straight line distance on the map, to the nearest millimetre, between

i. Wewak and Goroka

ii. Popondetta and Mt Hagen

d. Convert the map distances from part **c**. to actual distances in kilometres.

2. Convert the following simple scales to ratio scales:

a.

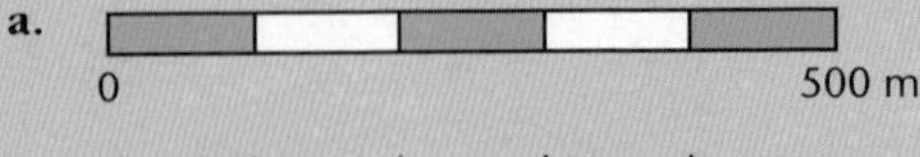

b.

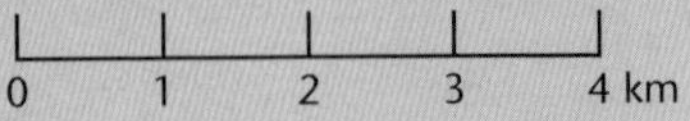

0 1 2 3 4 km

c. 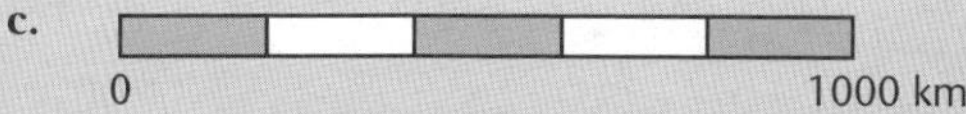

3. Draw a simple scale for each of the following ratio scales:

a. 1 : 2000

b. 1 : 150 000

c. 1 : 25 000

4. The map of Rabaul (below) has a ratio scale of 1 : 13 000.

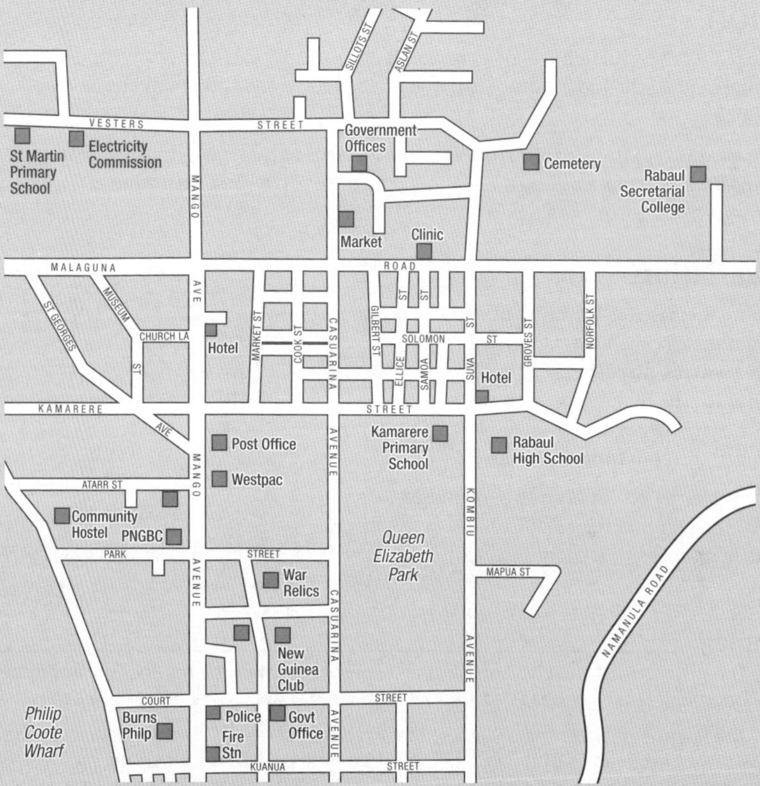

a. Use the ratio scale to find the actual straight line distance between:

i. The Police Station and the Post Office.

ii. The Community Hostel and the Market.

b. The straight line distance between two locations in Rabaul is 871 metres. How far apart are the two locations on the map?

c. Construct a simple scale for this map.

5. Two locations that are known to be 77 m apart are drawn on a map which has a ratio scale of 1 : 500. What is the distance between the locations on the map?

6. The straight line distance between two cities is known to be 560 km. If the simple scale on a map is:

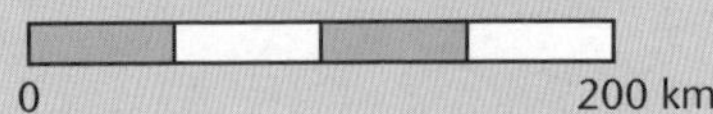

How far apart are the two cities on the map?

7. On a map two cities are 7.7 cm apart and the actual distance between the two cities is known to be 385km. What is the ratio scale of the map?

Variation

Variation describes the situation where a change in one quantity causes a change in another quantity. For example, the more cans of fish that are bought at the supermarket, the more money you pay.

The amount paid is said to be **proportional** to the number of cans bought. If the total cost paid is C kina and the number of cans bought is n then we can write $C \propto n$ where the $\propto$ symbol means 'is proportional to' or 'varies as'.

Direct variation

If one quantity increases as the other quantity increases then this is called **direct variation**. If the quantity y **varies directly** as (or **is directly proportional to**) the quantity $\boldsymbol{x}$, then we can write $y \propto x$ and hence $y = kx$ where $\boldsymbol{k}$ is a constant, called the **constant of variation**.

If there is direct variation between two quantities, x and y, then the graph of the relationship between the quantities is a straight line with gradient k.

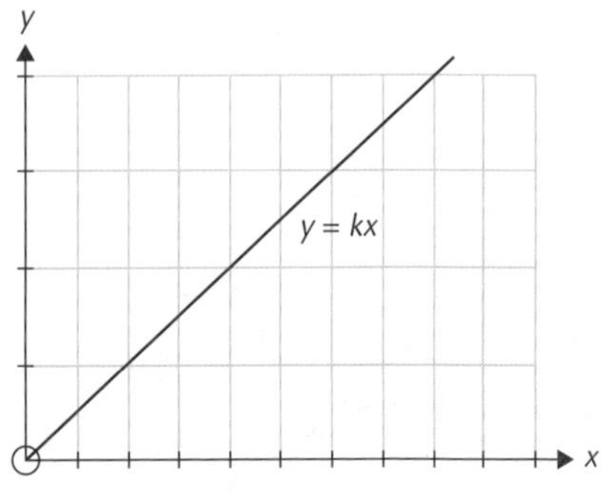

Example J

Q. 1. The price paid at the checkout is directly proportional to the number of cans of beans bought. If 6 cans of beans cost K3.78 then what is the cost of 15 cans of beans?

2. The volume of water that cylinders of a fixed height can hold varies directly as the square of the radius of the cylinder. If the volume is 3770 mL when the radius is 10 cm what is the volume of water in a cylinder of the same height when the radius is 6 cm?

A. 1. If b is the number of cans bought and P kina is the price paid at the checkout then we can say $P \propto b$ and $P = kb$ where k is a constant.

We know that when $b = 6$ then $P = 3.78$; substituting these values into the equation $P = kb$ will enable us to find a value for k.

$P = kb$

$3.78 = k \times 6$: divide both sides of the equation by 6

$k = 0.63$ (this is price of one can of beans)

so $P = 0.63b$

Substitute $b = 15$

$P = 0.63 \times 15 = 9.45$ You will pay K9.45 at the checkout for 15 cans of beans.

2. If V mL is the volume of water the cylinder can hold and r cm is the radius of the cylinder $V \propto r^2$ then and $V = kr^2$ where k is a constant.

Substituting V = 3770 and r = 10 into $V = kr^2$ to find k :

$3770 = k \times 10^2$

$\frac{3770}{10^2} = k$

$k = 37.70$

So $V = 37.70r^2$

Substituting $r = 6$

$V = 37.70 \times 6^2$

$= 1357.2$: the volume of the cylinder will be 1357.2 mL

Unit 11.1 Activity 9C: Variation

1. If y varies directly as x:

- **a.** Write a variation statement and an equation using the constant k.
- **b.** If $y = 34$ when $x = 1.7$ find the value of k.
- **c.** Find the value of y when $x = 3.2$.
- **d.** Find the value of x when $y = 56$.

2. If p is directly proportional to q and $p = 1.25$ when $q = 10.5$, find:

- **a.** p when $q = 63$
- **b.** q when $p = 30$

3. If $a \propto b^2$ and $a = 30$ when $b = 4$, find:

- **a.** a when $b = 12$
- **b.** b when $a = 187.5$

4. Which of the following pairs would you expect to be directly proportional?

- **a.** petrol consumption and distance travelled at a constant speed
- **b.** speed of travel and distance covered
- **c.** speed of travel and time taken for a journey

5. The exchange rate on a particular day is 1 Papua New Guinea Kina (PGK) = 0.421362 Australian dollars (AUD).

- **a.** Write down a rule connecting Australian dollars, A, and Papua New Guinea kina, P, on that day.
- **b.** How many Australian dollars, to the nearest cent, would you buy with K80?
- **c.** How many Papua New Guinea Kina, to the nearest toea, would you buy with $100 Australian?

6. Hooke's law states that the extension (**E**) of an elastic body, such as a spring, is directly proportional to the force (***f***), such as a weight, acting on it.

   When a weight of 2 kilograms is hung on a spring the extension is 5.2 centimetres. Find:

   a. the extension of the spring when a weight of 5 kg is hung on the spring.

   b. the weight required to extend the spring by 10 cm.

7. The value of a garnet (a type of gemstone) varies directly as the square of its weight. A garnet that weighs 4 carat is valued at K244. Find the value of a garnet that weighs 5 carat.

8. The surface area of a sphere, A, varies directly as the square of its radius, r. $A \propto r^2$. If a sphere of radius 6 cm has a surface area of approximately 452 cm^2, find the approximate radius (to the nearest mm) of a sphere of surface area 1000 cm^2.

9. When a stone falls freely, the time taken to hit the ground varies directly as the square root of the distance fallen. If it takes 4 seconds to fall 78.4 m, find how long it would take for a stone to fall 500 m.

10. The distance to the visible horizon at sea varies directly as the square root of the height of the observer's eye above sea level. If an observer's eye level is 5.4 metres then the distance to the visible horizon is 9 km. How far can an observer see to the horizon if his eye is at a height of 10 metres above sea level?

11. The period, ***T*** seconds, is the time that a pendulum takes to swing through one oscillation. The period is directly proportional to the square root of the length of the pendulum, ***l*** cm. $T \propto \sqrt{l}$.

    If a pendulum 12 cm long has a period of 0.7 seconds, find:

    a. the period of a pendulum that is 8 cm long

    b. the length of a pendulum that has a period of 1 second

Inverse variation

- If there is a decrease in one quantity corresponding to an increase in the other quantity then this is called inverse variation.
- For inverse variation $y \propto \frac{1}{x}$.
- If $y \propto \frac{1}{x}$ then $y = \frac{k}{x}$ where k is a constant (the constant of variation).
- A test for inverse variation is to find the product of the variables, xy, and show that it is a constant: $xy = k$.
- $y \propto \frac{1}{x}$ can be referred to as 'y varies inversely as x', 'y is inversely proportional to x' or 'y is proportional to $\frac{1}{x}$'
- If y varies inversely as x then the graph of y versus x has the shape shown at right.
- For inverse variation neither of the variables can be zero.

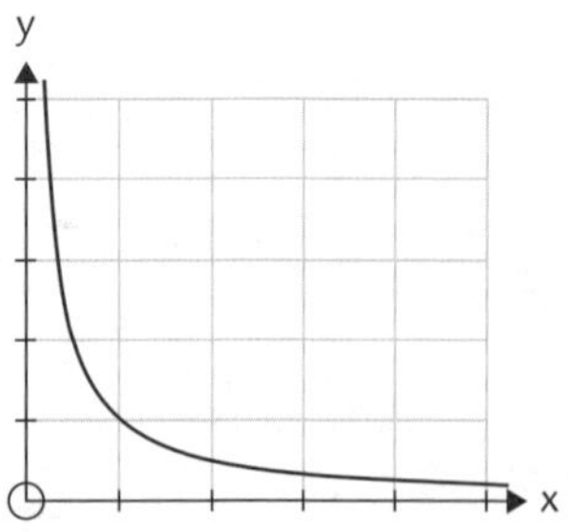

Example J

Q. 1. If y is inversely proportional to x and $y = 5$ when $x = 8$, find:

a. y when $x = 5$

b. x when $y = 4$

2. The amount of light that enters a camera lens depends on the area of the opening of the shutter. The relationship between the f- number on the camera settings and the area, A mm^2, of the opening of the shutter is shown in the table below.

f	2	4	8	16
A	480	120	30	7.5

a. Show that A and f^2 are inversely proportional

b. Find the value of A when $f = 2.8$; an actual setting on a manually operated camera.

A. 1. If y is inversely proportional to x then $y \propto \frac{1}{x}$ and $y = \frac{k}{x}$ where k is a constant.

a. If $y = \frac{k}{x}$ then $k = xy$. Substituting $y = 5$ and $x = 8$ gives $k = 5 \times 8 = 40$.

So $xy = 40$ is the general rule for this question.

Substituting $x = 5$ in $xy = 40 : 5 \times y = 40$ giving $y = 8$

b. Substituting $y = 4$ in $xy = 40 : x \times 4 = 40$ giving $x = 10$

2. a. We need to show that $A \propto \frac{1}{f^2}$ which means that $A = \frac{k}{f^2}$ or $Af^2 = k$ a constant.

Extending the table:

f	2	4	8	16
A	480	120	30	7.5
f^2	4	16	64	256
Af^2	1920	1920	1920	1920

$Af^2 = 1920$, a constant, so we can say that A and f^2 are inversely proportional.

b. Substituting $f = 2.8$ in $Af^2 = 1920$:

$Af^2 = 1920$

$$A = \frac{1920}{2.8^2} = 244.9 \text{ mm}^2$$

Unit 11.1 Activity 9D: Inverse variation

1. If y varies inversely as x and $y = 2.2$ when $x = 15$, find:

a. y when $x = 3$

b. x when $y = 5$

2. The variables in the table are inversely proportional.

i. Find the value of the constant of proportionality.

ii. Find the missing values.

a.

x	2	3	8	
y	60	40		10

b.

p	100		25	
q	0.75	1.5	3	10

3. Boyle's law states that 'for a given mass of gas at constant temperature, the volume, V, is inversely proportional to the pressure, P'.

a. Write an equation involving the variables V and P and a constant of proportionality, k.

b. A gas occupies 2 litres if it is at 740 mm of mercury pressure. Find the value of k.

c. What volume will the gas occupy if it is at 500 mm of mercury pressure?

4. The time taken, t, for a journey of fixed distance is inversely proportional to the average speed of travel, s. If the journey takes 2 hours 15 minutes travelling at an average speed of 50 km/hour, how long will it take if the average speed is 27 km/hr? Give your answer in hours and minutes.

5. On a guitar the frequency of vibration of a string, measured in Hertz (Hz), or vibrations per second, determines the note. The frequency of vibration, f, of a guitar string is inversely proportional to the length, l, of the string. The length of a string is shortened by pressing the string behind a fret. If the note A, which has a frequency of vibration of 440 Hz, is produced by a string of length 63 cm, find:

a. the length of a string that will produce the note C, which has a frequency of vibration of 523.2 Hz

b. the length of a string that will produce the note G, which has a frequency of vibration of 784 Hz

6. For a light with a constant output, the illumination, I units, at a distance d m from the light source is inversely proportional to d^2, ie. $I \propto \frac{1}{d^2}$.

a. If a light source has an illumination of 850 units at a distance of 1 metre, what will the illumination be at a distance of from the source?

i. 2 metres

ii. 5 metres

b. At what distance from the light source will the illumination be 68 units?

Unit 11.1 Number and Application
Topic 10: Basic algebra—revision

Basic algebra is a sub-section within Unit 11.1 of the Grade 11 syllabus (see Syllabus p. 14). This Topic deals with using straightforward algebraic methods and solving equations. It covers:

- Simplifying algebraic expressions involving exponents such as $(2x)^3$ and $\frac{12a^5}{8a^2}$.
- Manipulating and simplifying expressions such as $\frac{x}{4} + \frac{x}{3}$.

Algebraic terms

In **algebra**, letters such as x or a are used to represent unknown quantities or numbers. These letters are often called **pronumerals** or **variables**.

Algebraic terms are formed by combining numbers and variables. A **term** can be

- A **constant** such as 5 or $\frac{-1}{2}$.
- A **variable** such as y or A.
- A variable raised to a **power** or **exponent** eg y^3.
- Any product or quotient of constants and variables with exponents, eg $4r^2$ or $\frac{ab}{c}$.

In an algebraic term, the **coefficient** of the variable is the factor multiplying the variable, eg in the term $4x$, the coefficient of x is 4.

Algebraic expressions

Numerical expressions, such as $3 + \frac{4}{5}$ or $1 - 2 \times 3^2$, are formed using numbers along with the **operations** of addition, subtraction, multiplication and division etc. **Algebraic expressions** are formed in a similar way by adding and subtracting algebraic terms.

Example A

1. The **sum** of 8 and 4 is written 8 + 4. Similarly,
   - The sum of 8 and x is written $8 + x$.
   - The sum of y and x is written $y + x$.
2. 7 more than the **product** of 5 and 6 is written 5 x 6 + 7. Similarly,
   - 7 more than the product of 5 and a is written $5 \times a + 7$ or $5a + 7$.
   - c more than the product of a and b is written $ab + c$.
3. Nine less than the **square** of 7 is written $7^2 - 9$. Similarly,
   - 9 less than the square of x is written $x^2 - 9$.
   - y less than the square of x is written $x^2 - y$.

Simplifying algebraic expressions

Numerical expressions are simplified in the usual way using number rules, for example $1 - 2 \times 3^2 = 1 - 2 \times 9 = -17$. Some algebraic expressions may also be simplified, using the laws of algebra.

Since letters are representing numbers, the rules for arithmetic calculations hold true for algebraic calculations. These include:

Algebraic Rule	Numeric Example
$a + b = b + a$	$5 + 4 = 4 + 5$
$ab = ba$	$2 \times 3 = 3 \times 2$
$(a + b) + c = a + (b + c)$	$(3 + 4) + 5 = 3 + (4 + 5)$
$(ab)c = a(bc)$	$(3 \times 5) \times 6 = 3 \times (5 \times 6)$
$a(b + c) = ab + ac$	$7(3 + 2) = 7 \times 3 + 7 \times 2$
$a + 0 = 0 + a = a$	$2 + 0 = 0 + 2 = 2$
$a \times 1 = 1 \times a = a$	$5 \times 1 = 1 \times 5 = 5$
$a \times -1 = -1 \times a = -a$	$3 \times -1 = -1 \times 3 = -3$

Order of operations

Whether a problem is written in words or as a mathematical sentence, punctuation symbols are as important in mathematics as they are in English. In mathematics, the main punctuation symbols are **brackets**.

Example B

The messages 'No, cost too much' and 'No cost too much' have entirely different meanings, yet they differ by just one comma.

Likewise, the mathematical expressions $2 + 3 \times 4$ and $(2 + 3) \times 4$ give different answers:

$2 + 3 \times 4 = 2 + 12 = 14$ and $(2 + 3) \times 4 = 5 \times 4 = 20$.

The mnemonics **BEMA** or BEDMAS (brackets; exponents; multiplication/division; addition/subtraction) gives the convention for the correct order of operations.

Take care to include brackets in algebraic or numeric expressions wherever necessary to indicate the required order of operations.

Example C

Q. Write each of the following using a mathematical sentence:

1. 6 times the sum of 3 and 5.
2. 20 divided by the sum of p and q.
3. The product of 4, and another number that is increased by 5.
4. The square root of the difference between a and b (where $a > b$).

A.
1. $6(3 + 5)$ [the addition of 3 and 5 is done *before* multiplication by 6]
2. $\frac{20}{p + q}$ [the bracket around $p + q$ is understood, and need not be shown]
3. $4(x + 5)$ [x is the number, and it is increased by 5 before multiplication by 4]
4. $\sqrt{a - b}$ [the bracket around $a - b$ is understood, and need not be shown]

Unit 11.1 Activity 10A: Algebraic expressions

Write the following as algebraic expressions.

1. The sum of 9 and a.

2. b decreased by 4.

3. The product of 5 and c.

4. 6 more than x.

5. 7 less than y.

6. The product of 4 and the sum of x and y.

7. A number divided by 6.

8. The sum of x and y, divided by 10.

9. The difference between 12 and y (where $y < 12$).

10. x is increased by 4, then tripled.

11. y is decreased by 7, then halved.

12. The square root of the sum of x and 10.

13. The sum of two consecutive whole numbers.

14. The sum of a and 10 divided by the product of 7 and b.

15. A number which is 3 less than twice the value of y.

16. The cube of two less than p.

17. Jane is x years old. What was her age 5 years ago?

18. A trailer weighs t kg. It is loaded with 3 logs, each weighing l kg. Write down the weight of the loaded trailer.

19. The cost of x 10t stamps plus y 15t stamps.

20. I am a years old and my brother is b years old. Write down the sum of our ages in 2 years' time.

Like terms

Like terms have the same variables to the same powers. In **unlike terms** the variables or exponents (or both) are different.

Example D

1. $3a$, $4a$, and $-5a$ are *like* terms because they have the same variable, a, and the same exponent, 1 (ie $a = a^1$).
2. $3a$ and $4a^2$ are *unlike* terms because, although they have the same variable, a, they have different exponents (1 and 2 respectively).
3. $3ab$ and $4ba$ are like terms, but $3a$ and $4b$ are unlike terms.
   [the order of the variables (ab or ba) does not affect whether terms are/are not like terms]

Addition and subtraction of algebraic terms

Algebraic expressions containing *like* terms *can* be simplified by addition and subtraction. *Unlike* terms *cannot* be simplified by addition or subtraction. Gather like terms together and combine them by adding/subtracting their coefficients. (Numbers are combined with each other in the usual way.)

Example E

The following expressions contain like terms and can be simplified:

1. $3a + 4a = (3 + 4)a$ [adding coefficients]
   $= 7a$
2. $8 - 3x - 2 + x = 8 - 2 - 3x + x$ [gathering like terms]
   $= 6 + (-3 + 1)x$
   $= 6 - 2x$ [adding coefficients]
3. $x^2 + 3x + 4x^2 - 7x = x^2 + 4x^2 + 3x - 7x$ [gathering like terms]
   $= (1 + 4)x^2 + (3 - 7)x$ [adding/subtracting coefficients of like terms]
   $= 5x^2 + -4x$, or (better) $5x^2 - 4x$
4. $4a - 3b + 2ab - a + 2b = 4a - a - 3b + 2b + 2ab$ [gathering like terms]
   $= 3a - b + 2ab$

Note: In part **3**, the terms $5x^2$ and $-4x$ are unlike terms (unequal exponents) and so the expression $5x^2 - 4x$ will not simplify further. Similarly, in part **4**, $3a$, $-b$, and $2ab$ are all unlike terms.

Multiplication and division of algebraic terms

Algebraic expressions involving *like* or *unlike terms* can be multiplied together eg $a \times b \times c = abc$. Exponents are used to simplify answers eg $a \times a$ is written as a^2. Note that multiplication signs are not included in the final answer.

Example F

Q. Simplify **1.** $3a \times 2b$ **2.** $4w \times 3y$ **3.** $8ab \times 12ac$

A. **1.** $3a \times 2b = 3 \times a \times 2 \times b$
$= 3 \times 2 \times a \times b$
$= 6 \times a \times b$
$= 6ab$

2. $4w \times 3y = 4 \times 3 \times w \times y$
$= 12wy$

3. $8ab \times 12ac = 8 \times a \times b \times 12 \times a \times c$
$= 96a^2bc$ [$a \times a = a^2$]

Algebraic division is set out in fraction form and common factors are 'cancelled', using the fact that, if $x \neq 0$ then $\frac{x}{x} = \frac{1}{1} = 1$.

Example G

The following expressions involve division of unlike terms:

1. $\frac{8ab}{4a} = \frac{{}^{2}\cancel{8} \times {}^{1}\cancel{a} \times b}{{}_{1}\cancel{4} \times \cancel{a}_{1}}$ [dividing numerator and denominator by 4 and a]
   $= 2b$

2. $\frac{24abc}{16ac} = \frac{{}^{3}\cancel{24}\ {}^{1}\cancel{a}\ b\ {}^{1}\cancel{c}}{{}_{2}\cancel{16}\ {}_{1}\cancel{a}\ {}_{1}\cancel{c}}$ [dividing numerator and denominator by 8, a and c]
   $= \frac{3b}{2}$

Unit 11.1 Activity 10B: Combining algebraic terms

For questions 1–20, simplify the expressions where possible:

1. $5a + 3a$	**2.** $7a - a$	**3.** $3 + x + x$
4. $2x + 3 + x^2 - x - 2$	**5.** $x^2 - x + 5 + 3x^2 - 2x$	**6.** $6x^2y - 3xy^2 - 2yx^2$
7. $7x + 3x$	**8.** $5a - 2a$	**9.** $4 + x + x + x$
10. $5x - 2 + x^2 + 3x + 4$	**11.** $x^2 + 2x - 3 + 2x^2 - 3x$	
12. $2x^2 - 4x + 7 - x^2 - 5x + 2$	**13.** $8a \times 3b$	**14.** $4a \times 3ab$
15. $4p \times 3q \times 2r$	**16.** $5a \times 2ab^2$	**17.** $3x^6 + x^4 - 2x^4$
18. $\frac{9xy}{3y}$	**19.** $\frac{12abc}{8a}$	**20.** $\frac{6x}{9xy}$

Algebraic fractions

Algebraic fractions are fractions which have variables (letters) as well as numbers in them, eg $\frac{x}{3}$, $\frac{5a}{3b}$, $\frac{x^2 + 5}{x - 2}$, $\frac{1}{a - 7}$, $\frac{p + q}{4}$.

Since the letters stand for numbers, algebraic fractions obey the same rules as 'ordinary' fractions do when used in calculations.

Simplifying algebraic fractions

Algebraic fractions, like ordinary fractions, are usually written in their simplest form. To simplify an algebraic fraction, divide the numerator and denominator by the same factor (ie cancel common factors). The laws of indices are often required.

Example H

Q. Simplify: **1.** $\frac{12b^3}{30b}$ **2.** $\frac{5x^4y^2}{15x^3y^3}$

A. **1.** $\frac{12b^3}{30b} = \frac{2b^3}{5b}$ [dividing numerator and denominator by 6]

$= \frac{2b^2}{5}$ [using index laws, $\frac{b^3}{b} = b^{3-1} = b^2$]

2. $\frac{5x^4y^2}{15x^3y^3} = \frac{\overset{1}{\cancel{5}}\,\overset{x}{\cancel{x^4}}\,\cancel{y^2}}{\underset{3}{\cancel{15}}\,\cancel{x^3}\,\underset{y}{\cancel{y^3}}}$ [dividing numerator and denominator by 5 and using index laws for division]

$= \frac{x}{3y}$

Addition and subtraction of algebraic fractions

To add/subtract fractions when the denominators are the *same.*

- The numerators are added together (or subtracted).
- The denominator stays the same.

If the denominators are *different*, the fractions must first be changed to **equivalent fractions** with the *same* denominators before the rule is applied.

Example I

Q. Simplify: **1.** $\frac{a}{4} + \frac{3}{4}$ **2.** $\frac{b}{2} + \frac{b}{3}$ **3.** $\frac{3c}{4} - \frac{c}{8}$ **4.** $\frac{1}{x} + \frac{1}{y}$

A. **1.** $\frac{a}{4} + \frac{3}{4} = \frac{a + 3}{4}$ [denominators the same, add numerators together]

2. $\frac{b}{2} + \frac{b}{3} = \frac{b \times 3}{2 \times 3} + \frac{b \times 2}{3 \times 2}$ [common denominator is the lowest common multiple of the denominators 2 and 3, which is 6]

$= \frac{3b + 2b}{6}$

$= \frac{5b}{6}$ [adding numerators]

3. $\frac{3c}{4} - \frac{c}{8} = \frac{3c \times 2}{4 \times 2} - \frac{c}{8}$ [LCM of 4 and 8 is 8]

$= \frac{6c - c}{8}$

$= \frac{5c}{8}$

4. $\frac{1}{x} + \frac{1}{y} = \frac{1 \times y}{x \times y} + \frac{1 \times x}{y \times x}$

$= \frac{y + x}{xy}$

Take care with signs when expanding brackets in algebraic fractions.

Example J

$\frac{x + 1}{2} - \frac{2x + 1}{3} = \frac{3(x + 1)}{3 \times 2} - \frac{2(2x + 1)}{2 \times 3}$ [making equivalent fractions with denominator 6]

$= \frac{3x + 3 - 4x - 2}{6}$ [expanding brackets]

$= \frac{1 - x}{6}$ [simplifying]

Multiplication of algebraic fractions

To multiply algebraic fractions, all the numerators are multiplied together and all the denominators are multiplied together. Often the process can be made easier by cancelling first.

Example K

Q. Simplify: **1.** $\frac{a}{2} \times \frac{a}{3}$ **2.** $\frac{a^2}{b^3} \times \frac{b^2}{a}$ **3.** $\frac{3a^2}{4bc} \times \frac{8c^3}{9a}$

A. **1.** $\frac{a}{2} \times \frac{a}{3} = \frac{a \times a}{2 \times 3}$

$= \frac{a^2}{6}$

2. $\frac{a^2}{b^3} \times \frac{b^2}{a} = \frac{a^2b^2}{ab^3}$

$= \frac{a}{b}$ [using the laws of indices]

Alternatively, $\frac{\cancel{a^2}^{\,a}}{{}_{b}\cancel{b^3}} \times \frac{\cancel{b^2}}{\cancel{a}} = \frac{a}{b}$ [cancelling first]

3. $\frac{3a^2}{4bc} \times \frac{8c^3}{9a} = \frac{\overset{2}{\cancel{24}}\,\overset{a}{\cancel{a^2}}\,\overset{c^2}{\cancel{c^3}}}{\underset{3}{\cancel{36}}\,\cancel{a}\,b\,\cancel{c}}$ [multiplying numerators and multiplying denominators then simplifying by cancelling and index laws]

$= \frac{2ac^2}{3b}$

Division of algebraic fractions

To divide by an algebraic fraction, multiply by its **reciprocal** (the reciprocal of $\frac{a}{b}$ is $\frac{b}{a}$).

Example L

Q. Simplify: **1.** $\frac{a}{2} \div \frac{a}{3}$ **2.** $\frac{4a^2}{b} \div \frac{8ab}{c}$

A. **1.** $\frac{a}{2} \div \frac{a}{3} = \frac{a}{2} \times \frac{3}{a}$ [changing '$\div \frac{a}{3}$' to '$\times \frac{3}{a}$']

$= \frac{3}{2}$ [cancelling a's]

2. $\frac{4a^2}{b} \div \frac{8ab}{c} = \frac{4a^2}{b} \times \frac{c}{8ab}$ [$\frac{c}{8ab}$ is the reciprocal of $\frac{8ab}{c}$]

$= \frac{\overset{1}{\cancel{4}}\,\overset{a}{\cancel{a^2}}\,c}{\underset{2}{\cancel{8}}\,\cancel{a}\,b^2}$ [multiplying numerators and denominators then cancelling]

$= \frac{ac}{2b^2}$ [using laws of indices to simplify]

Order of operations in algebraic fraction expressions

Be sure to follow the correct order of operations (BEMA or BEDMAS) (see p. 5) when simplifying algebraic fraction expressions with more than one operation.

Example M

Q. Simplify $2x - \frac{x}{2} \times \frac{1}{3}$

A. $2x - \frac{x}{2} \times \frac{1}{3} = 2x - \frac{x}{6}$ [multiplying $\frac{x}{2} \times \frac{1}{3}$ first]

$= \frac{12x - x}{6}$ [subtracting]

$= \frac{11x}{6}$

More complicated algebraic fractions

When there is more than one term in either the denominator or numerator, simplifying is done by factorising first, then cancelling. This is discussed in Topic 11.

Unit 11.1 Activity 10C: Algebraic fractions

1. Simplify each of the following:

a. $\frac{x^{10}}{x^2}$ **b.** $\frac{3y^{12}}{y^4}$ **c.** $\frac{16a^6}{4a^3}$ **d.** $\frac{20b^2}{5b}$ **e.** $\frac{36x^2y}{12xy}$

f. $\frac{(3a^3)^2}{3a^2}$ **g.** $\frac{15a^4b^2}{30ab}$ **h.** $\frac{24x^3y^2}{16xy^3}$ **i.** $\frac{(2a^3b^2)^2}{10a^4b^3}$ **j.** $\frac{8x^6y^2z}{12xy^2z^3}$

2. Write each of the following as a single fraction:

a. $\frac{x}{3} + \frac{1}{3}$ **b.** $\frac{y}{9} - \frac{2}{9}$ **c.** $\frac{x}{3} + \frac{x}{4}$ **d.** $\frac{3a}{5} - \frac{a}{2}$

e. $\frac{4k}{5} - \frac{k}{3}$ **f.** $\frac{5y}{6} - \frac{2y}{3}$ **g.** $\frac{3}{x+1} + \frac{4}{x+1}$ **h.** $\frac{x}{2} + \frac{1}{x}$

i. $\frac{4}{a} + \frac{3}{b}$ **j.** $\frac{1}{3x} + \frac{1}{x}$ **k.** $\frac{1}{x^2} + \frac{2}{x}$ **l.** $\frac{1}{y^2} - \frac{2}{3y}$

3. Simplify each of the following:

a. $\frac{a}{3} \times \frac{a}{4}$ **b.** $\frac{a}{5} \times \frac{10}{a}$ **c.** $\frac{3a}{b} \times \frac{b}{3a}$ **d.** $\frac{x}{2} \times \frac{x}{5}$

e. $\frac{a}{5} \div \frac{a}{2}$ **f.** $\frac{a}{3} \div \frac{a}{4}$ **g.** $\frac{a}{3b} \times \frac{b}{3a}$ **h.** $\frac{4a^2}{3} \div \frac{15}{2a}$

i. $\frac{6x^2}{5} \times \frac{15}{4x}$ **j.** $\frac{12a^2b}{15ab} \times \frac{5ab^2}{4b}$ **k.** $\frac{6xy^2}{5x} \div \frac{4y}{3xy}$ **l.** $\frac{3xy^3}{4y} \div \frac{9x^2}{8xy}$

4. Simplify each of the following:

a. $\frac{x+1}{4} - \frac{x-1}{5}$ **b.** $\frac{x+1}{4} + \frac{x-1}{5}$ **c.** $\frac{x+1}{4} - \frac{x-1}{2}$

d. $\frac{x+1}{7} + \frac{x-1}{2}$ **e.** $\frac{x+2}{3} - \frac{3x}{4}$ **f.** $\frac{1}{4x} + \frac{1}{3}$

5. Use the correct order of operations to simplify the following.

a. $\frac{a}{2} + \frac{a}{3} \times \frac{1}{4}$ **b.** $\frac{3}{x} - \frac{2}{3} \div \frac{x}{6}$ **c.** $\left(\frac{a}{4} - \frac{a}{8}\right) \times \frac{a}{3}$

d. $\frac{y}{3} - \frac{y}{5} + \frac{y}{10}$ **e.** $\frac{2w}{3} \times \frac{1}{4} + \frac{w}{6}$ **f.** $\frac{5a^4}{12b} \div \frac{3}{4b} \times \frac{6}{a^2}$

Unit 11.1 Number and Application

Topic 11: Basic algebra—expanding and factorising

Basic algebra is a sub-section within Unit 11.1 of the Grade 11 syllabus (see Syllabus p. 14). This Topic deals with using straightforward algebraic methods and solving equations. It covers:

- Factorising and expanding.
- Manipulating and simplifying expressions such as $\frac{x^2 - 4}{x - 2}$.

Bracketed expressions

Brackets around an algebraic expression mean that the expression in the brackets is regarded as one **term**.

Example A

1. The expression $x + 4$ consists of two terms, 'x' and '4', but the bracketed expression $(x + 4)$ is considered to be only one term, '$x + 4$'.
2. An expression such as $3(x + 4)$ means 3 lots of $(x + 4)$, thus

$3(x + 4) = (x + 4) + (x + 4) + (x + 4)$

$= 3 \times x + 3 \times 4$ [3 lots of x plus 3 lots of 4].

$= 3x + 12$ [simplifying]

Expanding brackets

The process of removing brackets from an algebraic expression is called **expanding**.

- The number in front of the brackets (or sometimes behind) multiplies every term inside the bracket.
- The answer is then **simplified** (where possible) by combining like terms.

Example B

Q. Expand each of the following expressions and simplify where possible:

1. $2(x + 3)$ 2. $y(y - 2)$

A. 1. $2(x + 3) = 2 \times x + 2 \times 3$ [x and $+3$ are *each* multiplied by 2]

$= 2x + 6$

2. $y(y - 2) = y \times y + y \times -2$ [y and -2 are each multiplied by y]

$= y^2 - 2y$

By considering areas in the diagram alongside, a general rule for expanding brackets (called the distributive law) can be seen:

$$a(b + c) = ab + ac$$

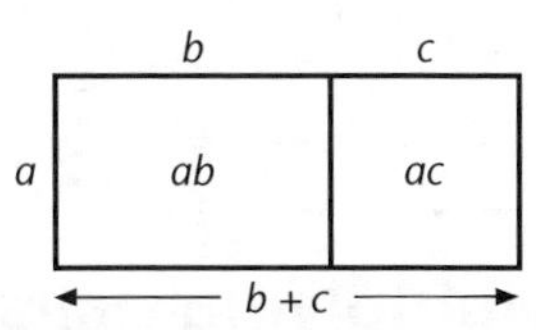

These rules also apply to expanding brackets with more than two terms.

Example C

1. $-3(a + 2b - c) = -3 \times a + -3 \times 2b + -3 \times -c$ [each term is multiplied by –3]

   $= -3a - 6b + 3c$

   Note: The product of two negatives is a positive, so $-3 \times -c = +3c$.

2. $(x + y - 2)y = y(x + y - 2)$ [swapping y to front of bracket since $a \times b = b \times a$]

   $= yx + y^2 - 2y$ [removing the bracket]

   $= xy + y^2 - 2y$ [since $yx = xy$]

3. $-3x^2(x - xy + 4y) = -3x^2 \times x + -3x^2 \times -xy + -3x^2 \times 4y$

   $= -3x^3 + 3x^3y - 12x^2y$ [simplifying]

Take care to follow the correct order of operations (BEMA or BEDMAS) when removing brackets.

Example D

1. $4 + 2(y - 5) = 4 + 2 \times y + 2 \times -5$ [y and –5 are each multiplied by +2]

   $= 4 + 2y - 10$

   $= 2y - 6$ [collecting like terms]

   Note: The order of operations (BEMA) applies when removing brackets. Therefore, in $4 + 2(y - 5)$, the brackets are expanded *before* the addition of 4.

2. $3 - (4 - x) = 3 + -1 \times 4 + -1 \times -x$ [$-(a + b)$ means $-1 \times (a + b)$]

   $= 3 - 4 + x$

   $= -1 + x$ or $x - 1$

Unit 11.1 Activity 11A: Expanding brackets

Expand each of the following expressions and simplify where possible:

1. $3(x + 1)$ **2.** $4(x + 2)$ **3.** $a(b - 2)$

4. $x(x - 3)$ **5.** $-4(x - 5)$ **6.** $2x(x^2 + 3x - 4)$

7. $5 - 3(2x + 1)$ **8.** $8 - (x - 3)$ **9.** $4 + 2(x + 3)$

10. $4(2x + 3y) + 3(x - 2y)$ **11.** $3(4a + 5b) - 2(a - 3b)$ **12.** $3(x - 2y) - (x - 5y)$

13. $x(x + 3) + 2(x - 1)$ **14.** $p(3p - q) - q(p - 2q)$ **15.** $x(x - 5) - 2x(x - 2)$

16. $2x(x - 3) - x^2$ **17.** $3x^2(1 - 2x - x^2)$ **18.** $-2xy(x - y - 3z)$

19. A square room of side length x is made 3 m longer (its width stays the same). What is the new area of the room?

x 3

x

20. Lucy has Ky. She spends K10.

a. How much money does Lucy have now?

b. Lucy's brother says, 'I've got three times as much money as you have now.' How much money does Lucy's brother have?

Multiplying pairs of brackets

When two bracketed expressions are multiplied together, each term in the first bracket multiplies the terms in the second bracket.

Example E

1. An expression such as $(x + 3)(x + 4)$ means $(x + 3)$ lots of $(x + 4)$, therefore

$(x + 3)(x + 4) = x(x + 4) + 3(x + 4)$ [x lots of $(x + 4)$ plus 3 lots of $(x + 4)$]

$= x^2 + 4x + 3x + 12$ [expanding]

$= x^2 + 7x + 12$ [simplifying]

2. $(x - 2)(x + 5) = x(x + 5) - 2(x + 5)$ [multiplying $(x + 5)$ by x and -2]

$= x^2 + 5x - 2x - 10$

$= x^2 + 3x - 10$

Note: The second bracket is multiplied by –2, *not* 2, because the second term in the first bracket is –2.

3. $(a + 2)(b + c - 3) = a(b + c - 3) + 2(b + c - 3)$

$= ab + ac - 3a + 2b + 2c - 6$

By considering areas in the diagram alongside a general rule for expanding pairs of brackets can be seen:

$$(a + b)(c + d) = a(c + d) + b(c + d) = ac + ad + bc + bd$$

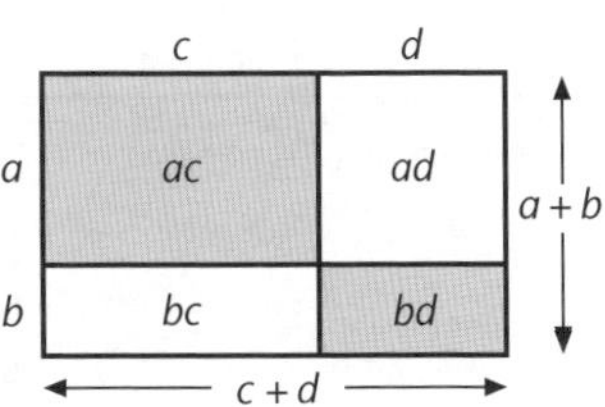

The mnemonic '**FOIL**' is a useful way to remember how to expand two pairs of brackets, each containing two terms.

F **F**irst terms in each bracket are multiplied together. $(a + b)(c + d) = ac$ ($a \times c$)

O **O**utside terms are multiplied together. $(a + b)(c + d) = ad$ ($a \times d$)

I **I**nside terms are multiplied together. $(a + b)(c + d) = bc$ ($b \times c$)

L **L**ast terms in each bracket are multiplied together. $(a + b)(c + d) = bd$ ($b \times d$)

Thus the **FOIL** method gives $(a + b)(c + d) = ac + ad + bc + bd$

Example F

$(3x + 2)(2x - 5) = 3x \times 2x + 3x \times -5 + 2 \times 2x + 2 \times -5$ [applying FOIL method]

$= 6x^2 - 15x + 4x - 10$

$= 6x^2 - 11x - 10$

Some important expansions

Take care when squaring a bracket. $(x - 3)^2$ means $(x - 3)(x - 3)$. It *does not* mean $x^2 - 3^2$ or $x^2 + (-3)^2$.

Example G

Q. Expand $(x - 3)^2$.

A. $(x - 3)^2 = (x - 3)(x - 3)$ [$(x - 3)^2$ means $(x - 3) \times (x - 3)$]

$= x^2 - 3x - 3x + 9$ [expanding using FOIL method]

$= x^2 - 6x + 9$ [adding like terms]

The expressions $(a + b)^2$ and $(a - b)^2$ are known as **perfect squares** and their expansions are:

$$(a + b)^2 = a^2 + 2ab + b^2$$
$$(a - b)^2 = a^2 - 2ab + b^2$$

For example, $(x + 5)^2 = x^2 + 2 \times x \times 5 + 5^2 = x^2 + 10x + 25$ and
$(3x - 4y)^2 = (3x)^2 - 2 \times 3x \times 4y + (4y)^2 = 9x^2 - 24xy + 16y^2$

Another important expansion gives the **difference between two squares**.

$$(a + b)(a - b) = a^2 - b^2$$

For example, $(x + 2)(x - 2) = x^2 - 2^2 = x^2 - 4$.

Unit 11.1 Activity 11B: Further expansions

Expand each of the following expressions and simplify where possible:

1. $(x + 4)(x + 5)$ **2.** $(x - 3)(x + 2)$ **3.** $(x + 4)(x - 5)$

4. $(x - 5)(x + 3)$ **5.** $(x - 7)(x - 4)$ **6.** $(x - 5)(x - 6)$

7. $(2x + 3)(x + 4)$ **8.** $(4x + 3)(3x - 4)$ **9.** $(2x - 3)(x - 6)$

10. $(3x + 2)(3 - x)$ **11.** $(2x + 7)(5 - x)$ **12.** $(2x - 5)(3x - 4)$

13. $(3x + 1)(x + 3)$ **14.** $(5x - 2)(3x + 4)$ **15.** $(3 - 2x)(4 + 3x)$

16. $(x + 6)^2$ **17.** $(x + 4)^2$ **18.** $(x - 7)^2$

19. $(x - 10)(x + 10)$ **20.** $(4x + 1)(4x - 1)$ **21.** $(2x - 5)(2x + 5)$

22. $(2x + 3)^2$ **23.** $(3x - 5)^2$ **24.** $(x + 2)(x^2 + x + 1)$

25. $(x - 3)(x^2 + 2x - 4)$ **26.** $(2x - 5)(x^2 - 2x - 3)$

27. Pythagoras' Theorem is a relationship involving the three sides of a right-angled triangle. For the triangle ABC, shown alongside, $a^2 = b^2 + c^2$.

Use Pythagoras' Theorem to find the length of the hypotenuse (ie side YZ) in terms of x.

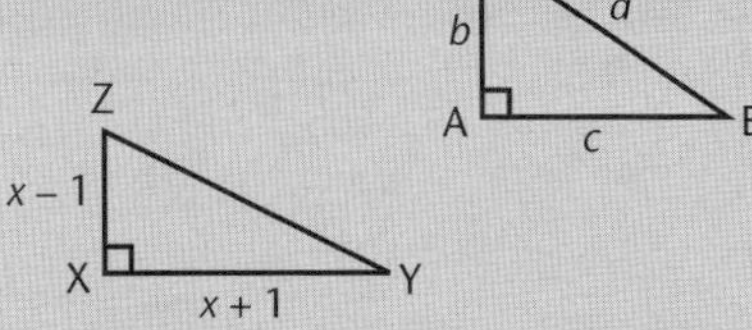

28. The diagram shows a plan of a garden. Find an expanded expression for the area of the garden in terms of x.

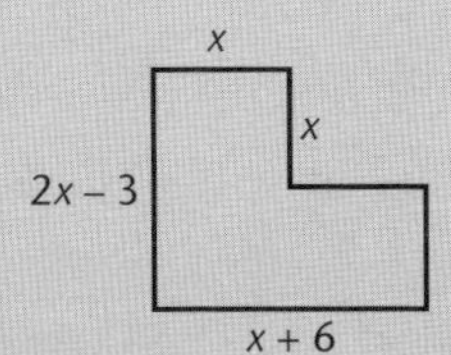

29. A square of side length x cm is cut from the centre of a square piece of card whose side length is 5 cm longer than the side length of the square that is cut out. Find an expression for the area of the card that remains.

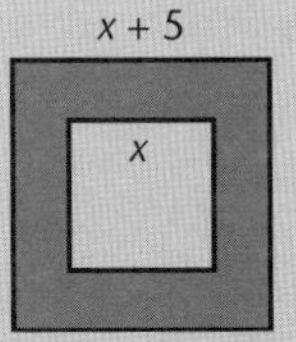

30. A cube has side length $(x + 2)$. Find an expanded expression for its volume $(x + 2)^3$.

Factorising

Factorising means writing an expression as a **product** of its **factors**. To do this the rules for expanding are used in reverse.

$$ab + ac = a(b + c)$$

Thus, the expression $ab + ac$ is expressed as a product of the two factors a and $(b + c)$.

To factorise an expression, the first step is taking out the **highest common factor** (HCF) of the terms in the expression.

Example H

Q. Factorise: **1.** $8x + 12y$ **2.** $3x + xy$ **3.** $x^2y^3 - xy^2$

A. **1.** In $8x + 12y$, the terms $8x$ and $12y$ have a highest common factor of 4.

$\therefore 8x + 12y = 4(2x + 3y)$ [since $8x = 4 \times 2x$ and $12y = 4 \times 3y$]

Note: 2 is also a common factor of $8x$ and $12y$, but it is not the *highest* common factor.

2. In $3x + xy$, the terms $3x$ and xy have a common factor of x.

$\therefore 3x + xy = x(3 + y)$ [since $3x = x \times 3$ and $xy = x \times y$]

3. In $x^2y^3 - xy^2$, the terms x^2y^3 and xy^2 have a highest common factor of xy^2.

$\therefore x^2y^3 - xy^2 = xy^2(xy - 1)$ [since $x^2y^3 = xy^2 \times xy$ and $xy^2 = xy^2 \times 1$]

Note: The HCF of two powers of the same variable is the lower power of the variable. For example, the HCF of x^7 and x^4 is x^4.

Some expressions may not have a common factor, but *groups of terms* within the expression may have.

Example I

If $pq - 3q + 3p - 9$ is considered as $(pq - 3q)$ plus $(3p - 9)$, it can be seen that $(pq - 3q)$ has a common factor of q, and $(3p - 9)$ has a common factor of 3. This gives:

$pq - 3q + 3p - 9$	$= (pq - 3q) + (3p - 9)$	[grouping terms]
	$= q(p - 3) + 3(p - 3)$	[factorising]
	$= (p - 3)(q + 3)$	[since $(p - 3)$ is a common factor]

Unit 11.1 Activity 11C: Basic factorising

Factorise each of the following completely by taking out the highest common factor:

1. $4a + 12$	**2.** $7x - 14$	**3.** $5y + 25$
4. $9x + 15y$	**5.** $xy - 3x$	**6.** $y^3 + 5y$
7. $a^2 + ab$	**8.** $x^2 - 7x$	**9.** $6x + 8y$
10. $ab - a^2$	**11.** $3p^2 - 6p$	**12.** $4x + 14y$
13. $3x^3 + 6x^2 - 3x$	**14.** $8x + 12y - 20z$	**15.** $15x + 10y - 25$
16. $8x^2y^3 + 12xy^4 - 10xy^2$	**17.** $8x - 24y + 20$	**18.** $3a^2b + 5a^3b^2 + ab$
19. $4x^2 + 8x - 10$	**20.** $2y - 6y^2 + y^3$	**21.** $3p^2q - 9pq^2$
22. $p^4 - 3p^2$	**23.** $4a^2 - 18a^3$	**24.** $a^2b^3 + a^3b^2$
25. $2x(x + 3) + 5(x + 3)$	**26.** $6x(x - 2) - 5(x - 2)$	**27.** $ab + xb + ac + xc$
28. $3x + 6y + ax + 2ay$	**29.** $5x + 15y - px - 3py$	**30.** $ax - bx + ay - by$

Factorising quadratic expressions of the form $x^2 + bx + c$

Quadratic expressions are expressions of the form $ax^2 + bx + c$.

When the **coefficient** of x^2 is 1 (ie $a = 1$), the quadratic expression is written as $x^2 + bx + c$. Such expressions are factorised by finding two numbers which

- *Multiply* together to give c.
- *Add* together to give b.

If both b and c are positive, then both the numbers required for the factorisation will be positive.

Example J

Q. Factorise: **1.** $x^2 + 7x + 12$ **2.** $x^2 + 8x + 12$

A. **1.** To factorise $x^2 + 7x + 12$, find two numbers which multiply to give 12 and add to give 7.

The numbers are 3 and 4, because 3 x 4 = 12 and 3 + 4 = 7.

$\therefore x^2 + 7x + 12 = (x + 3)(x + 4)$

2. To factorise $x^2 + 8x + 12$, find two numbers which multiply to give 12 and add to give 8.

The numbers are 2 and 6, because 2 x 6 = 12 and 2 + 6 = 8.

$\therefore x^2 + 8x + 12 = (x + 2)(x + 6)$

Note: Factorisations may be checked by **expanding** the answer. For example, in part **2,**

$(x + 2)(x + 6) = x^2 + 6x + 2x + 12 = x^2 + 8x + 12$ as required [using FOIL method]

If c is positive and b is negative in the expression $x^2 + bx + c$ then both numbers in the factorisation will be negative.

Example K

Q. Factorise $x^2 - 11x + 30$.

A. The two numbers required to factorise $x^2 - 11x + 30$ must multiply to give 30 and add to give –11.

The numbers are –5 and –6, because –5 x –6 = 30 and –5 + –6 = –11.

$\therefore x^2 - 11x + 30 = (x - 5)(x - 6)$

If c is negative in the expression $x^2 + bx + c$ then the two numbers used for the factorisation will have *different signs*, regardless of whether b is negative or positive. If b is positive, the bigger of the two numbers will also be positive and the second number will be negative.

Example L

Q. Factorise $x^2 + 2x - 48$.

A. The numbers required to factorise $x^2 + 2x - 48$ are +8 and –6, because –6 x 8 = –48 and –6 + 8 = +2.

$\therefore x^2 + 2x - 48 = (x + 8)(x - 6)$

Note: +8 and –6 are chosen so that the bigger number is positive. This gives a positive b as required.

If both c and b are negative in the expression $x^2 + bx + c$ then the two numbers will have *different signs* and the bigger number will be negative.

Example M

Q. Factorise $x^2 - 5x - 24$.

A. The numbers –8 and +3 are used to factorise $x^2 - 5x - 24$, because –8 x 3 = –24 and –8 + 3 = –5

$\therefore x^2 - 5x - 4 = (x - 8)(x + 3)$

Factorising the difference of two squares

Quadratic expressions of the type $x^2 - c^2$ are called the **difference of two squares**. These expressions factorise to give $(x + c)(x - c)$.

Example N

Q. Factorise: **1.** $x^2 - 36$ **2.** $4p^2 - 9q^2$

A. **1.** $x^2 - 36 = (x + 6)(x - 6)$ [since $\sqrt{36} = 6$]

2. $4p^2 - 9q^2 = (2p + 3q)(2p - 3q)$ [$\sqrt{4p^2} = 2p$, $\sqrt{9q^2} = 3q$]

When the coefficient of x^2 is a number other than 1, the method of factorising becomes more involved. In cases where a common *factor can be removed*, the problem still involves factorising a quadratic expression of the form $x^2 + bx + c$.

Example O

Q. Factorise: **1.** $2x^2 + 8x - 24$ **2.** $x^3 + 3x^2 + 2x$

A. **1.** $2x^2 + 8x - 24 = 2(x^2 + 4x - 12)$ [2 is a common factor]

$= 2(x + 6)(x - 2)$ [factorising $x^2 + 4x - 12$]

2. $x^3 + 3x^2 + 2x = x(x^2 + 3x + 2)$ [x is a common factor]

$= x(x + 2)(x + 1)$ [factorising $x^2 + 3x + 2$]

Factorising quadratic expressions of the form $ax^2 + bx + c$

A procedure for factorising expressions of the form $ax^2 + bx + c$ is illustrated in the following example.

Example P

Q. Factorise $3x^2 + 11x - 20$.

A. **a.** Multiply the coefficient of x^2 (3) by the constant (–20) to get –60.

b. Find the factors of –60 that add to the coefficient of x (11).

The pairs of numbers that multiply together to give 60 are:
(1,60), (2,30), (3,20), (4,15), (5,12), (6,10).

Thus the factors of –60 that add to 11 are –4 and 15 since $-4 \times 15 = -60$ and $15 - 4 = 11$.

c. Rewrite the term in x as two terms using these numbers, ie $11x = 15x - 4x$.

d. This results in the expression $3x^2 + 15x - 4x - 20$ to factorise:

$3x^2 + 15x - 4x - 20 = 3x(x + 5) - 4(x + 5)$ [factorise in pairs]

$= (x + 5)(3x - 4)$ [$x + 5$ is a common factor]

Example Q

Q. Factorise $4x^2 - 5x - 6$.

A. $4 \times -6 = -24$. Factor pairs of 24 are (1, 24), (2, 12), (3, 8), (4, 6).

The required numbers are 3 and −8 since $3 - 8 = -5$ and $3 \times -8 = -24$.

$\therefore 4x^2 - 5x - 6 = 4x^2 - 8x + 3x - 6$ [rewriting $-5x$ as $-8x + 3x$]

$= 4x(x - 2) + 3(x - 2)$ [factorising in pairs]

$= (x - 2)(4x + 3)$ [$x - 2$ is a common factor]

Unit 11.1 Activity 11D: Factorising quadratic expressions

Factorise each of the following completely:

1. $x^2 + 6x + 8$	**2.** $x^2 + 10x + 25$	**3.** $x^2 - 8x + 16$	**4.** $x^2 + 2x - 15$
5. $x^2 + 5x - 24$	**6.** $x^2 - 5x - 14$	**7.** $x^2 + 9x + 20$	**8.** $x^2 - 10x + 25$
9. $x^2 + 7x + 10$	**10.** $x^2 - 4x + 3$	**11.** $x^2 - 11x + 28$	**12.** $x^2 + x - 20$
13. $x^2 + 4x - 21$	**14.** $x^2 + 9x + 18$	**15.** $x^2 - 12x + 36$	**16.** $x^2 - 9$
17. $x^2 - 49$	**18.** $x^2 - 64$	**19.** $25x^2 - 16$	**20.** $4x^2 - 81$
21. $5x^2 - 20$	**22.** $3x^2 + 12x - 63$	**23.** $2x^2 - 2x - 4$	**24.** $4x^2 - 16x + 12$
25. $2x^2 + 5x - 3$	**26.** $2x^2 + 9x + 10$	**27.** $3x^2 - 7x + 2$	**28.** $5x^2 - 13x - 6$
29. $3x^2 + 4x - 7$	**30.** $6x^2 + x - 12$		

Simplifying rational expressions by factorising

Rational expressions are simplified by cancelling common factors. When a numerator or denominator of an algebraic fraction has more than one term, then factorisation must be done before cancelling.

Example R

Q. Simplify: **1.** $\frac{5a + 6a}{a}$ **2.** $\frac{x^2 + 5x + 6}{x + 2}$

A. **1.** $\frac{5a + 6a}{a} = \frac{11a}{a}$ [simplifying the numerator]

$= 11$ [cancelling by dividing top and bottom by a]

2. $\frac{x^2 + 5x + 6}{x + 2} = \frac{(x + 3)(x + 2)}{(x + 2)}$ [factorising the numerator and placing brackets around the denominator]

$= x + 3$ [cancelling by dividing top and bottom by $(x + 2)$]

Sometimes both the numerator and denominator can be factorised, as shown below.

Example S

Q. Factorise $\frac{x^2 - 9}{2x + 6}$.

A. $\frac{x^2 - 9}{2x + 6} = \frac{(x + 3)(x - 3)}{2(x + 3)}$ [factorising numerator and denominator]

$= \frac{x - 3}{2}$ [dividing top and bottom by $(x + 3)$]

Unit 11.1 Activity 11E: Simplifying rational expressions by factorising

Simplify each of the following rational expressions.

1. $\frac{6a + 3a}{a}$ **2.** $\frac{10b - 3b}{b}$ **3.** $\frac{4x + 12}{4}$ **4.** $\frac{3x + 15}{x + 5}$

5. $\frac{x^2 + 4x}{x + 4}$ **6.** $\frac{x^2 + 8x + 12}{x + 2}$ **7.** $\frac{x^2 + x - 12}{x - 3}$ **8.** $\frac{x^2 + 3x - 4}{x^2 - 3x + 2}$

9. $\frac{x^2 - 5x - 24}{x^2 + 5x + 6}$ **10.** $\frac{x^2 + 6x + 8}{x^2 - 4}$ **11.** $\frac{x^2 - x - 6}{x^2 - 9}$ **12.** $\frac{x^2 - x - 20}{2x^2 - 50}$

Unit 11.1 Activity 11F: Multiple choice – factorising quadratic expressions

1. In simplest form the expression $4ab + 3a + 3b + 5a \times -2b$ is equal to:

A. $3a + 3b - 6ab$

B. $-5ab$

C. $4ab + 8a + b$

D. $ab + 3a + 3b$

2. Jacob is x years old and his sister is two years younger. In three years time the sum of the ages of Jacob and his sister will be:

A. $2x + 6$

B. $2x + 4$

C. $x + 4$

D. $2x + 8$

3. $\frac{2a}{3} + \frac{3}{2} =$

A. $\frac{2a + 3}{6}$

B. $\frac{2a + 3}{5}$

C. $\frac{2a + 9}{3}$

D. $\frac{4a + 9}{6}$

4. $\frac{x}{2} + \frac{2}{x} =$

   A. 2

   B. $\frac{4x}{2 + x}$

   C. $\frac{x^2 + 4}{2x}$

   D. $\frac{x + 2}{x}$

5. $(2x - 3)(x + 1)$ expanded and simplified is equal to:

   A. $2x^2 - 3$

   B. $2x^2 + 5x - 3$

   C. $2x^2 - 3x + 1$

   D. $2x^2 - x - 3$

6. $(x - 1)^2 =$

   A. $x^2 - 1$

   B. $x^2 + 1$

   C. $x^2 - 2x + 1$

   D. $x^2 - 2x - 1$

7. An expression that represents the shaded area, at right, is:

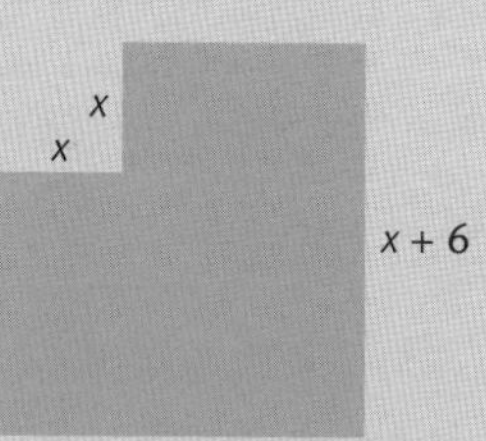

   A. $11x + 30$

   B. $x^2 + 10x + 30$

   C. $2x^2 + 11x + 30$

   D. $4x + 22$

8. When factorised, the expression $ab + ax + bc + cx$ is equal to:

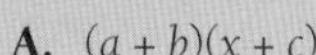

   A. $(a + b)(x + c)$

   B. $(b + x)(a + c)$

   C. $(a + x)(b + c)$

   D. $2(a + b + c + x)$

9. When factorised, the expression $x^2 + 3x - 4$ is equal to:

   A. $(x + 4)(x - 1)$

   B. $(x + 4)(x + 1)$

   C. $(x - 4)(x + 1)$

   D. $(x - 4)(x - 1)$

10. When fully factorised, the expression $4x^2 - 64$ is equal to:

   A. $(2x - 8)(2x + 8)$

   B. $4(x + 4)(x - 4)$

C. $4(x-1)^2$

D. $4(x-8)(x+8)$

11. When factorised, $3x^2 - 16x - 12$ is equal to:

A. $(3x+4)(x-3)$

B. $(3x-12)(x-1)$

C. $(3x+6)(x-2)$

D. $(3x+2)(x-6)$

12. When factorised, $4x^2 + 4x - 15$ is equal to:

A. $(2x-5)(2x+3)$

B. $(4x+5)(x-3)$

C. $(2x+5)(2x-3)$

D. $(4x+15)(x-1)$

13. The rational expression $\frac{x^2+3x-10}{x^2-4}$, when simplified, is equal to:

A. $\frac{x+5}{x+2}$

B. $\frac{x-5}{x+2}$

C. $\frac{x-5}{x-2}$

D. $\frac{x+5}{x-2}$

14. The rational expression $\frac{2x^2+x-1}{x^2-1}$, when simplified, is equal to:

A. $\frac{2x+1}{x+1}$

B. $\frac{2x-1}{x-1}$

C. $\frac{2x-1}{x+1}$

D. $\frac{2x+1}{x-1}$

Unit 11.1 Number and Application
Topic 12: Basic algebra—solving quadratic equations

Basic algebra is a sub-section within Unit 11.1 of the Grade 11 syllabus (see Syllabus p. 14). This Topic deals with using straightforward algebraic methods and solving. It covers:

- Solving factorised equations such as $(x - 1)(x + 3) = 0$.
- Solving simple quadratic equations such as $x^2 + 30x = 400$ and interpreting the results.
- Modelling by forming and solving appropriate equations.
- Interpretation in context.

Introduction

The general form of a quadratic equation in x is

$$ax^2 + bx + c = 0$$

In a **quadratic equation**, the highest power of the variable is 2.

For example: $x^2 - 4x + 1 = 0$, $y^2 = y + 1$, and $3a^2 = 4$ are all quadratic equations.

Solving quadratic equations

The following method for solving quadratic equations depends on an important property of zero:

If two real numbers multiply to make zero, then one or the other (or both) *of the numbers must be zero.*

This result is expressed using symbols as follows:

If $ab = 0$ then either $a = 0$ or $b = 0$ (or both)

This result makes the solving of quadratic equations straightforward if the quadratic is given in factored form (and is equal to zero).

Example A

Q. Solve the quadratic equations: **1.** $(x + 4)(x - 2) = 0$ **2.** $3x(2x + 3) = 0$

A. 1. If $(x + 4)(x - 2) = 0$ then

either $(x + 4) = 0$ or $(x - 2) = 0$ [setting each factor in turn to zero]

$x = -4$ or $x = 2$ [solving both equations for x]

2. $3x(2x + 3) = 0$

$\therefore 3x = 0$ or $(2x + 3) = 0$ [setting each factor in turn to zero]

$\therefore x = 0$ [dividing by 3] or $2x = -3$ [subtracting 3]

$\therefore x = 0$ or $x = -\frac{3}{2}$ [dividing second equation by 2]

If quadratic equations are not given in factored form then the quadratic must be factorised first.

Example B

Q. Solve: **1.** $x^2 + 5x - 6 = 0$ **2.** $2y^2 - 18 = 0$

A. 1.

$$x^2 + 5x - 6 = 0$$
$$(x + 6)(x - 1) = 0 \quad \text{[factorising]}$$
$$\therefore x + 6 = 0 \quad \text{or} \quad x - 1 = 0$$
$$\therefore x = -6 \quad \text{or} \quad x = 1$$

2.

$$2y^2 - 18 = 0$$
$$2(y^2 - 9) = 0 \quad \text{[factorising]}$$
$$2(y + 3)(y - 3) = 0 \quad \text{[factorising]}$$
$$y + 3 = 0 \quad \text{or} \quad y - 3 = 0$$
$$\therefore y = -3 \quad \text{or} \quad y = 3$$

If necessary, the quadratic equation must be **rearranged** before factorising, so that every term of the equation is on one side of the equals sign, leaving zero on the other side.

Example C

Q. Solve: **1.** $x^2 = 4x + 12$ **2.** $(x + 2)(x - 3) = 50$

A. 1. To solve $x^2 = 4x + 12$, the equation must be rearranged and then factorised.

$$x^2 = 4x + 12$$
$$x^2 - 4x - 12 = 0 \quad \text{[subtracting 4x and 12 from both sides]}$$
$$\therefore (x - 6)(x + 2) = 0 \quad \text{[factorising]}$$
$$x - 6 = 0 \quad \text{or} \quad x + 2 = 0$$
$$\therefore x = 6 \quad \text{or} \quad x = -2$$

2.

$$(x + 2)(x - 3) = 50$$
$$x^2 - 3x + 2x - 6 = 50 \quad \text{[expanding LHS]}$$
$$x^2 - x - 6 = 50 \quad \text{[simplifying]}$$
$$x^2 - x - 56 = 0 \quad \text{[subtracting 50]}$$
$$(x - 8)(x + 7) = 0 \quad \text{[factorising]}$$
$$x - 8 = 0 \quad \text{or} \quad x + 7 = 0 \quad \text{[solving]}$$
$$x = 8 \quad \text{or} \quad x = -7$$

Note: Although the quadratic in part **2** was factorised, the RHS was not zero, so the quadratic could not be solved using the techniques described previously. The quadratic needed to be expanded and the RHS made equal to zero before re-factorising and solving.

The solutions to a quadratic equation can be checked by **substituting** back into the original equation. (This can sometimes be done mentally.)

Example D

Q. Check that the solutions to $x^2 + 5x - 6 = 0$ are $x = -6$ or $x = 1$ (as found in Example B, part **1**).

A. $x = -6 \Rightarrow (-6)^2 + 5 \times -6 - 6 = 0$ [substituting $x = -6$ into $x^2 + 5x - 6$]

$36 - 30 - 6 = 0$

$36 - 36 = 0$ which is true.

$x = 1 \Rightarrow 1^2 + 5 \times 1 - 6 = 0$ [substituting $x = 1$ into $x^2 + 5x - 6$]

$1 + 5 - 6 = 0$

$6 - 6 = 0$ which is true.

$\therefore$ both solutions, $x = -6$ and $x = 1$, are correct.

Once it is recognised that an equation (in x say) is **quadratic** (the equation has terms in x^2 and/or x and/or constants) it is important to remember the correct strategy for solving such equations:

- **Z**ero: the right-hand side of the equation must equal zero.
- **F**actorise: express the quadratic as a product of its factors.
- **S**olve: set each factor in turn to zero and solve for the variable.

A useful reminder of this strategy is the mnemonic **ZFS**.

Example E

Q. Solve $(x - 1)(x + 3) = 0$.

A. A quadratic equation such as $(x - 1)(x + 3) = 0$ has already been made equal to zero (**Z**) and factorised (**F**) so that all that remains is to solve (**S**) the equation by setting each factor to zero.

Thus the solution is $x = 1$ or $x = -3$ [solving $x - 1 = 0$ and $x + 3 = 0$]

Note: A common error in solving equations such as $(x - 1)(x + 3) = 0$ is to expand the left-hand side of the equation, then attempt to isolate the x (as if the equation were linear).

Unit 11.1 Activity 12A: Solving quadratic equations

1. Solve the following quadratic equations given in factored form.

a. $(x + 2)(x + 3) = 0$ **b.** $(x - 2)(x - 4) = 0$ **c.** $(x + 1)(x - 3) = 0$

d. $(x - 5)(x + 2) = 0$ **e.** $x(x - 4) = 0$ **f.** $x(x + 5) = 0$

g. $(2x - 3)(x + 6) = 0$ **h.** $(2x + 1)(x - 1) = 0$ **i.** $(5x + 3)(2x + 5) = 0$

j. $(3 - 4x)(2x - 9) = 0$ **k.** $(3x - 4)(x - 5) = 0$ **l.** $(3x - 2)(3x + 2) = 0$

2. Factorise and solve the following quadratic equations.

a. $x^2 + 3x = 0$ **b.** $2x^2 - 10x = 0$ **c.** $x^2 + x - 6 = 0$

d. $x^2 + 5x + 4 = 0$ **e.** $x^2 - 3x - 40 = 0$ **f.** $x^2 - 6x + 9 = 0$

g. $x^2 - 8x + 12 = 0$ **h.** $x^2 + 3x - 10 = 0$ **i.** $x^2 - 4 = 0$

3. Solve the following quadratic equations.

a. $x^2 = 2x + 15$ **b.** $x^2 + 3x = 18$ **c.** $x^2 + 3 = 4x$

d. $x^2 = 6x$ **e.** $2x^2 = 5x$ **f.** $x^2 = 49$

g. $x^2 - 2x = 8$ **h.** $3x^2 = 12$ **i.** $2x^2 + x = 6$

4. Solve the following quadratic equations to find n.

a. $(n + 1)(n - 1) = 63$ **b.** $(n - 8)(n + 2) = 11$ **c.** $(n + 2)(n - 2) = 140$

d. $(2n + 3)(2n - 3) = 91$ **e.** $n(2n + 1) = 36$ **f.** $(3n - 1)(2n + 3) = 35$

Solving word problems with quadratic equations

When solving word problems, the solutions must be interpreted correctly so that they are meaningful within the given context. For example, if the unknown value, x, is a side length, then only positive values for x make sense. In word problems at the Merit level, the practical application is described and the quadratic equation to be solved is given.

Example F

Q. An aircraft dropped a box of emergency supplies to some villagers. The height of the box above the ground, h, at time, t, is given by the formula:

$h = 520 - 5t^2$, where h is in metres and t is in seconds.

To find how long the box takes to fall to the ground, the equation $520 - 5t^2 = 0$ must be solved. Find how long it takes for the box to fall to the ground.

A.

$520 - 5t^2 = 0$ [at ground level, $h = 0$]

$t^2 - 144 = 0$ [dividing by –5 and rearranging]

$(t + 12)(t - 12) = 0$ [factorising]

$t = -12$ or 12 [solving, giving both solutions]

The time taken is 12 seconds [ignore the negative solution, since time taken is positive].

At the 'very high achievement' level, the context of the problem will be described but the forming and solving of the quadratic equation will be left to the student.

Example G

Q. In an art shop, canvases are sold by the square centimetre. Two canvases in the shop have the same price. One canvas is a rectangle four times as long as it is wide. The other is a square canvas of side length 18 cm longer than the width of the rectangular canvas. Find the dimensions of the square canvas.

A. Let x be the width of the rectangular canvas. Thus, the length of this canvas is $4x$ and the side length of the square canvas is $(x + 18)$.

4x

x

x + 18

x + 18

The area of the rectangular canvas is $x.4x = 4x^2$ and the area of the square canvas is $(x + 18)^2$. Equating areas gives the following quadratic to be solved.

$(x + 18)^2 = 4x^2$

$x^2 + 36x + 324 = 4x^2$ [expanding]

$3x^2 - 36x - 324 = 0$ [collecting terms on one side]

$x^2 - 12x - 108 = 0$ [dividing by 3]

$(x + 6)(x - 18) = 0$ [factorising]

$x = -6$ or $x = 18$

Only $x = 18$ applies, since x is a length and therefore can't be negative.

The side length of the square is $18 + 18 = 36$ cm.

Unit 11.1 Activity 12B: Solving word problems with quadratic equations

1. Sam is the middle child of a 3-child family. His brother is 2 years older than Sam and his sister is 3 years younger than Sam. The product of their (Sam's brother and sister) ages is 176. The equation $(x + 2)(x - 3) = 176$ can be used to find the ages of the 3 children.

a. What does 'x' represent?

b. Solve the equation to find the ages of the 3 children.

2. In a village, the equation $L = -n^2 + 12n + 40$ modelled the number of pigs born each day for a period of time in the village. L = number of pigs born and n = day number.

To find n when 67 pigs were born in a day, the equation $-n^2 + 12n + 40 = 67$ needs to be solved. Solve this equation to find when 67 pigs were born.

3. A field is 20 m longer than it is wide. The area of the field is 384 m^2. In order to find the dimensions of the field, the quadratic equation $x(x + 20) = 384$ must be solved.

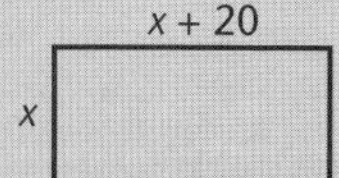

Solve this equation to find the length and width of the field.

4. A square room has side length x m. The room is made 2 m wider and 4 m longer so that its area is now 24 m^2.

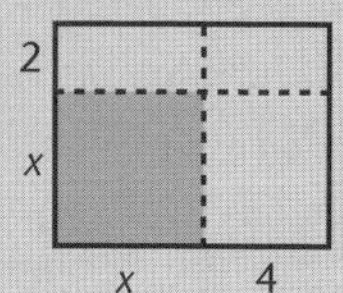

Solve the equation $(x + 2)(x + 4) = 24$ to find the original dimensions of the room.

5. William is playing a series of computer games in which he has to capture aliens. William scores one point for each alien captured. So far William has a total score of 45 points. At the end of his next game, William notices that if he squared and doubled the number of points he got for that game then he would get the same number of points as his new total score.

Solve the equation $2x^2 = x + 45$ to find how many aliens William captured in his latest game.

6. A box for soap has a square base and is 6 cm deep. Its top is open. The area of cardboard used for the box is 256 cm^2. The surface area of the outside of the box is $A = a^2 + 24a$, where a is the side length of the base of the box in cm.

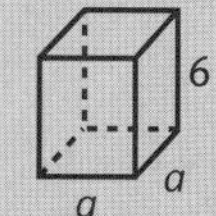

Solve the equation $a^2 + 24a = 256$ to calculate the side length of the base of the box.

7. Alice and Felix think of two numbers. The numbers are 10 apart and their product is 56. Solve the equation $x(x - 10) = 56$ to work out the numbers Alice and Felix thought of (there are two different possible pairs of numbers).

8. Two square pieces of cardboard are required as backing for a photo. The smaller square has side length 4 cm less than the side length of the larger square.

If the total area of cardboard required is 400 cm^2, find the dimensions of the larger square.

9. The product of two numbers is 360. If one number is 16 less than double the other number, find the two possible number pairs.

10. Three boys played a game in which they won points for skill. At the endof the game the winner has 6 points less than double the middle player's score. The loser has 4 points less than the middle player's score. The product of the winner's and loser's scores is equal to the square of the middle player's score. Find the three scores.

Unit 11.1 Activity 12C: Multiple choice – Solving quadratic equations

1. Which one of the following is not a quadratic equation?

   A. $2x^2 + 3x = 6x - 1$ **B.** $x^2 + 5x + 6 = 0$

   C. $(x + 2)(x - 5) = x^2$ **D.** $x^2 = 5x + 24$

2. The solution(s) of the equation $x(x - 4) = 0$ are:

   A. $x = 4$ **B.** $x = -4$

   C. $x = 2$ or $x = -2$ **D.** $x = 0$ or $x = 4$

3. The solutions of the equation $x^2 + 2x = 3$ are:

   A. $x = 1$ or $x = 3$ **B.** $x = 1$ or $x = -3$

   C. $x = -2$ or $x = 3$ **D.** $x = 2$ or $x = -3$

4. Which one of the following equations has only one solution?

   A. $x^2 + 9 = 6x$ **B.** $x^2 - 3x = 0$

   C. $x^2 = 16$ **D.** $x^2 + 4x + 16 = 0$

5. The solutions of the equation $x^2 + 2x - 8$ are:

   A. $x = 2$ or $x = 4$ **B.** $x = -2$ or $x = 4$

   C. $x = -2$ or $x = -4$ **D.** $x = 2$ or $x = -4$

6. Which one of the following equations has the solutions $x = -1$ or $x = 2$?

   A. $(x - 1)(x + 2) = 0$ **B.** $x^2 = x + 2$

   C. $x(x + 1) = 2$ **D.** $x^2 - 1 = 1$

7. The solutions to the equation $(3x - 1)(2x + 5) = 0$ are:

   A. $x = 1$ or $x = -5$ **B.** $x = 3$ or $x = -2.5$

   C. $x = \frac{1}{3}$ or $x = -2.5$ **D.** $x = -\frac{1}{3}$ or $x = 2.5$

8. The solutions of the equation $(x + 4)(x - 10) + 24 = 0$ are:

   A. $x = -4$ or $x = 10$ **B.** $x = 20$ or $x = 34$

   C. $x = 8$ or $x = -2$ **D.** $x = -6$ or $x = 2.4$

Unit 11.1 Number and Application
Topic 13: Basic algebra—graphs of quadratic functions

Basic algebra is a sub-section within Unit 11.1 of the Grade 11 syllabus (see Syllabus p. 14). This Topic deals with sketching and interpreting graphs. It covers:

- Sketch and interpret features of quadratic graphs.
- Determining and applying an appropriate model for a situation involving graphs.
- Writing equations from a graph to solve a problem (a combination of only two different types of transformation is expected, eg $y = 2x^2 + 3$ or $y = (x - 2)^2 + 1$).
- Drawing a graph to find the solution to a problem.

The parabola

The equation of a **quadratic function** can be written in the form

$$y = ax^2 + bx + c, \text{ where } a, b, c \text{ are constants, } a \neq 0$$

The following are all examples of equations of quadratic functions:

$y = 3x^2 - 2x + 4$ $\quad y = x^2 + 2x$ $\quad y = -x^2 + 3$ $\quad y = -2x^2$ $\quad y = (x - 3)^2$

The graph of a quadratic function is called a **parabola**. The parabola is a smooth curve, with one **axis of symmetry** and a **turning point** called the **vertex**.

The graph of $y = x^2$

The simplest quadratic function has the equation $y = x^2$. The graph is drawn by setting up a table of values and plotting points.

A table of values and graph for $y = x^2$ are shown.

x	–4	–3	–2	–1	0	1	2	3	4
$y = x^2$	16	9	4	1	0	1	4	9	16

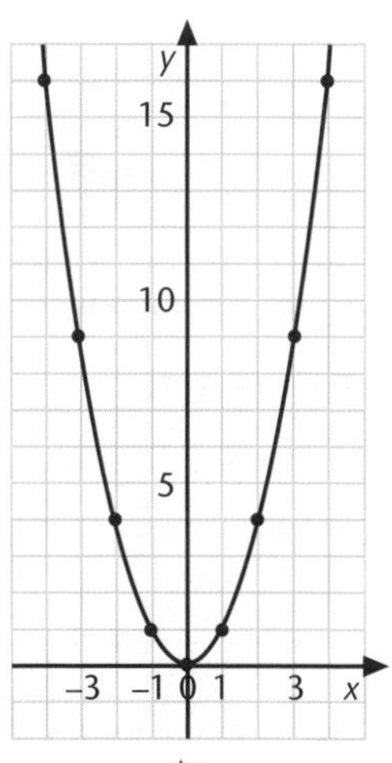

Note: 1. The graph is symmetrical about the y-axis, ie the line $x = 0$.

2. There is a vertex at (0, 0) giving a minimum value for y of 0 when $x = 0$.

3. The points are joined with a smooth curve. Notice how the curve 'flattens out' at the vertex; it is *not* a sharp point.

To see more clearly the behaviour of y between $x = -1$ and $x = 1$ consider the following table of values.

x	–1	$-\frac{3}{4}$	$-\frac{1}{2}$	$-\frac{1}{4}$	0	$\frac{1}{4}$	$\frac{1}{2}$	$\frac{3}{4}$	1
$y = x^2$	1	$\frac{9}{16}$	$\frac{1}{4}$	$\frac{1}{16}$	0	$\frac{1}{16}$	$\frac{1}{4}$	$\frac{9}{16}$	1

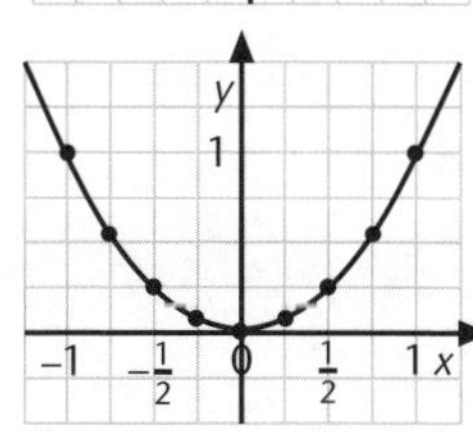

The 'magnification' of the part of the graph near the turning point shows the 'flattening out' of the curve much more clearly.

Transformations of the curve $y = x^2$

Changes to the basic equation of a parabola, $y = x^2$, cause various transformations of the curve. The general equation of a transformed parabola is

$$y = a(x - h)^2 + k \text{ where } a, h, k \text{ are constants}$$

Four transformations are now discussed. Their graphs are drawn by plotting points.

Graphs of the form $y = ax^2$

By changing the value of a in the equation $y = ax^2$, the steepness of the graph changes. The larger the size of a, the steeper the graph.

Example A

The graphs of $y = 2x^2$ and $y = \frac{1}{2}x^2$ are drawn below and compared with the graph of $y = x^2$.

x	-2	-1	0	1	2
x^2	4	1	0	1	4
$2x^2$	8	2	0	2	8
$\frac{1}{2}x^2$	2	$\frac{1}{2}$	0	$\frac{1}{2}$	2

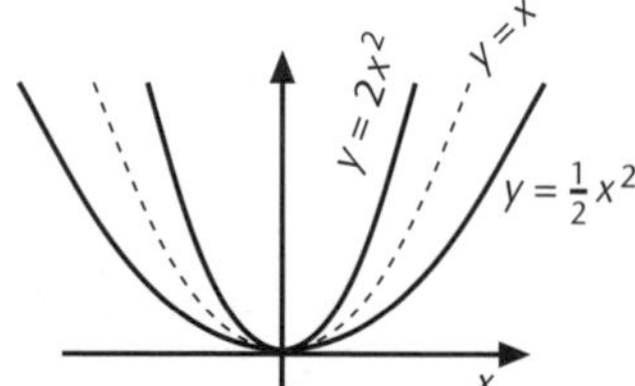

Note: 1. The graph of $y = ax^2$ is steeper than the graph of $y = x^2$ if $a > 1$. For example above, the graph of $y = 2x^2$ lies above the graph of $y = x^2$.

2. The graph of $y = ax^2$ is shallower than the graph of $y = x^2$ if $0 < a < 1$. For example above, the graph of $y = \frac{1}{2}x^2$ lies below the graph of $y = x^2$.
3. For each of these graphs, the vertex is (0, 0) and the axis of symmetry is the y-axis.

If a is negative, the parabola is **inverted** (upside-down). This gives a **maximum** value at the vertex, *not* a **minimum**. (Axis of symmetry remains the y-axis.)

Example B

The graphs of $y = -2x^2$ and $y = -\frac{1}{2}x^2$ are shown and compared with the graph of $y = -x^2$:
[Tabled values for x^2, $2x^2$, $\frac{1}{2}x^2$ (in Example A) are multiplied by -1.]

The graphs of $y = -2x^2$, $y = x^2$ and $y = -\frac{1}{2}x^2$ are *inverted* compared with the graphs of $y = 2x^2$, $y = x^2$ and $y = \frac{1}{2}x^2$.

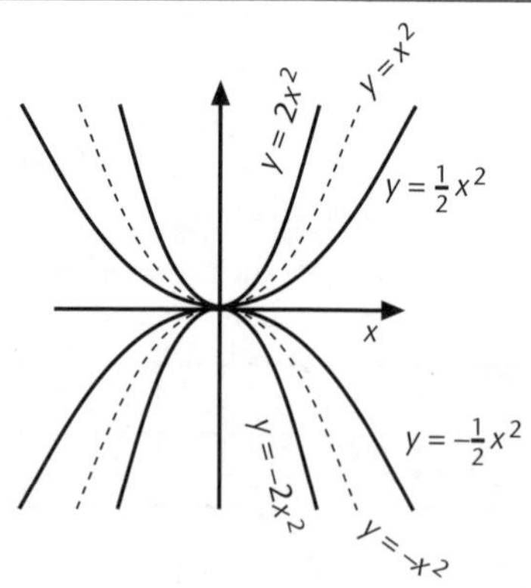

Graphs of the form $y = x^2 + k$

The graph of $y = x^2 + k$ is the graph of $y = x^2$ **transformed** by the **translation** $\begin{pmatrix} 0 \\ k \end{pmatrix}$.

In other words, it is the graph of $y = x^2$ shifted vertically k units.

- If k is positive, the shift is up.
- If k is negative, the shift is down.

Example C

By setting up a table of values, the graph of $y = x^2 + 2$ can be drawn and compared with the graph of $y = x^2$.

x	–3	–2	–1	0	1	2	3
x^2	9	4	1	0	1	4	9
$y = x^2 + 2$	11	6	3	2	3	6	11

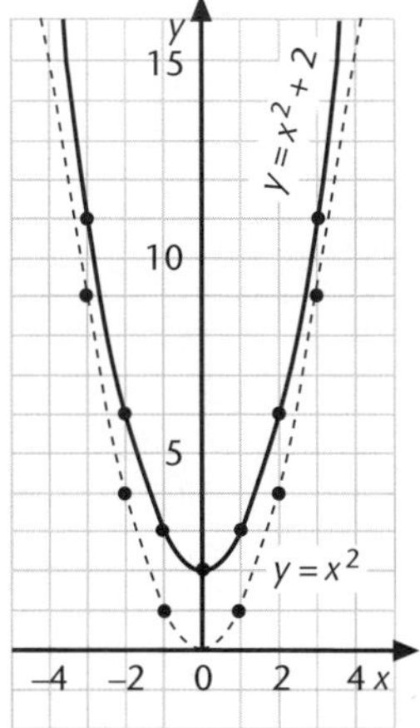

Note: The table and graph show that:

1. The graph of $y = x^2 + 2$ results from moving the graph of $y = x^2$ *up* 2 units.
2. The graph of $y = x^2 + 2$ is symmetrical about the y-axis.
3. The vertex is at (0, 2) and the minimum value for y is 2.

Graphs of the form $y = (x - h)^2$

The graph of $y = (x - h)^2$ is the graph of $y = x^2$ transformed by the translation $\begin{pmatrix} h \\ 0 \end{pmatrix}$ ie the graph of $y = x^2$ moved sideways h units.

- The axis of symmetry is $x = h$ and the vertex is $(h, 0)$.
- The y-intercept is $(0, h^2)$ and the x-intercept is $(h, 0)$.

Example D

Drawing up a table and plotting the graph of $y = (x - 2)^2$, and then comparing it with $y = x^2$, gives:

x	–1	0	1	2	3	4	5
$x - 2$	–3	–2	–1	0	1	2	3
$y = (x - 2)^2$	9	4	1	0	1	4	9

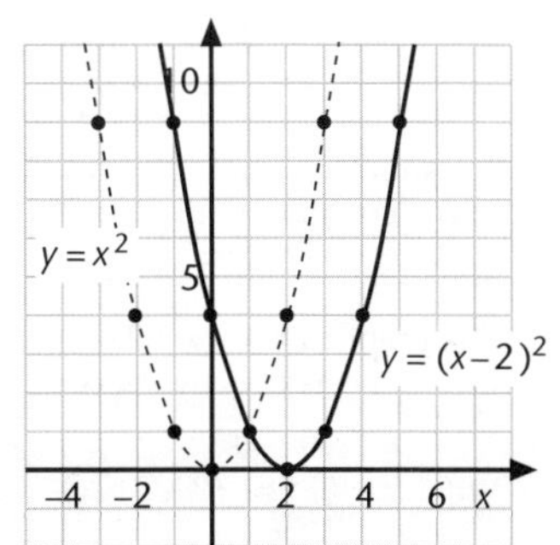

Note:

1. The axis of symmetry is $x = 2$ and the vertex is (2, 0).
2. The y-intercept is (0, 4) and the x-intercept is (2, 0).
3. The graph of $y = x^2$ has been shifted 2 units to the right.

Graphs of the form $y = (x - h)^2 + k$

The graph of $y = (x - h)^2 + k$ is the graph of $y = x^2$ transformed by the translation $\begin{pmatrix} h \\ k \end{pmatrix}$, ie the graph of $y = x^2$ is shifted horizontally h units and vertically k units.

- The axis of symmetry is the line $x = h$.
- The vertex is (h, k), so the minimum value of y is k.
- The y-intercept is $(0, h^2 + k)$.

Example E

Q. Draw the graph of $y = (x + 3)^2 - 2$.

A. Drawing up a table and plotting the points gives:

x	–6	–5	–4	–3	–2	–1	0
$x + 3$	–3	–2	–1	0	1	2	3
$(x + 3)^2$	9	4	1	0	1	4	9
$y = (x + 3)^2 - 2$	7	2	–1	–2	–1	2	7

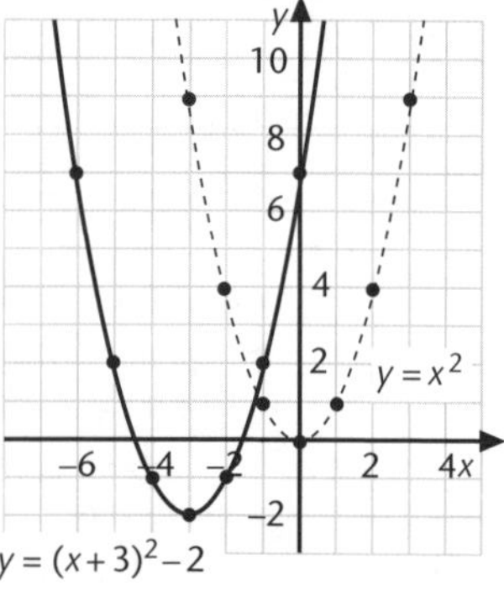

Note:
1. The axis if symmetry is the line $x = -3$.
2. The vertex is $(-3, -2)$, which is a minimum turning point.
3. The y-intercept is $(0, 7)$.

Sketching quadratic graphs by transformations

If the equation of a parabola is given in transformation form, $y = a(x - h)^2 + k$, its graph can be sketched by considering the various transformations involved.

Example F

1. The graph of $y = (x + 3)^2 - 2$ was drawn in Example E by plotting points. However, this graph can be drawn directly from the equation. The graph of $y = x^2$ is transformed as follows:

   $y = \underbrace{(x + 3)^2}_{\text{graph is moved 3 units left}} \underbrace{- 2}_{\text{graph is moved 2 units down}}$

   To find the y-intercept, substitute $x = 0$ in the equation.

   The graph is as in Example E.

2. The graph of $y = -(x + 3)^2 - 2$ is the graph in part **1** inverted (the negative outside the bracket causes this).

   When $x = 0$, $y = -(0 + 3)^2 = -11$

   The graph is as shown.

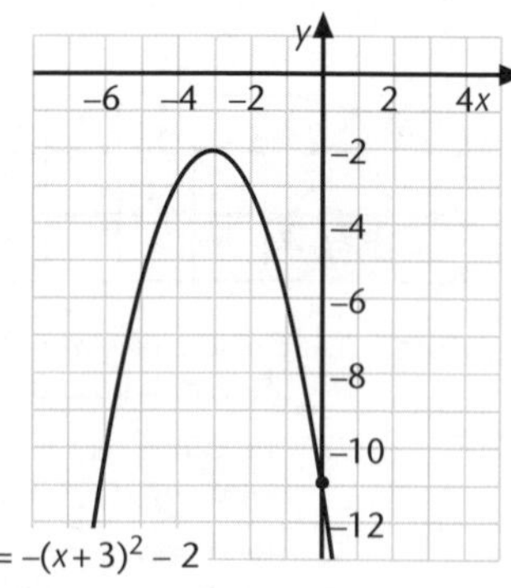

If the coefficient of x^2 is not ± 1, the steepness of the graph changes.

Example G

Q. Sketch the graph of $y = \frac{1}{2}(x - 4)^2 - 3$.

A. The graph of $y = x^2$ is transformed as follows:

$y = \underbrace{\tfrac{1}{2}}_{\text{shallower}} \underbrace{(x-4)^2}_{\text{4 units right}} \underbrace{-\,3}_{\text{down 3}}$

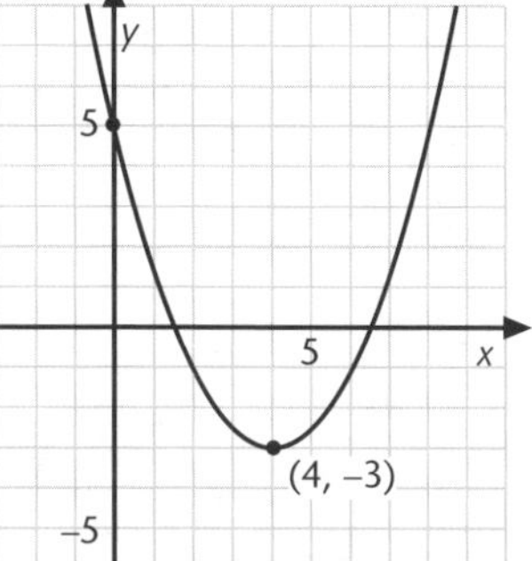

The vertex is at (4, –3) and the y-intercept is

$y = \frac{1}{2}(0 - 4)^2 - 3$ [substituting $x = 0$]

$= 5$

The graph is as shown.

Note: 1. The basic shape of the parabola is the same as that of $y = \frac{1}{2}x^2$, ie from the vertex travel out ± 1 and up $\frac{1}{2}$ to reach a point on the curve. Another point is out ± 2 up 2 from the vertex, etc.

2. The exact positions of the x-intercepts are not given. (If required, substitute $y = 0$ in the equation of the curve and solve the resulting equation.)

Unit 11.1 Activity 13A: Quadratic graphs

1. Complete the tables of corresponding values of x and y which fit the equations.

a. $y = x^2 - 2$

x	–2	–1	0	1	2
x^2	4				
$y = x^2 - 2$	2				

b. $y = (x - 2)^2$

x	0	1	2	3	4
$x - 2$					2
$y = (x - 2)^2$					4

c. $y = (x + 3)^2$

x	–5	–4	–3	–2	–1
$x + 3$	–2				2
$y = (x + 3)^2$	4				

d. $y = 2 - x^2$

x	–2	–1	0	1	2
x^2					
$y = 2 - x^2$					–2

e. $y = x^2 + 3$

x	–2	–1	0	1	2
x^2					4
$y = x^2 + 3$					7

f. $y = 2x^2$

x	–2	–1	0	1	2
x^2					4
$y = 2x^2$					8

2. Match the graphs below with the equations from Question 1.

a. 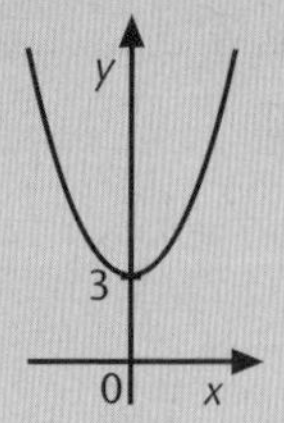

b. 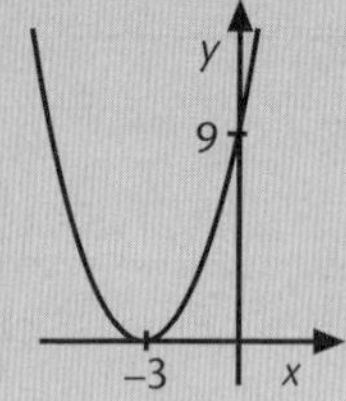

c. 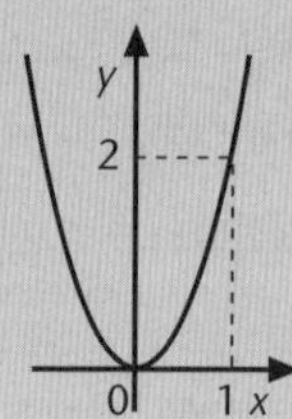

d. 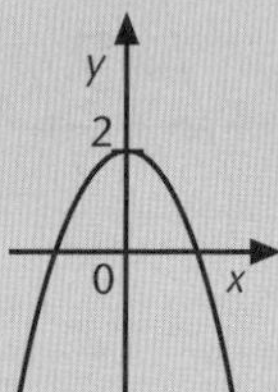

e. 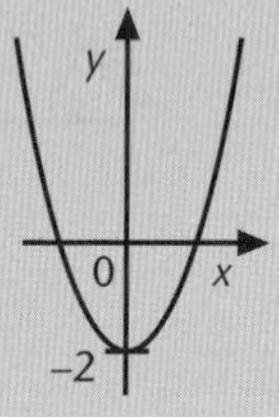

f. 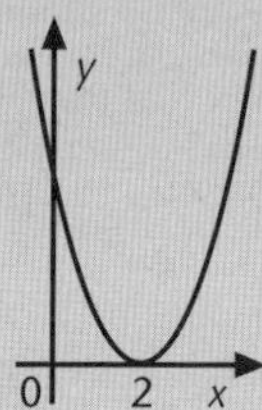

3. Sketch the graphs of the following equations, labelling carefully all intercepts (where possible) and the vertex for each graph.

a. $y = x^2 + 1$ **b.** $y = x^2 - 4$ **c.** $y = (x - 1)^2$

d. $y = (x + 1)^2$ **e.** $y = -2x^2$ **f.** $y = 4 - x^2$

4. Sketch the graphs of the following functions, labelling carefully all intercepts (where possible) and the vertex for each graph.

a. $y = (x + 1)^2 - 4$ **b.** $y = (x - 2)^2 + 3$ **c.** $y = (x + 2)^2 + 1$

d. $y = 2(x - 1)^2 - 3$ **e.** $y = -(x + 4)^2 + 1$ **f.** $y = -\frac{1}{2}x^2 - 2$

Sketching graphs of quadratic functions in factored form

When the equation of a quadratic function is given in factored form, eg $y = (x + 1)(x - 2)$, a different approach is used for drawing the graph. The steps are as follows:

1. Find the *x-intercepts* and the *y-intercept*.
2. Find the *axis of symmetry*.
3. Find the *vertex*.
4. Sketch the parabola.

Example H

Q. Sketch the graph of $y = (x - 4)(x + 2)$.

A. • Intercepts

For the *y*-intercept substitute $x = 0$

$y = (0 - 4)(0 + 2) = -4 \times 2 = -8$ $\therefore$ *y*-intercept is $(0, -8)$

For the *x*-intercepts substitute $y = 0$

$(x - 4)(x + 2) = 0 \Rightarrow x = 4$ or -2 $\therefore$ *x*-intercepts are $(4, 0)$ and $(-2, 0)$

- The axis of symmetry lies midway between the x-intercepts, so the axis of symmetry is:

  $x = \frac{4 + -2}{2}$ [averaging the values of the x-intercepts]

  $x = 1$ [shown by dotted line on graph]

- The vertex lies on the axis of symmetry ie on the line $x = 1$, and will have a y-coordinate of:

  $y = (1 - 4)(1 + 2)$ [substituting $x = 1$ in function]

  $= -3 \times 3$

  $= -9$

  $\therefore$ vertex is $(1, -9)$

The graph is as shown.

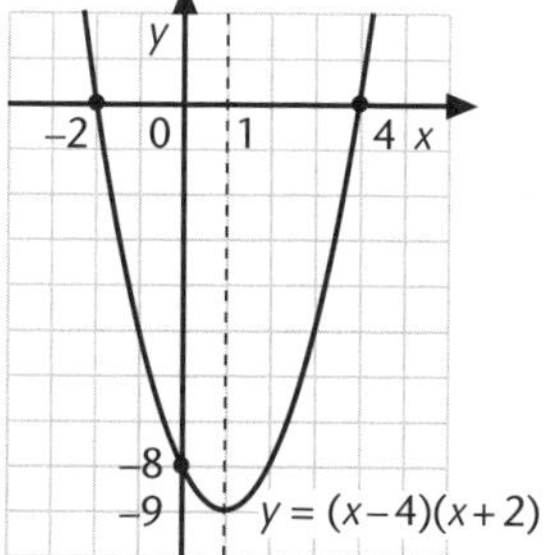

If the equation of the quadratic function is not in factored form, the quadratic will need to be factorised first.

Example I

Q. Sketch the graph of $y = x^2 + x - 6$

A. $y = (x + 3)(x - 2)$ [factorising]

- x-intercepts are -3 and 2

  y-intercept is $3 \times -2 = -6$

- Axis of symmetry is $x = \frac{-3 + 2}{2} = -\frac{1}{2}$
- Vertex is $(-\frac{1}{2}, -6.25)$

  [substituting $x = -\frac{1}{2}$ in equation]

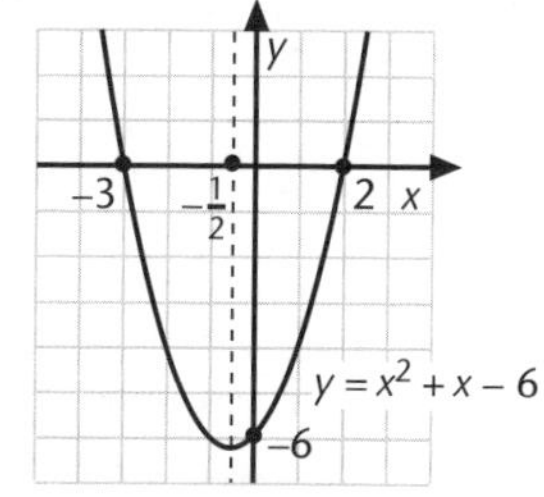

If the term in x^2 is negative then the graph will be inverted (upside down).

Example J

Q. Sketch the graph of $y = (2 - x)(4 + x)$

A. $y = (2 - x)(4 + x)$

- x-intercepts are 2 and -4

  y-intercept is $2 \times 4 = 8$

- Axis of symmetry is $x = \frac{-4 + 2}{2} = -1$
- Vertex is $(-1, 9)$

  [substituting $x = -1$ in equation]

Note: Expansion of $(2 - x)(4 + x)$ gives $-x^2 - 2x + 8$ so graph is 'upside down.' [coefficient of x^2 is negative]

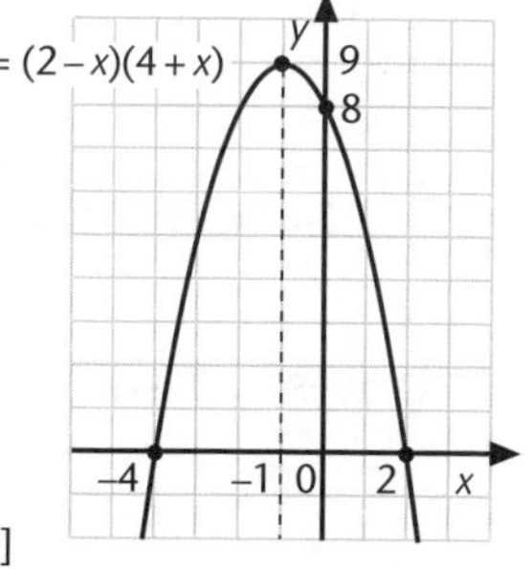

Writing equations of parabolas

By examining features of the graph of a parabola, such as intercepts, vertex, etc, the equation of a parabola can be written down.

Transformation method

If the vertex and one other point are known, the transformation method works well. The equation will be of the form

$$y = a(x - h)^2 + k$$

where h is the sideways translation, k is the vertical translation and a is a factor indicating the change in steepness.

Example K

Q. Write down the equations of the parabolas shown.

1.

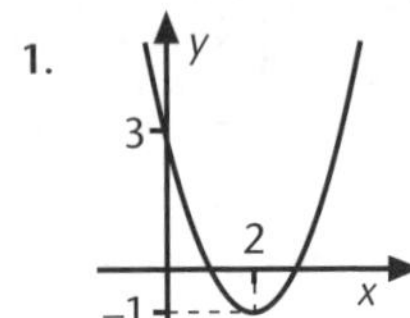

2.

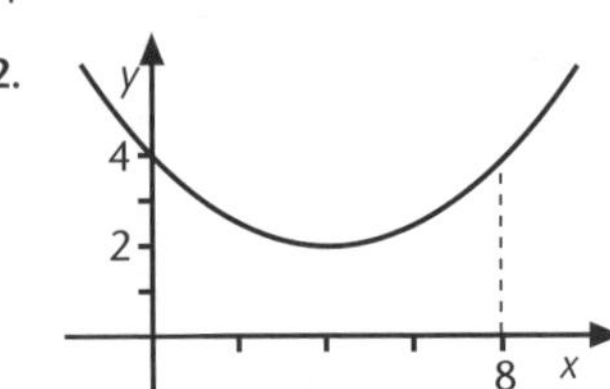

A. **1.** The parabola has been translated by the vector $\begin{pmatrix} 2 \\ -1 \end{pmatrix}$ so its equation is of the form

$y = a(x - 2)^2 - 1$ [general transformation equation with $h = 2$, $k = -1$]

Substituting the point (0, 3) gives

$$\begin{aligned} 3 &= a(0 - 2)^2 - 1 \\ 3 &= 4a - 1 \\ 4 &= 4a \\ a &= 1 \end{aligned}$$

So the equation of the parabola is $y = (x - 2)^2 - 1$.

2. The axis of symmetry of the parabola is $x = 4$. The minimum value is 2.

$\therefore$ the parabola $y = x^2$ has been translated by the vector $\begin{pmatrix} 4 \\ 2 \end{pmatrix}$ and so its equation is

$y = a(x - 4)^2 + 2$ [general transformation equation with $h = 4$, $k = 2$]

Substituting the point (0, 4) gives

$$\begin{aligned} 4 &= a(0 - 4)^2 + 2 \\ 4 &= 16a + 2 \quad \text{[simplifying]} \\ 16a &= 2 \\ a &= \frac{2}{16} = \frac{1}{8} \end{aligned}$$

So the equation of the parabola is $y = \frac{1}{8}(x - 4)^2 + 2$.

Factorised form method

If the x-intercepts are known, or can be worked out, then the quadratic function can be written down in factored form. The equation of the parabola will be of the form

$$y = a(x - b)(x - c)$$

where b and c are the x-intercepts and a is a factor affecting the steepness of the curve. Another known point is then substituted to evaluate a.

Example L

Q. Write down the equations of the parabolas shown.

1.

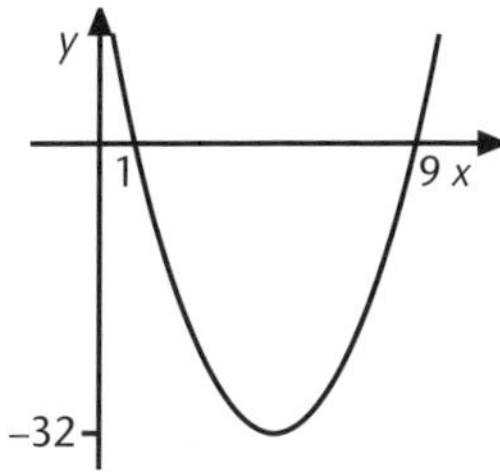

2.

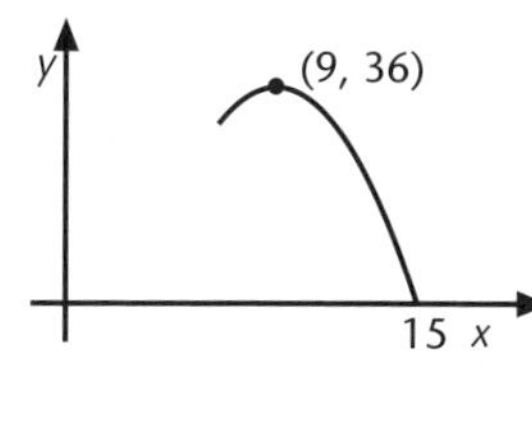

A. **1.** The x-intercepts are 1 and 9. Thus, the equation is of the form $y = a(x - 1)(x - 9)$.

The axis of symmetry of the graph is $x = 5$ [since $\frac{1 + 9}{2} = 5$]

From the graph $y = -32$, when $x = 5$:

$-32 = a(5 - 1)(5 - 9)$ [substituting $x = 5$, $y = -32$ in $y = a(x - 1)(x - 9)$]
$-32 = a \times 4 \times -4$
$-32 = -16a$
$a = 2$

$\therefore$ the equation is $y = 2(x - 1)(x - 9)$

The equation could also be written as $y = 2x^2 - 20x + 18$ after expanding.

2. Only one intercept is given, the x-intercept, 15; the maximum point is (9, 36). This means the axis of symmetry is $x = 9$ and the other x-intercept is 3 (by symmetry, $15 - 9 = 6$, $9 - 6 = 3$). So, the equation will be of the form:

$y = -a(x - 3)(x - 15)$ [$-a$ since the parabola is 'upside down']

When $x = 9$, $y = 36$:

$\therefore 36 = -a(9 - 3)(9 - 15)$ [substituting]
$36 = 36a$ [simplifying]
$a = 1$

$\therefore$ the equation will be $y = -(x - 3)(x - 15)$ (or $y = -x^2 + 18x - 45$ in expanded form)

Unit 11.1 Activity 13B: Quadratic graphs

1. Sketch the graphs whose equations are given.

a. $y = (x + 2)(x - 4)$ **b.** $y = (x - 5)(x - 1)$ **c.** $y = x(x - 3)$

d. $y = (x + 3)(x + 1)$ **e.** $y = (x - 4)(x - 2)$ **f.** $y = (3 - x)(1 + x)$

g. $y = x^2 + 6x + 5$ **h.** $y = x^2 + 4x - 12$ **i.** $y = 5x - x^2$

2. The sketch shows the graph of $y = x^2 + 2x - 8 = (x + 4)(x - 2)$.

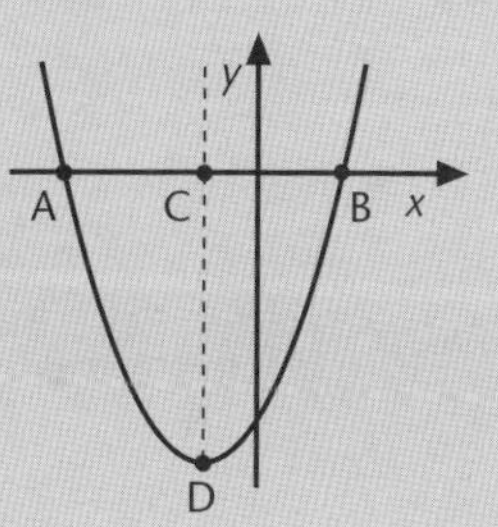

CD is the axis of symmetry.

a. Write the coordinates of:

i. A **ii.** B **iii.** C **iv.** D

b. What is the least possible value of $x^2 + 2x - 8$?

c. Write down the equation of the parabola in the form $y = a(x - h)^2 + k$.

3. Find the equations of the following parabolas (*not* drawn to scale):

a.

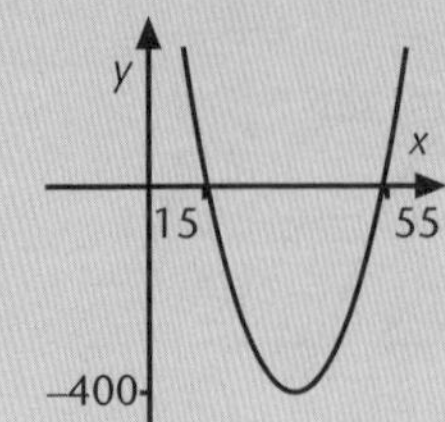

b.

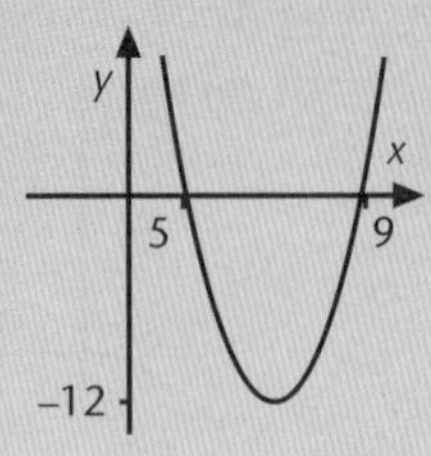

c.

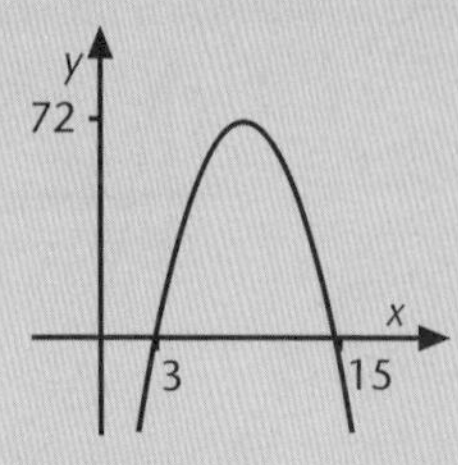

d.

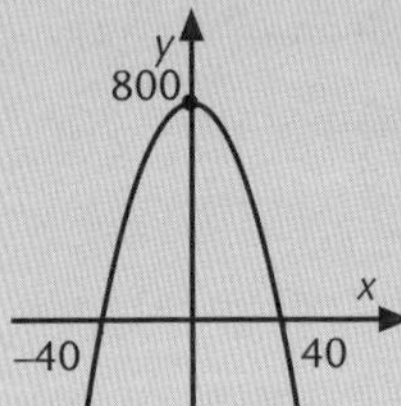

e.

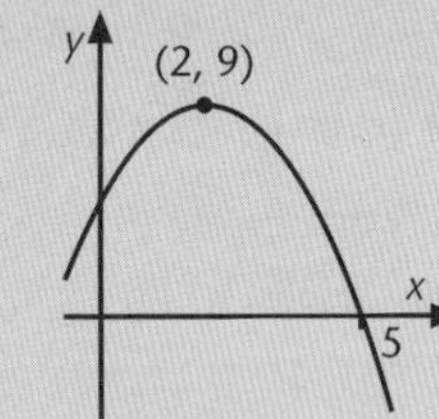

f.

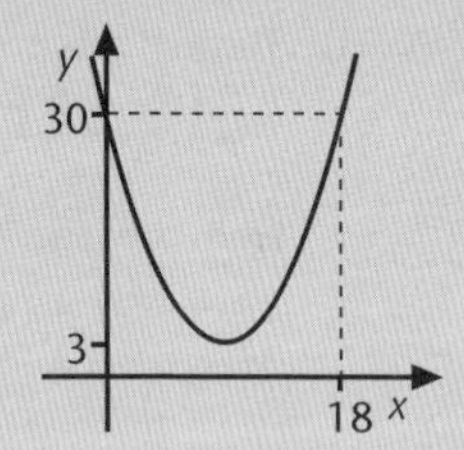

Using quadratic graphs to solve problems

Quadratic functions and their graphs can be used to model many different situations and to solve many problems in a practical context.

Example M

Q. Johnny's birthday present is a rocket launcher which projects a small toy in a parabolic arc.

The equation of the path of the toy is

$$h = \frac{(4 - d)(d + 2)}{5}$$

h (height)
d metres
Johnny marker landing point

where d is the distance in metres from a marker and h is the height in metres of the toy above the ground.

1. How far away from Johnny does the toy land?
2. What is the maximum height the toy reaches?
3. What is the height of the toy as it passes over the marker?

A. 1. When the toy lands, $h = 0$ [height of toy above ground is 0 metres]

Solving $\frac{(4 - d)(d + 2)}{5} = 0$

gives $(4 - d)(d + 2) = 0$ [multiplying by 5]

$d = 4$ or -2

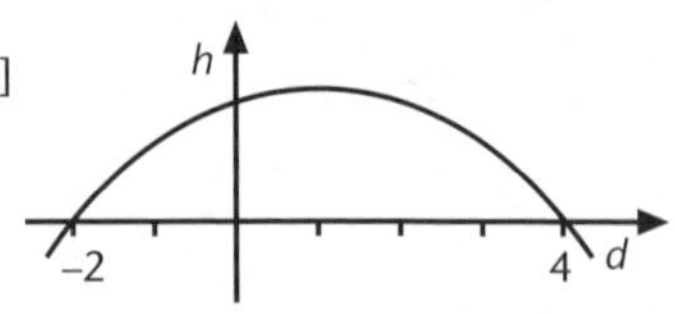

Relative to the marker, Johnny is at position –2 and the toy lands 6 m away at position 4.

2. Toy reaches maximum height halfway between Johnny and the landing point, ie at $d = \dfrac{-2 + 4}{2} = 1$

   When $d = 1$ $\quad h = \dfrac{(4 - 1)(1 + 2)}{5}$ $\quad$ [substituting $d = 1$ in equation of curve]

   $= 1.8$ m

   The maximum height reached is 1.8 m.

3. As the toy passes over the marker, $d = 0$

   $\therefore h = \dfrac{(4 - 0)(0 + 2)}{5} = 1.6$ m

The properties of the parabola can be used to find maximum or minimum values of quantities modelled by quadratic functions.

Example N

Q. Sue has 100 metres of fencing wire and wishes to build a rectangular garden. What are the dimensions of the garden that will give the rectangle the largest area?

A. Let the length of the rectangle be x and the width y.

Hence
$$\begin{aligned} x + y + x + y &= 100 && \text{[perimeter = 100 m]} \\ 2x + 2y &= 100 \\ x + y &= 50 && \text{[dividing by 2]} \\ y &= 50 - x && \text{[rearranging]} \end{aligned}$$

The area, A, of the garden is
$$\begin{aligned} A &= xy \\ &= x(50 - x) && \text{[substituting for } y\text{]} \\ &= 50x - x^2 \end{aligned}$$

From the table alongside, the graph of $A = 50x - x^2$ can be drawn.

x	0	5	10	15	20	25	30	35	40	45	50
A	0	225	400	525	600	625	600	525	400	225	0

The graph of $A = 50x - x^2$ shows that the maximum area is 625 m^2 when $x = 25$ m, and hence $y = 50 - x = 25$, ie a square garden of side 25 metres gives the maximum area.

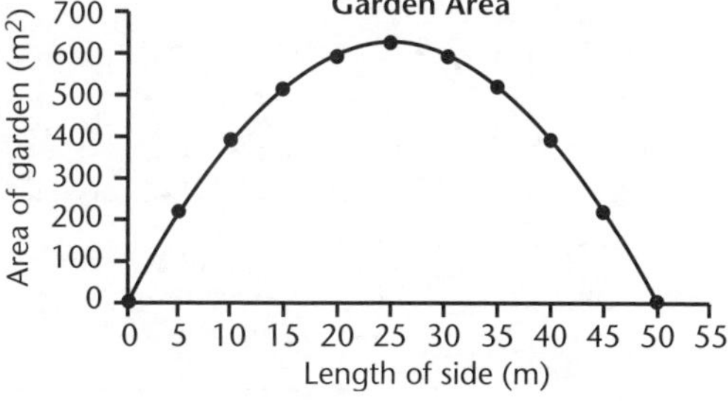

Note: The graph of $A = 50x - x^2$ can also be drawn by factorising first to give A = x(50 – x), then using factored form techniques (x-intercepts at (0, 0) and (50, 0), vertex at $x = 25$, $A = 625$ etc).

The equation of a curve may need to be found in order to solve a problem.

Example O

Q. A large trench is made so that its cross-section is in the shape of a parabola which is 3 m across. A pole inserted in the trench one metre out from one side of the trench shows a depth of 3 m. What is the depth of the trench at its deepest point?

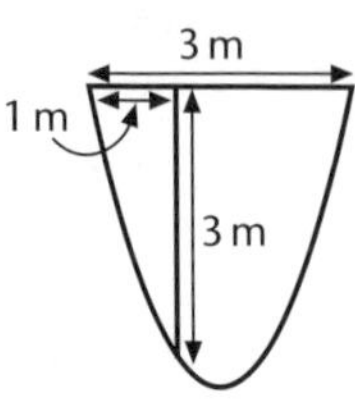

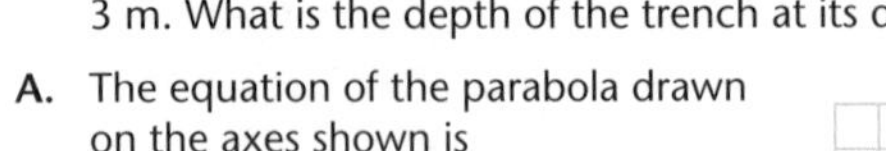

A. The equation of the parabola drawn on the axes shown is

$y = k(x + 1)(x - 2)$
[since $x = -1$ and $x = 2$ are the x-intercepts]

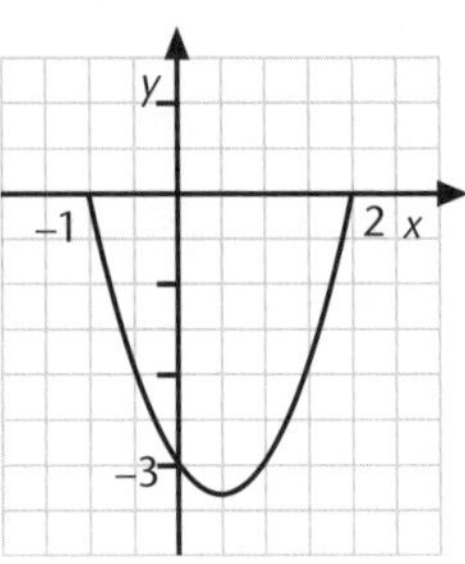

Since the point (0, –3) lies on the curve, it follows that

$-3 = k(0 + 1)(0 - 2)$ [substituting $x = 0$, $y = -3$ in equation of curve]

$-3 = -2k$ [simplifying]

$k = \dfrac{-3}{-2}$ [dividing by –2]

$= 1.5$

The equation of the parabola is $y = 1.5(x + 1)(x - 2)$.

The trench is deepest halfway between the intercepts, ie at $x = \dfrac{-1 + 2}{2} = 0.5$

At $x = 0.5$, $y = 1.5(0.5 + 1)(0.5 - 2)$ [substituting $x = 0.5$ in equation]

$= -3.375$

The trench is 3.375 m deep at its deepest point.

Unit 11.1 Activity 13C: Using quadratic graphs to solve problems

1. A villager has 24 m of fencing and builds a rectangular garden using an existing fence as one side. The area, A m^2, of the garden is given by $A = 24x - 2x^2$, where x is the width of the garden.

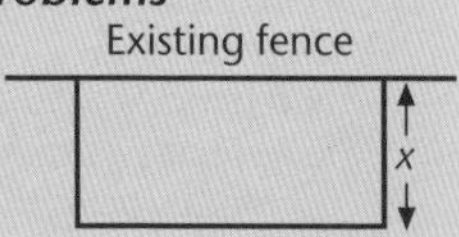

a. Copy and complete the following table:

x	0	1	2	3	4	5	6	7	8	9	10	11	12
A	0	22	40	54									

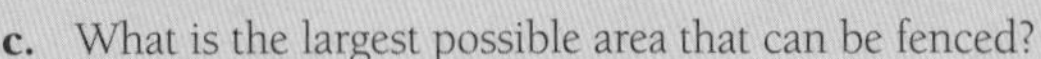

b. Draw the graph of $A = 24x - 2x^2$, using axes as shown.

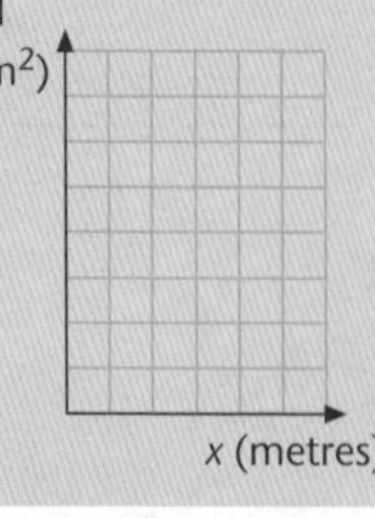

c. What is the largest possible area that can be fenced?

d. Why is $x \geq 0$?

e. Use your graph to find the width of the garden when its area is 60 m^2.

2. Hosea finds that the depth of water in a small drain in his village is related to the distance across the drain by the quadratic equation

$d = (w - 10)(w - 70)$

where d is the depth, in millimetres, of the water in the drain and w is the distance, in centimetres, measured from a post on one of its banks.

Depth (mm): 800, 600, 400, 200, 0, –200, –400, –600, –800, –1 000
Width (cm): 20, 40, 60, 80

a. Graph the equation for the depth of water in the drain. Draw the graph for values of w from 0 to 80. (Use axes as shown)

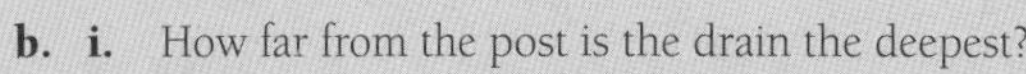

b. i. How far from the post is the drain the deepest?
ii. What is the depth at this distance?

c. i. How wide is the drain at water level?
ii. How is this shown on the graph?

d. What are the two distances from the post that the depth of water is –500 mm?

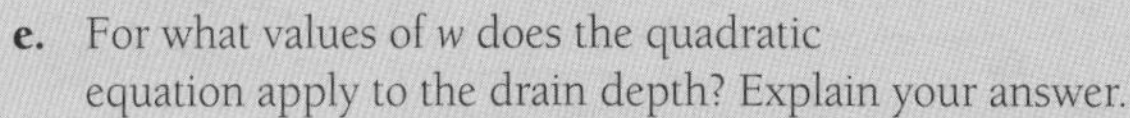

e. For what values of w does the quadratic equation apply to the drain depth? Explain your answer.

f. Give some practical reasons why the *actual* depth of water in the drain may be different from that shown by the graph.

3. In Gabriel's village there is an electrical wire suspended between two poles across a gully. The poles are on the edges of the gully and the wire is at the same height on both sides of the gully. The curve of the wire can be modelled by the equation

$h = (d - 6)^2 + 3$

where h is the height, in metres, of the wire above the floor of the gully and d is the distance, in metres, from the left-hand post.

a. Draw the graph of $h = (d - 6)^2 + 3$ for the values of distance, d, from 0 to 12.

b. What is the minimum height that the wire is above the floor of the gully?

c. i. How far apart are the two poles?
ii. How is this shown on the graph?

4. Judy has a piece of string hung across a wall in her room that has her birthday cards on it. The string takes on a shape which is approximately that of a parabola. The ends of the string are 2 m above the floor of each end, and the string is 1.5 metres above the floor at the lowest point. The ends of the string are 4 metres apart.

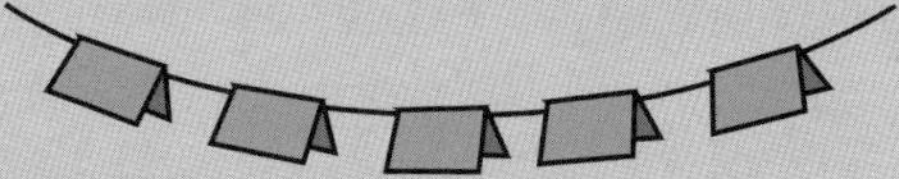

a. Find an equation that represents the shape of the string and draw its graph.

b. Judy is 1.65 m tall. She stands under the string and 1.2 m to the right of the lowest point of the string. Will her head touch the string?

5. Josie has a stack of yams built into a small pit near her village. She measures the height, h cm, from the floor of the pit, every metre from the left-hand end.

Josie's data are shown in the table below. A graph of these data is shown

Distance d (m)	0	1	2	3	4	5	6	7	8	9	10
Height h (cm)	10	60	110	160	210	260	310	360	320	200	0

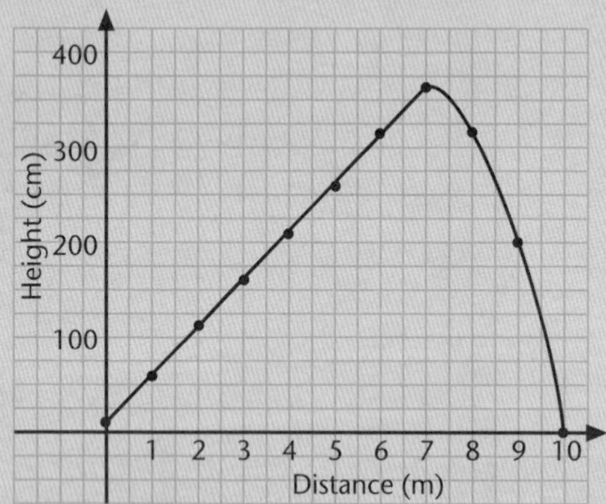

a. Write the equation for h in terms of d that models the **first seven metres** of the stack.

b. Write the equation for h in terms of d that models the **last three metres** of the stack. (Assume (7, 360) is the turning point of this quadratic graph).

6. The path of a bouncing ball was recorded by multiple flash photography.

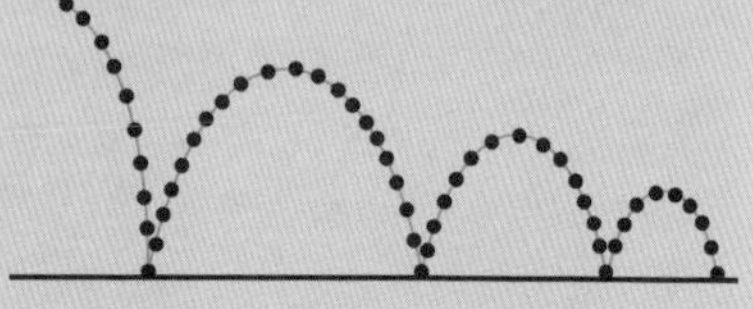

The flash operated every 0.1 seconds and the height of the ball from the floor was measured. The measurements of the path of the first bounce are shown in the table below.

Time (secs)	1.0	1.1	1.2	1.3	1.4	1.5	1.6	1.7	1.8	1.9	2.0
Height (m)	0	0.45	0.80	1.05	1.20	1.25	1.20	1.05	0.80	0.45	0

a. Graph the results, joining the points with a smooth curve.

b. What was the maximum height reached by the ball on its first set of bounces?

c. Write an equation for the curve.

d. How high would the ball have been above the floor after 1.25 seconds?

7. A gutter for the roof of a shed is to have a rectangular cross-section and to be made from a long strip of sheet metal 30 cm wide. The sides are to be turned up at right angles to the base and the sides are to be of equal height.

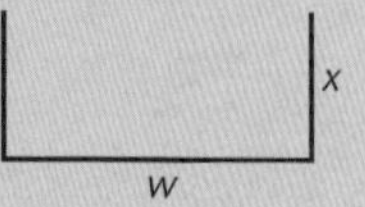

For what value of x is the cross-sectional area the greatest?

Hint: Express w in terms of x then sketch the graph of the quadratic function which models the area of the cross-section.

Unit 11.1 Number and Application

Topic 14: Basic algebra—simultaneous equations

In the Grade 11 syllabus (see Syllabus p.14), the basic algebra sub-section requires students to solve simultaneous equations using elimination, substitution and graphical methods. This topic covers:

- Solving pairs of simultaneous linear equations.
- Informal methods such as 'guess and check' for solving simultaneous equations.
- Solving problems involving modelling by forming and solving appropriate equations.

Introduction

Simultaneous equations are two or more equations which are true for the *same values* of their unknowns or variables. In this chapter, pairs of simultaneous equations with *two unknowns* are solved.

Solving simultaneous equations requires finding the value of each of the unknowns so that *both* equations are true.

- 'Guess and check' methods take a numerical approach, and may offer a quick solution if the answers are whole numbers.
- Algebraic techniques usually offer a more direct path to the answer.

The solutions to problems involving simultaneous equations should be checked by substituting the values in both equations. (This can often be carried out mentally.)

A graphical calculator, such as the *fx*-975, will find solutions to a pair of simultaneous linear equations immediately, providing the equations are entered in a a certain form.

Guess and check method

Some simultaneous equations can be solved using repeated substitutions, improving these estimates as you proceed.

Example A

Q. Two numbers a and b are four apart in size. If I triple the smaller number and double the bigger number, the sum of these new numbers is 73. This gives the following pair of simultaneous equations to solve:

$$b = a + 4$$
$$3a + 2b = 73$$

where a is the smaller number. Solve these equations to find the values of a and b.

A. A starting value of $a = 10$ is chosen.

Setting out estimates for a and b in a table:

a	$b = a + 4$	$3a + 2b$	Compare answer with 73
10	14	30 + 28 = 58	Numbers too small
16	20	48 + 40 = 88	Numbers too large
14	18	42 + 36 = 78	Numbers a little too large
13	17	39 + 34 = 73	Numbers correct

The numbers are 13 and 17.

Algebraic solutions to simultaneous equations

Two commonly used methods for solving simultaneous equations follow.

Substitution method

The **substitution method** is used when one of the unknowns is the subject of one (or both) equation(s). This expression is then substituted for the unknown in the other equation.

Example B

Q. Solve the following pairs of simultaneous equations using the substitution method.

1. $x + 2y = 8$ and $y = 3x - 10$ **2.** $x = y - 4$ and $2y - x = 6$

A. 1.

$x + 2y = 8$ ... (1)

$y = 3x - 10$ ... (2) [labelling the equations]

Substituting $3x - 10$ for y in (1) gives:

$x + 2(3x - 10) = 8$ [writing $3x - 10$ in place of y]

$x + 6x - 20 = 8$ [expanding]

$7x - 20 = 8$ [simplifying]

$7x = 28$ [adding 20]

$x = 4$ [dividing by 7]

Substituting $x = 4$ in (2) gives $y = 3 \times 4 - 10$

$= 2$

The solution is $x = 4$ and $y = 2$.

To check the answer $x = 4$ and $y = 2$, substitute into equation (1) to give: $4 + 2 \times 2 = 8$ which is true, so the solutions are correct.

2.

$x = y - 4$ ... (1)

$2y - x = 6$ ... (2) [labelling the equations]

Substituting $x = y - 4$ in (2) gives:

$2y - (y - 4) = 6$ [replacing x with $y - 4$]

$2y - y + 4 = 6$ [expanding bracket]

$y + 4 = 6$ [simplifying]

$y = 2$ [subtracting 4]

Substituting $y = 2$ in (1) gives $x = 2 - 4$

$= -2$

The solution is $x = -2$, $y = 2$

Elimination method

The **elimination method** is best used when both unknowns in both equations are on the same side of the equals sign. (This is the form required by graphical calculators.)

In the elimination method, the two equations are added together (or subtracted from each other), so that one of the two variables is eliminated. This leaves one equation to solve, in which there is only one unknown.

Example C

Q. Solve the simultaneous equations $2x + y = 8$ and $x - y = 1$ using the elimination method.

A. The equations $2x + y = 8$ and $x - y = 1$ have 'y' in one equation and '$-y$' in the other. If the equations are *added* together, the two y's will be eliminated, because $y + -y = 0$.

	$2x + y = 8$	... (1)	
	$x - y = 1$	... (2)	[labelling the equations]
Adding (1) and (2)	$3x + 0 = 9$	[eliminating the y's]	
	$x = 3$	[dividing by 3]	

Substituting $x = 3$ in (1) gives $2 \times 3 + y = 8$

$y = 2$ [subtracting 6]

The solution is $x = 3$ and $y = 2$.

Check: Substituting $x = 3$ and $y = 2$ in (2) gives $3 - 2 = 1$, which is true.

Sometimes one (or both) of the equations must be 'scaled up' first (ie multiplied through by a constant) so that addition or subtraction will eliminate one of the two variables.

Example D

Q. Solve the equations $2x + 3y = 14$ and $3x + y = 10.5$ using elimination.

A. If the equation $3x + y = 10.5$ is multiplied through by 3, an equation with the term $3y$ results. One equation can then be subtracted from the other to eliminate the term in y.

	$2x + 3y = 14$	... (1)	
	$3x + y = 10.5$	... (2)	[labelling the equations]
3 x (2) gives	$9x + 3y = 31.5$	... (3)	[labelling the new equation]
(3) – (1) gives	$7x = 17.5$		[subtracting, (3) – (1)]
	$x = 2.5$		[dividing by 7]

Substituting $x = 2.5$ in (1) gives:

$2 \times 2.5 + 3y = 14$

$3y = 9$ [subtracting 5]

$y = 3$ [dividing by 3]

The solution is (2.5, 3).

Check: Substituting $x = 2.5$ and $y = 3$ in (2) gives $7.5 + 3 = 10.5$, which is true.

Note: **1.** In the elimination method, the two equations to be subtracted or added together are best written one above the other (with variables lined up) before doing the addition or subtraction.

2. It is usually best to use the two *original* equations for finding the value of the second unknown and for the check.

The better algebraic method for solving simultaneous equations depends on the form in which the equations are given, as the following example shows. The substitution method is used in the first example, while the elimination method is used in the second example.

Example E

Q. Solve the simultaneous equations **1.** $y = x + 4$, $y = 2x - 1$ **2.** $4x + 3y = 1$, $5x - 2y = 30$

A. **1.** $y = x + 4$ ... (1)

$y = 2x - 1$ ... (2)

Substitute for y in (1) $2x - 1 = x + 4$

$x - 1 = 4$ [subtracting x]

$x = 5$

Substitute for x in (1) $y = 5 + 4 = 9$

Solution is (5, 9)

2. $4x + 3y = 1$ ... (1)

$5x - 2y = 30$ ... (2)

Multiply (1) by 2 $8x + 6y = 2$ ... (3)

Multiply (2) by 3 $15x - 6y = 90$ ... (4)

Add (3) and (4) $23x = 92$

$\therefore x = 4$ [dividing by 23]

Substitute $x = 4$ in (1) $16 + 3y = 1$

$3y = -15$

$y = -5$

Solution is $x = 4$, $y = -5$.

Graphical solution to simultaneous linear equations

The graphical solution to a pair of linear equations is *the point of intersection of the graphs of the two lines*. This is the point that satisfies both equations at the same time, or simultaneously.

Example F

Q. Find the simultaneous solution to the linear equations $y = 4x - 5$ and $y = -2x + 7$

A. The line $y = 4x - 5$ goes through the point (0, –5) and, substituting $x = 1$, the point (1, –1).

The line $y = -2x + 7$ goes through the point (0, 7) and, substituting $x = 1$, the point (1, 5).

Graphing both lines on a set of axes:

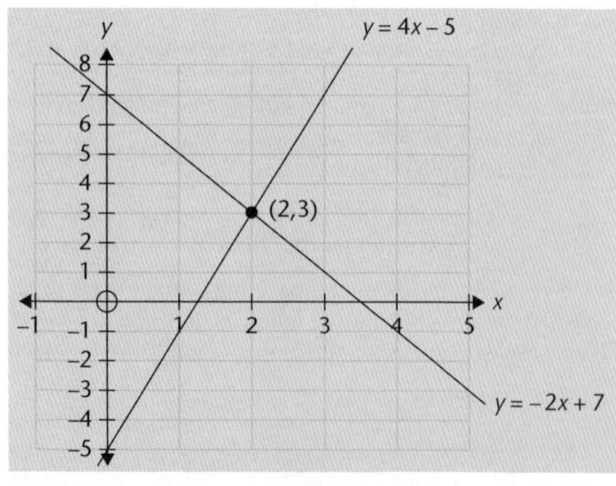

Accurate graphing of the linear equations reveals the simultaneous solution to be $x = 2$ and $y = 3$

Unit 11.1 Activity 14A: Simultaneous equations

1. Solve the following pairs of simultaneous equations using 'guess and check' techniques.

a. $y = x + 5$, $y + 2x = 47$ **b.** $y = x + 4$, $3x - y = 18$ **c.** $y = 2x + 3$, $x + 3y = 58$ **d.** $y = 2x + 3$, $y - x = 15$

2. Solve the following pairs of simultaneous equations by elimination.

a. $2x + y = 11$, $x + y = 7$ **b.** $4x - 3y = 8$, $2x + 3y = 22$ **c.** $x + y = 8$, $3x + y = 14$ **d.** $2x + y = 11$, $x - 2y = 13$

e. $2x + y = 6$, $x - 2y = \frac{1}{2}$ **f.** $3x - 2y = 13$, $5x + y = 0$ **g.** $2x + 3y = 23$, $5x - 2y = 10$ **h.** $2x - 3y = 10$, $3x + 5y = 15$

3. Solve the following pairs of simultaneous equations by substitution.

a. $x + y = 7$, $y = 2x - 2$ **b.** $y = 2x + 3$, $x + 2y = 16$ **c.** $y = 2x - 5$, $3y - 2x = 5$ **d.** $y = x + 5$, $x - 2y = -7$

e. $2y + x = 14$, $x = y + 5$ **f.** $y = 2x + 3$, $2x + 3y = 5$ **g.** $y = x - 3$, $2y = 3x - 8$ **h.** $x = 9 - y$, $y - 2x = 12$

4. Solve the following pairs of simultaneous equations using any appropriate technique.

a. $3x - 4y = 6$, $2x + 3y = 4$ **b.** $3x - 4y = 4$, $y = x - 1$ **c.** $3x + 2y = 7$, $4x + 3y = 9$ **d.** $1.1x + 3.2y = 22$, $1.5x - 4y = -16$

5. On a set of axes, accurately graph each of the following pairs of linear equations and find the simultaneous solution.

a. $y = 1 - 2x, y = 10 + x$ **b.** $y = 3x, y = 8x - 6$

c. $y = 2x + 7, y + 4x = 16$ **d.** $c = 30n, c = 126 + 12n$

Solving problems using simultaneous equations

You may be required to form and solve a set of simultaneous equations in order to find the solution to a problem.

Example G

Q. A trip to a rugby game is arranged. The total number of children and adults going on the trip is 20. Each adult's ticket is K25 and each child's ticket is K15. The total price of the tickets is K370. How many adults and children went to the game?

A. Let a = number of adults and c = number of children.

$a + c = 20$ ... (1) [20 people go altogether]

Adults' tickets cost a total of $a \times 25 = 25a$ kina

Children's tickets cost a total of $c \times 15 = 15c$ kina

$25a + 15c = 370$ ... (2) [total cost is K370]

$15a + 15c = 300$ ... (3) [multiplying equation (1) by 15]

Subtracting (3) from (2) gives

$10a = 70$

$a = 7$ [dividing by 10]

Substituting a = 7 in (1) gives $7 + c = 20$ and $c = 13$.

So 7 adults and 13 children went to the rugby game.

Unit 11.1 Activity 14B: Word problems with simultaneous equations

1. Linus bought some takeaway burgers for his family. The hamburgers cost 80 toea less than the cheeseburgers. The total cost for three hamburgers and four cheeseburgers was K28.40. Solve the following pair of simultaneous equations to find the price of a cheeseburger.

$$h = c - 0.80$$
$$3h + 4c = 28.40$$

2. Greg spends ten hours each weekend at his two part-time jobs. His gardening job pays K8 per hour and his shop job pays K9 per hour. Altogether Greg earns K86.50 per weekend. Solve the following pair of simultaneous equations to find how long Greg spends gardening each weekend.

$$g + s = 10$$
$$8g + 9s = 86.50$$

3. A company employed two consultants, Artemius and Brown, for a total of 50 hours. Artemius charged K120 per hour and Brown K90 per hour. The company paid a total of K5 460 for the two consultants. Solve the following pair of simultaneous equations to find how many hours each consultant worked:

$$120A + 90B = 5\,460$$
$$A + B = 50$$

4. Vavine bought some vegetable plants for her garden. She bought beans at K2.95 a box and aibika at K1.45 a box. Vavine bought 3 more boxes of aibika than she bought boxes of beans, and she spent a total of K30.75.

Solve the simultaneous equations $295x + 145y = 3\,075$ and $y = x + 3$ to find the number of boxes of beans (x) and the number of boxes of aibika (y) that Vavine bought.

5. A shop offered a special on socks and handkerchiefs for Father's Day. For K50 you could buy 7 pairs of socks and 3 boxes of handkerchiefs or for K24 you could buy 3 pair of socks and 2 boxes of handkerchiefs.

a. Form a pair of simultaneous equations for this information.

b. Solve your equations to find the cost of a pair of socks.

6. Tom decides to sell flowers and vegetables at the market, so he makes two rectangular gardens. The flower garden is 3 metres longer than it is wide. The length of the vegetable garden is 4 metres more than double its width. The combined perimeters of the two gardens is 126 metres.

Tom decides to fence both gardens to keep out his pigs. The fencing for the flower garden costs K4 per metre and the fencing for the vegetable garden costs K5 per metre. The total fencing cost for the two gardens is K584.

Find the dimensions of each garden.

Unit 11.1 Number and Application

Topic 15: Basic algebra—inequalities

This Topic covers the last bullet point listed under 'Basic algebra' (see p. 14 of the Grade 11 syllabus):

- Solve inequality and plot on number line or plane.

Inequalities

Solving linear inequations

A linear **equation** has the form $mx + c = 0$ where m and c are constants.

A linear **inequation** has a similar form but the equals sign (=) is replaced by one of the inequality signs:

- $>$, greater than.
- $\geq$, greater than or equal to.
- $<$, less than.
- $\leq$, less than or equal to.

The process for solving linear inequations is the same as that for solving linear equations except for the following rule:

The inequality sign is reversed when multiplying or dividing an inequation by a negative number

An inequality can be graphed on a number line or written in interval notation.

For example : $x \geq 3.5$.

On a number line:

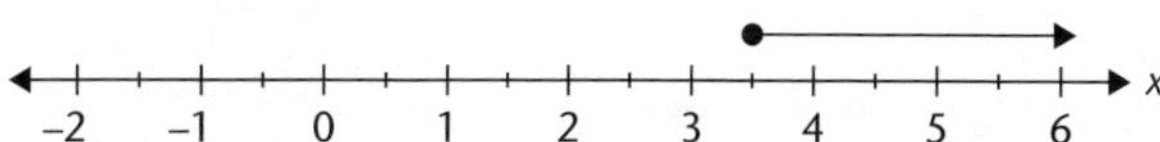

● means the value is included and ○ means the value is not included

This inequality can be expressed in **interval notation** as $[3.5, \infty)$ where the square bracket, [, means the value is included and the round bracket,), means the value is not included.

Note: Round brackets are always used with ∞ and $-\infty$.

Example A

Q. For each of the following inequations:

1. $3x + 2 \leq 14$

2. $4x - 3 > 5x - 7$

i. Solve for x.

ii. Show the solutions on a number line.

iii. Write the answer in interval notation.

A. 1. i. $3x + 2 \leq 14$: subtract 2 from both sides

$3x \leq 12$: divide both sides by 3

$x \leq 4$

ii. $x \leq 4$

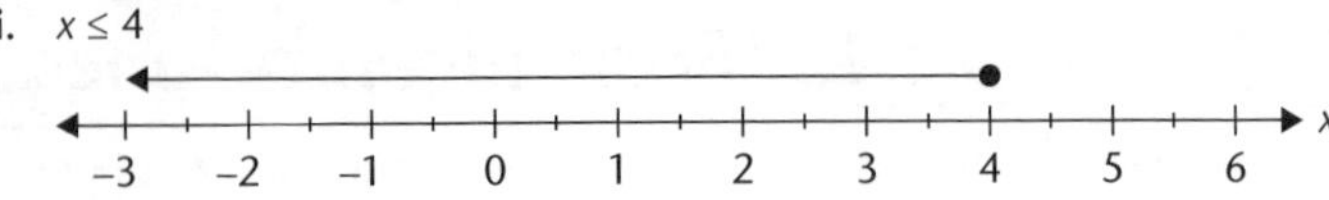

iii. $(-\infty, 4]$

2. i. $4x - 3 > 6x - 7$: subtract $6x$ from both sides

$-2x - 3 > -7$: add 3 to both sides

$-2x > 4$: divide both sides by -2; we need to reverse the inequality sign

$x < 2$

ii.  $x < 2$

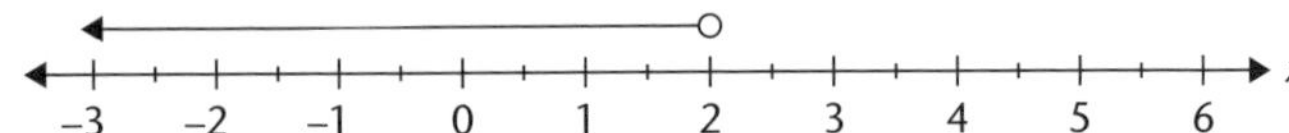

iii. $(-\infty, 2)$

Unit 11.1 Activity 15A: Inequations

1. For each of the following inequations:
   - **i.** Solve for x.
   - **ii.** Show the solutions on a number line.
   - **iii.** Write the answer in interval notation.

   a. $5x \leq 10$ **b.** $-7 > 6x + 5$ **c.** $9 \geq 5 - x$

   d. $\frac{x}{7} \geq \frac{3}{14}$ **e.** $\frac{x}{3} + 4 > 1.5$ **f.** $15 - 2x < 8$

2. Solve for x, giving your answer in interval notation.

   a. $3x + 5 > 0$ **b.** $-3x \geq 9$ **c.** $\frac{x}{6} \leq \frac{7}{2}$ **d.** $5 - 6x \leq -7$

   e. $\frac{2x+1}{2} \geq 5$ **f.** $\frac{x-3}{-2} < 7$ **g.** $\frac{4-x}{5} \geq -2$ **h.** $\frac{3-2x}{2} \leq -1$

3. Solve for a.

   a. $3a - 2 > a - 5$ **b.** $2a - 3 > 5a - 7$ **c.** $5 - 2a \geq a + 4$

   d. $7 - 3a \leq 5 - a$ **e.** $3(a-1) > a + 2$ **f.** $3(a + 2) > 4 - a$

   g. $4(a + 1) < 3(2 + a)$ **h.** $3 + 2(a - 5) \leq 5 - 3(a + 1)$

4. Solve for b.

   a. $5 + \frac{b+3}{2} > 1$ **b.** $3 - \frac{3-b}{4} \leq -1$ **c.** $\frac{b}{2} - \frac{b}{6} < 4$

   d. $\frac{b+2}{3} + \frac{b-3}{4} \leq 1$ **e.** $\frac{1-b}{2} + \frac{b+2}{3} \leq 1$ **f.** $\frac{b+1}{3} - \frac{b}{6} \geq \frac{2b-3}{2}$

Regions

On a set of Cartesian axes (the Cartesian plane) a **linear equation** defines a set of points that are in a straight line. A **linear inequation** defines a **region** of the Cartesian plane.

Example B

Q. 1. For each of the following, graph the region defined by the inequation

a. $x \geq 2$ **b.** $x + y > 3$

c. $2x + 3y \geq 6$

2. Find the equation of the boundary line and hence state the inequality that defines the following shaded region.

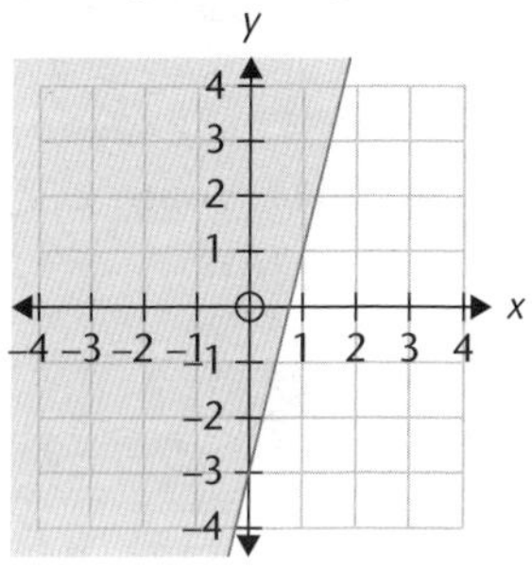

A. 1. a. We need to graph all the points that have an x-coordinate that is **greater than or equal to 2.**

The points that have an x-coordinate **equal to 2** will be the points on the straight line $x = 2$ and this line forms a boundary for the region.

The points that have an x-coordinate greater than 2 will be all the points to the right of the line $x = 2$, shaded opposite.

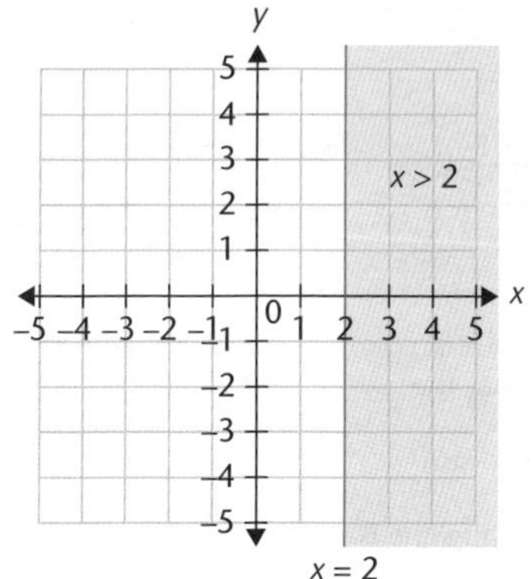

The region defined by $x \geq 2$ includes all points on the line $x = 2$ and the shaded region $x > 2$.

b. Two points on the line $x + y = 3$ are (0, 3) and (3, 0).

The straight line $x + y = 3$ can be drawn through these two points.

For the region $x + y > 3$ the points on the line $x + y = 3$ are **not** included so we draw this line as a dotted line.

Testing that the origin is in the region: substituting (0, 0) into $x + y > 3$, gives $0 + 0 > 3$ which is **not** true. The region is the side of the line not containing the origin.

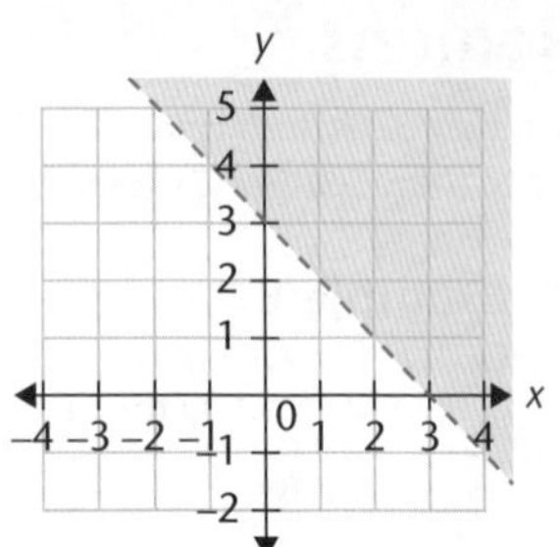

c. To graph the inequality $2x + 3y \leq 6$ we first graph the equality part as the boundary; the straight line $2x + 3y = 6$.

The line $2x + 3y = 6$ has intercepts with the axes $x = 3$ and $y = 2$.

One side of the line $2x + 3y = 6$ will contain all the points where $2x + 3y < 6$ and the other .

$2x + 3y > 6$.

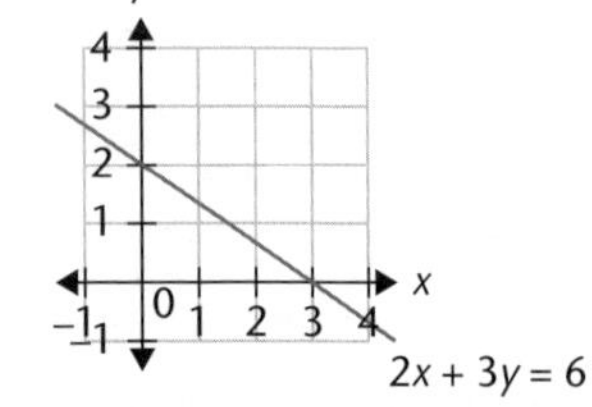

To find which side of this line is the 'less than' side we choose a point that is not on the line and test the inequation with the x and y values.

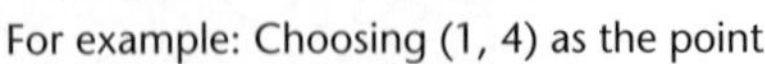

For example: Choosing (1, 4) as the point.

Testing: substituting in $2x + 3y$ gives $2 \times 1 + 3 \times 4 = 14$. Is this less than 6?

Note. So the point (1, 4) is not on the side of the line that includes all the points $2x + 3y < 6$.

The side required will be the other side of the line.

The region $2x + 3y \leq 6$ is the shaded region and the line.

Note: If the boundary line of an inequation does not go through the origin then it is usually convenient to choose the point (0, 0) as the testing point.

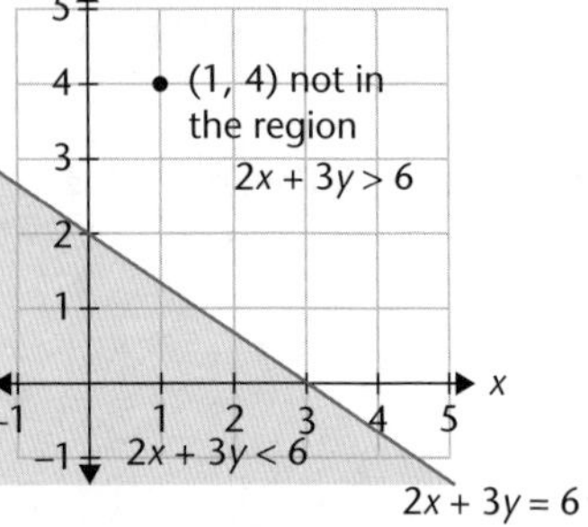

2. Looking at the boundary line of the shaded region we can see that it goes through the points (0, –3) and (1, 1). This means that the

line has a gradient of $\dfrac{1-(-3)}{1-0} = \dfrac{4}{1} = 4$ and

the y-intercept is –3 so the equation of the line is $y = 4x–3$.

The origin (0, 0) is in the region so testing $y \leq 4x–3$ gives $0 \leq 4 \times 0–3$ or $0 \leq –3$ which is **not true**. So the region must be $y \geq 4x–3$.

Unit 11.1 Activity 15B: Regions

1. Sketch the region defined by each of the following inequalities:

a. $x \geq 0$ **b.** $y < -2$

c. $x + y \leq 5$ **d.** $2x + y \geq 6$

e. $4x - 3y \leq 24$ **f.** $y \geq x$

g. $y \leq 5x - 3$

2. Find the equation of the boundary line and hence state the inequality that defines each of the following shaded regions:

a.

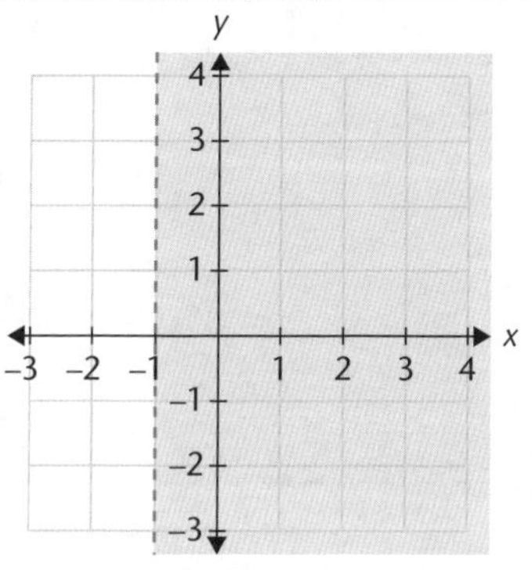

b.

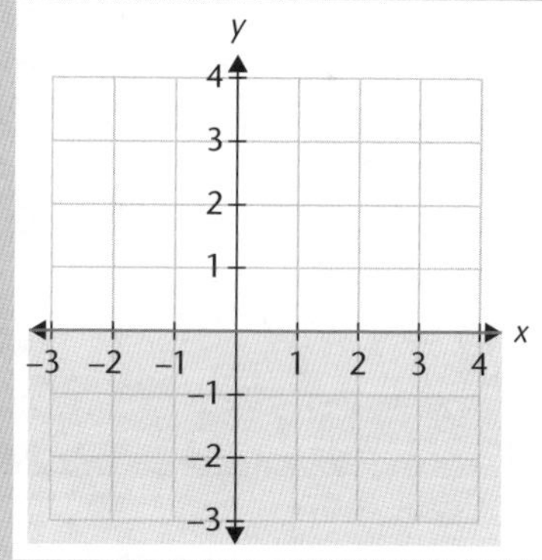

c.

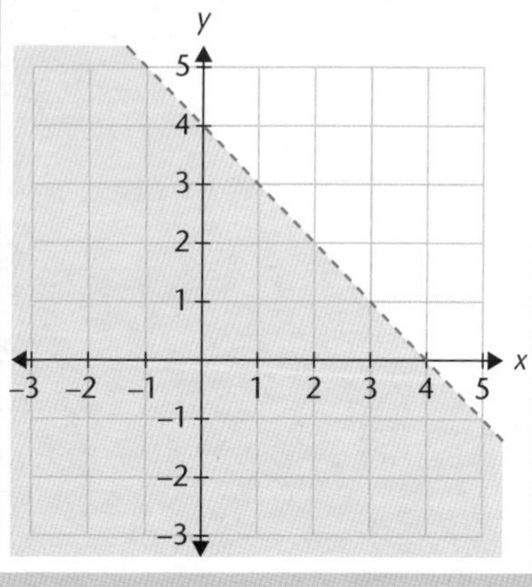

d.

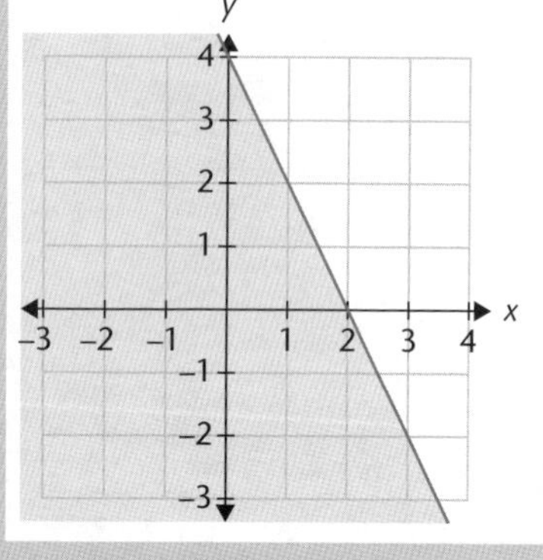

e.

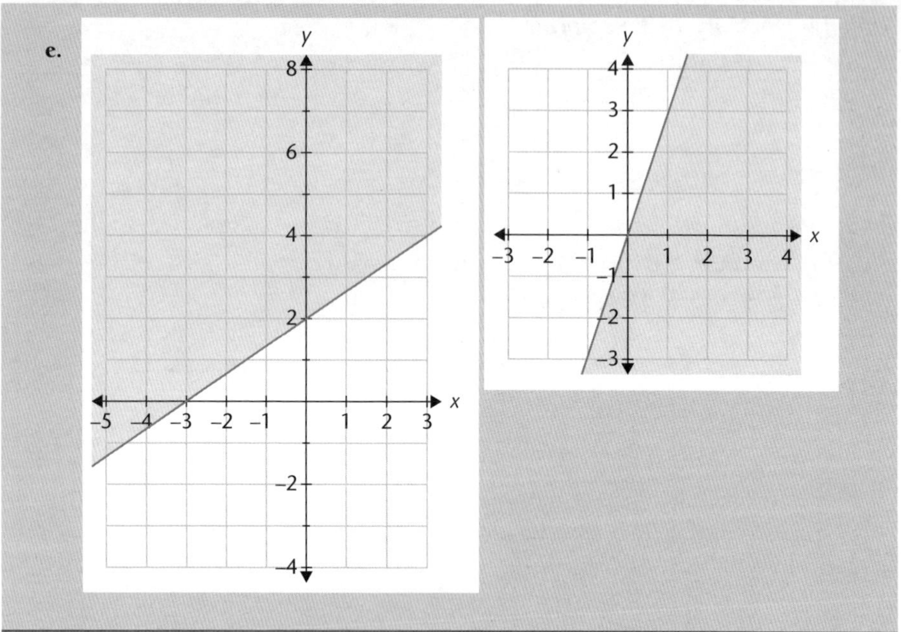

Unit 11.2 Managing Money 1
Topic 1: Working with percentages

Unit 11.2 Managing Money focuses on the mathematics that deals with money – spending, earning and lending – with a particular emphasis on the development of the applying mathematical skills. An ability to work with percentages is an essential prerequisite. Topic 1 focuses on working with percentages in a real-life context. It covers:

- Percentage increase or decrease.
- Expressing one quantity as a percentage of another.
- Finding the original quantity given a percentage change.
- Devising a strategy involving multiple and sequential calculations to solve a number problem.

Introduction

Percentages are so much a part of everyday life, that it is often overlooked that they are simply fractions with a denominator of 100. Thus, a percentage such as 51% means the fraction $\frac{51}{100}$ or the decimal 0.51, etc.

Since percentages are fractions with a common denominator of 100, percentages are very useful for making comparisons, eg a 10% increase in a person's rates is clearly going to be less than a 15% increase in that person's rates.

Using percentages

Many processes with percentages should by now be familiar. Some of these are summarised below.

- **To change a fraction or a decimal to a percentage**, multiply by 100.
  For example, $\frac{1}{2} = \frac{1}{2} \times 100\% = 50\%$ $\quad$ $0.17 = 0.17 \times 100\% = 17\%$
- **To change a percentage to a fraction or a decimal**, divide by 100.
  For example, $47\% = 47 \div 100 = \frac{47}{100}$ $\quad$ $95\% = 95 \div 100 = 0.95$
- **To find a percentage of a quantity**, multiply the percentage by the quantity.
  For example, to find w% of \$150 the calculation is $0.25 \times 150 = \$37.50$.

Note: It is useful to learn certain common conversions such as:

$$\frac{1}{20} = 5\% \quad \frac{1}{10} = 10\% \quad \frac{1}{5} = 20\% \quad \frac{1}{4} = 25\% \quad \frac{1}{2} = 50\%$$
$$\frac{3}{4} = 75\% \quad \frac{1}{3} = 33\frac{1}{3}\% \quad \frac{2}{3} = 66\frac{2}{3}\%.$$

One quantity as a percentage of another

To express one quantity as a percentage of a second quantity:

- Express the first quantity as a fraction of the second.
- Then change the fraction to a percentage.

The *same* unit must be used for both quantities.

Example A

Q. Express 1. K45 as a percentage of K112.50

2. 15 minutes as a percentage of 2 hours.

A. 1. The fraction is $\frac{45}{112.5}$. Changing this to a percentage gives

$\frac{45}{112.5} = \frac{45}{112.5} \times 100\% = 40\%$ [multiplying by 100 to change to a %]

So K45 is 40% of K112.50

2. The fraction is $\frac{15}{120}$ [since there are 120 minutes in 2 hours]

and $\frac{15}{120} = \frac{15}{120} \times 100\% = 12\frac{1}{2}\%$ [changing the fraction to a percentage]

So 15 minutes is $12\frac{1}{2}$ % of 2 hours.

When calculating the percentage by which a quantity has increased or decreased, the comparison is always made with the *original* amount.

Example B

Q. A school's roll dropped from 1 240 in 2011 to 1 124 in 2012. What was the percentage decrease in the roll?

A. The decrease in the roll is: 1 240 – 1 124 = 116.

The percentage decrease in the roll is:

$$\frac{\text{decrease in roll}}{\text{original roll}} \times 100\% = \frac{116}{1\,240} \times \frac{100}{1}\%$$

$$= 9.4\% \text{ (1 dp)}$$

Once again, ensure that the same units are being used in the comparison.

Example C

Q. Kerrie's pocket money of K1.50 is increased by 40t. What percentage increase is this?

A. Increase as a fraction = $\frac{40}{150}$ [increase divided by original amount (in cents)]

Increase as a percentage = $\frac{40}{150} \times 100\%$ [multiplying by 100]

= 26.666...%

= 27% increase (to the nearest unit)

Increasing or decreasing a quantity by a percentage

A common technique for increasing (or decreasing) a quantity by a percentage is to find the percentage amount, then add it to (or subtract it from) the original quantity.

Example D

Q. 1. Increase K140 by 15%. 2. Decrease 150 by 20%.

A. 1. 15% of K140 is 0.15 x 140 = K21

K140 is to be increased by K21.

∴ final amount = 140 + 21 = K161 [final amount = original amount + increase]

2. 20% of 150 is 0.2 x 150 = 30

150 is to be decreased by 30.

∴ final amount = 150 – 30 = 120 [final amount = original amount – decrease]

An alternative technique for calculating percentage increases and decreases is now shown. This is a quick and versatile method that is worth practising.

Example E

Q. 1. Increase K140 by 15% **2.** Decrease 150 by 20%

A. 1. final amount = original amount + increase

= 100% of K140 + 15% of K140

= 115% of K140 [since 100% + 15% = 115%]

= 1.15 x K140 = K161 [changing 115% to 1.15]

2. final amount = original amount – decrease

= 100% of 150 – 20% of 150

= 80% of 150 [since 100% – 20% = 80%]

= 0.8 x 150 = 120

Pricing is a common application of percentage increases and decreases. Prices may be

- **Marked up** by a certain percentage, eg to include sales tax or profit, or
- **Discounted** by a certain percentage, eg reduced in a sale.

Example F

Q. A shop adds a mark-up of 16% to the cost price of goods.

1. What would something costing K48 be marked up to?

The shop has a pre-Christmas sale in which all prices are discounted by 12%.

2. What would the sale price of a K50 item be?

A. 1. final amount = (100 + 16)% of 48 [original amount + increase]

= 1.16 x 48 [changing 116% to a decimal]

= K55.68

2. final amount = (100 – 12)% of 50 [original amount – decrease]

= 0.88 x 50 [changing 88% to a decimal]

= K44.00

Unit 11.2 Activity 1A: Calculations with percentages

1. Calculate each of the following:

   a. 20% of 60 **b.** 15% of 120 **c.** 78% of 260 **d.** 120% of 240

   e. 0.5% of 126 **f.** 3% of 240 **g.** $12\frac{1}{2}$% of 96 **h.** $87\frac{1}{2}$% of K248

   i. $33\frac{1}{3}$% of 81 **j.** $66\frac{2}{3}$% of K180

2. Increase:

   a. 80 by 20% **b.** 60 by 18% **c.** 400 by 16% **d.** K84 by 15%

   e. K14.50 by 30% **f.** K126 by $12\frac{1}{2}$%

3. Decrease:

   a. 80 by 10% **b.** 120 by 15% **c.** 240 by 25% **d.** K76 by 12%

   e. K12.60 by 5% **f.** K69.95 by $12\frac{1}{2}$%

4. Write the first quantity as a percentage of the second (give answers to the nearest whole number):

   a. 18 out of 24 **b.** 48 out of 150 **c.** K32 out of K80

   d. 49 L out of 76 L **e.** K26.50 out of K48 **f.** 15 minutes out of $2\frac{1}{4}$ hours

   g. 50c out of K40 **h.** 98 cm out of 5 m

5. Write each of the following changes as a percentage increase or decrease:

   a. from 60 to 75 **b.** from 160 to 240 **c.** from 38 to 49.4

   d. from 96 to 84 **e.** from 24 to 18 **f.** from 36.4 to 30.94

   g. from K15.25 to K17.69 **h.** from K88.56 to K77.49

6. In an electorate, 8 360 of the 38 000 electors did *not* vote.

   a. What percentage of the electors did vote?

   b. Of the votes cast, 45% went to Abel. How many votes did Abel receive?

   c. If Bisaun received 12 402 votes, what percentage of the votes cast did Vigil, the only other candidate, receive? (Give answer to the nearest per cent.)

7. Mark used to live 2 kilometres from school. When the family shifted house his new distance from school was $3\frac{1}{2}$ kilometres. What percentage increase is this?

8. When Feinga was five he was 1.10 m tall. At six he was 1.18 m tall. What percentage increase was this?

9. A DVD player sells for K295. The shop has a sale and the price is discounted by 25%. What is the selling price?

10. A new road reduces the distance between John's house and his work by 2.5%. If John used to travel 24 km to work, how long is his journey now?

Finding original quantities given a percentage change

Suppose a quantity has been increased by a certain percentage to a new value. It may now be required to find the original quantity, before the increase was made. To do this you need to work out what percentage of the original quantity is represented in the increased value given.

Example G

Q. A property increases in value by 25%. If it is now worth K375 000, what was its previous value?

A. New value = 100% of old value + 25% of old value

$\therefore$ new value = 1.25 x old value [new quantity represents 125% of original quantity]

375 000 = 1.25 x old value [substituting new value = 375 000]

$\therefore$ old value $= \dfrac{375\ 000}{125}$ [dividing by 1.25]

= K300 000

Finding an original quantity after a percentage decrease is done in a similar way.

Example H

Q. A car dealer offered 30% off the price of a car. If the reduced price is K3 640, what was the previous price?

A. Let the original price be x.

Reduced price = 100% of x – 30% of x

$= 0.7x$ [reduced price represents 70% of the previous price]

$3\ 640 = 0.7x$

$x = \dfrac{3\ 640}{2.7}$ [dividing by 0.7 and swapping sides]

= K5 200

The previous price was K5 200.

Unit 11.1 Activity 1B: Finding original quantities

1. Find the original price when:

a. a 10% increase results in K55

b. a 20% increase results in K18

c. a 15% discount results in K102

d. a 20% discount results in K14.76

e. an 18% increase results in K28.91

f. a 30% discount results in K417.20

g. a $12\frac{1}{2}$% increase results in K58.95

h. a $12\frac{1}{2}$% discount results in K94.71

2. An article in a shop is marked to sell at K156. If this represents a profit of 30%, what did the article cost the shopkeeper?

3. In a '35% off' sale, Kate buys a skirt for K54.60. What was the price of the skirt before the sale?
4. The material Miwako buys to make curtains loses 5% of its length when washed. If Miwako needs 33.25 m after washing, how much material should she buy?
5. Jason paid K104.50 for goods which include a 10% late payment fee. What was the price of the goods before the fee was added?
6. If a house is sold for K187 000 there is a loss of 15% on the cost price. For how much should it be sold to make a gain of 15% on the cost price?
7. A shopkeeper adds a total of 40% to the cost price of goods to determine the selling price of the goods. She decides to have a sale in which she offers 35% off the selling price of all goods in her store.
   - **a.** Will she make a loss on the goods in her shop (ie will her selling price be below her cost price) or will she still make a profit?
   - **b.** What percentage profit or loss will she make?

Unit 11.2 Managing Money 1
Topic 2: Earning and spending

'Earning and spending' is the first of two sub-sections in Unit 11.2 of the Grade 11 syllabus. Topic 2 deals with:

- Calculating wages and salaries, casual and piece work.
- Calculating commission and overtime.

Taxation and foreign exchange are covered in the next two Topics.

Earning money

Before people can spend money, they must earn it. So how is money earned? Money is earned by providing services, selling goods and working for an employer.

Some ways a person might receive an income are from **wages**, from a fixed **salary** or from doing **casual** or **piece work**.

Wages

A person working for **wages** would be someone who works in private industry, such as a shop assistant, a construction worker or a hotel worker. A position where wages are paid is usually ongoing and workers are guaranteed a set number of hours of work each week.

These workers can be paid a set amount for each hour that they work, and overtime is paid if they work extra hours in a paying period. Overtime is usually paid at a higher rate per hour, often quoted as $1\frac{1}{2} \times$ *hourly rate* (time-and-a-half) or $2 \times$ *hourly rate* (double time)

Alternatively, a worker in sales may receive a retainer each pay period, and in addition receive a percentage, in commission, of the sale price of the products that they sell. The retainer can be a set amount for a set number of hours worked or it can be given as a rate per hour for the number of hours worked.

Example A

Q. 1. Tomitom works a 40-hour week from Monday to Friday and he is paid fortnightly at a rate of K6.25/hour. How much does he earn in a fortnight?

A. 1. Rate = K6.25/hour   Fortnightly pay = $2 \times 40 \times 6.25 = 500$

Tomitom earns K500 a fortnight.

Q. 2. Meki worked as a trainee storeman for a heavy equipment spare parts company at an hourly rate of K3.25 for a 40 hour week. The supervisor asked him to stay back an extra 3 hours per day every day in a fortnight. Overtime is paid at double the normal rate. How much did Meki earn in that fortnight?

A. 2. Normal pay = $2 \times 40 \times 3.25 = 260$ Overtime pay = $3 \times 10 \times (3.25 \times 2) = 195$

Total pay = $260 + 195 = 455$

Meki earned a total of K455 in that fortnight.

Q. 3. Sam sells trucks and receives a base pay rate of K6.50/hour, plus 2% commission of the selling price of the trucks that he sells. In one fortnight he works 60 hours and sells one truck for K32 500. How much does he earn in that fortnight?

A. 3. Normal pay – $60 \times 6.25 = 375$ Commission = $\frac{2}{100} \times 32\,500 = 650$

Total pay = $375 + 650 = 1\,025$

Sam earns a total of K1 025 in that fortnight.

Unit 11.2 Activity 2A: Calculating wages

1. The cashiers at a supermarket are paid K4.50 an hour. How much would the following cashiers get in a fortnight if they worked the hours shown?
   - **a.** Priscilla: 7 hours
   - **b.** Jolene: 10 hours
   - **c.** Betty: 23.5 hours
   - **d.** Kuri-Ezi: 75.5 hours
2. Gabi is paid at K7.25 an hour for the working hours from 8 am to 5 pm with one hour lunch break in a day. She is paid double time if she works on Saturday and Sunday. In a fortnight Gabi works every weekday plus one Saturday and two Sundays. How much does she earn altogether in this fortnight?
3. Complete the table below by filling in the blank cells. The usual working week is 40 hours and any hours over 40 are considered overtime and are paid at 1.5 times the normal hourly rate.

Total Hours Worked	Hourly Rate (K)	Regular Wages(K)	Overtime Wages(K)
30	6.50		
50	12.00		
40		0	600.00
25	8.00		
35		306.25	

4. Poli operates a backhoe for 5 hours a day for a building contractor and is paid K50 in a working day. If he is paid at time-and-a-half on a Sunday, how much does he get for 6 days of work from Sunday to Friday?
5. Nella works an 8-hour day at a set rate but receives pay at time-and-a-half on Saturday and double time on Sunday.

   In a particular fortnight she can work the following days. Which option gives the most pay?
   - **A.** Two Mondays, one Wednesday and two Saturdays.
   - **B.** Two Tuesdays, one Thursday, one Saturday and one Sunday.
   - **C.** Two Saturdays and two Sundays
   - **D.** One Wednesday, one Saturday and two Sundays
6. An insurance sales person earns K300 per week plus 5% commission on sales of policies. If he sells K3 000 worth of insurance policies in a week, how much will he earn in that week?
7. Boni and Tolo wanted to build a house at Kuruvina and got quotes from two local timber millers. Their fixed costs, like fuel and oil for the chainsaws, were the same, but Miller 1 quoted his labour costs at K30 an hour for 7 hours a day for 4 days. Miller 2 quoted his labour cost as K150 per day for 4 days.

   Help Boni and Tolo decide on the cheaper quote.

8. Ana works at a women's fashion clothing store. She is paid K8.00 an hour plus a 5 percent commission on what she sells. She usually sells around K1 000 worth of clothing in a day and gets paid every 2 weeks. She works 5 hours a day for 3 days a week.

   a. How much is she earning in a fortnight?

   b. What is her commission in a fortnight?

9. Petrus works at a furniture store. He is paid a K220 retainer per week, plus a 3% commission on sales over K3 000. In a particular week he sold K6 250 worth of furniture.

   a. How much commission did he make in that week?

   b. How much was his pay in this week?

10. A salesman at Kawage Motors is on a retainer of K250 per week, plus 4% commission on any sales he makes in that time. If he manages to sell two reconditioned Honda CRVs, at K26 800 and K24 300 in a fortnight, what will be his total pay for the fortnight?

11. Taasi earns 7% commission on the first K7 000 worth of items sold in a month and 11% commission on all sales over that. What are her gross earnings for K38 431 in sales in a month?

12. If Joe earned a commission of K2 500 on sales of K40 000, find his rate of commission.

13. Mouvo gets a retainer of K350 per week and the remainder of his earnings come from commission on the sales that he makes. In a particular week he sold K24 400 worth of goods and he received K1 936 in pay. What is his rate of commission?

Salary

A person who is receiving a **salary** would be someone who, for example, works for the government, such as a teacher, a nurse or a postal worker. Their salary is usually expressed as a yearly amount but they are paid weekly, fortnightly or monthly, for the whole year including their holidays. Government tax, superannuation, and some subsidised allowances can be subtracted from the regular amount that they are paid. If work is done out of the regular working hours then a salary worker will not usually be paid for this work; for example, a teacher might do some correction at home in the evening but they are not paid extra for this work.

Example B

Q. Luke earns K34 600 per year. How much does he earn:

a. per month? b. per fortnight?

A. a. Assuming that there are 12 monthly payments per year:

$\frac{37\,440}{12} = 3\,120$ Luke earns K3 120 per month

b. Assuming that there are 26 fortnightly payments per year:

$\frac{37\,440}{26} = 1\,440$ Luke earns K1 440 per fortnight

Q. Sioni earns K45 600 per year base salary. Each fortnight 6 % of Sioni's fortnightly base salary is paid into a superannuation fund and then he has K86 for housing and K90 for tax subtracted from the remaining amount. How much does Sioni take home each fortnight?

A. Fortnightly gross pay $\frac{45\ 600}{26} = 1\ 753.85$

Deductions:

Superannuation : 6% of K1 753.85 = $\frac{6}{100} \times 1\ 753.85 = 105.23$

Housing : 86.00

Tax : 90.00

Total deductions: K281.23

Fortnightly take home pay = 1 753.85 – 281.62 = 1 472.62

Sioni takes home K1 472.62 each fortnight.

Unit 11.2 Activity 2B: Calculating salaries

1. Taraisi is paid an annual salary of K36 800. How much does he earn in:

a. a month? **b.** a fortnight?

2. The PNG Parliament in 2011 voted itself a 52% pay rise. How much would an MP who is currently getting K21 052.63 get after the pay rise?

3. Wari works for an LNG-related company where he earns K895 per week. He is paid an annual child allowance of K500 per child for his two children and K650 for his spouse. Calculate Wari's:

a. total payments for the year

b. take home pay each week

4. An academic staff member at the University of Papua New Guinea is paid an annual base salary of K32 000. She is paid for an equivalent of 73.5 hours per fortnight. She is also paid a domestic market allowance (DMA) of K11 000 per annum and an academic teaching allowance (ATA) at 20% of the base salary.

If she is paid fortnightly, compute the following;

a. the hourly rate on the base salary

b. the amount of DMA in a fortnight

c. the amount of ATA in a fortnight

d. her gross pay per fortnight.

5. Natalie, who had a salary of K32 000 per annum, was given a raise of K1 000 a year.

a. What was the percentage increase in salary for Natalie?

b. By what amount did Natalie's fortnightly pay increase?

c. An employee in the same company, earning K23 450, was given the same percentage salary increase. What was his increased fortnightly pay?

6. Gimot's salary increased from K400 a week to K424.
   a. By what percentage did Gimot's salary increase?
   b. What is the annual salary at this new rate?
7. James is working on a yearly salary and his take-home pay each fortnight is K1 486. Included in his fortnightly take-home pay is a housing allowance of K400 and K145 has been taken out in tax. What is James's annual gross salary?
8. Selina, a teacher, is to receive a 6% increase in salary in each of the next three years. If she has a present annual salary of K18 720:
   a. What will her annual salary be next year?
   b. What will her annual salary be after the three increases?
   c. How much extra will she get in her gross fortnightly pay after the three increases have been made?
9. Margaret has a part-time position. The salary for working full-time in this position is K32 578. If Margaret works six days in a fortnight instead of the full-time ten days, how much is her gross fortnightly salary, to the nearest kina?
10. Robert has been given a pay rise of 7.5% and his fortnightly pay is now K1 333. Calculate:
   a. his fortnightly pay before the increase
   b. the value of this pay rise

Casual and piece work

Many workers do not have permanent employment and rely on casual work or piece-work to earn money.

A person who does casual work could be someone who works as a driver, an office assistant, a gardener, a cleaner or a builder's laborer. Casual workers usually work for an agreed amount of time for a specified amount.

A piece-worker is paid per item that they process. Piece-workers do plantation picking and production of goods such as clothes or items that are sold at markets.

Example C

Q. Luther is a builder's labourer and he has been contracted to work 4 hours a day for 8 days at a rate of K7.40 per hour. How much in total will he earn in these 8 days?

A. Luther will work $4 \times 8 = 32$ hours in these eight days.
He will earn a total of $32 \times$ K7.40 = K236.80 in the eight days.

Q. Salome is paid K5.60 for each basket of coffee beans that she picks. Over the last five days she has picked 13, 15, 6, 12 and 11 baskets per day. How much in total will she be paid for these five days?

A. Salome has picked $13 + 15 + 6 + 12 + 11 = 57$ baskets of coffee beans.
She will be paid $57 \times$ K5.60 = K319.20 for the five days of picking.

Unit 11.2 Activity 2C: Calculating piece work and casual work

1. Kali varnishes chairs at a local furniture manufacturer's workshop. For each chair that he varnishes during the work day he receives 75 toea. During one week Kali varnished 825 chairs. How much did Kali earn that week?
2. Boni recently sold 40 kg of cocoa wet beans at K1.25 per kilo. How much was he paid?
3. Paraka owns a dump truck with which he delivers gravel and sand to customers by the truckload. If he is paid K75 for each load, how much would he earn if he delivers one hundred loads?
4. Ruby does casual work in an office and is paid K6.75 per hour if she works 3 or more hours in a day. She is paid K8.20 per hour if she works less than 3 hours in a day. In a particular fortnight she worked five days for 5, 2, 7, 3, and 1.5 hours. How much did she earn in that fortnight?
5. Joan works as a cleaner one day a week for each of three families. The first family pays her K80 for a 7-hour day, the second pays her K9.45 per hour for the hours 8 am to 4 pm, and the third family pays K6.40 per hour for the hours 12 noon to 4 pm then double time for the hours 4 pm to 8 pm.
   a. Which family pays the highest rate of pay?
   b. Which family pays the most in a day?
   c. How much will Joan make in a fortnight if she works the usual six days?
6. A woman is paid K2.00 for every bag of coconut that she packs up to 50 bags. She is paid a bonus of 30 toea for each bag that she packs over 50 bags. How much will she earn if she packs 120 bags?
7. Gari drives a PMV and charges K1.05 for a trip from his village to the nearest town. He charges K1.25 for the return trip. How much did he make in a day if he drove a total of 45 villagers into town and had a total of 39 return passengers?
8. Tomas works as a security guard. He charges K3.00 per hour for the hours 8 am to 4 pm, K3.60 for the hours 4 pm to 12 midnight and K6.40 for the hours 12 midnight to 8 am. In a particular week he worked the following times: 8 am to 3 pm, 2 pm to 10 pm, 8 pm to 6 am, 4 pm to 10 pm, 10 pm to 3 am, 12 noon to midnight. How much did he earn in this week?
9. A women's co-operative is contracted to make packing sacks for plant material. They are paid K3.55 for each sack up to the first 1 000 sacks in a week and are paid a bonus of 80 toea for each sack that they make over 1 000 in that week. In a particular week they make 1 645 sacks. How much do they earn in this week?

Unit 11.2 Managing Money 1

Topic 3: Taxation

Calculating taxable income, goods and services tax (VAT, GST), deductions, rebates, levies, superannuation, are all part of the 'Earning and spending' section of the Grade 11 syllabus (see Syllabus p. 15). This Topic covers:

- Income tax.
- Superannuation.
- Calculating taxable income.
- Calculating rebates.
- Goods and services tax (GST).

Taxation

Taxation is the means by which governments raise money to fund the services they provide to the citizens in their countries.

Some of the more familiar taxes are:

- **Income tax**, the tax paid by individuals according to their income. People who have a very low income will pay no tax. The percentage of their income that people pay as tax increases with the amount of income that they earn.
- **Goods and Services Tax (GST)**, also called Value Added Tax (VAT). This tax is added to all goods sold, and services provided, at the rate of 10%. Prices given in shops already have the GST added to them and it is the responsibility of the shop-owner to collect the tax and pay it to the government.
- **Company tax**, the amount that a business is taxed on the profit that they make.

Income tax

Taxation of salary and wage earners

Most **salary and wage earners** have income tax deducted from their pay each fortnight. The employer of a salary or wage earner calculates the amount of tax and sends it to the government.

There is a taxation table that employers use to calculate the tax payable (Table 1, below, shows extracts from the taxation table) or they can use an automatic calculator available on the Internal Revenue Commission (IRC) website.

The amount deducted depends on whether the employee is a resident or non-resident, whether he/she has completed a *Salary or Wages Declaration form*, and on the employee's number of dependents.

A **salary or wages declaration** provides the government with details of the employee's work, dependants and any extra benefits that they receive from their work.

Income tax return

An income tax return is a form that gives all the income that an employee has earned, including allowances and benefits, and outlines the deductions to which they are entitled. The Internal Revenue Commission determines the tax liability of the taxpayer from the information supplied in this form.

An employee who earns K100 or less in **other income** (eg bank interest) for the year, and whose tax is deducted from their fortnightly pay, does not need to submit an income tax return for that year.

Salary and wage earner allowances and benefits

Some employees have allowances or benefits paid for by their employer, in addition to their salary amount, and tax has to be paid on these allowances or benefits. However, certain benefits are exempted from taxation or taxed at less than their full value:

- Housing; the amount of taxation depends on the value of the housing and the area in which it is located.
- Motor vehicles: taxed at K125.00 per fortnight if the employee has unrestricted use and the fuel is provided and at K95.00 per fortnight without fuel.
- Meals: taxed at the exact cost of the meal except if they are mess-type meals, in which case K30.00 is added to the income per fortnight.
- School fees for employees' children are **exempt** from tax.
- Leave fares: one return fare for the taxpayer and his family to their home is exempt from tax. Additional fares are taxable.
- Payment of expenses for electricity, gas, telephone, domestic servants: these are taxable and the full value of the allowance should be added to the income for the fortnight.

Superannuation

Superannuation is the money that an employee puts aside from their income during their working life so that they will have some income to live on when they retire from work.

Superannuation contributions by an employee are not subject to taxation and are deducted from the gross income before the fortnightly taxation liability is calculated.

It is compulsory for employees to contribute 6% of their income to a superannuation fund.

Employers also have to pay a compulsory amount into a workers superannuation fund. At the moment this is 8.4% of the employee's gross wage or salary. Employers get a tax deduction for this contribution.

In general:

Taxable income = fortnightly salary or wage + allowances and benefits – deductions

The table below gives the tax payable on the fortnightly taxable income of an employee. It incorporates the deduction that is given for dependants.

Table 1: Tax payable per fortnight for Salary and Wage earners

Column 1		Column 2	Column 3	Column 4			
Taxable income per fortnight		Non resident tax payer	Where no declaration is lodged	Where a declaration is lodged			
Exceeding	Not exceeding			Number of Dependants			
				None	1	2	3 or more
:	:	:	:	:	:	:	:
277	279	61.38	117.18	0.46	0.00	0.00	0.00
279	281	61.82	118.02	0.90	0.00	0.00	0.00
281	283	62.26	118.86	1.34	0.00	0.00	0.00
283	285	62.70	119.70	1.78	0.05	0.00	0.00
285	287	63.14	120.54	2.22	0.49	0.00	0.00

:	:	:	:	:	:	:	:
327	329	72.38	138.18	11.46	9.73	8.57	7.42
329	331	72.82	139.02	11.90	10.11	8.92	7.73
331	333	73.26	139.86	12.34	10.49	9.25	8.02
:	:	:	:	:	:	:	:
373	375	82.50	138.18	11.46	9.73	8.57	7.42
375	377	82.94	139.02	11.90	10.11	8.92	7.73
377	379	83.38	139.86	12.34	10.49	9.25	8.02
:	:	:	:	:	:	:	:
491	493	108.46	207.06	47.54	40.41	35.65	30.90
493	495	108.90	207.90	47.98	40.78	35.98	31.19
495	497	109.34	208.74	48.42	41.15	36.31	31.47
497	499	109.78	209.58	48.86	41.53	36.64	31.76
499	501	110.22	210.42	49.30	41.90	36.97	32.04
501	503	111.10	212.10	50.18	42.65	37.63	32.62
:	:	:	:	:	:	:	:
547	549	120.78	230.58	59.86	50.88	44.89	38.91
549	551	121.22	231.42	60.30	51.25	45.22	39.19
551	553	121.66	232.26	60.74	51.63	45.55	39.48
:	:	:	:	:	:	:	:
563	565	124.30	237.30	63.38	53.87	47.53	41.20
565	567	124.74	238.14	63.82	54.24	47.86	41.48
567	569	125.18	238.98	64.26	54.62	48.19	41.77
:	:	:	:	:	:	:	:
599	601	132.22	252.42	71.30	60.60	53.47	46.34
601	603	132.66	253.26	71.74	60.98	53.80	46.63
603	605	133.10	254.10	72.18	61.35	54.13	46.92
:	:	:	:	:	:	:	:
617	619	136.18	259.98	75.26	63.97	56.44	48.92
619	621	136.62	260.82	75.70	64.34	56.77	49.20
621	623	137.06	261.66	76.14	64.72	57.10	49.49
:	:	:	:	:	:	:	:
651	653	143.44	273.84	82.71	70.30	62.03	53.76
653	655	143.88	274.68	83.15	70.68	62.36	54.05
655	657	144.32	275.52	83.59	71.05	62.69	54.33
:	:	:	:	:	:	:	:
705	707	156.59	296.52	95.25	80.96	71.44	61.91

(*continued*)

Table 1: Tax payable per fortnight for Salary and Wage earners (*continud*)

Column 1		Column 2	Column 3	Column 4			
Taxable income per fortnight		Non resident tax payer	Where no declaration is lodged	Where a declaration is lodged			
Exceeding	Not exceeding			Number of Dependants			
				None	1	2	3 or more
741	743	167.39	311.64	106.05	90.14	79.54	68.93
743	745	167.99	312.48	106.65	90.65	79.99	69.32
745	747	168.59	313.32	107.25	91.16	80.44	69.71
:	:	:	:	:	:	:	:
757	759	172.19	318.36	110.85	94.22	83.14	72.05
759	761	172.79	319.20	111.45	94.73	83.59	72.44
761	763	173.39	320.04	112.05	95.24	84.04	72.83
:	:	:	:	:	:	:	:
791	793	182.52	333.06	120.98	103.67	92.13	80.59
793	795	183.12	333.90	121.58	104.27	92.73	81.19
795	797	183.72	334.74	122.18	104.87	93.33	81.79
797	799	184.32	335.58	122.78	105.47	93.93	82.39
799	800	184.62	336.00	123.08	105.77	94.23	82.69

Example A

Q. John has a taxable income of K500 per fortnight, has made a declaration, and has a dependent wife and one child.

a. Use Table 1 to calculate the amount of tax that will be deducted from John's pay each fortnight.

b. How much will John take home each fortnight.

c. What percentage of his fortnightly wage is paid in tax?

A. a. John has made a declaration and has two dependants. Moving across the table from the K499–K501 category (marked on the table) to the second-last column gives the amount of tax as K36.97.

b. John will take home K500 – K36.97 = K463.03.

c. The percentage of his wage that is paid as tax = $\frac{36.97}{500} \times \frac{100}{1} = 7.4\%$ (to 1 d. p.).

Q. Sarah is a non-resident worker who earns K605 per fortnight. She is provided with housing, valued at K100 a fortnight, and a car. She buys her own fuel for the car and contributes K40 to superannuation per fortnight.

a. Calculate the amount that Sarah is taxed on each fortnight, ie her taxable income.

b. How much tax is deducted from her fortnightly pay?

c. What percentage of her fortnightly taxable income is paid in tax?

A. a. Sarah is taxed on her fortnightly wage plus amounts for her benefits minus the superannuation contribution.

Wages	K605.00
Benefits:	
Car	+K95.00
Housing	+K100.00
Total wages and benefits	K800.00
Deductions:	
Superannuation	–K40.00
Taxable income	K760.00

b. From the table: Tax on K760.00 for a non-resident is K172.79.

c. The percentage of her income that is paid as tax $= \frac{172.79}{760} \times \frac{100}{1} = 22.7\%$ (to 1 d. p.).

Unit 11.2 Activity 3A: Calculating taxable income, superannuation and allowances

1. Use Table 1 to complete the table below. Assume the employees are all resident employees and that the fortnightly taxable pay is the amount after they have made their superannuation contribution.

Employee	Fortnightly taxable income	Declaration?	No. of Dependents	Taxation	Tax as a percentage of taxable income
Lore	K330	Yes	1		
Mary	K552	No	0		
Laki	K620	Yes	5		
Hiari	K800	No	3		

2. A cleaner in a factory earns K4.25/hour and works a 70-hour fortnight.

a. How much will the cleaner earn each fortnight?

b. How much will the cleaner contribute to superannuation? What is his fortnightly taxable income?

c. What amount of tax will be deducted from his fortnightly wage if he does not make a declaration?

d. How much tax will be deducted from his wages if he makes a declaration and has two dependants?

e. How much extra will he take home in a year if he makes a declaration?

3. Therese earns K346.00 per fortnight. She pays 6% of her wage into a superannuation fund and her employer pays 8.4% of her wage into the fund.

a. How much does Therese contribute to the fund each fortnight?

b. How much does her employer contribute to the fund each year?

4. Flora and Martin work in a factory. Flora has a taxable income of K284 a fortnight and Martin, who is a foreman, has a taxable income that is twice as much as Flora's per fortnight. Both employees have made a declaration and have no dependants.

a. Complete the following table:

Employee	Income	Tax/fortnight	Tax as a % of income
Flora			
Martin			

b. Is the amount of tax paid in the same proportion as their taxable incomes? Discuss your result.

5. Robert has worked his usual 78 hours, at a rate of K8.35/hour. In addition he has worked overtime of 4 hours at time-and-a half and 3 hours at double-time in the fortnight.
   a. Calculate Robert's fortnightly pay.
   b. Calculate the amount that Robert will be contributing to his superannuation fund.
   c. Robert has made a declaration and has 3 dependants. Find the amount of tax that Robert will pay on his taxable income in this fortnight.
   d. What percentage of his taxable income is the tax that he pays?
   e. Use Table 1 to calculate the tax that Robert would pay if he did not work any overtime in a fortnight.
   f. What percentage of his pre-tax wage would the tax from part d. represent?
6. Gabriel has two jobs. In his main job he has a taxable income of K496 per fortnight and he has made a tax declaration to this employer declaring his two dependents. In Gabriel's second job he has a taxable income of K278 per fortnight and he has not made a declaration with this employer.
   a. Use Table 1 to find the fortnightly tax in:
      i. Gabriel's main job.
      ii. Gabriel's second job.
   b. How much does Gabriel pay in tax per fortnight?
   c. What percentage of Gabriel's fortnightly combined taxable income is paid in tax?
   d. If Gabriel earned his taxable income in one job where he had made a tax declaration he would have to pay K86.90 in tax. How much income does he lose per fortnight by not being able to make a declaration in his second job?
7. Taki has an income, after he has contributed to superannuation, of K612 per fortnight and is provided with a car-and-fuel and a housing benefit valued at K55.00 per fortnight. Education fees of K300 per year for each of his three children are paid by his employer. He has made a tax declaration and has four dependants. How much tax is deducted from his wage each fortnight?
8. Lemape has an income of K308 per fortnight after his superannuation contribution is made. He is provided with housing valued at K40.00 per fortnight and mess-type meals. He has made a tax declaration but has no dependants.
   a. How much tax is deducted from Lemape's pay per fortnight?
   b. What is Lemape's take-home pay?

9. Bagua's income is K485 per fortnight after he has contributed to superannuation. His employer provides him with a car-and-fuel, a housing allowance to the value of K131 per fortnight and a K50 allowance for electricity.

Bagua is slow to fill in his tax declaration form.

a. How much tax is deducted from Bagua's pay per fortnight if he has not completed the declaration form.

b. What is Bagua's take-home pay if he has not completed the declaration form?

c. Bagua has a dependent wife and child.

What is Bagua's take home pay after he completes the tax declaration form?

Taxation on income earned in other situations

When income is earned by means other than salary and wages then the individual earning the income is required to submit a **tax return.**

An individual who would submit a tax return would be a person whose income comes from renting a property, a tradesman, a contract worker or someone who receives income from investments.

A tax return should also be submitted by a salary and wage earner who earns extra income that is more than K100.00.

Personal income tax rates

The personal income tax rates for residents and non-residents as defined by law in PNG in 2010 are given in Tables 1 and 2, respectively, below. These tax tables are updated every year.

Table 1: Tax Rates for Residents

Taxable income (K)	Cumulative tax up to maximum (K)	Tax rate of income in bracket (%)
0–7 000	0	
7 001–18 000	2 420	22
18 001–33 000	6 920	30
33 001–70 000	19 870	35
70 001–2 50 000	91 870	40
250 001+	No limit	42

Table 2: Tax Rates for Non-Residents

Taxable income (K)	Cumulative tax up to maximum (K)	Tax rate of income in bracket (%)
0–18 000	3 960	22
18 001–33 000	8 460	30
33 001–70 000	21 410	35
70 001–2 50 000	93 410	40
2 50 001 +	No limit	42

Allowances and benefits

Tax has to be paid on the same allowances and benefits as those for salary and wage earners. Taxpayers submitting a taxation return form must include these benefits as income.

Deductions

Allowable deductions from taxable income include work expenses, superannuation and school fees of dependants.

Rebates

A rebate is a deduction on the gross tax that is payable.

In Papua New Guinea in 2010 the dependant rebates for the first three dependants are as set out below:

Table 3

First dependant	15% of gross tax with a maximum of K450 and a minimum of K45
Second and third dependants	10% of gross tax with a maximum of K300 and a minimum of K30

Rebates to salary and wage earners

Allowable rebates on the taxation that salary and wage earners have paid can be obtained if they complete a taxation return (form).

A salary or wage earner can receive a tax rebate on his/her tax if they have expenses that exceed K200 associated with their employment. The amount of the rebate is 25% of the allowable expenditure in excess of K200.

Salary and wage earners can also receive a tax rebate for school fees paid for their dependants. The rebate is 25% of school fees paid with a maximum of K750 per child.

In general, to calculating tax payable:

> **Taxable income** = income + allowances and benefits – allowable deductions
>
> Taxation tables are used to calculate the **gross tax** based on the taxable income.
>
> **Tax payable** = gross tax – rebates

Example B

Q. Dorothy is a resident who has a taxable income of K21 456. She has two dependants.

a. Use Table 1 to calculate the tax that Dorothy is liable to pay on her taxable income.

b. Calculate the tax rebate that Dorothy will receive for her two dependants.

c. Calculate Dorothy's tax liability after rebates.

A. a. Using Table 1, Dorothy's income is in the K18 001 – K33 000 bracket. This means that she will have to pay the maximum tax for the previous bracket (K2 420) plus 22% of the income that is in excess of K18 000.

$$\text{Tax payable} = 2\,420 + \frac{22}{100} \times (21\,456 - 18000) = 2\,420 + 760.32 = 3\,180.32$$

Dorothy is liable to K3 180.32 in tax, before any rebates. This figure is her gross tax.

b. Dorothy has two dependants. For the first dependant she will receive a rebate of 15% of her gross tax with a minimum of K45 and a maximum of K450.

15% of 3 180.32 = $\frac{15}{100} \times 3180.32 = 477.05$ which is more than the maximum of K450.00.

So Dorothy receives a rebate of K450 for the first dependant.

For the second dependant she will receive a rebate of 10% of her gross tax with a minimum of K30 and a maximum of K300.

10% of 3 180.32 = $\frac{10}{100} \times 3180.32 = 318.03$ which is more than the maximum of K300.00.

So Dorothy receives a rebate of K300 for the second dependant.

c. Dorothy is liable to pay K3 180.32 – K450 – K300 = K2 430.32 in income tax.

Q. Kema is a paid a salary and has the required tax taken from his salary each fortnight. During 2010 he has incurred deductable work expenses of K480 and he has paid school fees of K2 300 for his first child and K900 for his second child.

Calculate the tax rebate that Kema can expect.

A. The rebate for work expenses is 25% of the expenses in excess of K200.

Kema will receive a work-related rebate of 25% of (K480 – K200).

Rebate $= \frac{25}{100} \times 280 = 70$ kina

The rebate for school fees is 25% of school fees paid with a maximum of K750 per child.

The rebate for the first child is 25% of K2 300 = K575.

The rebate for the second child is 25% of K900 = K225.

Kema can expect a total rebate of K(70 + 575 + 225) = K870.

Q. Larson has earned K44 588 in the year 2010 but has allowable work expenses of K5 045 and he has made a superannuation contribution of K2 675. He has a dependent wife and two dependent children. He has paid K1 066 and K975 for school fees for his children.

Calculate the tax that he has to pay for the year 2010.

A. Larson has no allowances and benefits and his deductions are the work related expenses, superannuation and school fees.

Using *Taxable income = income + allowances and benefits – allowable deductions*

Larson's taxable income = 44 588 + 0 – (5 045 + 2 675 + 1 066 + 975) = 34 827 (K).

Using Table 1 to calculate his gross tax: Larson's income is in the 33 001–70 000 bracket so he pays the cumulative tax for the previous brackets plus 35% for any income in excess of 33 000.

Gross tax = 6 920 + 35% of (34 827 – 33 000)

= 6 920 + 0.35 × 1 827

= 7 559.45

Larson's gross tax is K7 559.45 before any deductions.

Larson had deductions for his three dependants:

The deduction for the first dependant is 15% of the gross tax to a maximum of K450.

15% of K7 559.45 = 0.15 × 7 559.45 = K1 133.92: the deduction is the maximum of K450.

The deduction for the other two dependants is 10% of the gross tax to a maximum of K300.

10% of K7 559.45 is K755.95; the deduction is K300 for each of the 2nd and 3rd dependants.

Larson is liable to pay K7 559.45 – K450 – 2 × K300 = K6 509.45 in tax.

Unit 11.2 Activity 3B: Calculating tax payable

1. Using Table 1 (for residents), calculate the tax payable on the following taxable incomes:
   a. K6 945
   b. K78 550
   c. K43 850
   d. K115 000
   e. K18 990
2. For a taxpayer who completes a taxation return, classify the following quantities as either income or deductions:
   a. superannuation
   b. housing allowance
   c. dependants
   d. work-related costs
   e. school fees
   f. use of a car
   g. electricity allowance
   h. interest from savings
   i. commission
   j. meals supplied to the taxpayer.
3. Gabi is a resident who earned K55 600 in consultant fees in 2010. In addition she received K11 960 in income from a rental property that she owns. She has costs of K8 640 associated with her work and costs of K8 620 related to the rental property. She contributes K5 000 to her superannuation fund.
   a. Calculate Gabi's taxable income for the year.
   b. Calculate the amount of tax that Gabi will have to pay for 2010.
4. Marcus is a non-resident who is working as a consultant. He has earned K97 560 in fees for the year but has K18 650 in work-related costs. He has two dependants and pays school fees of K12 400 for his child.
   a. Calculate Marcus's taxable income.
   b. Use Table 2 to calculate Marcus's gross tax.
   c. Calculate Marcus's tax payable for 2010.

5. Ben is a non-resident, working as an IT consultant. He earns K126 680 from his consultancy and K3 560 interest from some investments. He has work-related expenses of K15 270, three dependants and school fees for his children of K8 360.

a. Calculate Ben's taxable income.

b. Use Table 2 to calculate Ben's gross tax.

c. Calculate Ben's tax payable for 2010.

6. Solomon is a resident and has five dependants. Solomon has the following information ready to complete his tax return:

Income from contracts	K68 340
Income from publications	K2 560
Income from seminars	K3 000
Rental income	K7 280
Work-related expenses	K8 635
Expenses related to rental income	K5 400
Housing allowance	K3 600
Electricity allowance	K400
Car and fuel allowance	K1 800
Meals provided	K764
Superannuation contribution	K5 000
School fees	K6 400

Using the tax tables for the year 2010 calculate Solomon's:

a. taxable income

b. gross tax

c. tax payable

Goods and services tax (value added tax)

A **Goods and Services Tax (GST),** also known as a value added tax (VAT), is a tax applied to the sale of goods and services in PNG, or the importation of goods into Papua New Guinea. The rate of GST is 10% of the value of goods sold, services provided or goods that are imported. The GST was introduced to PNG on 1 July 1999 and it replaced a complex system of sales tax and import duties.

GST is not applied to all goods and services. Some of the **exempt items** are listed below:

- Market goods.
- Supplies of medical services by medical doctors, hospitals, nurses or dentists.
- PMV and taxi fares.
- School fees.
- Retail sales of newspapers.
- Sale of postage stamps.
- Supply of financial services.

Applying the goods and services tax

Businesses that sell goods or services collect GST and pay it to the Internal Revenue Commission (IRC) in a monthly GST return. The amount that a business pays is reduced by the amount of GST that they have paid to the their supplier.

For example, a manufacturer of T-shirts charges a retailer K8.40 for each T-shirt. To the cost of the T-shirt the manufacturer must add GST of 10% which is K0.84. The total charge to the retailer is K8.40 + K0.84 = K9.24.

The retailer sells the T-shirts for K22.00 which includes the retailer's 10% GST of K2.00 (K20.00 + 10% of K20.00 = K22.00).

For each T-shirt sold the retailer would return to the IRC:

(GST collected from sales) – (GST paid to the manufacturer) = K2.00 – K0.84

= K1.16

Working with the GST

GST-inclusive price = retailer's price + GST

= retailer's price + 10% of retailer's price

so

$$\text{GST-inclusive price} = 1.10 \times \text{retailer's price}$$

Rearranging this formula gives:

$$\text{retailer's price} = \frac{\text{GST inclusive price}}{1.1}$$

To find the GST that has been applied to a selling price:

$$\text{GST} = \frac{\text{selling price}}{11}$$

Example C

Q. Maisy, a retailer, wants a return of K3.40 for her item for sale.

a. What is the GST-inclusive price for this item?

b. If Maisy already has a GST credit of K0.16 for this item, how much GST does she return to the IRC in her monthly return?

A. **a.** Using GST-inclusive price = 1.10 × retailer's price gives

GST inclusive price = 1.10 × K3.40 = K3.74

b. The GST added to the retailer's price is K0.34 so Maisy will return

K0.34 – K0.16 = K0.18 to the IRC.

Q. Frank bought a mobile phone for K239.95. How much GST did he pay?

A. Using $GST = \frac{selling\ price}{11}$

$$\text{GST} = \frac{239.95}{11} = 21.813$$

Frank has paid K21.81 in GST.

Unit 11.2 Activity 3C: Calculating GST

Give prices to the nearest toea.

1. Calculate the GST that a retailer will need to add to these GST-exclusive prices:
   a. A chair at K42.00
   b. A jacket at K36.25
   c. A ring at K2 350.00
2. Calculate the amount of GST that is included in these retail prices:
   a. K46.20
   b. K399.95
   c. K2 045.00
3. A couch costs K1 243.00 (GST inc.) for a retailer to buy from the manufacturer. The retailer adds K200.00 to this cost for profit, then adds GST.
   a. How much GST has been included in the manufacturer's price of K1 243.00?
   b. How much GST does the retailer add to his price?
   c. What is the retailer's selling price?
   d. How much GST does the retailer include in his monthly GST return?
4. A light fitting costs K27.61 (GST inc.) for a retailer to buy from the manufacturer. The retailer adds K15.00 to this cost, for profit, then adds GST.
   a. How much GST has been included in the manufacturer's price of K27.61?
   b. How much GST does the retailer add to his price?
   c. What is the retailer's selling price?
   d. If the retailer sell 6 of these light fittings in a month, how much GST in total does the retailer include in his monthly GST return?
5. A retailer has added 80%, for profit, to the cost price (GST inc.) of K504.00 for a television. He then added the 10% GST to give his selling price of the television.
   a. Calculate the selling price of the television.
   b. What is the GST included in the cost price of K504.00?
   c. How much GST does the retailer include in his monthly GST return?
   d. How much money, considering cost and GST, did the retailer make from selling this item?
6. After adding 50% for profit and then the 10% GST, a retailer has K56.55 as the selling price of the item.
   a. What is the retailer's price before she adds GST?
   b. What is the original cost price to the retailer before she adds her 50% profit and GST?
   c. How much GST is included in the original cost price to the retailer?
   d. How much GST does the retailer include in her monthly GST return?

e. How much money, considering cost and GST, did the retailer make from selling this item?

7. Tommy works in a butcher's shop. Tommy's rugby club is having a barbecue and Tommy is arranging the purchase of the meat. The club requires:

- 12 kg of sausages (normal selling price of K5.25 per kg)
- 5.5kg of steak (normal selling price of K11.80 per kg)
- 8.5 kg of chicken (normal selling price of K7.50 per kg)

All meat in the shop is marked up from the cost price by 40%, then GST is added. The manager says that Tommy can have all the meat at cost price plus GST.

a. Calculate the cost price of each of the types of meat.

b. How much in total will Tommy pay for the meat?

c. What percentage deduction will Tommy receive over what he would have paid if he bought the meat at the normal selling price?

Unit 11.2 Managing Money 1

Topic 4: Foreign exchange

Using foreign exchange rates and performing calculations using exchange rates are part of the 'Earning and spending' section of the Grade 11 syllabus (see Syllabus p. 15). This Topic covers:

- Currency exchange rates.
- Calculations using exchange rates.

Currency exchange rates

Exchange rates reflect the value of one country's currency in terms of another currency. These exchange rates can change at any time depending on global financial factors.

For example, the goods that you could buy with 1 Australian dollar would require K2.38 to buy on one day but K2.35 on another day. Banks and money exchange companies will provide current exchange rates and these can change (usually by very small values) many times in a day.

On a particular day the following exchange rates were given for converting PNG Kina (PGK) to the stated currency:

Table 1

Country and currency	Exchange rate
Australian dollar (AUD)	0.42066
United States dollar (USD)	0.4391
British pound sterling (GBP)	0.26802
Euro(EUR)	0.30719
New Zealand dollar (NZD)	0.52147
Indonesian Rupiah (IDR)	3 808.3035
Singapore dollar (SGD)	1.87086

On the same day the following exchange rates were given for converting the stated currency to PNG Kina:

Table 2

Country and currency	Exchange rate
1 Australian dollar (AUD)	2.37724
1 United States dollar (USD)	2.2774
1 British pound sterling (GBP)	3.731
1 Euro(EUR)	3.25529
1 New Zealand dollar (NZD)	1.91765
1 Indonesian Rupiah (IDR)	0.00026
1 Singapore dollar (SGD)	0.53451

Note: the rates given in Table 2 are $\frac{1}{\textit{The rate given in Table 1}}$ eg $2.3772 - \frac{1}{0.42066}$

Example A

Q. Keke wants to convert K1 000 to Euro. How many Euro have the same value as K1 000?

A. To convert PNG kina to Euro we use the exchange rate in Table 1:

1 kina = 0.30719 Euro

so 1 000 × 1 kina = 1 000 × 0.30719 Euro

K1 000 ≡ €307.19

Keke could exchange PGK1 000 for EUR307.19.

Q. Gure has USD356 to change to kina.

a. Use Table 2 to determine the equivalent amount in kina.

b. If the bank charges a fee of 3.5% how much will Gure get when he exchanges his money?

A. a. 1USD ≡ PGK2.2774

so USD(1 × 356) ≡ PGK(2.2774 × 356) ≡ PGK810.75

b. For the amount to be reduced by 3.5% then the multiplying factor is 1 – 0.035 = 0.965.

K810.75 × 0.965 ≡ K782.37

Gure will receive K782.37 when he exchanges his USD.

Unit 11.2 Activity 4A: Calculations using foreign exchange

1. Use the exchange rates in Table 1 to complete the table, changing the given amounts in PNG Kina to the stated currency.

PNG Kina	Currency	Equivalent amount
300.00	USD	
50 900.00	EUR	
15.50	IDR	
234 800.00	AUD	
86.75	SGD	

2. Use the exchange rates in Table 2 to complete the table, changing the given amounts in the stated currency to PNG Kina.

Amount	Currency	Equivalent amount in Kina
45.50	USD	
568.95	EUR	
10 000	IDR	
5 650.00	AUD	
450.00	SGD	

3. Larson bought a camera for SGD$450.00 in Singapore. What is the camera's equivalent price in Kina?

4. A book costs AUD$20 plus AUD$9.80 postage. The same book can be bought for K65.00 in Papua New Guinea. Which is the better buy and by how much?

5. Koriak buys a pump online for USD$360.00 including postage. He pays with his credit card which charges a currency exchange fee of 3%. What amount will Koriak be charged, in Kina, on his credit card?

6. Rosalyn buys a handbag online for AUD$165. Postage is AUD$12 and her credit card charges a foreign exchange fee of 3.5%. What is the total cost of the handbag in Kina?

7. Stella buys three dress patterns online for USD$5.65 each. She is charged a total of USD$15 for postage and a credit card fee of 3%.
   a. What is her credit card fee in Kina?
   b. How much is she charged in total, in Kina, on her credit card?

8. On his holiday to New Zealand, Nixon takes K2 500 in spending money. He exchanges this amount at a bank for New Zealand dollars. If the bank charges an exchange fee of 2%:
   a. How much is the bank fee in Kina?
   b. How much spending money will Nixon have in New Zealand dollars?
   c. On his return to PNG Nixon has NZD$285 left. He exchanges this amount at the airport where they charge a fee of 6%. How much will he get back in kina?

9. a. Why do you think exchange rates are given with accuracy to five decimal places?
   b. In what circumstances would you need more decimal places in the quoted exchange rates?
   c. Using the exchange rate in Table 1, determine the amount that you could exchange in Australian dollars to PNK so that the exchange is accurate to the nearest toea.

10. a. Is it possible to find the exchange rate between two of the other countries mentioned in Tables 1 and 2?
   b. Find the exchange rate that could be used to convert AUD to USD.
   c. What is the value of AUD$100 in USD?
   d. Find the exchange rate that could be used to convert GBP to EUR.
   e. What is the value of EUR1 000 in GBP?

Unit 11.2 Managing Money 1

Topic 5: Budgeting

This Topic deals with the budgeting section of Unit 11.2 Managing Money in the Grade 11 syllabus (see Syllabus p. 15). It covers:

- Basic income and expenses.
- Cash flow.
- Budgets.

Basic income and expenses

An income is a sum of money obtained by an individual, a group of people or an organisation through employment, sales of goods and services, fees for membership or borrowing, etc. The amount of money earned from the same activity or by doing the same job is called basic income.

It is wise to plan how you will spend your income so that you do not use up every toea at one time. Money repeatedly spent on the same thing/item becomes a basic or regular expense.

A budget is a financial plan that is used to guide an individual, organisation or enterprise on how they should spend their income. Or we can say that a budget is basically the recording of how the basic/regular income is spent on basic needs or expenses.

To avoid debts and bankruptcy, it is convenient for individuals to make weekly or fortnightly budgets, while business firms or organisations make a monthly budget. The monthly budget should contain monthly income and expenses. The monthly income and expenses are the calculated total expected to be earned or spent in a month.

An individual, organisation or enterprise estimates their current income or expenses using the previous income and expenses.

Cash flow

Cash flow refers to the way money is coming in and going out of a household or business.

Individuals and businesses are encouraged to prepare a cash-flow budget for their household or business so that they do not run out of money at certain times of the year and have to rely on expensive loans to keep afloat.

A cash-flow budget:

- Records the amount of money you expect to receive and spend over a period of time.
- Identifies times when you have large expenses and may be short of money.
- Identifies times when your expenses are less than your income.
- Allows you to make considered decisions about savings and spending.

Households need to make a yearly cash-flow budget because some expenses, eg insurance, are only paid as a yearly amount. This budget can then be broken down to a monthly, fortnightly or weekly budget.

Businesses also need to make a yearly cash-flow budget as this will be needed if the business wants to borrow money to expand

Financial budget for a business or an organisation

A financial budget or balance sheet contains two columns and normally lists income on the left and expenses on the right. Balance sheet means that the budget has equal totals of income and expenses.

Example A

The table below is an example of how a school budget should look.

Income		Expenses	
Type of income	**Amount (K)**	**Type of expenses**	**Amount (K)**
School fees	895 000	Food	400 000
Projects	90 000	Power bill	100 000
Government subsidy	200 000	Water bill	100 000
Canteen	50 000	Stationery	50 000
		Uniform	20 000
		Ancillary staff wages	300 000
		Departments allocation	100 000
		Other expenses	165 000
Total	K1 235 000	Total	1 235 000

Activity budget

An activity budget does not have income shown since it is a plan for events that require only spending. For this particular budget the total amount spent depends on the number and the unit price for items, therefore it should include columns for unit cost and the number of items.

Example B

Titi's mum plans for his 9th birthday party and the costs are as follows:

Birthday cake: K79

Lollies: K3.50/packet × 4

Sweet biscuits: K2.30/packet × 6

Balloons and decorations: K3/packet × 3 packs

Cordial: K8.00/container × 2 containers

Ice-cream: K80

Chicken: K16/pack × 10 packet

Lamb flaps: K20

Sausages: K30

Rice: K40

Garden food: K50

Fruits: K20

Prepare a budget for the birthday party.

Expenses			
Item	Unit price (K)	Number	Amount (K)
Birthday cake	79	1	79
Lollies	3.50	4	14
Biscuits	2.30	6	13.80
Balloons and decorations	3	3	9
Cordial	8	2	16
Ice – cream	80	1	80
Chicken	16	10	160
Lamb flaps	20	1	20
Sausages	30	1	30
Rice	40	1	40
Garden food	50	1	50
Fruits	20	1	20
Total			531.80

Personal budget

A personal budget includes income and expenses columns and the sum of the total expenses and savings must be the same as the income. The expenses can be divided into two parts: necessities and entertainment. The amount that is not spent can be saved.

Example C

Rita's net wage is K359 and this is how she plans to spend her money. The rest will be saved.

Food and toiletries: K100

PMV fare: K20

Clothes: K20

Power: K30

Water bill: K20

Video CDs: K20

Games: K20

Newspaper: K10

Prepare Rita's personal budget.

Income		Expense	
	Amount (K)		Amount(K)
Net wage	359	Necessities:	
		Food and Toiletries	100
		PMV fare	20
		Clothes	20
		Power	30
		Water bill	20
		Entertainment:	
		Video CDs	20
		Games	20
		Newspaper	10
Total	359	Total	240
Savings: K119			

Unit 11.2 Activity 5A: Budgeting

1. Nagitum plantation employs five workers and owns cattle, pigs, chickens and a hectare of coconut palms. The market for the livestock is K1 000 per cow, K800 per pig, K25 per chicken and K150 per 50 kg bag of coconut. Within a fortnight the plantation sells 3 cows, 2 pigs, 100 chickens and 50 bags of coconuts and spends K450 for chicken feed, K500 on fuel, K3 000 for workers' pay and K1 000 on other expenses.

- **a.** Prepare a fortnightly budget for Nagitum plantation.
- **b.** What is the plantation's profit?
- **c.** What is its monthly estimated income?
- **d.** What is its monthly estimated expense?
- **e.** Calculate:
  - **i.** Its estimated annual profit.
  - **ii.** Its estimated annual income.
  - **iii.** Its estimated annual expenses.

2. Unitech Bulls basketball club organises a barbecue to celebrate its win at the end of the Lae Amateur Basketball Association's proper season. The club treasurer gets the following prices from Prima butcher shop and Pelgens small goods:

Sausages: K20 per 25

Tomato sauce: K9.00 per 1-litre bottle

Bread loaves: K3.70 per loaf

Butter: K6.90 per container

Cordial: K9.00 per 2-litre bottle

Cooking oil: K5.95 per litre

Onion: K4.00 per plastic

a. Prepare an activity budget for the club if it is going to provide 1 000 sausages, 20 loaves of bread, 2 bottles of 1 litre tomato sauce, two containers of butter, two 2-litre bottles of cordial, 1 litre of cooking oil and 1 plastic round onion.

b. What is the total cost for the barbecue?

3. Mondo is a sales representative for a car dealer company and earns a net wage of K350 per fortnight plus 3% commission of every vehicle he sells. In a fortnight he sells two cars worth K20 000 each. Mondo's fortnightly expenses are:

Food: K150

Toiletries: K50

PMV fare: K20

Power: K50

Water bill: K20

Clothes: K50

Rent: K150

a. What is Mondo's total income?

b. What are Mondo's total expenses?

c. What is his savings?

People also budget for holidays or trips which involve extra expenses. The total cost includes transport, accommodation, admission, meals and spending money and it is more convenient to find out about the costs and prepare in advance.

4. The arts students of a particular school are planning to attend an arts exhibition. The admission fee for each person is K5/student and K7/adult, the cost for hiring a 25-seater bus is K100 each way, a lunch pack costs K10 and a can of soft drink costs K3. There will be 140 students and 3 teachers going to the exhibition. Draw up a budget for the trip and work out:

a. how much each person will pay to the nearest Kina

b. the total contribution if teachers will pay K10 more than the students.

5. Students from a church agency school are to attend a three-day Easter camp. It is arranged that 40 girls and 60 boys are to travel. The students will be staying at the mission station where a cabin will cost K150. The six teachers accompanying the students will sleep in one cabin and the boys and girls will use separate cabins. There are five bunks in each cabin. Meals for the weekend are K40 per person and 25-seater buses will be hired for the cost of K200 per bus one way. The same buses are to pick up the students at the end of the camp.

a. Prepare a budget to determine the total cost for the trip.

b. If students are to meet the cost of the camp themselves then how much will each student contribute, to the nearest Kina?

c. An extra 5% of the total cost is allowed for other expenses and the amount each student will be contributing is rounded off to the next multiple of K5. How much will each student be asked to contribute?

Unit 11.2 Managing Money 1

Topic 6: Loans

This Topic deals with the loans section of Unit 11.2 Managing Money in the Grade 11 syllabus (see Syllabus p. 15). It covers:

- Types of loans.
- Calculating interest.
- Credit.

Introduction

A loan refers to money borrowed from a lender such as a bank, a financial institution or an individual lender.

People, organisations or business firms need money to make a plan work, to put up projects or extend or expand their operations. If the amount of money they have in hand is less than the total amount required then they can seek financial assistance from banks or other finance companies or institutions.

The bank or finance companies lend money to their customers provided that the amount borrowed is repaid with interest. Institutions have their own terms and conditions which must be met by the borrowers for them to be eligible for a loan.

Banks or finance institutions that lend money charge their customers for borrowing their money. The charge for using someone else's money is called **interest**.

The interest rate, a percentage, depends on the type of loan, the security of the loan, the borrower's credit history and the purpose of the loan.

The amount of interest is a certain percentage of the **principal** (amount borrowed) and depends on principal, **term** (length of time taken to repay the loan) and the **interest rate**.

Types of loan

Various types of loan are offered according to the borrower's loan purpose. The three common types of loan are:

Personal loans

People apply to a financial institution for a loan for their personal use and normally repay it with interest in equal fortnightly or monthly instalments. Personal loans are often short-term loans (eg a loan to pay for a computer could have a term of three months) or they can be for several years (eg a loan to buy a car could have a term of three years).

Housing loans

A loan obtained to purchase a house or property is called a mortgage and is often paid off using a reducing balance loan, as it is a long-term loan. With a reducing balance loan the total loan balance reduces and so the amount of interest reduces after each payment, even though the interest rate remains constant.

The mortgage can be 100% of the purchase price for some customers approved by the bank or finance institution but it is usually less than 100% for those who pay a deposit of a certain percentage.

With housing loans, a deposit of a certain percentage of the purchase price of the property is paid by the customer to the bank or finance institution as a guarantee that they can afford to pay back the loan.

Like any other loan the finance institution would consider the security of the loan, the borrower's credit history, the stability of the borrower's job and their commitment to make regular repayments before they agree to the loan.

If a borrower is unable to pay off the loan as agreed, the bank or the finance institution repossesses the property and sells it to recover the amount it is owed.

School fee loans

Customers who earn a regular income and who wish to take out a school fee loan are asked to provide valid documents to prove that their children are truly students of any particular schools. A school fee loan is a short-term loan and the money is not put into the borrower's account but put straight into the school's account. If the customer has children attending different schools then the bank or finance company does the school fee break-up before putting the respective amounts into the respective school accounts. School fee loans can be repaid in equal fortnightly or monthly instalments.

Calculating interest

Interest charged on loans can be calculated in different ways and the total amount to be paid will be different depending on the method used to calculate the interest.

Simple interest

For a loan that charges simple interest the interest is calculated on the original amount borrowed. A simple interest rate is sometimes referred to as a flat interest rate.

Simple interest is relatively easy to calculate as it is just a percentage of the principal:

Interest = Principal × Rate × Time

A formula for calculating the simple interest, I, on a principal of P Kina, for a term of T years at an interest rate of r% p.a. is:

$$I = \frac{P \times r \times T}{100}$$

Loans that charge simple interest are short-term loans and hire-purchase agreements.

Compound interest

Most home loans are calculated using compound interest. With each payment made on a home loan the amount owed is reduced and so the interest on the next repayment is less. These types of loans are called reducing balance loans.

There is a complex formula for calculating repayments for reducing balance loans and this is investigated further in Grade 12 General Mathematics.

Example A

Q. A man borrowed K2 500 and is charged interest of K100 a quarter. Calculate his annual interest and hence determine the interest rate.

A. Annual interest = 4 x K100 [there are four quarters in a year]

= K400

$$\text{Interest rate} = \frac{400}{2\ 500} \times 100$$

= 16%

Q. Freda and Maria are partners of a hair salon and they decided to get a loan to purchase stock. They were given a 100-day loan of K9 500 after which time they were to pay back K10 025. What is the interest rate charged for this loan?

A. Loan period is 100 days.

Interest for loan period = K10 025 – K9 500

= K525

In order to find the interest rate:

i. Find the interest for one day.

Interest for 1 day = interest for loan period divided by loan term

= K525 ÷ 100 days

= K5.25

ii. Find the interest for one year.

Interest for 1 year = interest for one day × 365 days

= K5.25 × 365 = K1 916.25

iii. Calculate the interest rate.

Interest rate = [annual interest] ÷ [Principle (amount borrowed)] × 100%

= 1 916.25 ÷ 9 500 × 100

= 20.17%

Q. Morea has organised a personal loan of K1 200, borrowed for 90 days, and charged 14.5% p.a. interest.

a. Calculate the interest that will be charged on the loan.

b. How much will Morea have to repay at the end of 6 months?

A. **a.** Using the formula $I = \frac{P \times r \times T}{100}$ where $P = 1\ 200$, $r = 14.5$, $T = \frac{90}{365}$

$$I = \frac{1\ 200 \times 14.5 \times \frac{90}{365}}{100} = 42.90$$ Morea will owe K42.90 in interest.

b. Morea will need to repay K1 200 + K42.90 = K1 242.90 after 90 days.

Unit 11.2 Activity 6A: Interest and loans

1. Lasa obtained a K7 000 loan and was charged interest of K87.50 per month. Calculate the yearly interest rate.

2. Manu borrows K1 500 and is to repay it in 31 days. Work out the yearly interest rate if he is to pay K1 512 on the 31st day.

3. Amos borrows K5 000 to buy a used vehicle and he pays quarterly interest of K150. What is the annual interest rate?

4. Anduo obtained a personal loan of K3 000 and pays fortnightly interest of K17. Calculate the yearly interest rate.

5. Bapi borrows K40 000 to purchase a property and pays a weekly interest of K70. What is the interest rate?

6. Cynthia borrowed K30 000 from her best friend and promised to pay it back in 120 days. At the end of the 120 days she gave K31 578 to her friend. Cynthia then asked her friend if she could borrow back K15 000 for another 60 days. Cynthia repaid K15 444 for the second loan on the last day of the loan. Calculate the interest rates for the two loans.

7. Ako lends Gima K41 000. Gima agrees to repay the loan after 10 years at a simple interest rate of 10% per year.

a. What would the total amount of interest be for those ten years?

b. When the ten-year total interest is added to the principal of K41 000, what would the total amount be?

8. A K11 500.00, two-year loan has interest due of K3 234.38. What is the rate of simple interest?

9. Kati borrows K10 000. She borrowed K3 000 from her friend at 8% annual interest, K6 000 from her bank at 9% p.a., and the remainder from her insurance company at 5% p.a. How much does she pay in total interest for the first year?

10. Semi borrowed some money at 15% simple interest. He paid off the loan in one year, paying a total amount of K1 199.45. What was the original amount borrowed?

11. To finance her community college education, Sara takes out a loan for K3 600. After a year Sara decides to pay off the interest, which is 7% of K3 600. How much will she pay?

12. Grace borrows K1 000 from a small-loan lender. The loan is to be repaid in 12 fortnightly payments of K115 per fortnight.

a. How much in total will Grace be repaying?

b. How much interest is Grace paying for the loan?

c. Hw much interest is Grace paying per fortnight?

d. What is the yearly rate of interest?

13. Kimen borrows K3 000 to pay for his daughter's wedding. He is required to pay back the loan in 30 fortnightly repayments of K179.67 per fortnight.

a. How much in total will Kimen be repaying?

b. How much interest is Kimen paying for the loan?

c. How much interest is Kimen paying per fortnight?

d. What is the yearly rate of interest?

14. Kete and Eva have taken out a reducing balance housing loan for K124 000. The term of the loan is 25 years and the interest charged is 7.6% compounding monthly. It is calculated that they will pay K924.43 per month (12 months per year) for the term of the loan.

a. i. How much in total will Kete and Eva pay to the bank over the 25 years?

ii. What is the total amount of interest that they will pay?

iii. If Kete and Eva had a simple interest loan, what would be the yearly rate of interest?

b. If Kete and Eva were paying off their loan in yearly repayments they would be making a payment of K11 221.87 per year

i. Copy and complete the following table showing the balance owing and the interest payment for the first four years.

Year	7.6% interest charged (K)	Balance after interest (K)	Repayment (K)	Amount owing at the end of the year (K)
0				1 24 000
1	0.076 × 124 000 = 9 424	124 000 + 9 424 = 1 33 424	11 221.87	133 424 – 11 221.87 = 122 202.13
2	0.076 ×122 202.87 =		11 221.87	
3			11 221.87	
4			11 221.87	

ii. What is happening to the interest payment each year?

Credit

Credit can be thought of as a short term loan with a variable repayment plan.

Credit can be provided by a financial institution issuing a **credit card**, by a retailer offering a **store loyalty card** or by a finance company offering a **time-payment plan** (also known as a hire-purchase agreement).

Credit cards and store-loyalty cards

Buying an item using a credit card is similar to obtaining a mini loan. It is easy to make a purchase with a credit card but it is also easy to overspend and then have to pay large amounts in interest.

Interest on credit cards is charged in various ways:

- Cash is usually charged interest from the day the cash is borrowed.
- Some cards have an interest-free period but charge an annual fee for the use of the card. If the amount that is owed is not paid within the interest-free period then the cardholder is charged interest from the date of the purchase.
- Some cards have no interest-free period but do not charge an annual fee.
- A minimum repayment amount (about 3%) is set each month. If this is the only amount that is repaid then the amount owed can continue to increase and interest rates of 16–18% interest will charged on the balance.

To obtain a credit card from a financial institution, customers need to have a good credit history and also have an income that means they can repay the amount that they owe. Financial institutions assess a person's ability to repay and set a limit on the amount of credit they can obtain.

Time-payment plans (hire-purchase agreements)

If a purchaser cannot pay the full price of an item then a time-payment plan may be offered by the retailer. A time-payment plan usually requires the purchaser to pay a deposit, then the remainder of the purchase price, plus interest, is paid in an agreed number of equal instalments.

With a time-payment plan (also called a hire-purchase agreement) the purchaser is hiring the item from the retailer until the final payment is made. If the purchaser defaults on any of the payments then the item will be repossessed by the retailer without any return of the payments.

The interest paid on a time payment plan is quoted as a simple interest rate (flat rate) per annum and even though the repayment amount can seem affordable sometimes the amount of interest paid is large.

Example A

Q. Violet has a 'no interest-free period' credit card that charges 16.6% interest. She purchased a dress for K62.00 on the 5th of April and took out a cash advance of K50.00 on the 12th of April. Her credit card statement arrived on the 25th of April. How much interest would be included in Violet's monthly statement?

A. Violet will owe interest on K62.00 for 25 – 5 + 1 = 21 days and interest on the K50.00 for 25 – 12 + 1 = 14 days. Note : The number of days include both the purchase date and the end date.

$$\text{Interest} = \frac{62 \times 16.6 \times \frac{21}{365}}{100} + \frac{50 \times 16.6 \times \frac{14}{365}}{100} = 0.91$$

K0.91 will be included in Violet's monthly statement.

Q. Marisa is offered a time-payment plan, charging a flat rate of interest of 16%, to buy a refrigerator which has a purchase price of K1 450. If she pays a deposit of K250 and will pay the remainder plus interest in 18 monthly payments, how much will she pay each month?

A. The amount owing after the deposit is paid is K1 450 – K250 = K1 200.

Interest at 16% on K1 200 for 18 months = $\frac{1\,200 \times 16 \times 1.5}{100} = 288$ Interest is K288.

Marisa will pay a total of K1 200 + K288 = K1 488 in 18 equal instalments.

Marisa will pay $\frac{\text{K}1\,488}{18}$ = K82.67 each month.

Q. Nick is using a time-payment plan to purchase a sound system with a purchase price of K980. He has agreed to pay K180 as a deposit and the remainder in six monthly payments of K150.

a. How much interest is Nick paying with this contract?

b. What is the flat rate of interest being charged?

A. **a.** The amount owing after the deposit is paid = K980 – K180

= K800

Six monthly payments of K150 = 6´ K150

= K900

Nick is paying K900 – K800 = K100 in interest.

b. Nick is paying K100 in interest for 6 months and this would be K200 for 12 months (1 year).

The yearly interest as a percentage of the amount owed = $\frac{200}{800} \times \frac{100}{1} = 25$

Nick is being charged a flat rate of interest of 25% p.a. This is a very high rate of interest which was not immediately apparent from the amount of interest being charged or the amount of the repayments.

Unit 11.2 Activity 6B: Calculating interest

1. Joe was charged K26 interest for 1 month on a K1 200 credit card balance. What was the monthly interest rate?
2. Helen obtains a K500 cash advance on her credit card on the 8th of September. Her credit card statement is dated the 26th of September and she is being charged 17.5% p.a. interest. If the cash advance is the only item on the statement, how much interest is Helen charged on the credit card statement?
3. Morea has a 'no interest-free period' credit card that charges 16.5% interest p.a. On the 12th of August he purchases a phone costing K295.00. His August statement is dated the 28th of August and the phone is the only item on it.
   - **a.** How much interest is charged on the August credit card statement?
   - **b.** Morea pays K50.00 off the amount owed and makes no more purchases or payments before the next credit card statement arrives 31 days later. How much will the statement show that Morea owes?
4. Find the interest paid on a hire-purchase contract that is used to pay off a principal of K2 000 over 2 years and is charging a flat rate of 12% p.a.
5. Find the amount and the flat rate of interest when a hire-purchase agreement is made to pay off a principal of K1 000, plus interest, with 12 monthly payments of K100.
6. Calculate the monthly instalments paid for a hire-purchase agreement that pays off a principal of K900, plus interest charged at a flat rate of 9.6%, in 6 monthly instalments.
7. James has agreed to a hire-purchase contract, charging a flat rate of interest of 8%, to buy a car costing K5 400. He is paying a deposit of K1 400 and he has a choice of the number of monthly payments he can make. Calculate the amount he will be paying per month if he makes:
   - **a.** 12 payments
   - **b.** 18 payments.
8. Flora has taken out a time-payment plan to buy household goods to the value of K3 865. She is paying a deposit of K500 and the contract requires 20 monthly payments of K185. Find the flat rate of interest charged.
9. A time-payment plan requires the purchaser to pay a deposit of K60 and then 8 monthly payments of K32.85 on goods that cost a total of K300. Calculate the flat rate of interest per annum being charged.

10. Somin has bought a scooter costing K3 500. He has paid a deposit of K500 and has agreed to pay K240 per fortnight for the next 6 months to pay off the balance.

a. How many payments will he make?

b. How much interest is he paying?

c. What is the flat rate of interest per annum?

Unit 11.3 Statistics

Topic 1: Exploring data

Unit 11.3 focuses on everyday data and how it is preserved, analysed and interpreted. This Topic looks at the use of statistical methods and information and covers:

- Answering a given question or hypothesis, communicating statistical ideas clearly.
- Posing a question or hypothesis and giving a justified answer.
- Calculating statistics (where appropriate) and interpreting statistical information.
- Evaluating the statistical process used to respond to the question or hypothesis.

Introduction

Statistics is the collection, display and **analysis** of **data** (information) in order to answer a question or provide evidence for decision making. This Topic involves the use of statistical methods to respond to a question or **hypothesis**, and the interpretation of statistical information.

Statistical variables

Statistical **variables** are features of the data such as height, age, mass, eye colour and so on. More than one variable may be studied at a time:

- **Univariate data** involves only one variable (such as height) for each member of the survey. For example, the heights of boys may be compared with heights of girls.
- **Bivariate data** involves a pair of variables for each member of the survey – it is the relationship between the two variables that is of interest, eg the relationship between rainfall levels and hours of sunshine per year for a province.

There are various types of data:

- **Quantitative data** are numerical quantities such as population or distance. Quantitative data may be **discrete** (often involving counting, eg number of eggs laid per week by a hen) or **continuous** (obtained by measuring, eg the time taken to run a race).
- **Qualitative data** require word descriptions, such as gender, season or colour.

In this course only quantitative variables will be considered.

Stages of a statistical investigation

The steps in a **statistical investigation** of the relationship between two variables are:

1. Posing the question

A suitable statistical question must be posed. A good question is clearly worded and must:

- Be open – it does not have a one-word or one-number answer.
- Ask about relationships and trends, differences and similarities, and other important features of the variables.
- Allow for the calculation of statistics and the drawing of graphs to investigate the relationship between the variables.
- Allow the reporting of the investigation and the justification of the answer.

Good questions are often of the form, 'What is the relationship between variable A and variable B?' or 'How does variable A compare with variable B?'

There may be several possible choices of question for a data set.

Example A

Sama, a secondary school General Maths student, obtained a list of the overall percentage marks of Grade 11 students who studied Applied English, General Maths and Biology the previous year.

Part of the list is shown below:

Name	Applied English %	General Maths %	Biology %
Andrew, Joshua	67	72	68
Gaima, Benori	75	52	57
....			
Tailasa, Yaun	69	60	51
Tiba, Apeng	88	76	48
...			

She poses some questions:

'Is there a relationship between English and Maths marks for this group of students?'

[this is a question about *bivariate* data (two marks from one student are compared)]

'How do the Biology marks of boys compare with the Biology marks of girls in this group?'

[this is a question involving *univariate* data (the same feature of two different groups is involved in the comparison)]

Avoid posing questions that ask whether changes in one variable *cause* changes in the other variable, eg 'Does a high maths mark cause a high science mark?' Just because high maths marks are *associated* with high science marks does not mean that high maths marks *cause* high science marks (or vice versa). In fact, the marks are probably both responding to other external factors, such as good study habits, or a natural talent for maths/science-type subjects.

Note: It is important to practise the skill of posing suitable questions.

2. Selecting a sample

To choose a **sample** to investigate, it is necessary to define:

- The **population** (the entire group(s) of interest), about which the question is posed. This is to ensure that sampling is carried out from the correct group.
- The **variables** of interest, including their unit of measurement. It is important to be aware of the nature of the variables (univariate or bivariate).

Raw data (initial, unsorted data) are gathered using surveys, experiments, etc.

- If every member of a population is included in a statistical investigation, then the survey is known as a **census**.
- Usually only a sample (a portion of the population) is included in a survey. A sample size of at least 30 is recommended for reliable results.

The aim is to have **unbiased data**, which are truly **representative** of the population from which they are drawn.

- In a **random sample** every member of the population is equally likely to be selected. This should mean that the sample is **unbiased**, ie the characteristics of the sample are similar to the characteristics of the population from which it is selected.
- There are various techniques for obtaining an unbiased sample. These include the **lottery method** (eg drawing names from a hat), **systematic sampling** (eg selecting every 10th name on a list), or **stratified sampling** (eg if a school has twice as many boys as girls, then twice as many boys as girls are included in the sample). In **cluster sampling**, the sample is drawn from a representative subgroup of the population (rather than from the population as a whole).

3. Analysing the data

The raw data are organised using **tables** and **graphs** (see Topics 2 and 3). Various summary statistics such as measures of centre (**mean**, **median**, **mode**), **proportions** and measures of spread (**range**, **quartiles**) are calculated. These are discussed later in the chapter.

The graphs and the summary statistics will allow the variables to be compared, in order to answer the question posed concerning their relationship.

4. Conclusions

The question posed is answered, with justification (evidence backing up the conclusion that has been reached). Some indication should be given as to how much confidence you have in your conclusion.

There should also be a discussion of any other aspects of the investigation which could affect your conclusion.

5. Evaluation

Comments should be made about the usefulness and effectiveness of the investigation. This may include a discussion of:

- What has been done, and why.
- The effectiveness or accuracy of the data collected, and the resulting display and analysis.
- Any limitations of the statistical process (possible inaccuracies and sources of bias).
- Any improvements that could be made.

Carrying out a statistical investigation

In practice, the data set in a statistical investigation consists of the specific information needed to answer the question that has been posed (as in the example below). In an assessment, data suitable to the posing of a question are usually supplied (students may, however, be required to collect their own data).

Example B

Peter chooses two schools in Lae, Lae Secondary School and Bumayong Lutheran Secondary School. He wants to know whether students at Bumayong Lutheran Secondary School get more homework than the students at Lae Secondary School. Peter decides to carry out a statistical investigation.

Posing the question

Peter wants to know, 'How does the number of hours of homework per week done by students at Bumayong compare with the number of hours of homework per week done by students at Lae Secondary?'

Selecting a sample

- The population involves students at Bumayong and students at Lae Secondary.
- The variable to be measured is the number of hours of homework completed per week. (Comparing the hours of homework at one school with the hours at another is an example of univariate data.) Another variable could be the year level of each student surveyed.

Peter decides to select 100 student names randomly (using the lottery method) from the school roll at each school. A simple questionnaire could then be used to gather the information required. A key question might be, 'How many hours of homework did you do last week?'

Analysing the data

Peter draws up tables and graphs to summarise the data collected. He also calculates summary statistics measuring central values and spread for each data set (these techniques are covered later in the chapter).

Peter compares the various statistics he has calculated and graphs he has drawn for each data set.

Conclusions and evaluation

From the above analysis, in which the hours spent doing homework at Lae Secondary and Bumayong were compared using graphs and summary statistics, Peter drew conclusions in order to answer his question.

Peter noted some possible limitations and sources of error in his investigation:

- ***Poor choice of sampling method***

  Peter noticed that there was a strong relationship between the grade level of a student and the number of hours of homework done (a relationship involving bivariate data). Inclusion of more seniors in one sample than the other affected the total homework hours overall for that school. Peter decided that *stratified sampling* would have been a better choice (to get similar proportions of seniors in each sample).

- ***Inaccurate data***

  Peter noticed that some students seemed unsure of the number of hours of homework they had completed, and the figure they quoted was more of a guess than an exact number. This problem could be overcome by requiring that students record their homework hours for one week, then hand in the record.

- ***Non-representative or biased data***

  Peter found out that some Lae Secondary students in his survey had had an overnight school trip during the week of the survey. He concluded that he should have checked the week choice more carefully so that it fairly represented a typical week's homework (it should not be a short week, involving holidays, trips or exams, etc).

Note: Here the two sample sizes were equal in size. It is not necessary for this to be true in order to compare two variables and form valid conclusions.

Unit 11.3 Activity 1A: Statistical variables

1. Classify the following variables as either discrete or continuous.
   a. Number of peas in a pod.
   b. The height of a student.
   c. The length of a person's arm.
   d. The number of students in a class.
   e. The time spent watching TV.
   f. A person's age in years completed.
2. Classify the following as involving univariate or bivariate data:
   a. Comparing the weight of ginger cats with the weight of black cats.
   b. Comparing the weight of an orange with its circumference.
   c. Comparing girls' times with boys' times in a 100 m race.
   d. Comparing the size of sections for sale in Mount Hagen with their price.
3. Decide whether the following questions are 'good' or 'bad' questions, according to the statistical investigation criteria previously discussed. Explain your answer.
   a. Is there a difference between the head circumference of boys and girls?
   b. What is the median head circumference of Grade 11 boys and girls?
   c. Does low rainfall cause increased pest levels in orchards?
   d. How do athletes' 100 m times compare with their 200 m times?
4. Suggest problems which may be encountered in surveys of the following:
   a. The hours spent on homework by students in Grade 11.
   b. How much a student spends in the school tuck shop each week.
   c. Popularity of dog breeds.
   d. The ranking of the popularity of the subjects Mathematics, English and Geography.
5. The Personal Development teachers at Goroka Secondary School wanted to investigate the smoking habits of the pupils at their school compared with the smoking habits at nearby Bena Bena Secondary. The question posed was 'Do students at Bena Bena Secondary School smoke more (or less) than students at Goroka Secondary?'
   a. What is the population for this problem?
   b. What are the variables for the problem defined? Are they univariate or bivariate?
   c. What other aspects of smoking could be investigated in this survey?
   d. Discuss possible sources of error in this problem and how to overcome them.
6. A mathematics student is to conduct an investigation into weights of school bags carried by students at Malahang Technical School. The question asked is 'What is the relationship between the mass of a school bag and the grade level of a student at Malahang Technical School?'
   a. What is the population for this problem?
   b. What are the variables for this problem? Are they univariate or bivariate?
   c. Discuss possible sources of error and how they could be overcome.
   d. Who would possibly use the results of this investigation?

Organising raw data

Raw data can be sorted and displayed in a number of different ways. The most elementary form of data display is that of a *table*.

Frequency distribution tables

The **frequency** of a score is the number of times it occurs. A **frequency table** or **frequency distribution** lists scores and the frequency of each of those scores. A frequency table shows the **distribution** of scores for a set of data, in a way that a large collection of raw data cannot.

A frequency table is often combined with a **tally chart**. Tallies are a simple record of the number of times a particular outcome occurs. The tallies are then totalled and the results put in the frequency column of the table.

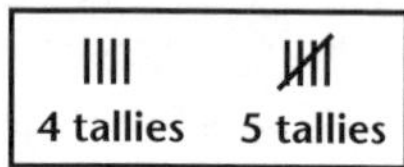

Example C

Q. The following scores are the results of a test marked out of 5:

1, 3, 4, 2, 3, 5, 0, 1, 5, 4, 3, 4, 5, 5, 2, 5, 2, 5, 3, 4, 5, 3, 0, 1, 5, 3, 2, 4

Set up a frequency table for these results.

A.

Score	Tally	Frequency
0	||	2
1	|||	3
2	||||	4
3	卌 |	6
4	卌	5
5	卌 |||	8

The table shows **ungrouped data** (each possible score is listed separately). Small numbers of data values are usually ungrouped.

Note: When using tallies to construct a frequency table from data, mark off each score as it occurs, eg a stroke is put by 1, then by 3, then by 4, etc. Do *not* count all the 0's then all the 1's, etc.

When there is a large number of possible data points, data values are grouped. With **grouped data**, scores are placed into groups or classes, and the *group frequency* is recorded.

Class intervals

Frequency distribution of a large data set is done using class intervals. Data values are put into groups of 5, 10, 15, etc. and each group starts with a multiple of the size. For instance, if the proposed class size is 5, then the consecutive class intervals should start with 0, 5, 10, 15 and so on; that is, 0–4, 5–9, 10–14, 15–19, etc. The limited number of scores for each class interval is 5.

The frequency of each class interval is the number of scores that fall in that category.

A frequency distribution for a large data set can be constructed as follows:

- Work out the range of the data set.
- Choose the appropriate class width or size.
- The number of classes is the quotient of the range and the class width.

Example D

Q. Following are the results of an examination. Construct a frequency table for the data.

34 59 48 54 57 50 63 58 52
50 54 71 53 81 56 54 32 57
30 65 41 58 57 61 53 43 43
64 46 73 55 77 61 60 54 55
80 67 61 30 50 59 69 82 67
61 65 43 63 34

A. Step 1: Find the range.

Range = Highest score – the lowest score
= 82 – 30
= 52

Step 2: Choose the class size.
Class size is five and since the lowest score is 30, the first class interval should start with 30.

Step 3: Determine the number of class intervals.

$$\text{Number of class intervals} = \frac{52}{5} = 10.4$$

When determining the number of class intervals, do not round the quotient off to the nearest whole number but always round it up. That means the number of class intervals for this example is 11.

Marks	Tally	Frequency
30–34	𝍸	5
35–39		0
40–44	||||	4
45–49	||	2
50–54	𝍸 𝍸	10
55–59	𝍸 𝍸	10
60–64	𝍸 |||	8
65–69	𝍸	5
70–74	||	2
75–79	|	1
80–84	|||	3

Continuous data are always grouped in classes, since there is an infinite number of possible values.

Example E

Some Elementary students were timed doing a task. The table shows that 3 took less than 10 seconds to complete the task, 5 took at least 10 seconds but less than 20 seconds, etc.

Note: 10– means $10 \leq \text{time} < 20$ seconds.

Time (sec)	Frequency
0–	3
10–	5
20–	6
30–	4
40–50	2

Analysing data

An analysis of statistical data would be expected to include finding **measures of central tendency** (often referred to as **averages**), and simple **measures of spread**.

Averages

An **average** is a number that indicates the *central value* of a set of numbers or data.

There are three common types of average – median, mode and mean.

Median

The **median** is the middle number when the numbers are **ranked** (listed in order of size).

If there is an *odd* number of data values, there will be a single middle value (the median). If there is an *even* number of data values, there will be two middle values. In this case, the median is the number midway between these two values.

Example F

1. The median of the numbers 1, 8, 6, 4 and 5 is found by placing the numbers in order of size (here, largest to smallest): 8, 6, 5, 4, 1.

   The middle of the data is 5 (there are two numbers 'above' and two numbers 'below' 5), therefore the median is 5.
2. The median of the numbers 7, 3, 5 and 4 is found by placing the numbers in numerical order (here, smallest to largest): 3, 4, 5, 7.

   In this case, the middle of the data is halfway between 4 and 5. The median is found by taking the number midway between the middle pair of numbers:

   $$\frac{4+5}{2} = 4.5$$

Mode

The **mode** is the *most frequent* score; ie the score which occurs most often.

Example G

The set of numbers 3, 7, 8, 4, 2, 6, 5, 4, 3 and 9 has two modes (*bimodal*), namely 3 and 4, because they both appear twice in the data set.

Note: There can be either 0, 1, or 2 modes. If there are 3 or more numbers which occur the same number of times, there is no mode.

Mean

The **mean** (or **arithmetic mean**) is commonly referred to as the average.

$$\text{Mean} = \frac{\text{total of scores}}{\text{number of scores}}$$

Example H

The mean of the set of numbers 3, 7, 8, 4, 2, 6, 5, 4, 3 and 9 is:

$$\text{Mean} = \frac{3 + 7 + 8 + 4 + 2 + 6 + 5 + 4 + 3 + 9}{10}$$
$$= \frac{51}{10}$$
$$= 5.1$$

Finding averages from frequency tables with ungrouped data

Consider the following data, presented in order of size:

1 1 1 1 2 2 2 5 5 5 5 5 5 5 5

The number of scores is 15, so the *median* is the 8th score, 5. The most commonly occurring score is 5, so the *mode* is also 5. Adding the scores and dividing by 15, gives *mean* = $\frac{50}{15} = 3\frac{1}{3}$.

If the data were put in a frequency table it would appear as alongside.

Score	Frequency
1	4
2	3
5	8
Total	15

To use the table to find these averages, proceed as follows.

- The total frequency is 15, so the middle score, or *median*, is the 8th score (7 below and 7 above). Adding the frequencies, it is found that 4 + 3 = 7 scores have been included once the 1's and 2's are counted. The 8th score is therefore a 5, so the median is 5.
- The *mode* is readily found by locating the highest frequency (8) and reading off the corresponding score (5). The mode is 5.
- The *mean* is calculated from the table by adding all of the 1's (there are four, so total is 4 × 1 = 4), all the 2's (there are 3, so total is 3 × 2 = 6) and all the 5's (there are 8, so total is 8 × 5 = 40). These products are denoted *f.x*, where x is a score and *f* its frequency.

Score	Frequency, *f*	*fx*
1	4	4
2	3	6
5	8	40
Total	15	50

  Summing these values for *fx* (ie calculating the grand total of all scores) and dividing by the total frequency (ie the number of scores), the mean = $\frac{50}{15} = 3\frac{1}{3}$. Often the symbol $\bar{x}$ is used to denote the mean of a sample. Thus $\bar{x} = 3\frac{1}{3}$.

Example I

Q. The following results are scores from a test:

Mark, *x*	0	1	2	3	4	5	6	7	8	9	10
Frequency, *f*	2	1	4	5	6	8	12	10	7	3	2

Find the mean, the median and the mode.

A. • To find the mean, calculate the extra column of $f.x$ values and total them. Total the frequencies (f).

$$\text{Mean} = \frac{\text{Total of } fx}{\text{Total of } f} = \frac{333}{60} = 5.55 \text{ (2 dp)}$$

• The median is the average of the 30th and 31st scores. Add the frequencies (f) until the total is in the neighbourhood of 30 and 31.

$2 + 1 + 4 + 5 + 6 + 8 = 26$ (26 marks are of size 5 or less).

The next mark is the 27th, and its value is 6. Since there are 12 sixes, the 30th and 31st scores are also both 6. Hence the median is 6.

• The mode is also 6, since this is the score with the highest frequency (a frequency of 12).

x	f	fx
0	2	0
1	1	1
2	4	8
3	5	15
4	6	24
5	8	40
6	12	72
7	10	70
8	7	56
9	3	27
10	2	20
Totals	60	333

Finding averages from frequency tables with grouped or continuous data

The **modal class** in a frequency table with grouped or continuous data is the class whose frequency is greatest.

The mean of grouped or continuous data can be estimated by finding the mean of the middle scores of each group. Since the actual values of the underlying data are unknown, this is an *approximate mean* only.

Example J

A student collected data about the weights of pupils in her class. The results are shown in the frequency table. The student wishes to estimate the mean of the data. Her working is shown.

Weight, x	Midpoint of group, m	Frequency, f	$f.m$
$45 \le x < 50$	$\left(\frac{45+50}{2}\right) = 47.5$	2	95.0
$50 \le x < 55$	52.5	6	315.0
$55 \le x < 60$	57.5	9	517.5
$60 \le x < 65$	62.5	8	500.0
$65 \le x < 70$	67.5	4	270.0
$70 \le x < 75$	72.5	1	72.5
	Totals	30	1770

$$\text{Mean} \approx \frac{\text{Total of } f.m}{\text{Total } f} = \frac{1770}{30}$$

Note: The class $55 \le x < 60$ is the modal class (its frequency of 9 is the greatest frequency).

Comparing averages

The *arithmetic mean* is probably the best known of all the measures of averages. It seems 'inherently fair', in that all of the data is used in its calculation and an exact result can be produced from its mathematical formula. However, there are disadvantages:

- Extreme values can affect a mean quite significantly. For example, the inclusion of one or two unusually large or small pieces of data (ie **outliers**) can affect the value of the mean markedly, eg the mean of 2, 3, 4, 5, 6, 150 is 28.7, which doesn't seem relevant to the data.
- The exact result for a mean is often difficult to interpret directly, eg a mean class size of $24\frac{2}{3}$ students.

The *median* is less affected by extreme values, and often gives very relevant results in that the median is a data point (or the average of two data points). It too has drawbacks however:

- For a small set of data the median can give a poor picture of the behaviour of the data, eg the median of 1, 2, 3, 99, 100 is 3.
- Since the median has no mathematical formula (unlike the mean), it is difficult to apply mathematically, and hence of limited use in more advanced statistical work.

The *mode* is easily calculated, and yields the most common or most popular result, which can be of interest to voters or manufacturers. However:

- The mode can alter rapidly and radically as new scores are included, eg 2, 2, 2, 8, 8, 8, 8 has mode 8; while 2, 2, 2, 2, 8, 8, 8 has mode 2.
- The mode is of little importance in further statistical study.

Proportions

It is often of interest to find the **proportion** of a population which has a certain property or characteristic eg the proportion of voters for a political party in an election, A good estimate can be made of a population proportion by taking a representative sample (of at least thirty members) and calculating the sample proportion.

$$\text{Sample proportion} = \frac{\text{Number in sample with the property}}{\text{Total number in sample}}$$

For example, in order to determine the percentage of overweight bags checked in at an airport, a random sample of 100 bags is taken and each bag weighed. If 13 out of the 100 bags are found to be overweight, then $\frac{13}{100} = 13\%$ is a reasonable estimate of the overall percentage of overweight bags.

Unit 11.3 Activity 1B: Averages

1. For the following sets of scores, find:

i. The median. **ii.** The mode. **iii.** The mean.

a. 3, 5, 3, 4, 2

b. 7, 6, 8, 3, 4, 6, 5, 2, 4, 1

c. 12, 14, 10, 17, 16

d. 13, 14, 16, 12, 11, 13, 12, 15, 16, 17

e. 103, 105, 106, 104, 105, 107

f. 205, 208, 203, 204, 203, 201

2. The table shows the number of goals scored per match by the Hekari United football team for a season:

Goals per match	0	1	2	3	4	5	6
Number of matches	3	5	6	4	4	2	1

a. Calculate the mean number of goals scored per match by the Hekari team.

b. Find the median number of goals per match.

3. The table shows the number of wickets taken in an innings by a bowler in a cricket season:

No. of wickets per innings	0	1	2	3	4	5	6	7
Frequency	3	6	9	7	5	3	2	1

a. Calculate the mean number of wickets per innings for the bowler.

b. What is the median number of wickets per innings?

4. A discus thrower recorded the distance (in metres) of her throws in a training session:

36.8	42.3	41.4	37.8	39.1	40.3	32.3	41.5	38.6	37.8
43.1	37.8	33.4	42.1	37.6	39.3	40.2	37.6	41.3	43.5
41.3	41.2	41.5	39.8	35.4	41.2	43.1	39.3	36.2	41.2
36.5	33.4	42.6	33.2	42.5	32.8	37.6	41.2	35.4	37.6

a. Copy and complete the table:

Distance, metres, x	**Tally**	**Frequency, f**	**Midpoint, m**	**$f.m$**
$32 \le x < 34$				
$34 \le x < 36$				
$36 \le x < 38$				
$38 \le x < 40$				
$40 \le x < 42$				
$42 \le x < 44$				
	Totals			

b. What is the modal class?

c. Estimate the mean distance.

d. What proportion of throws went further than 40 m?

Measures of spread

Data can be spread out between their lowest and highest values in various patterns. The **spread** of data can be measured in different ways. One method of describing spread uses the **range**.

Range = highest score – lowest score

Example K

The range of the set of scores 3, 5, 9 and 10 is $10 - 3 = 7$.

Quartiles, along with the median, divide a set of scores into four equal parts.

- The **upper quartile** (UQ) can be viewed as the *median of the upper half of the values.*
- The **lower quartile** (LQ) can be viewed as the *median of the lower half of the values.*

The **interquartile range** is the difference between the upper and lower quartiles.

Example L

The following set of scores: 3, 5, 8, 9, 7, 6, 4, 3, 5, 6, 1, 3, 4, 2 and 5 can be ranked to give:

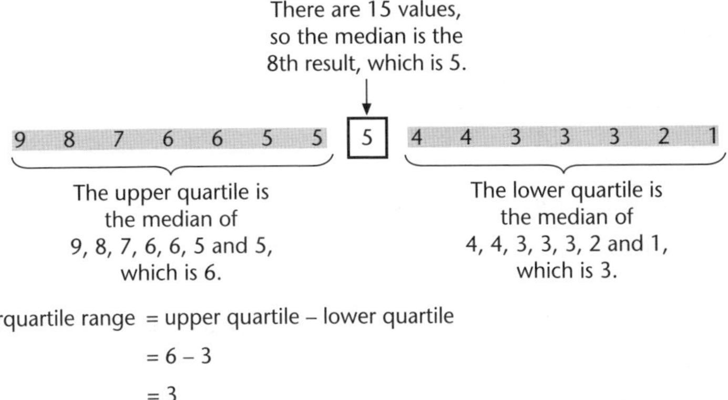

The interquartile range = upper quartile – lower quartile

= 6 – 3

= 3

Note: When calculating quartiles, the median is not included in the upper or lower half of the data.

The problem with the range and IQR as a measure of spread is that both only use two data values in their calculation. Some data sets can have their characteristics hidden when the IQR is quoted. It would be helpful if we could have a measure of spread that used all of the data values in its calculation. One such statistic is the **variance**. Variance measures the average of the squared deviations of each data value from the mean. The deviation of a data value, x, from the mean, $\overline{x}$, is given by $x - \overline{x}$.

The other method of describing the spread of scores is the **standard deviation**.

Standard deviation is the measure of how clustered or scattered the data values (scores or observations) are around the mean. The further away the scores are from the mean, the greater the standard deviation. For instance, the standard deviation of 1.9 signifies that the distance of each score from the mean is short, and the standard deviation of 20 signifies that the scores are widely spread around the mean.

The standard deviation is the positive square root of the variance. The standard deviation of a population is denoted by the symbol σ (sigma) and σ^2 (sigma squared) denotes the variance of a population.

The **variance of a sample** (symbol s) is the mean of the squares of the distance of the data values from the centre (the mean). If x refers to the values in the sample, f refers to the respective frequencies of those values and $\overline{x}$ is the mean of the sample, then the formula for the variance of grouped data is:

$$s^2 = \frac{\sum f(x - \overline{x})^2}{\sum f - 1}$$

The variance for the ungrouped data can be calculated using:

$$s^2 = \frac{\sum (x - \overline{x})^2}{n - 1}$$

where n is the number of data values.

With practice, the variance can be easily found from the formula. In the following example the variance calculation is done directly by first finding the mean then using the variance formula. In the second method, a table is used to calculate the variance.

Example M

Q. The following table is a sample taken from a much larger population.

x	1	2	3	4	5
f	4	3	2	3	4

Find the variance of the sample values.

A. Direct solution

$$\text{Mean} = \frac{4\times1+3\times2+2\times3+3\times4\times4\times5}{4+3+2+3+5} \qquad \left[\text{substituting in } \bar{x}=\frac{\sum fx}{\sum f}\right]$$

$$=\frac{48}{16}$$

$$=3$$

$$\text{Variance} = \frac{\sum f(x-\bar{x})^2}{\sum f-1}$$

$$=\frac{4(1-3)^2+3(2-3)^2+2(3-3)^2+3(4-3)^2+4(5-3)^2}{(4+3+2+3+4)-1}$$

$$=\frac{4(-2)^2+3(-1)^2+2(0)^2+3(1)^2+4(2)^2}{15}$$

$$=\frac{16+3+0+3+16}{15}$$

$$\approx 2.53$$

A. Table solution

x	f	fx	$(x-\bar{x})$	$(x-\bar{x})^2$	$f(x-\bar{x})^2$
1	4	4	1 – 2 = –2	4	4.4 = 16
2	3	6	2 – 3 = –1	1	3
3	2	6	3 – 3 = 0	0	0
4	3	12	4 – 3 = 1	1	3
5	4	20	5 – 3 = 2	4	16
Totals	**16**	**48**			**38**

$$\bar{x}=\frac{\sum fx}{\sum f}=\frac{48}{16}=3 \qquad s^2=\frac{\sum f(x-\bar{x})^2}{\sum f-1}=\frac{38}{16-1}\approx 2.53$$

Note:

- The calculation of $\bar{x}$ must be done before the table can be completed.
- In the exercises or exams, do not show the calculation of the deviation.

- The deviation from the mean of an observation is the difference of an observation, x, and the mean, $\bar{x}$, of a sample.
- Deviation = $x - \bar{x}$
- The mean deviation of a sample average of the positive values of the deviations from the mean.

  Mean deviation $= \frac{\Sigma |x - \bar{x}|}{n}$

  $= \frac{\text{Sum of positive deviations}}{\text{The number of scores}}$

In the example above, the standard deviation is:

$s = \sqrt{2.53} \approx 1.59$ (3 dp)

Unit 11.3 Activity 1C: Measures of spread

1. Find: **i.** The range. **ii.** The upper quartile. **iii.** The lower quartile.

iv. The interquartile range for each of the following data sets.

a. 1 2 3 3 4 7 8 9 9 10 12

b. 17 17.2 17.5 17.6 17.8 17.9 18.1 18.5 19.6

c. 104 107 108 108 108 117 123 134 146 152

d. 23.42 23.46 23.48 23.48 24.62 25.04 25.16 26.93

2. Jack recorded times in the 25 m freestyle competition at his swim club. The times were as shown, in seconds (to the nearest tenth of a second).

15.6 20.3 14.7 20.3 21.1 17.4 17.8 19.4 18.2 22.3 19.4 18.6

Find:

a. The range of times. **b.** The upper quartile.

c. The lower quartile. **d.** The interquartile range.

3. In the annual report about daily visitor numbers at a children's fun park, the report included the following information. 'On 75% of the days at least 85 children visited. On the least successful day we had only 24 children visit, but on half of the days there were at least 120 children visiting. Our best day saw 246 children visiting and 25% of the time we had at least 174 children visiting.' Using this information, state:

a. The range. **b.** The interquartile range.

4. a. Find the mean and the standard deviation of 1, 2, 5, 5, 7, 10.

b. i. Find the new mean and standard deviation if 10 is added to each score.

ii. How are the new mean and standard deviation related to the original mean and standard deviation?

c. i. Find the new mean and standard deviation if the scores are multiplied by 10.

ii. How are the new mean and standard deviation related to the original mean and standard deviation?

5. A set of scores has a mean of 6 and a standard deviation of 2. Find the new mean and standard deviation if:

a. 5 is added to all of the scores.

b. All the scores are multiplied by the five.

c. 20 is added to all of the scores.

d. All of the scores are multiplied by 20.

e. 2 is subtracted from all of the scores.

f. All of the scores are divided by 2.

g. The frequencies of the scores are trebled.

6. Following are the marks obtained by Grade 10 students of a school in their science assignment: 2, 9, 3, 1, 4, 7, 2, 3, 4, 5.

Work out: **a.** The mean.

b. The mean deviation.

c. The standard deviation.

7. In a new guava plantation, the number of guava fruits on 7 trees were counted and recorded as follows: 23, 43, 16, 40, 18, 27, 29.

Work out: **a.** The mean.

b. The mean deviation.

c. The standard deviation.

8. Eight students were interviewed on the number of phone calls they receive weekly from their friends and the following data was comprised: 3, 7, 8, 2, 7, 9, 7, 5.

a. What is the average number of calls?

b. Determine the standard deviation.

9. A Grade 12 student gave his 10-page argumentative essay to one of his teachers to edit and the following data contains the number of grammatical errors: 4, 3, 6, 7, 0, 1, 5, 2, 2, 0.

a. Calculate the mean error.

b. Calculate the standard deviation.

Unit 11.3 Statistics
Topic 2: Analysis of data—using statistics

Unit 11.3 focuses on everyday data and how it is collected, analysed and interpreted. This Topic looks at the use of statistical methods and information and covers:

- Drawing appropriate graph(s) and commenting on their features.
- Using statistical information to answer straightforward questions.

Introduction

Graphical displays can show trends and patterns in data more clearly than raw data or tables of data can. When investigating relationships between variables, the drawing of appropriate graphs provides a visual and very effective method of comparison.

Statistical graphs

Every statistical graph should have the following features, where appropriate:

1. A title that clearly states what the graph is about.
2. Clearly labelled axes with scale and units shown.
3. A key of symbols used.
4. Clarity – it should be easy to read and understand.

Pie graphs

Pie graphs or **pie charts** show **proportions** by using sectors of a circle.

- The whole pie graph represents the total number of data points.
- Each sector represents the number of data points for a particular category.

To work out the angle size of a sector, the *proportion in each category* is calculated by dividing the *number of data points in the category* by the *total number of data points*. This proportion is then multiplied by 360, the number of degrees in the circle.

Example A

Fifty students were given a choice of sport to take part in for a phys-ed option. The results are shown in the following table:

Sport	Athletics	Basketball	Tennis	Volleyball	Weightlifting
Number of students	20	12	4	9	5

Calculating sector angles for each category (as for athletics) gives the following table, from which a pie graph can be drawn (using a protractor to measure each angle).

Category	Frequency	Sector angle (°)
Athletics	20	$\frac{20}{50} \times \frac{360}{1} = 144.0$
Basketball	12	86.4
Tennis	4	28.8
Volleyball	9	64.8
Weightlifting	5	36.0
Total	50	360.0

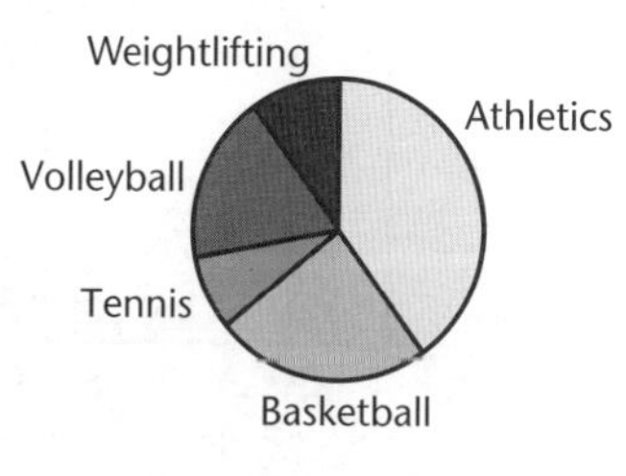

Using pie graphs to compare sets of data

Since pie graphs show the proportion (or percentage) of the total quantity in each category, a pair of pie graphs can be used to compare proportions (the bigger the sector, the larger the proportion).

Bar graphs

A **bar graph** (sometimes called a **column graph**) is a common method of displaying *discrete data*. The frequencies of the data can be easily compared, because the length or height of the bars is **proportional** to the *frequency of the data.*

Example B

The phys-ed choices from Example A are shown in the graphs below:

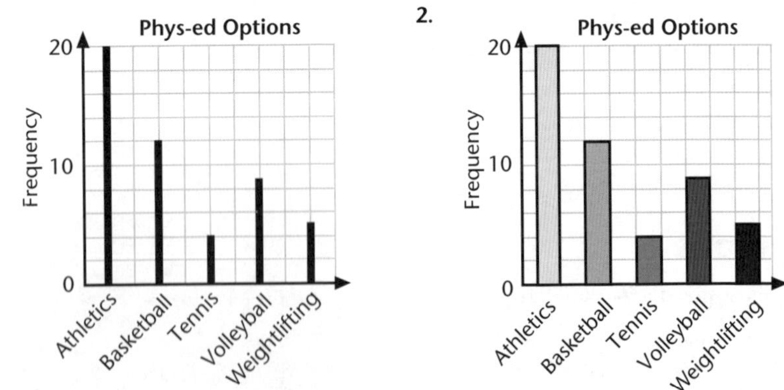

Note: 1. The frequency is usually on the vertical axis, but bar graphs may be turned sideways so that frequency is on the horizontal axis.

2. The bars should be the same width and should not be touching.

When drawing bar graphs for grouped discrete data, bars are labelled on the horizontal axis with the class interval. The vertical height of the bar shows the class frequency. Since the classes have 'gaps' between them, the bars do not touch (although they are sometimes drawn touching).

Example C

Q. Draw a bar graph for the data in the frequency table.

A. The graph is as shown.

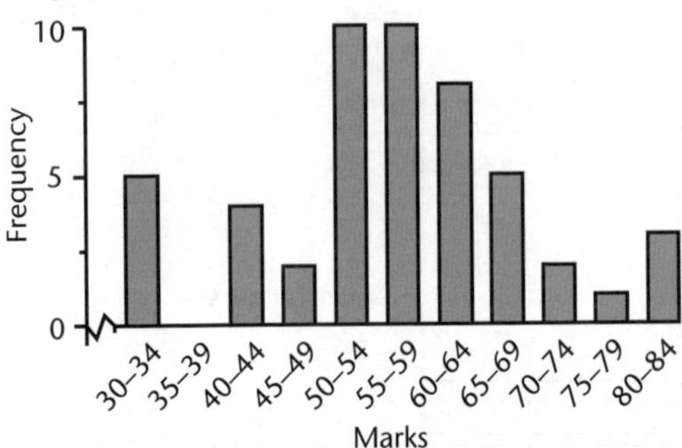

Marks	Tally	Frequency
30–34	卌	5
35–39		0
40–44	IIII	4
45–49	II	2
50–54	卌 卌	10
55–59	卌 卌	10
60–64	卌 III	8
65–69	卌	5
70–74	II	2
75–79	I	1
80–84	III	3

Using bar graphs to compare sets of data

Composite bar graphs can be used for comparing two or more sets of univariate data.

Example D

The table shows the number of pupils, by gender, and their phys-ed option choices.

Sport	Athletics	Basketball	Tennis	Volleyball	Weightlifting
Number of girls	8	7	3	6	1
Number of boys	12	5	1	3	4

This information can be displayed by means of a composite bar graph, as follows:

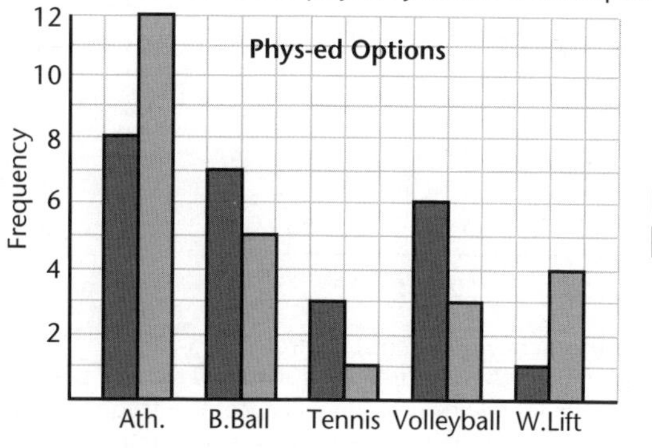

From the graph it can easily be seen that girls outnumber boys in basketball, tennis and volleyball, but not in athletics or weightlifting.

Histogram

A **histogram** is a bar graph of **continuous data**. The bars of a histogram touch and the area of the bars are proportional to the frequency of the data.

Example E

A student collected data about the weight of pupils in her class. The results are displayed below in both a table and using a histogram.

Weight (kg) x	Frequency
$45 \le x < 50$	2
$50 \le x < 55$	6
$55 \le x < 60$	9
$60 \le x < 65$	8
$65 \le x < 70$	4
$70 \le x < 75$	1

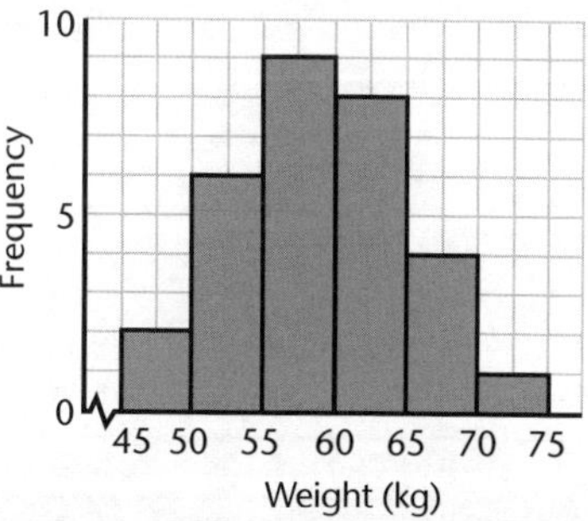

Notes: 1. Concertinaing (⩘) of the horizontal axis is an often-used technique in statistics, and it indicates that there is nothing graphed between 0 and the left-hand value of the first bar. In general, vertical axes should *not* be treated like this.

2. The end points of each interval are shown on the horizontal axis.

Frequency polygon

The **frequency polygon** is a straight line graph obtained by joining the centre of the tops of the bars of a histogram.

Example F

The frequency polygon for the data in the example above is illustrated below:

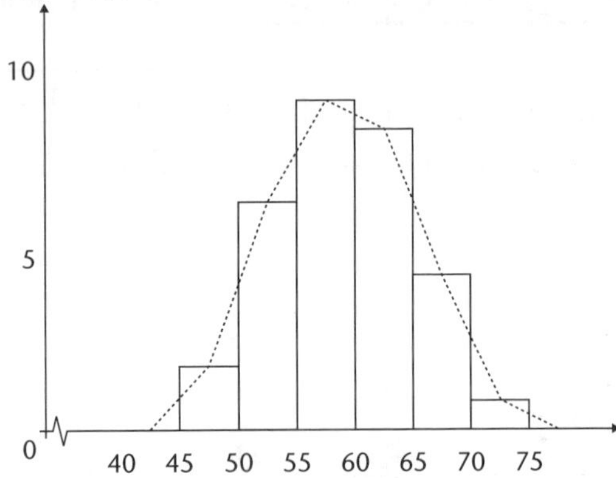

You can see from the graph that the **frequency polygon** starts from the *x*-axis and ends at the *x*-axis. The **starting point** is the **midpoint** of the class interval 40–45 and the end point is the **midpoint** of the class interval 75–80. The frequency polygon will become a **frequency curve** if the number of scores and the number of class intervals only increase. A frequency polygon can also be drawn by plotting the **frequencies** against the **class centres**.

Unit 11.3 Activity 2A: Bar graphs

1. A class of 36 students listed their favourite subject:

Subject	English	Mathematics	Science	Social Science	Phys-ed
No. Students	7	10	4	9	6

Display this information using a bar graph.

2. The following scores are the results of a test marked out of 10:

3	4	5	1	10	1	6	3	5	2	7	8	6	5	9
9	8	0	2	3	6	5	2	1	8	4	10	3	1	3

Set up a frequency table for these results and use it to draw a suitable graph for these data.

3. The frequency table below shows the results of a survey of the heights of students in a class.

Height (cm)	150–	154–	158–	162–	166–	170–	174–178
Frequency	2	4	6	3	4	7	4

a. Draw a histogram to display these data.

b. Construct a frequency polygon on the same set of axes.

4. A student collected the following data about the height (in cm) of the pupils in his class:

173	162	173	181	159	164	169	177	152	160
154	180	176	172	168	163	171	174	162	169
169	178	181	158	166	168	179	166	164	173

Using class intervals of width 5 cm:

a. Draw up a frequency table.

b. Draw an appropriate graph.

c. Construct a frequency polygon on the same set of axes.

5. A group of 72 students listed their means of transport to school:

Means of transport	PMV	Car	Cycle	Boat	Walk
Number of girls	12	3	3	3	18
Number of boys	8	5	9	3	8

a. Display this information using a composite bar graph.

b. Compare the boy/girl data sets.

6. The algebra results of two classes (both of size 30) were compared. The results are shown in the table.

	Low achievement	Satisfactory achievement	High achievement	Very high achievement
Class A	10	9	7	4
Class B	13	11	4	2

a. Draw a composite bar graph for this information. b. Compare the data sets.

Cumulative frequency graphs and percentiles

Cumulative frequency graphs are constructed for grouped data and are used to give estimates of the median, the upper and lower quartiles and percentiles.

For the following data collected about the weight of pupils in a class:

Weight (kg)	Frequency
$45 \leq x < 50$	2
$50 \leq x < 55$	6
$55 \leq x < 60$	9
$69 \leq x < 65$	8
$65 \leq x < 70$	4
$70 \leq x < 75$	1

A separate table has to be constructed to show the cumulative frequency:

Weight (kg)	Cumulative frequency	
< 45	**0**	There are no data values less than 45 kg
< 50	**2**	There are 2 data values less than 50 kg
< 55	2 + 6 = **8**	There are 8 data values less than 55 kg
< 60	2 + 6 + 9 = **17**	There are 17 data values less than 60 kg
< 65	2 + 6 + 9 + 8 = **25**	There are 25 data values less than 75 kg
< 70	2 + 6 + 9 + 8 +4 = **29**	There are 29 data values less than 70 kg
< 75	**30**	There are 30 data values (all) that are less than 75 kg

Note: The number of data values in the last row of the second column should be the sum of all the frequencies.

Below is the cumulative frequency curve for the frequency distribution above. The variable (weight values) is placed on the horizontal axis and the cumulative frequencies on the vertical axis. The points (45, 0), (50, 2), (55, 8) ... (75, 30) are plotted and straight lines are drawn to join the points. The graph should always have the 'stretched- s' shape.

Graph 1

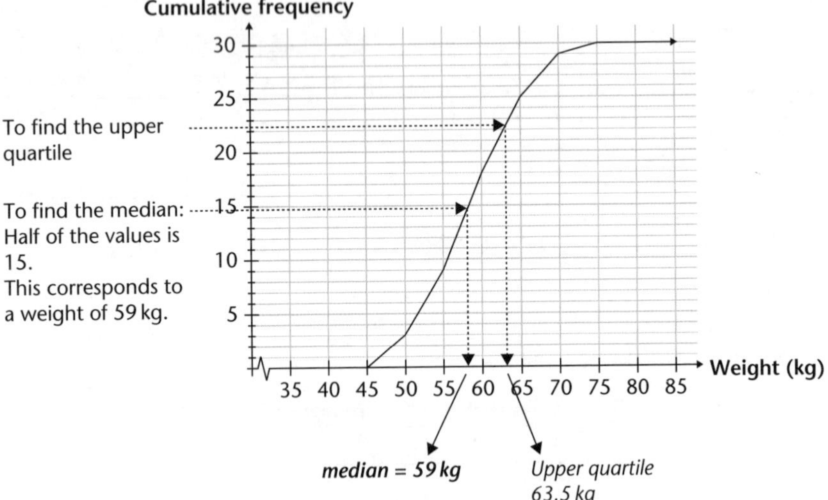

A percentage cumulative frequency table and curve can also be drawn:

Table

Weight (kg)	Cumulative frequency	Percentage cumulative frequency (%)
< 45	0	$\frac{0}{30} \times 100 = 0$
< 50	2	$\frac{2}{30} \times 100 = 6.7$
< 55	8	$\frac{8}{30} \times 100 = 26.7$
< 60	17	56.7
< 65	25	83.3
< 70	29	96.7
< 75	30	100

Graph 2

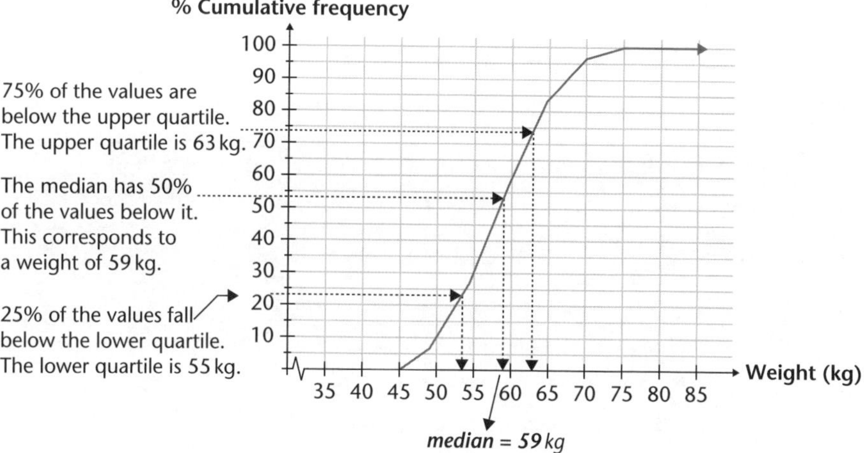

Reading statistics from a cumulative frequency curve

We can read estimates of statistics from the cumulative frequency curve that we would not be able to calculate if we do not have the actual data values.

The **median** is the middle value in a set of ordered data. For our example, using Graph 1, 15 values would have a value less than the median so going across to the graph at 15 and then down to the *weight* axis gives the median as 59 kg.

The **upper and lower quartiles** can also be read from the graphs. 75% of the data values are less than the upper quartile, so moving horizontally at 22.5(Graph 1) or 75% (Graph 2) across to the graph then down to the *weight* axis gives an upper quartile of about 63.5 kg. Similarly, the lower quartile can be read as approximately 55 kg.

This data set could be described as: 'The weight of students in this class was centred around 59 kg (median) with the lightest 25% having a weight of 55 kg or less (lower quartile) and the heaviest 25% having a weight of 63.5 kg or more (upper quartile)'.

Percentiles

Percentiles are frequently used to describe sets of data.

For the data graphed above we could make statements such as: '90% of the students in the class had a weight less than 67 kg. This figure is called the **90th percentile**'; or 'the lightest 10% of the class had a weight of 51kg or less.' Here we are quoting the **10th percentile**.

- The median is also called the 50th percentile.
- The lower quartile is also called the 25th percentile.
- The upper quartile is also called the 75th percentile.

Unit 11.3 Activity 2B: Cumulative frequency graphs

1. A percentage cumulative frequency curve, shown below, has been constructed for the heights of a basketball squad.

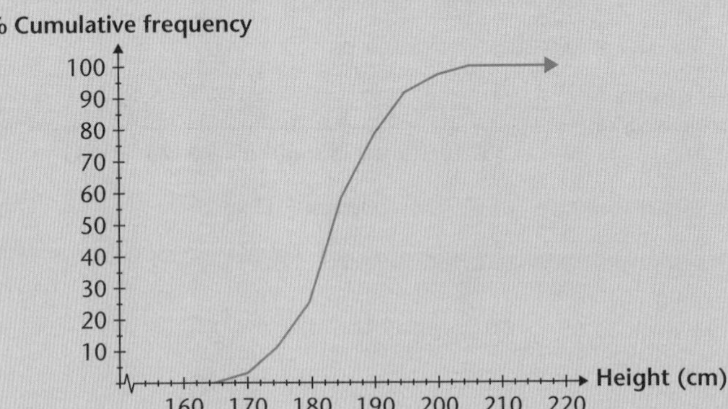

a. From the graph find estimates of:

i. The median height.

ii. The upper quartile.

iii. The lower quartile.

iv. The 80th percentile.

v. The 10th percentile.

b. Copy and complete the following statements:

i. Half of the heights of the basketballers are greater than …

ii. 25% of the basketballers have a height that is less than …

iii. … of the basketballers have a height that is less than 190 cm.

iv. …% of the basketballers have a height that is greater than 186 cm.

2. a. Copy and complete the cumulative frequency table and graph for the data given below:

Maximum daily temperature (°C)	Frequency
15 –	1
20 –	5
25 –	14
30 –	5
35 –	3
40 –	2

Maximum daily temperature (°C)	Cumulative frequency
< 15	0
< 20	
< 25	
< 30	
< 35	
< 40	
< 45	30

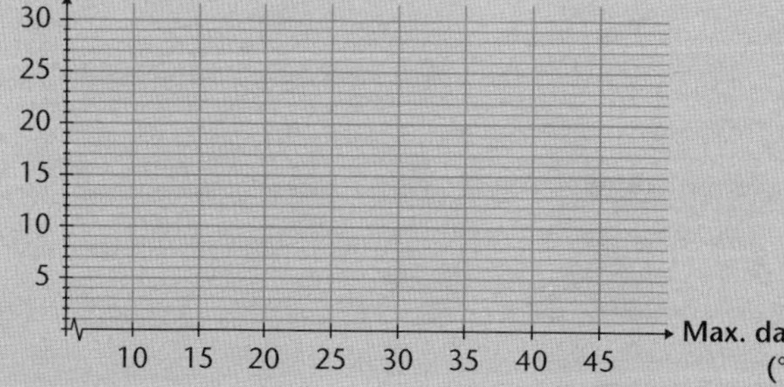

b. From your graph estimate:

i. The median maximum daily temperature.

ii The 80th percentile.

iii The 10th percentile.

iv The interquartile range.

c. Write a sentence that interprets each of the statistics from **b** parts **i** to **iv**.

3. The speed of vehicles travelling along a section of highway where the speed limit is 100km/hr has been recorded and given in the table below:

Speed (km/hr)	Number of cars (Frequency)
50 –	8
60 –	18
70 –	84
80 –	116
90 –	212
100 –	150
110 –	10
120 – < 130	2

a. How many vehicles were included in this survey?

b. What percentage of the vehicles were travelling at speeds equal to or greater than 100 km/hr?

c. Construct a percentage cumulative frequency table for the data.

d. Construct a percentage cumulative frequency graph from your table.

e. From your graph find estimates of the given statistic, and write a sentence to interpret:

i. The median.

ii. The lower quartile.

iii. The upper quartile.

iv. The 90th percentile.

v. The 10th percentile.

vi. The interquartile range.

4. The following frequency table gives the time it takes to swim 100 m for the forty members of a swimming squad.

Time to complete 100m swim (secs)	Number of swimmers
50 –	1
55 –	7
60 –	17
65 –	11
70 –	2
75 – < 80	2

a. Construct a percentage cumulative frequency table for the data.

b. Construct a percentage cumulative frequency graph from your table.

c. From your graph find estimates of, and write a sentence to interpret:

i. The median.

ii. The lower quartile.

iii. The upper quartile.

iv. The 90th percentile.

v. The 10th percentile.

vi. The interquartile range.

Stem-and-leaf plots

Stem-and-leaf plots are used to organise data into order as the data are collected. The first significant figure(s) of a data value are placed on the **stem**, in order of size (smallest to largest or vice versa), and the remaining significant figure is placed as a **leaf** alongside.

Example G

The data 124 116 118 103 125 117 109 112 126 105, can be shown on a stem-and-leaf plot by putting the numbers 10, 11, 12 on the stem first.

Stem	Leaves	
12	4 5 6	[this includes 124, 125, 126]
11	6 8 7 2	[ie 116, 118, 117, 112]
10	3 9 5	

An **ordered stem-and-leaf plot** has leaves arranged in order of size.

Example H

The distances of the best twelve throws by the competitors in the men's hammer at the 1990 Commonwealth Games were (in metres rounded down to 3 sf):

70.9 72.7 70.7 71.2 70.1 73.5 70.7 69.8 70.1 75.6 71.1 70.9

Distances can be plotted on a stem-and-leaf graph with the whole metres on the stem.

Men's Hammer Throw (unordered)

Stem	Leaves
75	6
74	
73	5
72	7
71	2 1
70	9 7 1 7 1 9
69	8

Men's Hammer Throw (ordered)

Stem	Leaves
75	6
74	
73	5
72	7
71	1 2
70	1 1 7 7 9 9
69	8

Key:
75 | 6 means 75.6 m

Various statistics, such as median, upper and lower quartiles, and range, can now be found easily.

Comparison of variables using back-to-back stem-and-leaf plots

By using **back-to-back stem-and-leaf plots**, two sets of univariate data can be compared.

Example I

Compare the efforts of the men hammer throwers at the 1974 and 1990 Commonwealth Games.

From the graph, it can easily be seen that the standard of performance in the men's hammer at the Commonwealth Games has improved from 1974 to 1990.

For example, the 1994 median is

$$\frac{67.3 + 67.4}{2} = 67.35 \text{ m},$$

whereas the 1990 median is 70.9 m.

Men's Hammer Throw

1974		1990
	75	6
	74	
	73	5
	72	7
	71	1 2
	70	1 1 7 7 9 9
5 4	69	8
9 0	68	
9 4 3 1 0	67	
8 8 6	66	

Key:
4 | 69 | means 69.4 m (1974)
| 69 | 8 means 69.8 m (1990)

Box-and-whisker plots (boxplots)

Another method of displaying the spread of data is by means of a **box-and-whisker plot**. Five points are graphed on a box-and-whisker plot. They are shown on the diagram to the right.

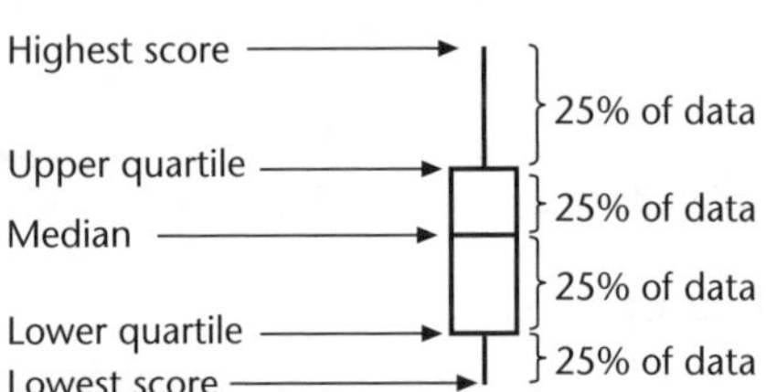

Example J

Q. Draw a box-and whisker plot for the data 3, 5, 8, 9, 7, 6, 4, 3, 5, 6, 1, 3, 4, 2, 5.

A. Ranking the scores:

1, 2, 3, 3, 3, 4, 4, 5, 5, 5, 6, 6, 7, 8, 9

Highest score 9
Upper quartile....... 6
Median 5
Lower quartile....... 3
Lowest score 1

or

The graph shows that 25% scored less than 3, 25% scored between 3 and 5, 25% scored between 5 and 6 and 25% scored between 6 and 9.

Comparing two variables using box-and-whisker plots

Box-and-whisker plots are very useful for comparing two (or more) sets of univariate data.

Example K

Q. Use the back-to-back stem-and-leaf plots of the men's hammer throw in Example H to draw a pair of box-and-whisker plots for the data.

A. From the graphs the following statistics are calculated.

Score \ Year	1974	1990
Highest	69.5	75.6
Upper quartile	68.45	71.95
Median	67.35	70.9
Lower quartile	66.9	70.4
Lowest	66.6	69.8

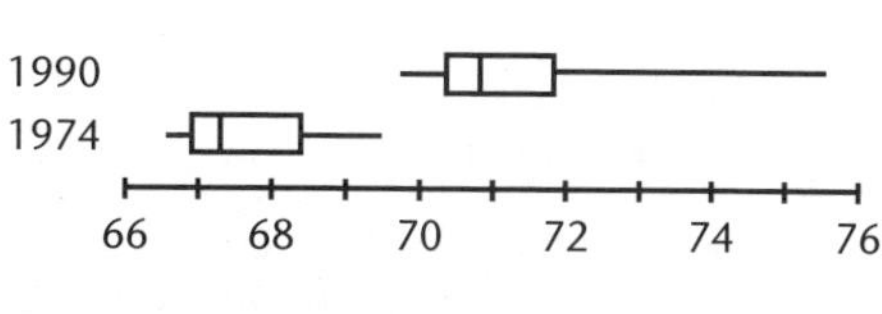

The superiority of the 1990 data can clearly be seen from the graph. The interquartile range is similar for both years, but all 1990 competitors threw better than the best throw for the 1974 competitors. Improved throw techniques and training methods are some possible reasons for the superior distances of the 1990 results.

Unit 11.3 Activity 2C: Further data displays

1. For a science project, Sara measured the lengths of 20 leaves of an indoor pot plant. The lengths, to the nearest mm, were:

38 41 44 51 42 29 33 36 42 37 34 42 51 30 53
43 41 28 31 38

a. Display these data using a stem-and-leaf plot.

b. Find the: **i.** Median. **ii.** Upper quartile. **iii.** Lower quartile lengths.

c. Display these data using a box-and-whisker plot.

2. The test marks of 30 students are shown below:

32 44 72 52 69 49 32 68 53 71 55 51 33 46 74
35 48 61 31 47 47 68 54 36 48 70 52 50 63 51

a. Plot these values using a stem-and-leaf plot.

b. Find the: **i.** Median. **ii.** Upper quartile. **iii.** Lower quartile scores.

c. Display these data using a box-and-whisker plot.

3. The number of points scored by the PNG Senior Girls Netball team in their twenty games last season were:

76 84 92 53 65 82 76 45 91 78 63 79 81 64 58 71
93 81 99 92

a. For these scores find:

i. Median. ii. Upper quartile. iii. Lower quartile.

iv. Range. v. Interquartile range.

b. Draw a box-and-whisker graph to display these data.

4. Jason set out to do a comparison between the times of competitors in the men's decathlon 400 m and the women's decathlon 400 m at the 1990 Commonwealth Games. The times, measured to the nearest 0.1 second, were:

Men: 47.8 51.2 47.7 49.9 49.8 49.6 51.3 49.5 50.5 51.4 51.8 50.6

Women: 51.7 52.8 53.8 54.8 54.8 55.9 56.4 51.2 52.4 52.8 53.3 53.9

a. Plot these values on a back-to-back stem-and-leaf graph.

b. Display these data using box-and-whisker plots.

c. Jason felt that a woman 400 m runner should be given a 4-second start against the men decathletes over 400 m. Comment on this.

5. The data alongside relates to the maximum daily temperatures (°C) in January of two provinces, A and B.

a. Draw box-and-whisker plots for the two data sets.

b. Comment on the statement 'Province A is hotter than Province B'.

	Province A	Province B
Mean	19.6	20
Median	22	21
Maximum	28	26
Minimum	12	15
Lower quartile	17	18
Upper quartile	24	22

Time series graphs

When numerical data is collected over **consecutive time intervals** a line graph joining the data points is the appropriate type of graph. This is called a **time series graph.**

The consecutive time intervals can be seconds, minutes, hours, days, weeks, months, years, etc.

The following time series graph shows the exchange rate for one Papua New Guinea kina against the Australian dollar for the years 1988 to 2008.

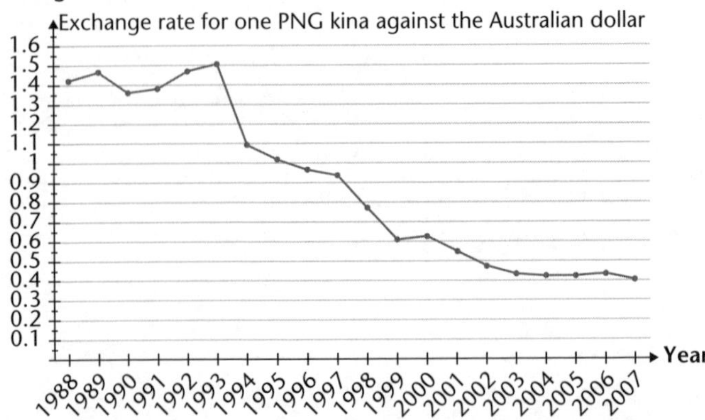

A continuous line joins the points on a time series graph so that any upward or downward pattern, called a **trend**, can be seen.

The graph above shows a slight upward trend for the years 1988 to 1993 then a steady downward trend from 1993 to 2003. After 2003 the exchange rate has remained about the same at approximately 0.42.

If there is a **repeated pattern** apparent on a graph then the graphed data is said to be **seasonal**. Seasonal data is often seen on graphs for sales.

Sales (K1000s)

60 50 40 30 20 10

1 2 3 4 5 6 7 8 9 10 11 12

Quarter

The graph at right shows a seasonal pattern that is repeated every four quarters. There also appears to be a downward trend to the data.

Time series graphs are used to **predict** into the future. A **trend line** can be applied to the data so that predictions can be made. Using the trend line to predict within the given data set is called **interpolation.**

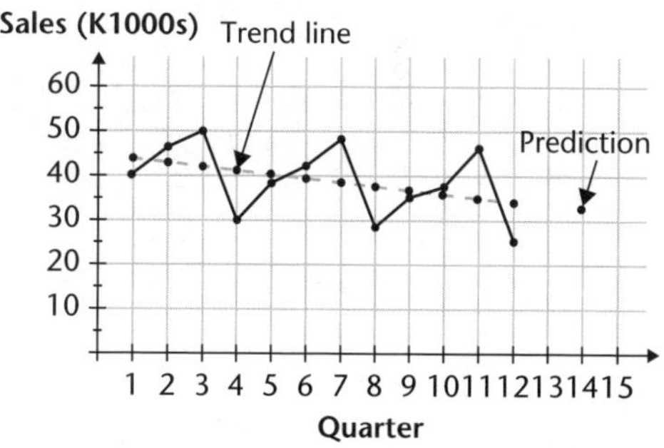

For example, if the trend line is used to predict the sales for quarter 4 then the prediction would be K41 000.00. This would be a prediction that ignores the seasonal fluctuations; called a de-seasonalised prediction.

Using the trend line to predict sales outside the given data set is called **extrapolation**. For example, using the trend line to predict the sales (de-seasonalised) for quarter 14 gives the value K33 000.

Some time series graphs show no trend or seasonal pattern; these graphs are described as showing **random fluctuations**. On graphs that show random fluctuations any particular highs or lows should be mentioned.

Example L

Kendra recorded the number of hours that she worked on each of 12 days.

The graph shows only a random pattern (no trend or seasonal fluctuations). The most hours that she worked in a day was 12 hours and on two days she worked the minimum of 4 hours.

Hours spent working

14 12 10 8 6 4 2

1 2 3 4 5 6 7 8 9 10 11 12 13 **Day**

Unit 11.3 Activity 2D: Time series graphs

1. **a.** Construct a time series graph for the data given below:

   Estimated volume (tonnes) of cured vanilla bean exports in PNG for the years 1997–2006.

   b. Comment on the graph.

Year	Volume (tonnes)
1997	1
1998	4
1999	8
2000	24
2001	50
2002	160
2003	200
2004	215
2005	68
2006	26

2. The time series graph below shows the estimated rice production in PNG for the years 1980–2000.

Estimated production of rice (tonnes)

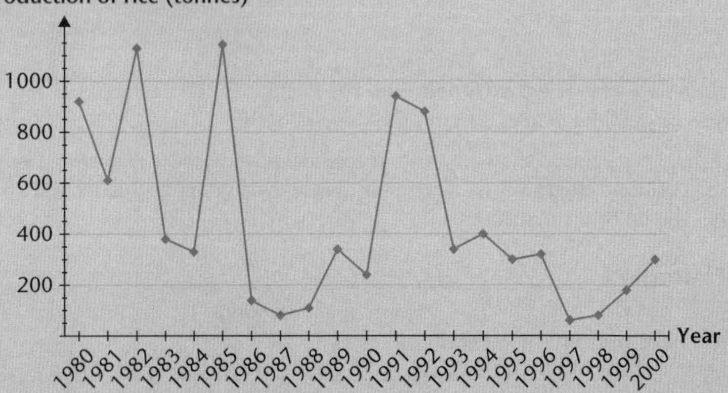

   a. Does the graph show: **i.** A random pattern?
   ii. A seasonal pattern?
   iii. A trend?

   b. Determine the year of: **i.** Highest production.
   ii. Lowest production.

   c. Describe the trend over the 21 years.

   d. Use the trend over the last four years to predict the rice production in the year 2001.

3. The graph below shows the volume (tonnes) of fruit and vegetable imports from Australia and New Zealand for the years 1983–2003.

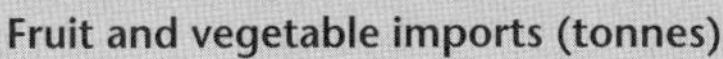

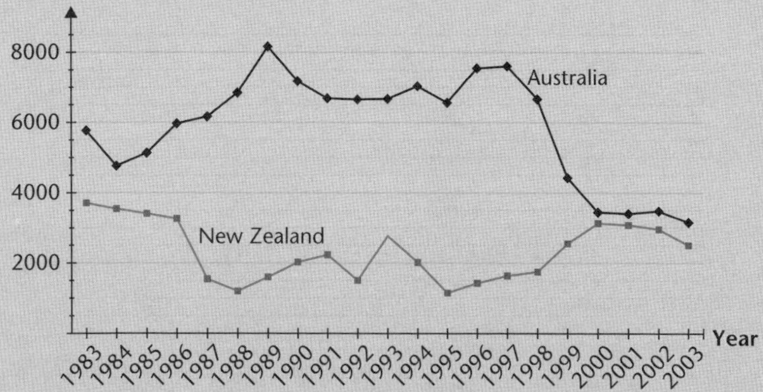

a. Describe the pattern of the data for: **i.** Australia.

ii. New Zealand.

b. Compare the pattern for each of the countries over the years 2000 to 2003.

c. Use the data to predict the fruit and vegetable imports from Australia and New Zealand in 2004.

4. The following graph shows small holder coffee production and sales of food in the Kainantu area, Eastern Highlands Province, 1982–1984.

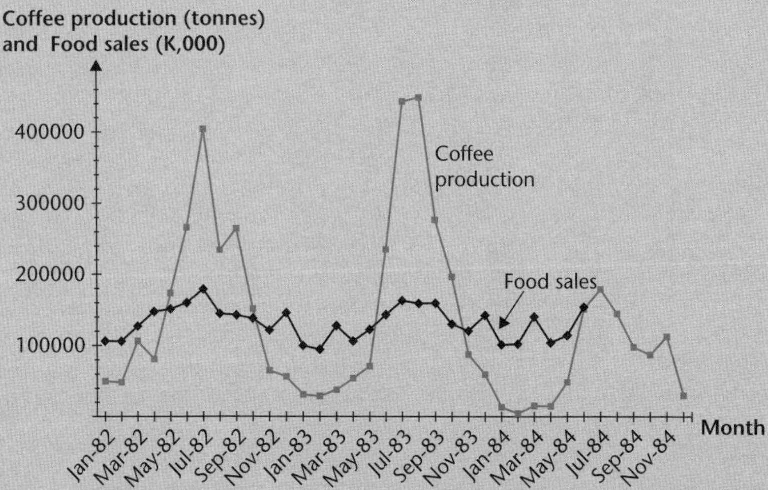

a. Describe the pattern in the coffee production graph.

b. Describe the pattern in the food sales graph.

c. What do you notice when comparing the coffee production pattern and the food sales pattern?

5. The following graph shows the average price per year, in toea, of sweet potato in Port Moresby and Madang.

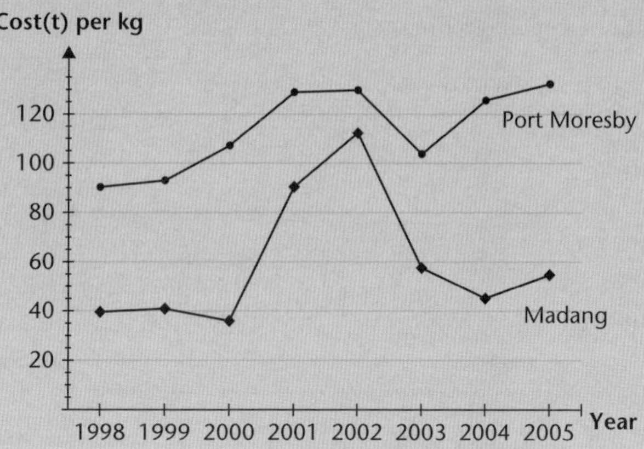

a. Describe the pattern in the Port Moresby data.

b. Describe the pattern in the data from Madang.

c. Compare the data for Port Moresby and Madang.

6. The graph below shows the distance travelled by a PMV owner in the first four weeks of operation. Day 1 is a Sunday.

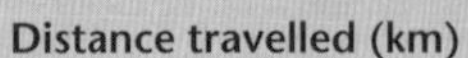

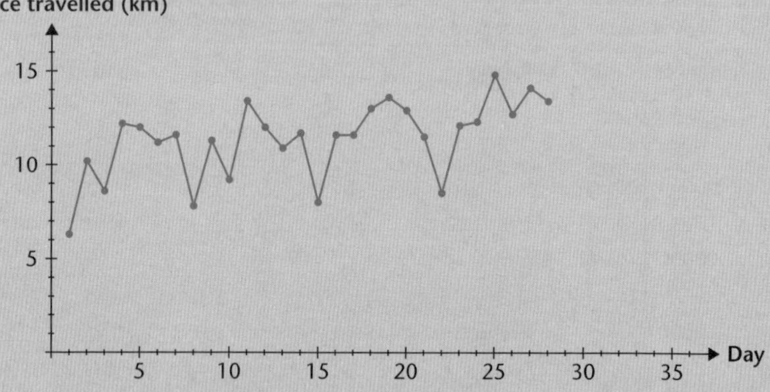

a. Describe the pattern in this time-series graph.

b. A line that can be used to predict the trend in this data goes through the points (5, 10) and (15, 11.5). Locate this line on the graph using a ruler.

c. Use this line to predict the distance the PMV will travel on:

i. Sunday of the 5th week.

ii. Wednesday of the 5th week.

iii. Saturday of the 5th week.

d. Which of the predictions from part c is likely to be closer to the actual value? Why?

Unit 11.3 Statistics

Topic 3: Analysis of data—interpreting statistics

Unit 11.3 focuses on everyday data and how it is collected, presented, analysed and interpreted. This Topic looks at the use of statistical methods and information and covers:

- Interpreting statistical information and answering straightforward questions.
- Commenting on the validity of a statement with respect to features of statistical information.

Introduction

In everyday life statistical information is presented in many different forms, such as tables, graphs and summary statistics (averages, spread, etc). This chapter deals with techniques for interpreting such information and commenting on its features.

Interpreting graphs and statistics

Reading tables

Often data is presented in a table, and, although a suitable graph can be drawn from the table, the data can be analysed just as easily from the figures in the table.

Example A

A. The life expectancy at birth for the years 1995 and 2011 is given in the table below for 15 countries.

Life expectancy at birth (years)

Country	1995	2011
PNG	61	65
Australia	78	82
Indonesia	66	71
Solomon Islands	70	74
Fiji	66	71
Malaysia	69	74
India	60	67
Thailand	70	74
China	69	75
Russia	65	66
Turkey	68	73
South Africa	64	49
Samoa	68	72
Uganda	42	53
Cuba	75	78

a. What does 'life expectancy at birth' mean?

b. Which of the countries had the lowest life expectancy at birth in:

i. 1995?

ii. 2011?

c. For which country did the life expectancy at birth decrease from 1995 to 2011?

d. Which of these countries had the greatest increase in life expectancy at birth from 1995 to 2011?

e. Which of these countries had an increase in life expectancy at birth of more than 5 years in the period from 1995 to 2011?

f. What is the percentage change in life expectancy at birth for China?

g. List some factors that you think might affect the life expectancy at birth.

A. a. Life expectancy at birth means the number of years you are expected to live before you die. A child born in 2011 in PNG is expected to live for 65 years.

b. i. Uganda had the lowest life expectancy at birth in 1995 and 2011.

ii. South Africa had the lowest life expectancy at birth in 2011.

c. The life expectancy at birth decreased in South Africa from 1995 to 2011.

d. Uganda had the greatest increase (9 years).

e. India, China and Uganda.

f. Percentage change for China = $\frac{75-69}{69} \times \frac{100}{1} = 8.7\%$

g. Nutrition, access to medical treatment, education, immunisation.

The distribution of a set of data

If a set of data is graphed we can comment on the overall shape, called the distribution of the data. We use the terms symmetric, positively skewed and negatively skewed.

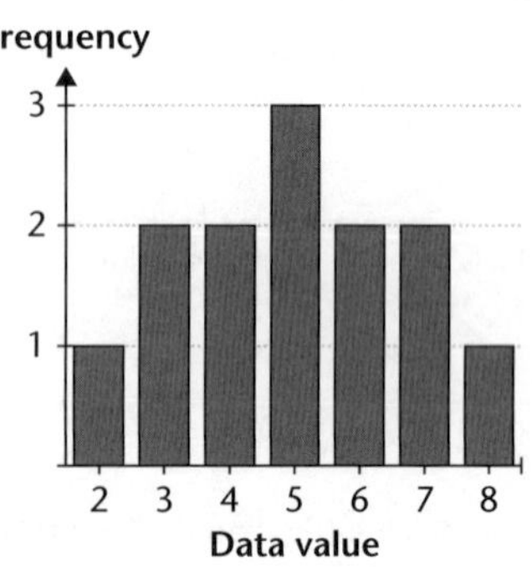

A **symmetric** distribution of a set of data appears equally spread on either side of a central value.

For this **symmetric data set** the mean, median and mode are all the same value (5). The graph is **symmetric** about the value 5.

A **positively skewed** set of data would appear on a graph to be stretched to the right and the mean (19) would be greater than the median and mode (both in the 10–15 class interval). This set of data has an extreme value in the 60–65 interval.

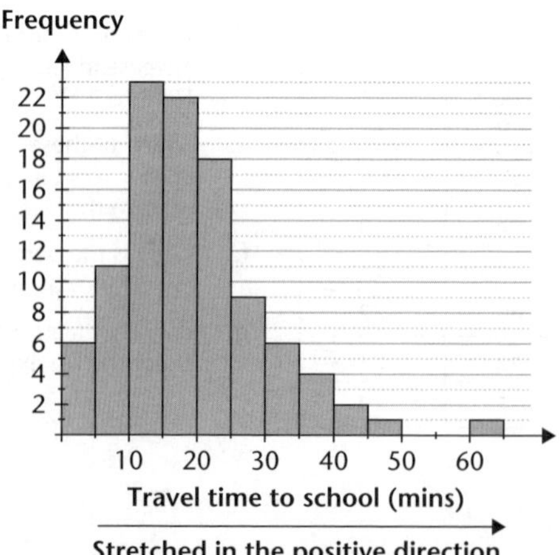

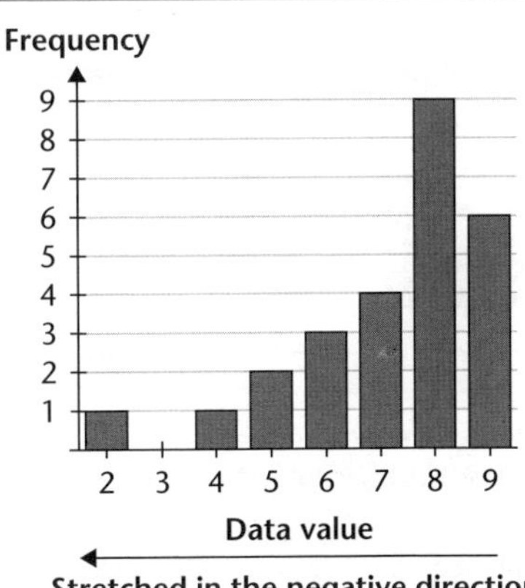

A **negatively skewed** set of data appears stretched to the left.

For this **negatively skewed data set** the mean is less than the median and mode because the mean is dragged towards the extreme small values (4 and 2).

In cases where there are extreme data values and the distribution of the data is skewed the most suitable measure of centre to use is the **median** or the **mode**.

The distribution of a set of data can also be seen from a **stem-and-leaf plot.**

Looking sideways at the stem-and-leaf plot below, we can see a frequency graph:

Stem	Leaf
1	6 7 8 8
2	2 3 3 4 4 6 7 8 9
3	1 2 5 8
4	0 4 6
5	1

The distribution of this set of data can be described as **positively skewed**; it is stretched towards the larger values.

Boxplots can also display the fact that a distribution is symmetric or skewed.

A symmetrically distributed set of data will have a symmetric boxplot; the median will be in the centre of the 'box' and the whiskers will be the same length.

The boxplot for a **positively skewed set of data** has a longer right whisker and the median is not in the centre of the 'box'; the whole boxplot appears stretched to the right:

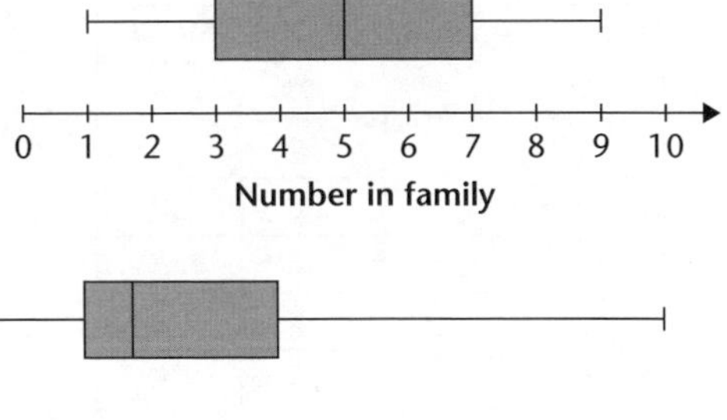

The boxplot for a **negatively skewed set of data** has a longer left whisker and generally appears stretched to the left.

When comparing two sets of data any statements made should be supported by the relevant statistic.

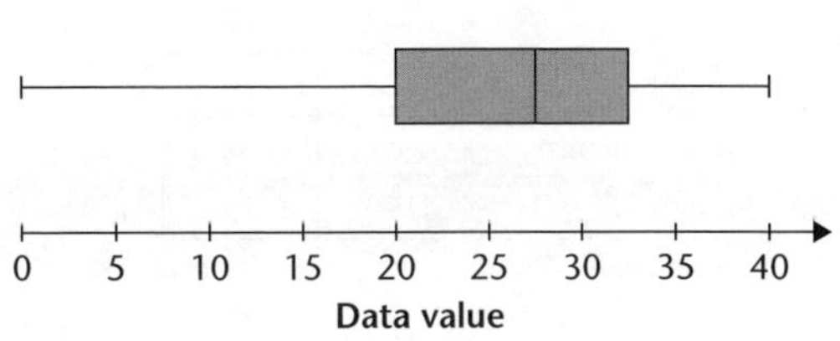

Example B

Q. The mathematics results for two classes are displayed on the boxplots below. Comment and compare the two sets of data.

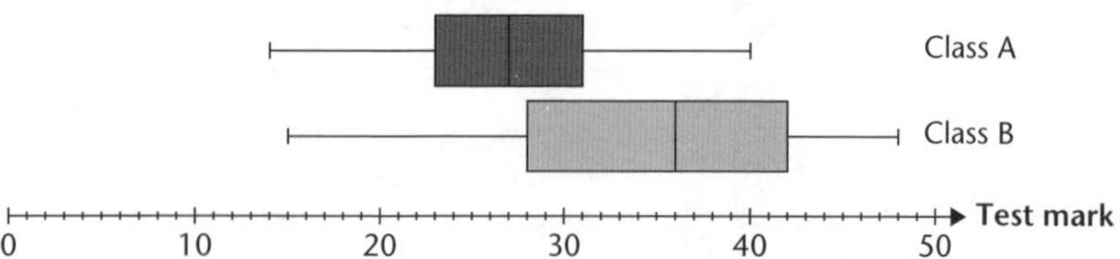

A. The distribution of test marks in class A appears to be symmetric about a median of 27. The middle 50% of test results are spread over 8 marks (IQR); the lowest mark is 14 and the highest is 40, giving a range of 26.

The distribution of test marks in class B appears to be negatively skewed with a greater spread of marks below the median mark. Half of the students in class B had a test mark of 36 or more (median) and the middle 50% of marks were spread over 14 marks (IQR). The lowest mark in class B was 15 and the highest was 48, giving a range of 33.

In general the marks were lower in class A than class B (comparing the medians: 27 compared to 36) but the students in class A had less varied results (comparing IQR's: 8 compared to 14. Also comparing ranges; 26 compared to 33).

More than 75% of the students in class B had a mark greater than the median of class A.

More than 25% of class B had a mark greater than the highest mark in class A.

Unit 11.3 Activity 3A: Interpreting graphs and statistics

1. The data given below gives the under-5 mortality rate for the years 1990 and 2009, for thirteen countries.

Under-5 mortality rate (deaths per 1000 live births)

Country	1990	2009
PNG	91	68
Australia	9	5
Indonesia	86	39
Solomon Islands	38	36
Fiji	22	18
Malaysia	18	6
India	118	66
Thailand	32	14
China	46	19
Russia	27	12
Turkey	84	20
South Africa	62	62
Samoa	50	25

a. What does the figure 68, for PNG in 2009, mean?

b. Which of these countries had the greatest decrease in their under-5 mortality rate from 1990 to 2009?

c. Which country had no change in their under-5 mortality rate from 1990 to 2009?

d. Which countries had an under-5 mortality rate greater than 50 in:

i. 1990?

ii. 2009?

e. What is the percentage change in under-5 mortality for PNG from 1990 to 2009?

2. In a village the number of chickens per household was counted to give the following data:

4 4 4 5 5 5 5 5 6 6 6 7 7 7 8 8 9 10 11 13

   a. Construct a suitable graph for the data.

   b. Calculate the median and the upper and lower quartiles.

   c. Construct a boxplot for the data.

   d. Use your graph and statistics to comment on the distribution of the data.

3. The following graph gives the PNG production of export tree crops in the years 1975 and 2005.

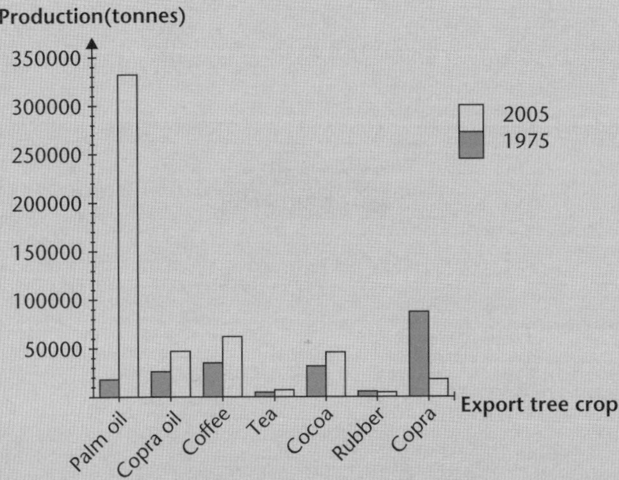

Comment on the change, from 1975 to 2005, in production for:

   a. Palm oil.

   b. Coffee.

   c. Copra.

4. The number of avocados per tree was counted and the data displayed on the graph below:

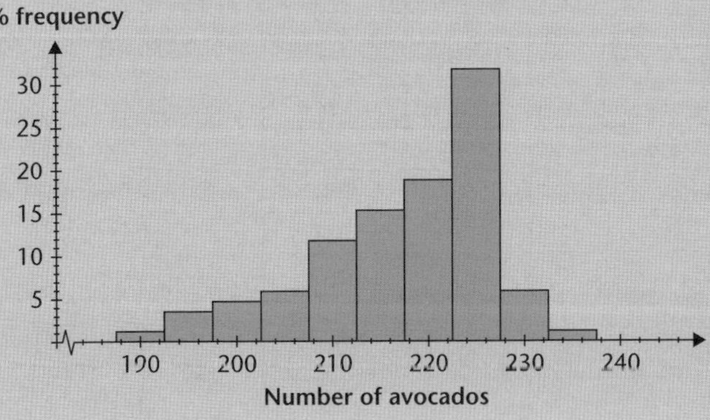

A boxplot showing the median, upper and lower quartiles of the data is given below:

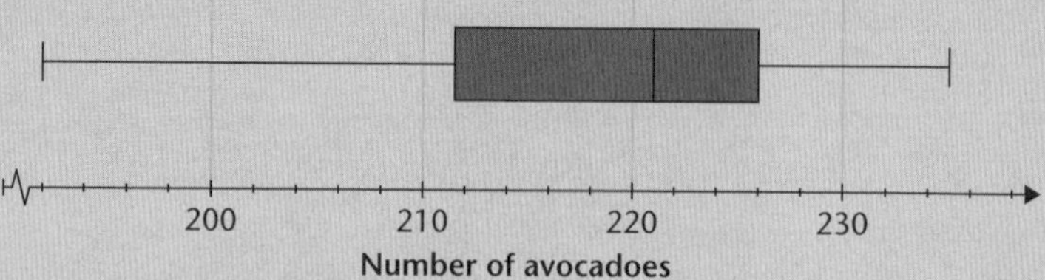

Comment on the distribution of the data, quoting statistics from the boxplot.

5. A team has recorded the number of points scored in each game of rugby over the last three years. The relevant statistics are displayed on the boxplots below:

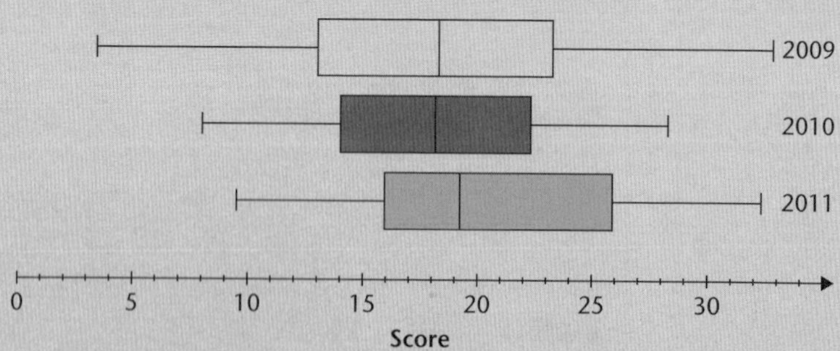

Comment on the team's scoring over the last three years.

6. The table below gives the male and female literacy rate (%) for 14 countries for the period 2004–2008:

Literacy (%)	Male	Female
PNG	65	69
Indonesia	97	96
India	88	74
China	99	99
Malaysia	98	99
Russia	100	100
Turkey	99	94
South Africa	96	98
Samoa	99	100
Cuba	100	100
Uganda	89	86
Afganistan	49	18
Egypt	88	82
Philippines	94	96

a. What does a male literacy rate of 65% for PNG mean?

b. In which of these countries is the male literacy rate less than the female literacy rate?

c. In which countries is there a difference in male/female literacy of more than 5%?

d. Which country has the greatest difference in male/female literacy?

e. Construct boxplots for male literacy and female literacy on the same axis.

f. Comment and compare the boxplots from part e, supporting your statements with the relevant statistic.

Unit 11.4 Geometry

Topic 1: Geometry of lines and angles

Unit 11.4 focuses on mathematics that deals with shapes and properties of planes and solid figures. This Topic covers:

- Properties of parallel and intersecting lines.
- Finding unknowns using two-step processes.
- Investigating a conjecture, presenting a proof or solving a problem using at least three steps of reasoning.

(Note that the definition of a plane, calculating angles between a line and a plane, and angles between two planes, are covered in Unit 11.5 Topic 3.)

Introduction

In the assessment of performance standards, the focus is on using the geometric properties of polygons and lines, to find unknowns and analyse shapes. This will involve at least two steps of reasoning.

- At the 'high achievement' level, all relevant geometric reasons must be included with your solution. Reasons are often given in abbreviated form in brackets.
- At the 'very high achievement' level, investigations or proofs of geometric results involve at least three steps of reasoning.

When solving geometric problems, writing geometric reasons for each step of your working will help clarify your thinking and improve your learning of the necessary geometrical facts.

Types of angle

An angle is a turning. Angles are measured in **degrees**. The angle shown is referred to as angle ABC or ∠ABC (or ∠CBA), or simply ∠B.

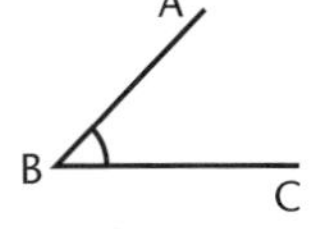

Acute Angle	Right Angle	Obtuse Angle
An angle between 0° and 90°	An angle which is 90°	An angle between 90° and 180°
Straight Angle	**Reflex Angle**	**Full Turn**
An angle which is 180°	An angle between 180° and 360°	An angle which is 360°

Angles on a straight line

The sum of the adjacent angles on a straight line is 180°. (Abbreviation: ∠s on a line.)

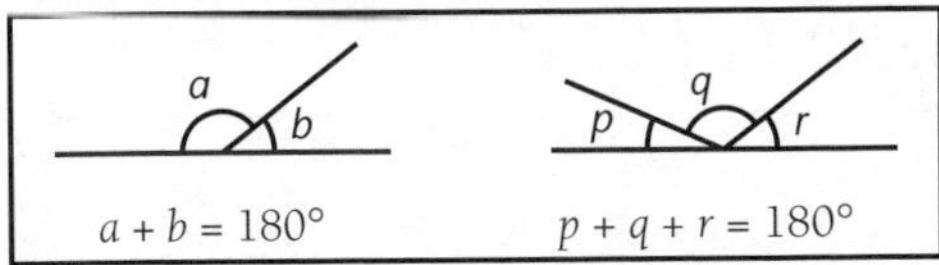

Example A

Q. Find the size of the angles shown, with reasons.

1.

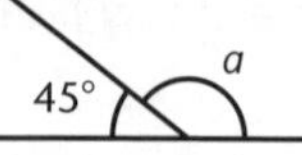

2.

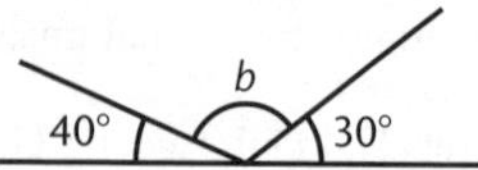

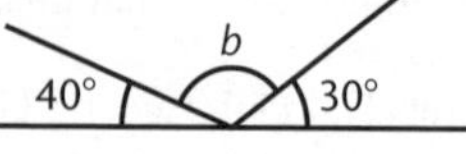

A. 1. $a + 45 = 180$ (∠s on a line)

$a = 135°$ [subtracting 45]

2. $40 + 30 + b = 180$ (∠s on a line)

$b = 110°$ [solving]

Angles at a point

The sum of the angles formed at the same point is 360°. (Abbreviation: ∠s at a point.)

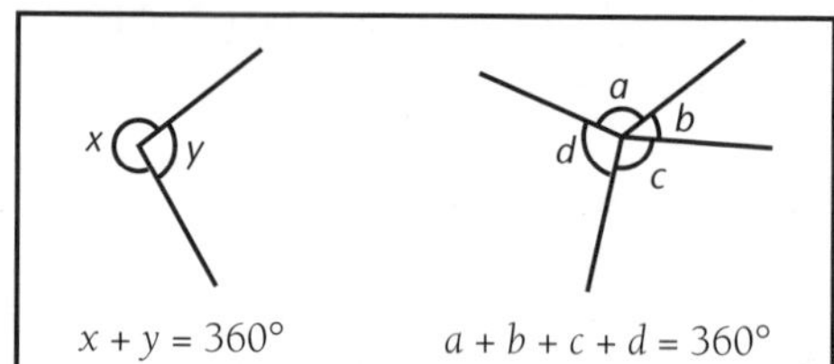

Example B

In the diagram, the size of the angle labelled x is given by:

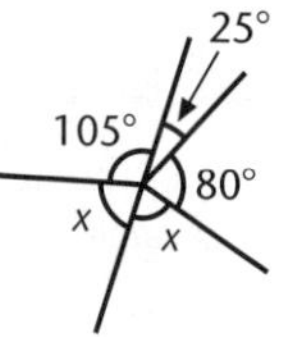

$x + x + 80 + 25 + 105 = 360$ (∠s at a point)

$2x + 210 = 360$ [simplifying]

$2x = 150$ [subtracting 210]

$x = 75°$ [solving]

Vertically opposite angles

Vertically opposite angles form when two straight lines intersect.

Vertically opposite angles are equal. (Abbreviation: vert opp ∠s.)

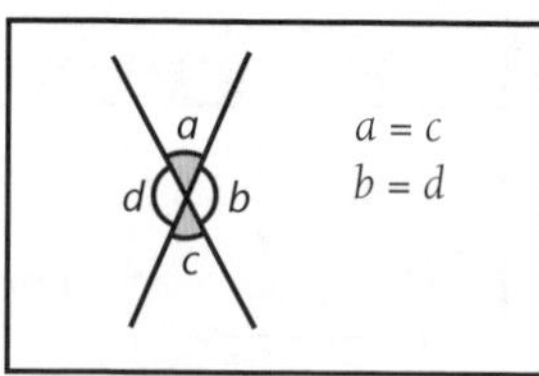

Example C

In the diagram:

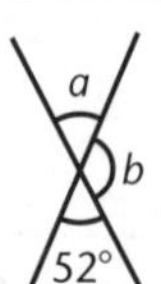

$a = 52°$ (vert opp ∠s)

$b = 180° - 52°$ (∠s on line)

$= 128°$

Unit 11.4 Activity 1A: Angles

Calculate the size of the angles shown by the letters. Give reasons with each answer.

1. a.

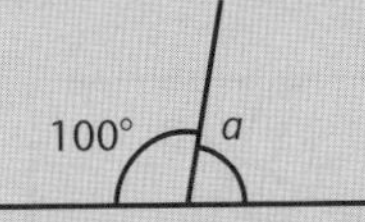

b.

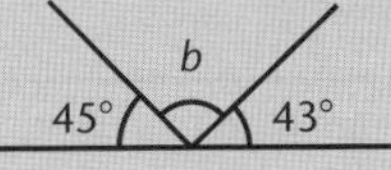

c.

d.

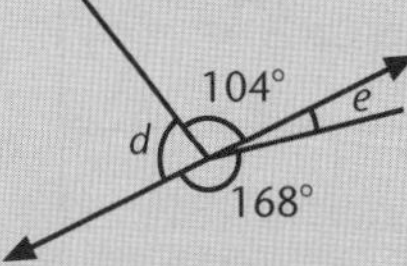

e.

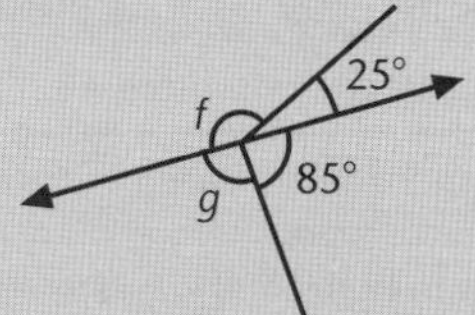

f.

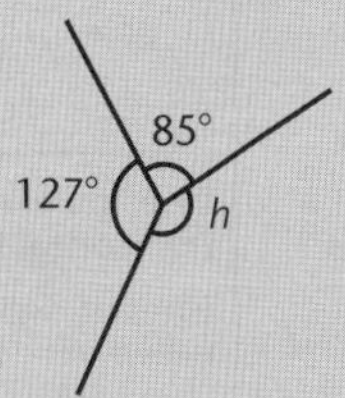

g.

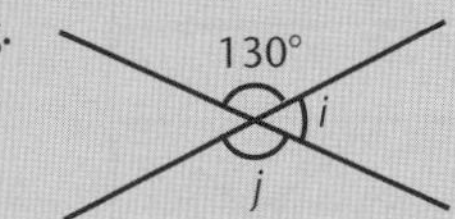

h.

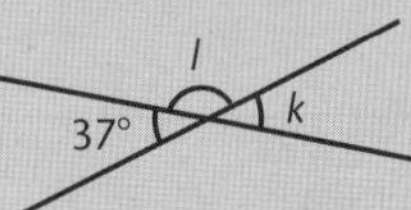

i.

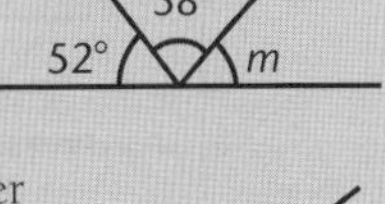

2. Tom wants to divide a straight angle into two angles, so that the larger of the two angles is double the size of the smaller angle. He draws the diagram alongside where x is the smaller angle and $2x$ is the larger angle.

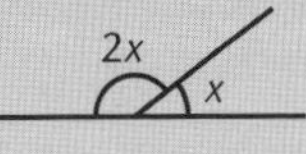

Solve the equation $2x + x = 180°$ in order to find the size of the smaller angle.

3. Form an equation and solve it to find the values of the unknown letters.

a.

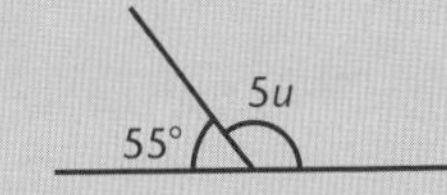

b.

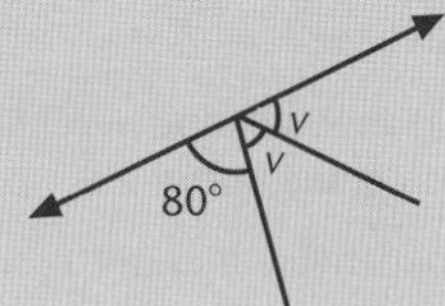

c.

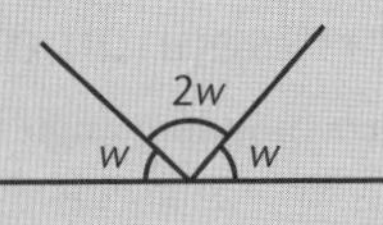

d.

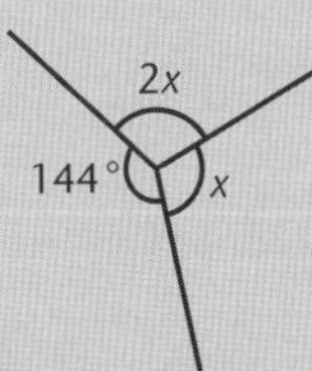

e.

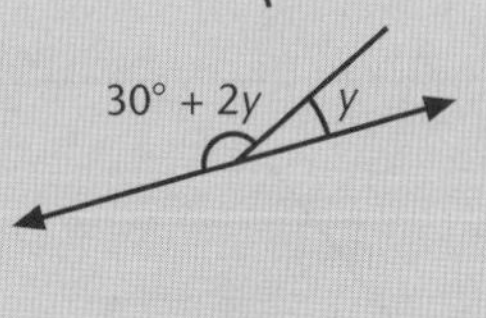

f.

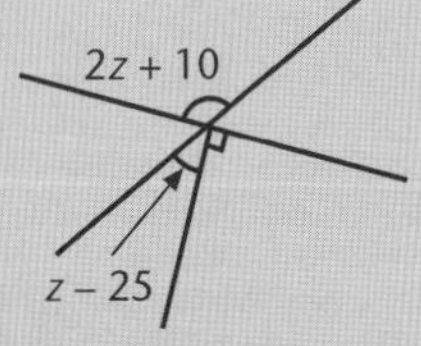

4. Write down an equation linking s, t and u using the geometrical properties of the figures. Simplify your answers.

a.

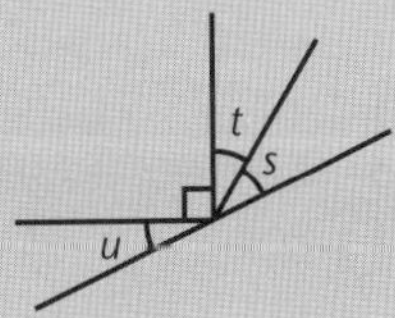

b.

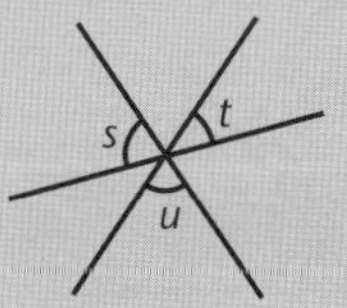

c.

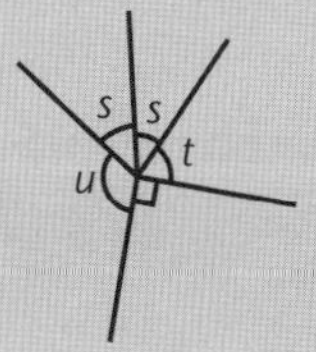

Parallel lines

Lines in the same plane that do not intersect are called **parallel lines**. Parallel lines are always the same distance apart.

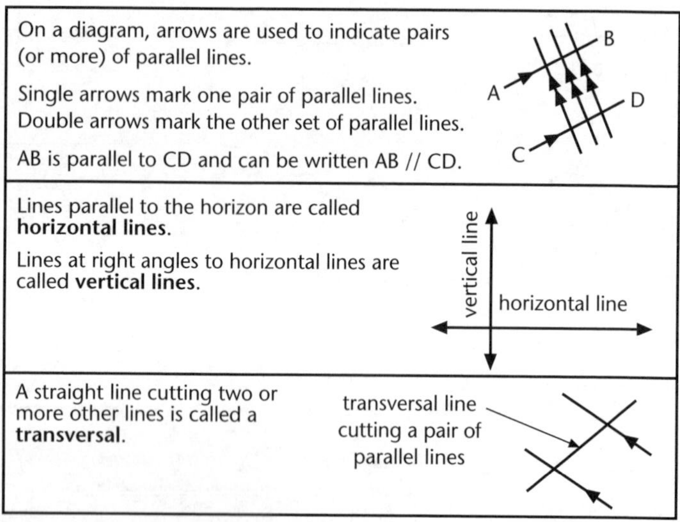

Angles and parallel lines

Three types of angles are formed when a transversal crosses a pair of parallel lines.

Corresponding angles

Corresponding angles are the two angles on the *same side* of the transversal and the lines. They are both 'above' or both 'below' each of the lines.

When a transversal crosses a pair of parallel lines, the corresponding angles are equal. (Abbreviated to: corr ∠s // lines.)

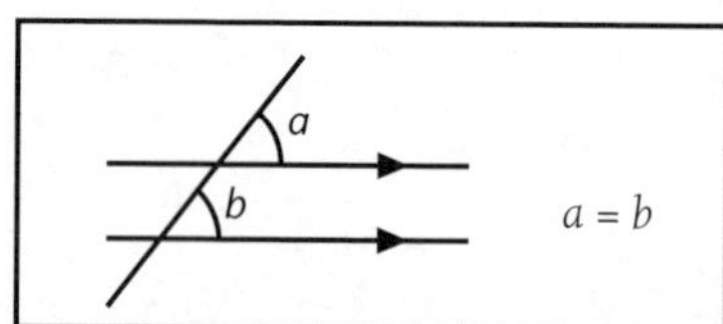

Note: There are four pairs of corresponding angles formed when a transversal cuts a pair of parallel lines. (Check these pairs of angles are equal with a protractor.)

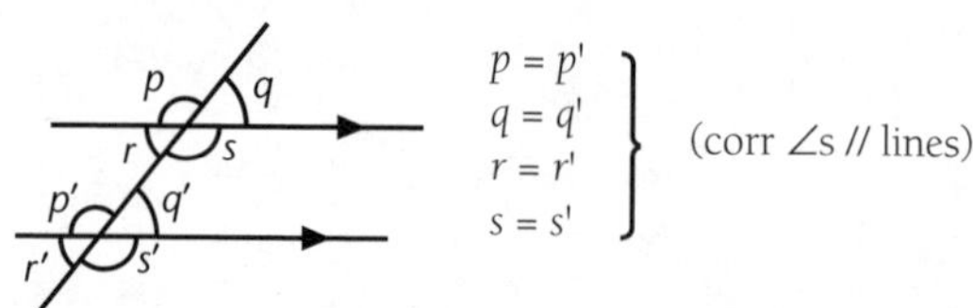

Example D

In the diagram: $a = 180° - 115°$ ($\angle$s on line)

$= 65°$

$b = 65°$ (corr $\angle$s // lines)

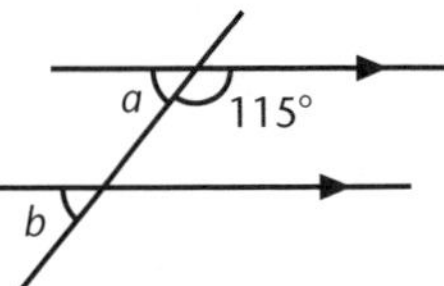

If a transversal cuts two lines such that corresponding angles are equal, then the lines are parallel. If the corresponding angles are *not equal*, then the lines are *not parallel*.

Example E

Q. In the diagram (not drawn to scale) which pairs of lines are parallel?

A. CD is parallel to AB since corresponding angles are equal (120°).

EF is not parallel to GH since corresponding angles are not equal (115° ≠ 120°).

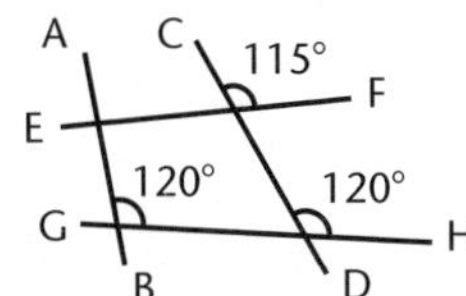

Alternate angles

The two angles between lines, but on different sides of the transversal, are called **alternate angles**.

When a transversal cuts a pair of parallel lines, the alternate angles formed are equal. (Abbreviated to: alt $\angle$s // lines.) Check this with a protractor.

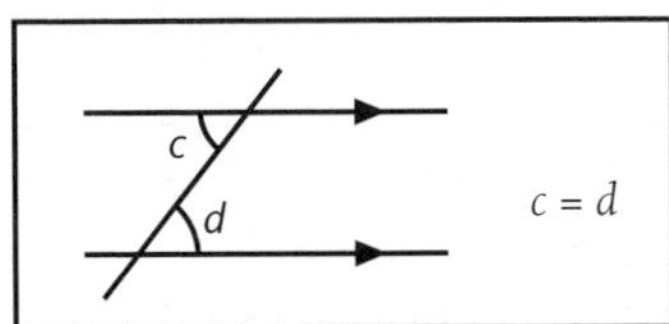

Note: There are two pairs of alternate angles formed when a transversal cuts a pair of parallel lines.

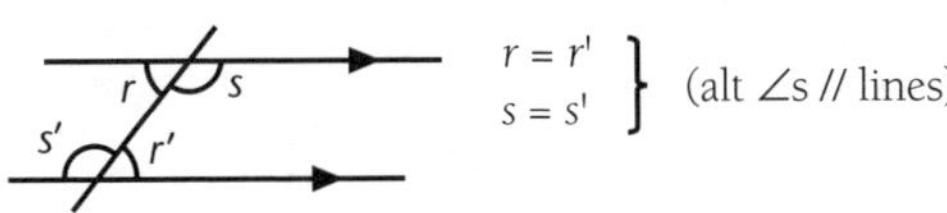

Example F

In the diagram: $c = 110°$ ($\angle$s on line)

$d = 110°$ (alt $\angle$s // lines)

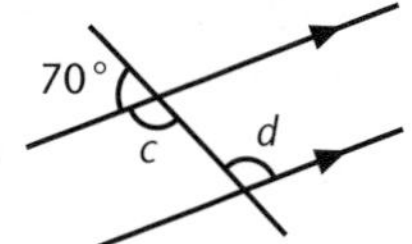

If a transversal cuts two lines such that the alternate angles are equal, then the lines are parallel. If the alternate angles are *not equal*, then the lines are *not parallel*.

Example G

Q. In the diagram (not drawn to scale) which pairs of lines are parallel?

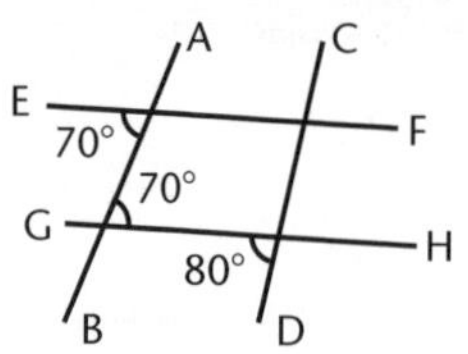

A. EF is parallel to GH since alternate angles are equal (both 70°).

AB is not parallel to CD since alternate angles are not equal ($70° \neq 80°$).

Cointerior angles

The two angles formed between lines and on the *same* side of the transversal are called **cointerior** (or **allied**) angles.

When a transversal cuts a pair of parallel lines, the sum of the cointerior angles is 180°. (Abbreviated to: coint ∠s // lines.) Check this with your protractor.

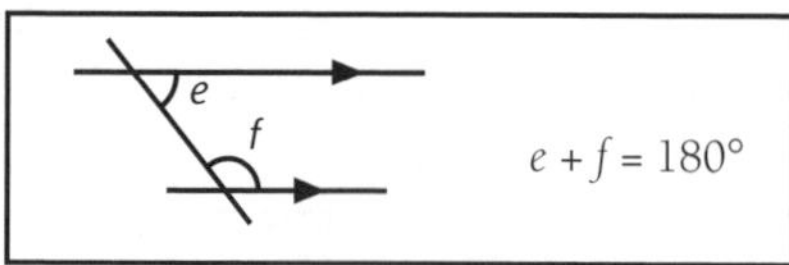

There are two pairs of cointerior angles formed when a transversal cuts a pair of parallel lines.

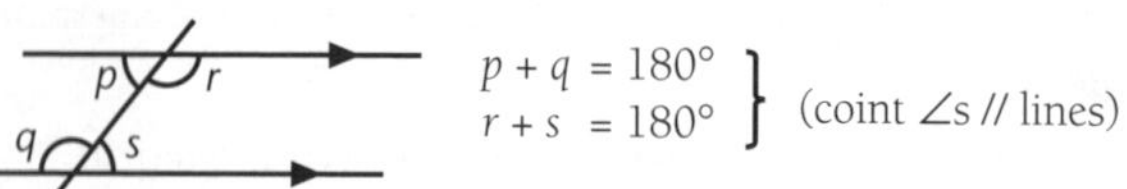

Note: Cointerior angles are not equal unless each one is 90°.

Example H

In the diagram:

$e = 130°$ (vert opp ∠s)

$e + f = 180°$ (coint ∠s // lines)

$\therefore 130° + f = 180°$ [substituting $e = 130°$]

$\therefore f = 50°$ [subtracting 130 from each side]

If a transversal cuts two lines such that the cointerior angles add to 180° then the two lines are parallel. If the cointerior angles do *not have a sum of 180°*, then the lines are *not parallel*.

Example I

In the diagram, the line AB is not parallel to the line CD since cointerior angles do not add to 180° ($110° + 80° = 190°$).

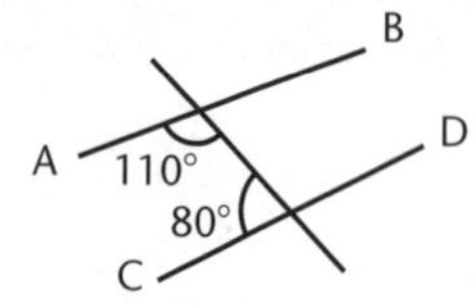

Problem solving using angles and parallel lines

Some problems involve several unknown angles. Labelled problems are usually designed so that it is best to find the unknown angles in alphabetical order. Angle rules of all types may be involved; more than one may apply to finding an unknown.

Example J

Q. Find the angles labelled with letters, giving reasons.

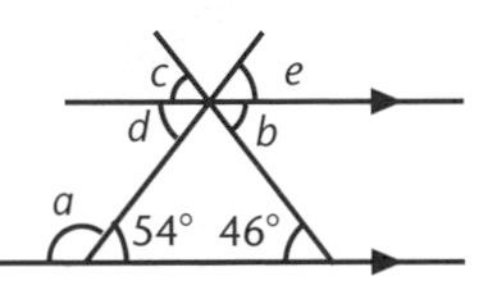

A. $a = 126°$ (∠s on line)

$b = 46°$ (alt ∠s // lines)

$c = 46°$ (vert opp ∠s) or (corr ∠s, // lines)

$d = 54°$ (alt ∠s // lines) or (coint ∠s // lines)

$e = 54°$ (corr ∠s // lines) or (vert opp ∠s)

Unit 11.4 Activity 1B: Angles and parallel lines

1. Calculate the size of the angles shown by the letters. Give reasons with each answer.

a.

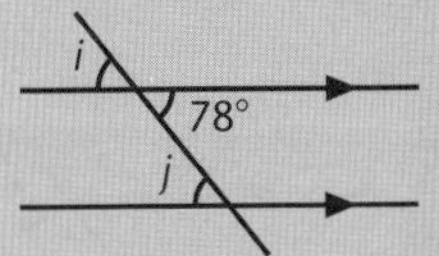

b.

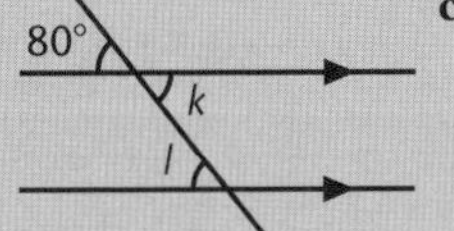

c.

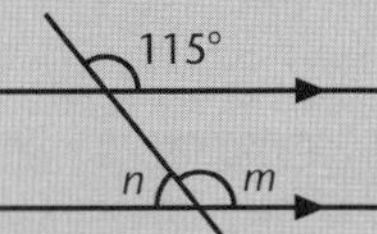

d.

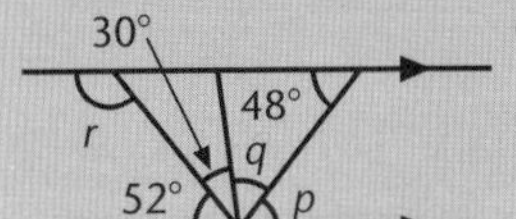

e.

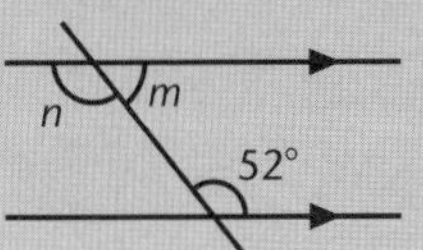

f.

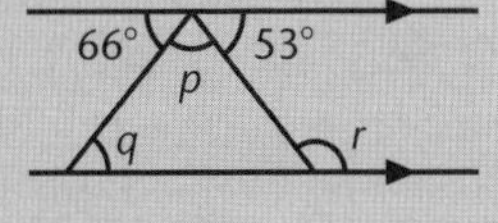

2. Two vertical poles AB and CD are put in sloping ground. A rope joins A to C and is parallel to the ground. Support ropes join A to the ground at E, and C to the ground at F, as shown.

Find ∠AEB, the angle the support rope AE makes with the ground.

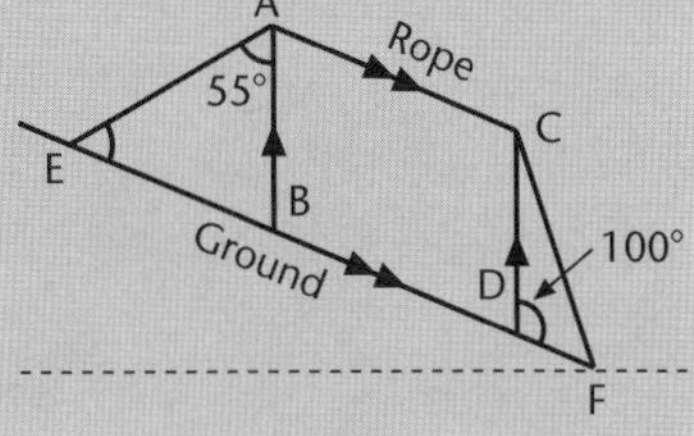

3. In each diagram below find x, y and z and decide whether the lines cut by the transversal are parallel. Give reasons.

a.

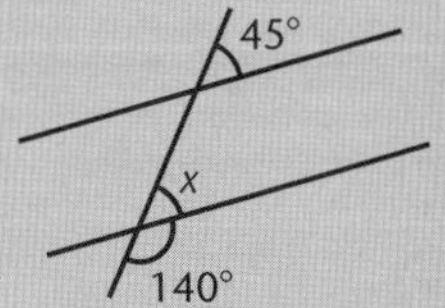

b.

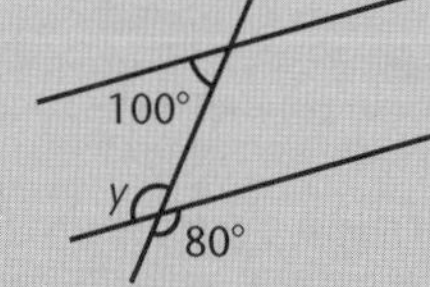

c.

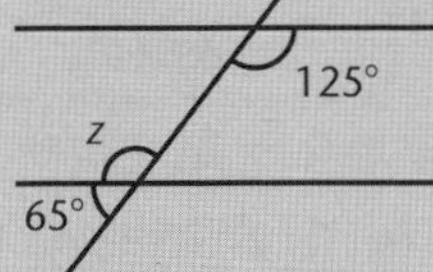

4. In the diagram alongside:

a. Find x and y.

b. Name any pairs of parallel lines in the diagram, giving reasons for your answer.

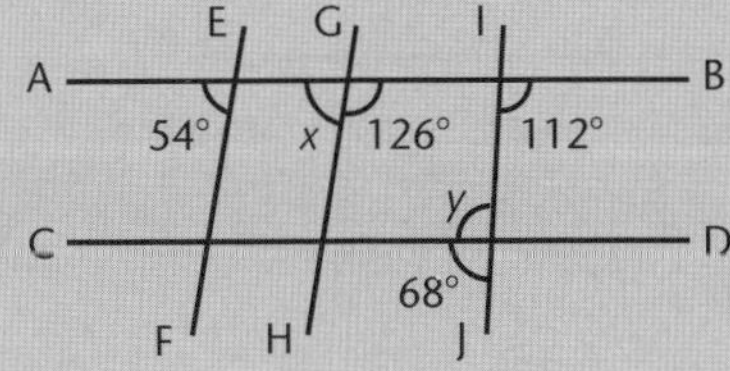

5. **a.** Find, with reasons, the value of x.

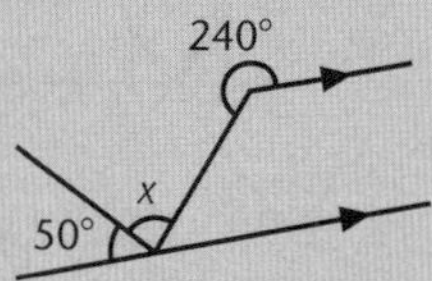

b. Find, with reasons, the value of a.

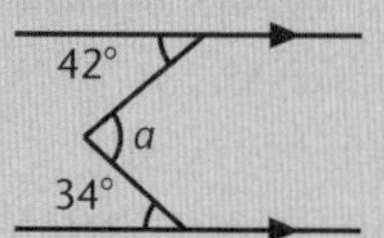

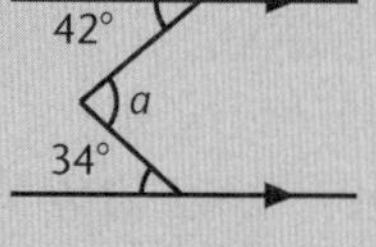

6. In the diagram, PQ and RS are parallel.

Fill in the blanks to prove that $a = b$.

$\angle CBQ = a$ (________________)

$\angle CBQ =$ _____ (corr ∠s // lines)

$\therefore\ a = b$

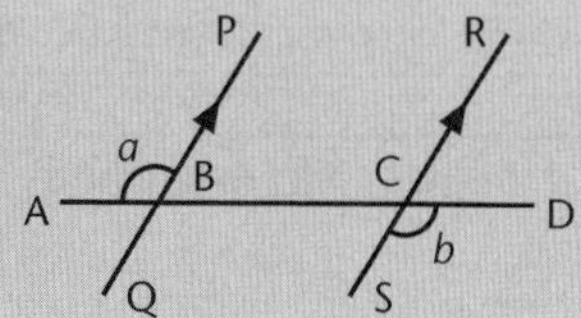

7. AB is parallel to DC and AE is parallel to FC

Fill in the blanks to prove that $x = y$.

$\angle BAD = x$ (________________)

$x = y$ (________________)

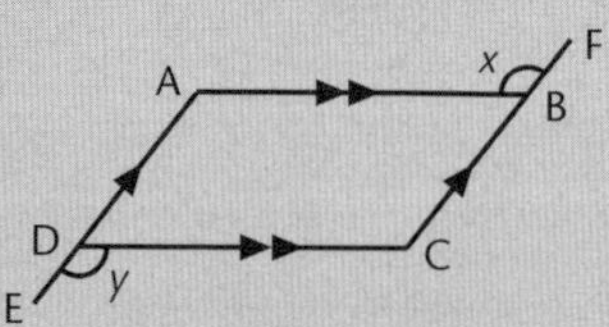

8. PR is parallel to ST and $\angle SQR = \angle STR = x$.

Prove that SQ is parallel to TR.

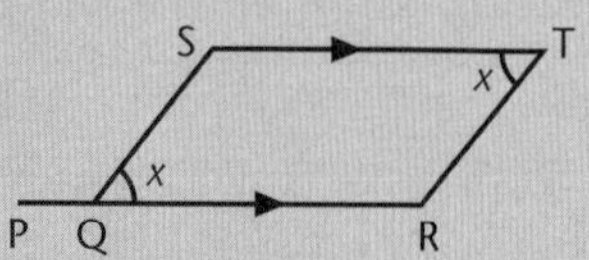

9. AB is parallel to CE and $x = y$.

Prove that BG is parallel to EF.

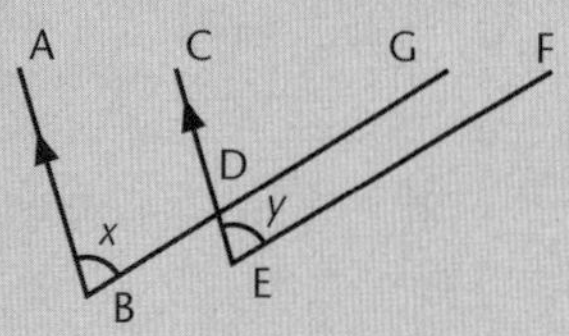

10. Each diagram has an equation linking s, t and u. Prove that each equation is true.

a.

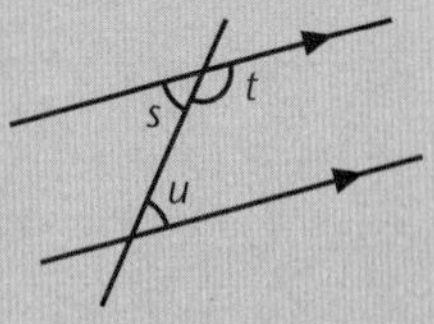

$s + 2t + u = 360°$

b.

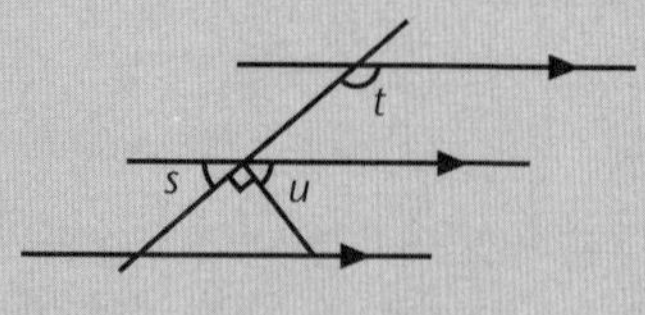

$2s + t + u = 270°$

c.

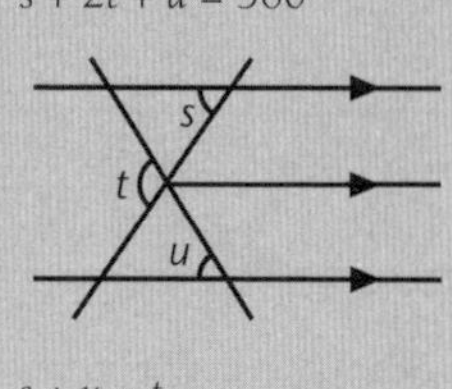

$s + u = t$

Unit 11.4 Geometry

Topic 2: Triangles, quadrilaterals and other polygons

This Topic deals with the use of geometric reasoning to solve problems, building on concepts covered in Topic 1 and introducing polygons and their properties. It covers:

- Properties of polygons.
- Similar polygons.
- Finding unknowns in polygons using two-step processes.
- Investigating a conjecture, presenting a proof or solving a problem using at least three steps of reasoning.

Polygons

A **polygon** is a closed *many-sided* figure whose sides are straight line segments. The names of some important polygons are shown in the table alongside.

A **regular polygon** has all sides of equal length and all angles of equal size.

In a polygon, a **diagonal** is any line segment that joins two **vertices** (corners) and is not a side.

Number of sides	Name of Polygon
3	triangle
4	quadrilateral
5	pentagon
6	hexagon
8	octagon
10	decagon
12	dodecagon

Example A

The pentagon PQRST has 5 diagonals: diagonals PR and QT are shown.

The remaining diagonals are the line segments PS, QS and RT (not shown).

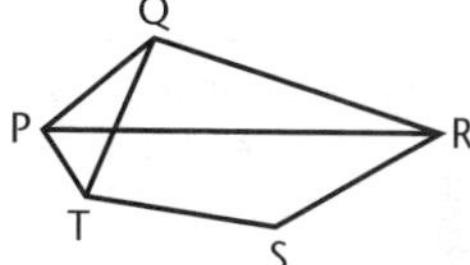

The angles inside a polygon are called **interior angles**. **Exterior angles** are formed by extending the sides of the polygon. An interior angle and an exterior angle are adjacent angles on a straight line (they add to 180°).

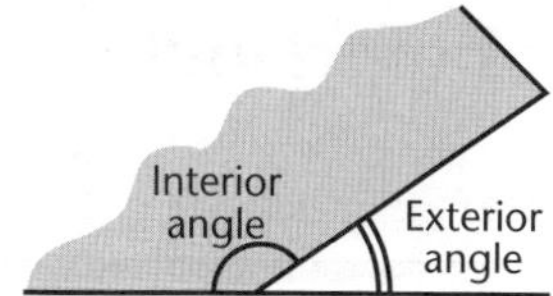

The interior and exterior angles of a triangle, quadrilateral, and hexagon are shown below:

Triangle	Quadrilateral	Hexagon
Interior angles: *a*, *b*, *c* Exterior angles: *p*, *q*, *r*	Interior angles: *a*, *b*, *c*, *d* Exterior angles: *p*, *q*, *r*, *s*	Interior angles: *a*, *b*, *c*, *d*, *e*, *f* Exterior angles: *p*, *q*, *r*, *s*, *t*, *u*

Triangles

A triangle is a three-sided polygon. Some of the important geometrical properties of a triangle are discussed below.

The sum of the interior angles of any triangle is 180°. (Abbreviated to: $\angle$ sum Δ.)

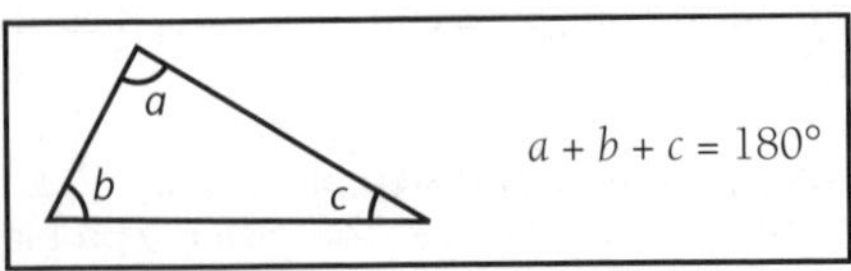

Example B

The size of the angle labelled x is given by:

$x + 70 + 30 = 180$ ($\angle$ sum Δ)

$x + 100 = 180$

$x = 80°$

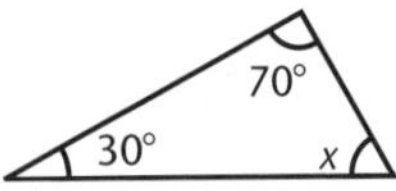

The size of an exterior angle of any triangle is equal to the sum of the two opposite interior angles. (Abbreviated to: ext $\angle$ Δ.)

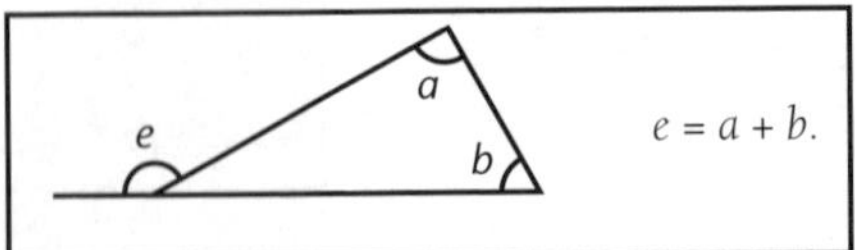

Example C

The angle x shown in the diagram is given by:

$x + 80 = 150$ (ext $\angle$ Δ)

$x = 70°$

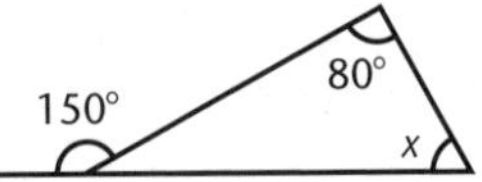

Types of triangle

Triangles can be classified into different types according to the relative lengths of the sides and the sizes of the interior angles. Most of these triangles have different **axes of symmetry**.

Equilateral triangle	Isosceles triangle	Scalene triangle
Three sides of equal length. Each interior angle is 60°. Three axes of symmetry.	Two equal sides, other side called **base**. Two equal angles called **base angles**. One axis of symmetry.	No equal sides. No equal angles. No axes of symmetry.

Note: **1.** Matching marks on sides indicate line segments of equal length.

2. The longest side of a triangle is always opposite the largest angle, and the shortest side is always opposite the smallest angle.

Example D

The size of the angle labelled x is given by:

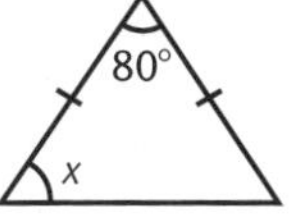

$x + x + 80 = 180$ (sum $\angle$s isos Δ)

$2x = 100$ [subtracting 80]

$x = 50°$

Note: Stating that the triangle is an 'isos Δ' in the solution shows that the unmarked angle is equal to x (since the base angles of an isosceles triangle are equal).

Triangles can also be classified according to the size of the angles:

Acute-angled triangle	Right-angled triangle	Obtuse-angled triangle
62°, 45°, 73°	30°, 60°	35°, 27°, 118°
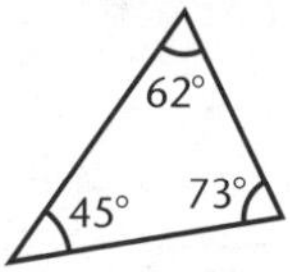 Each angle is less than 90°.	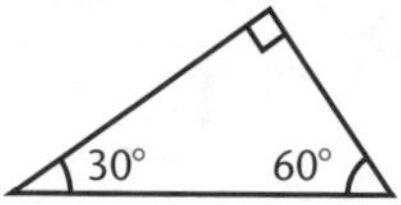 One angle is equal to 90°.	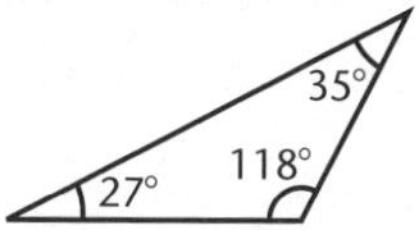 One angle is greater than 90°.

Unit 11.4 Activity 2A: Angles and triangles

1. Find, with reasons, the size of the labelled angles.

a. a, 46°, 42°

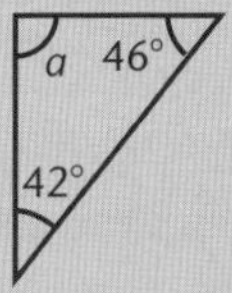

b. 88°, b

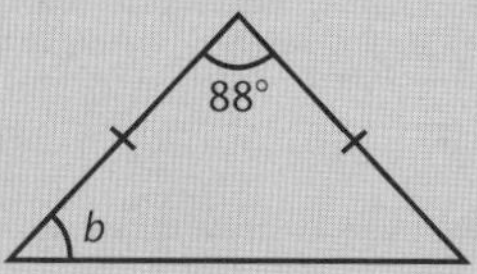

c. c, 54°, 128°

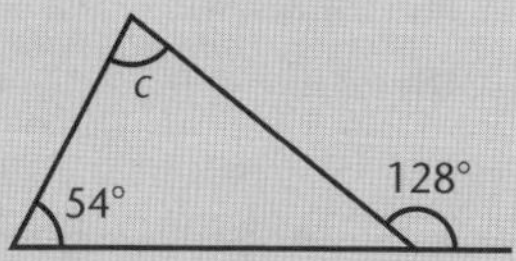

d. e, d, 55°

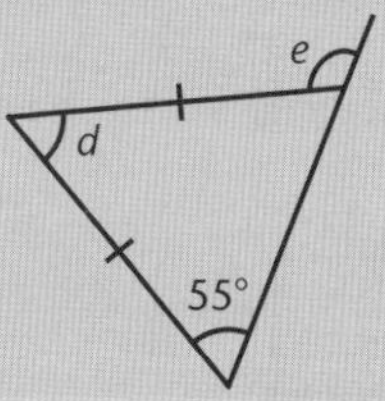

e. 56°, g, f, 43°, 37°, 51°

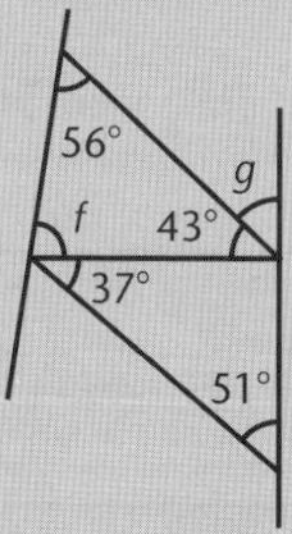

f. 106°, i, h, 84°

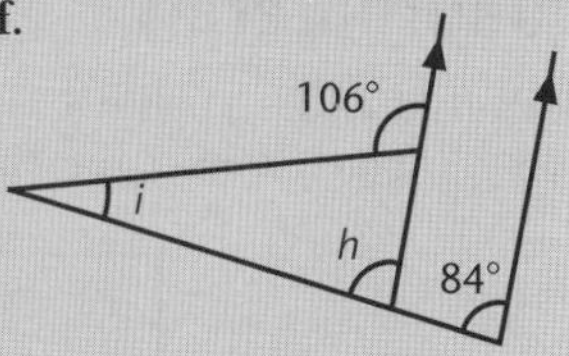

g. d, 40°, e

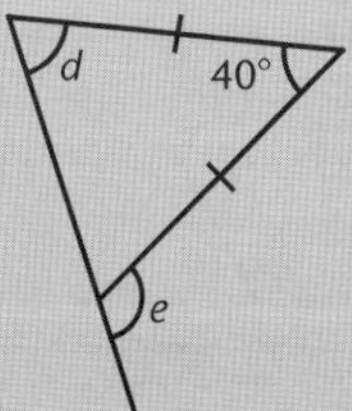

h. f, g, 28°, 81°, 32°, 47°

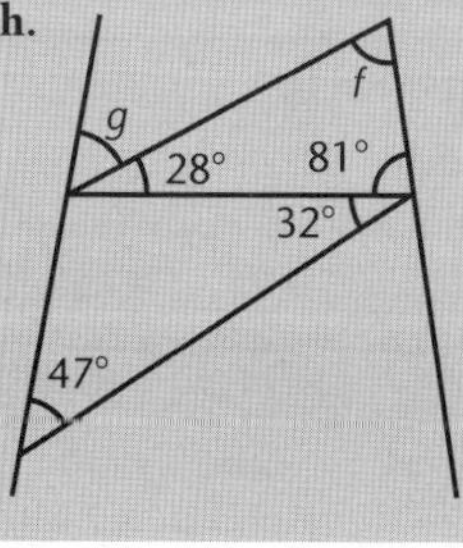

i. 108°, i, h, 40°

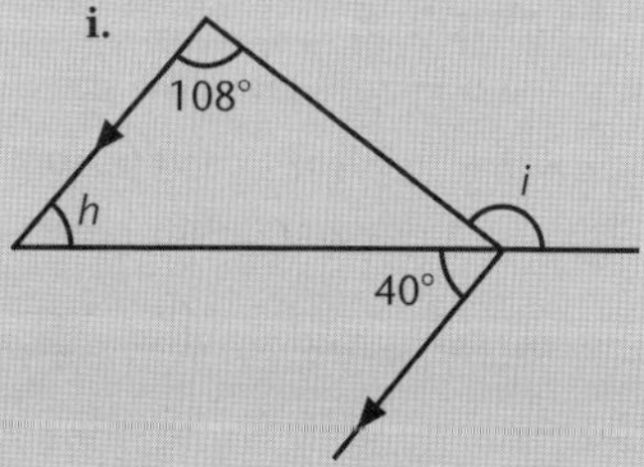

2. Prove that PR is parallel to ST, given that length PQ = length QR and the angle sizes are as marked on the diagram (// means 'is parallel to').

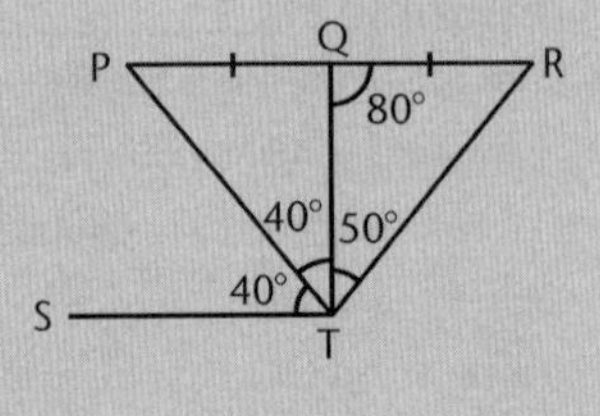

Copy the following and fill in the blanks to complete the proof.

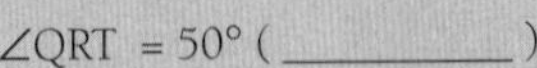

$\angle QRT = 50°$ (____________)

$\therefore$ length QR = length _____ (ΔQRT is isosceles)

$\therefore \angle QPT =$ _____° (base $\angle$s isos Δ)

$\therefore$ PR // ST (________________)

3. AB is parallel to CD. Lengths AC, AD and BD are equal.

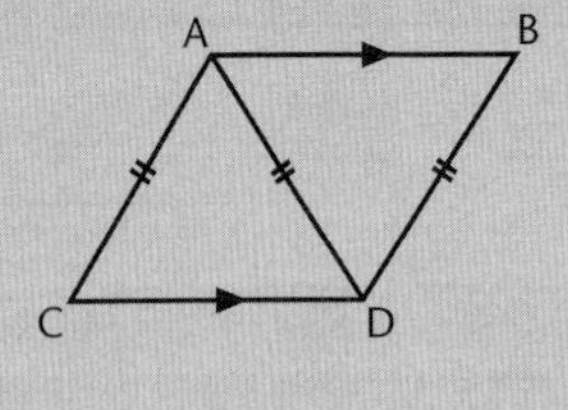

Copy the following and fill in the blanks to prove that $\angle ACD = \angle ABD$.

Let $\angle ACD = x$

$\angle ADC =$ __ (base $\angle$s isos Δ)

$\angle BAD =$ __ (alt $\angle$s // lines)

$\angle ABD = x$ (________________)

4. ABC and ADE are straight lines and $a = b$.

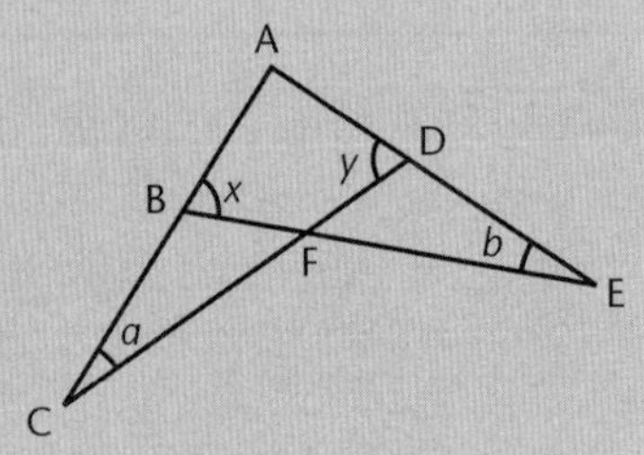

Prove that $x = y$.

Hint: Let $\angle BFC = c$.

Quadrilaterals

A **quadrilateral** is a polygon with four sides. The sum of the interior angles of any quadrilateral is 360°. (Abbreviation: $\angle$ sum quad.)

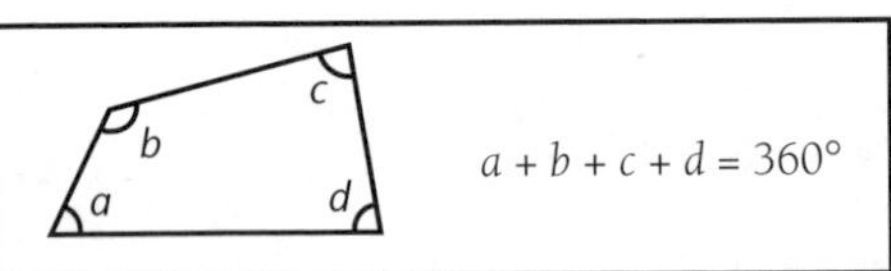

Example E

In the diagram shown, the size of y is given by:

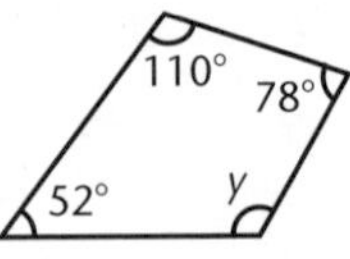

$y + 78 + 110 + 52 = 360$ ($\angle$ sum quad)

$y + 240 = 360$

$y = 120°$

Types of quadrilateral

The table shows some quadrilaterals with special properties. (Note that a **mediator** is a perpendicular bisector.)

Name and Shape	Properties	Symmetry
Trapezium	One pair of parallel sides.	No axes of symmetry.
Isosceles Trapezium	A trapezium with one pair of non-parallel sides which are equal in length. Diagonals of equal length.	One axis of symmetry – mediator of parallel sides.
Kite	Two pairs of equal adjacent sides. Diagonals which intersect at right angles.	One axis of symmetry – one diagonal.
Arrowhead	Two pairs of equal adjacent sides. One interior angle is a reflex angle.	One axis of symmetry – one diagonal.
Parallelogram	Opposite sides are parallel and equal. Opposite angles equal. Diagonals bisect each other.	No axes of symmetry. Point symmetry only.
Rhombus	Opposite sides are parallel. All sides are equal in length. Opposite angles equal. Diagonals bisect each other at right angles.	Two axes of symmetry – each diagonal.
Rectangle	Opposite sides parallel and equal. All angles are right angles. Diagonals equal and bisect each other.	Two axes of symmetry – mediators of the sides.
Square	Opposite sides are parallel. All sides are equal in length. All angles are right angles. Diagonals have equal length and bisect each other at right angles.	Four axes of symmetry – mediators of the sides and each diagonal.

Unit 11.4 Activity 2B: Angles and quadrilaterals

1. An equilateral triangle BCD is joined to an isosceles triangle ABD to form a kite, as shown.

If ∠ABC = 90°, find the size of ∠BAD.

2. Find, with reasons, the size of the labelled angles.

a. ABEF is a square.
ABC and FED are straight lines.

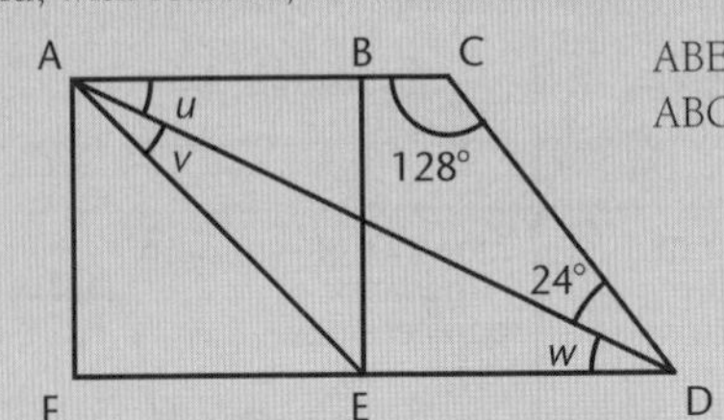

b.

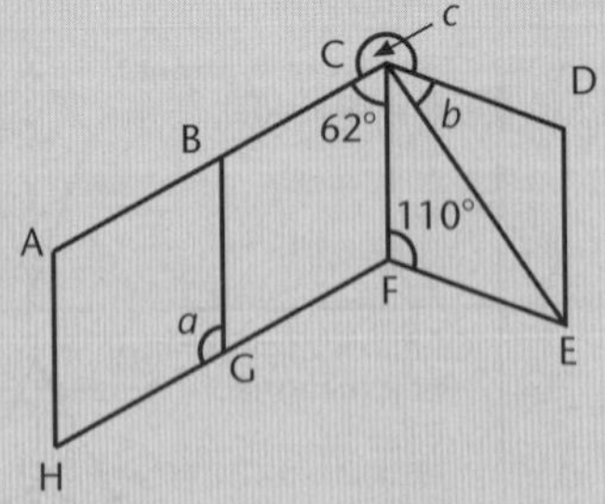

ABGH, BCFG and CDEF are rhombuses.

c. EL is an axis of symmetry.

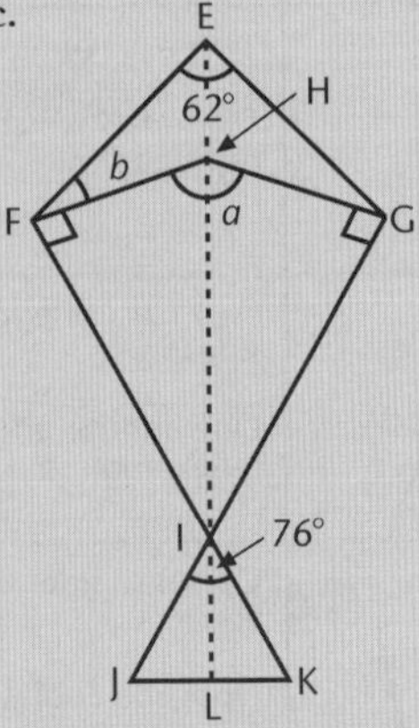

d.

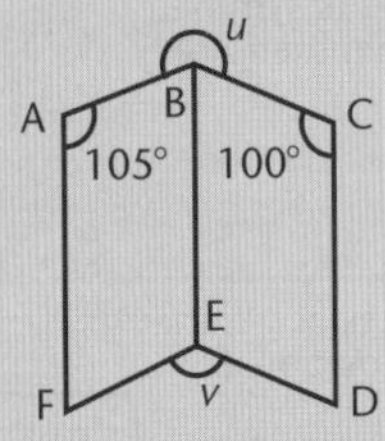

ABEF and BCDE are parallelograms.

3. Given that $b + d = 180°$, fill in the blanks to prove that $c = e$.

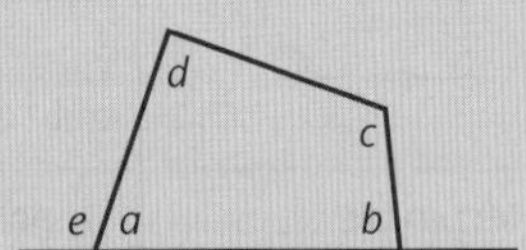

$a + b + c + d =$ ______ (∠ sum quad)

$a + c + 180° =$ ______

$a + c =$ ______

$a + e =$ ______ (∠s on a line)

$\therefore a + c = a + e$

$\therefore c = e$

4. Prove that $a + b + c + d = 360°$ in the diagram alongside.

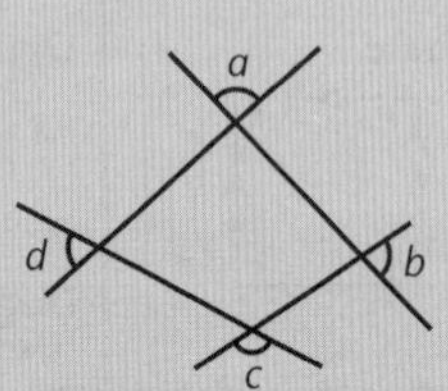

5. ABCD is a parallelogram. Length AB = length BE.

Prove that triangle BCF is isosceles.
Hint: Consider ∠s BEA and EAB.

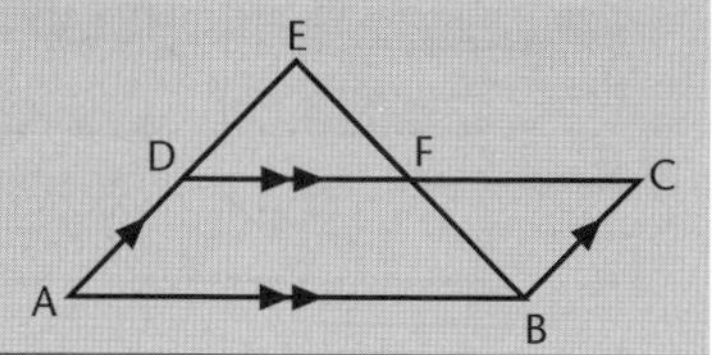

Interior angles of polygons

A polygon with n sides has n **interior angles**.

The four polygons shown below have their interior angles shown. The polygons have also been divided into triangles by drawing all possible diagonals from any one vertex. The sum of the interior angles of each polygon can be found by adding together the interior angles of each triangle.

Triangle	Quadrilateral	Pentagon	Hexagon
One triangle Sum int ∠s = 1 x 180° = 180°	Two triangles Sum int ∠s = 2 x 180° = 360°	Three triangles Sum int ∠s = 3 x 180° = 540°	Four triangles Sum int ∠s = 4 x 180° = 720°

In general, if a polygon has n sides, then $(n - 2)$ triangles can be formed, and the sum of the interior angles is:

Sum of interior angles of an n-sided polygon = $(n - 2) \times 180°$

Example F

The sum of the interior angles of an octagon is

$(8 - 2) \times 180° = 1\,080°$ [using above formula with $n = 8$]

If the octagon is regular, then

each interior angle $= \dfrac{1\,080}{8}$ [since all angles equal size]

$= 135°$

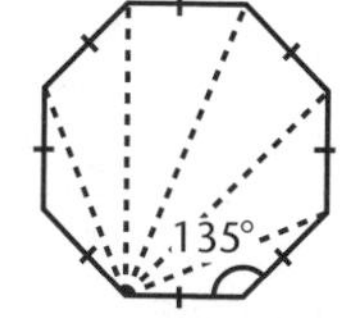

Exterior angles of polygons

A polygon with n sides has n **exterior angles**. For any polygon (regular or not) the following rule holds:

Sum of the exterior angles of a polygon is 360°

The exterior angles of a **regular polygon** are the same size as each other (since interior angles are equal).

The size of each exterior angle of a regular n-sided polygon is $\left(\frac{360}{n}\right)^{\circ}$

Example G

A regular pentagon has five equal exterior angles.

Therefore, $a + a + a + a + a = 360°$ (sum ext ∠s polygon)

$5a = 360°$

$a = 72°$

Each exterior angle is 72°.

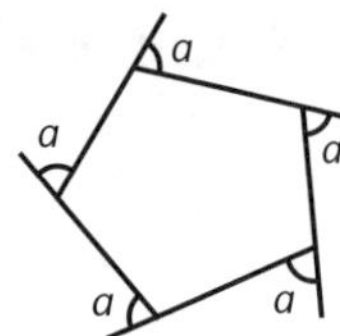

Another method for calculating the size of the interior angle of a regular polygon uses the relationship:

Interior ∠ size = 180° – (exterior ∠ size)

Since the angle sum of an n-sided polygon (whether regular or not) is constant, this gives another formula for determining the angle sum of an n-gon (a polygon with n sides):

The sum of the interior angles of an n-gon is $n \times$ (int ∠ size of a regular n-gon)

(Abbreviation: sum = $n \times$ int ∠.)

Example H

Q. Find the exterior angle of a regular decagon and use this to find the sum of the interior angles of a decagon.

A. The size of the exterior angles is $\frac{360}{10} = 36°$ [exterior angle of regular n-gon = $\frac{360}{n}$]

Since the exterior angle and the interior angle form a straight angle, the size of the interior angle of a regular decagon is 180 – 36 = 144°.

The sum of the interior ∠s of a decagon = 10 x 144° [sum = n x int ∠]

= 1 440°

The properties of the interior and exterior angles of polygons can also be used in inverse problems.

Example I

Q. If each interior angle of a regular polygon is 165°, how many sides does the polygon have?

A. Each interior angle is 165°, so each exterior angle is 180 – 165 = 15°.

∴ number of ext ∠s = $\frac{360}{15}$ (sum ext ∠s polygon is 360°)

= 24

∴ the polygon has 24 sides.

Unit 11.4 Activity 2C: Angles and polygons

1. Find, with reasons, the sizes of the labelled angles.

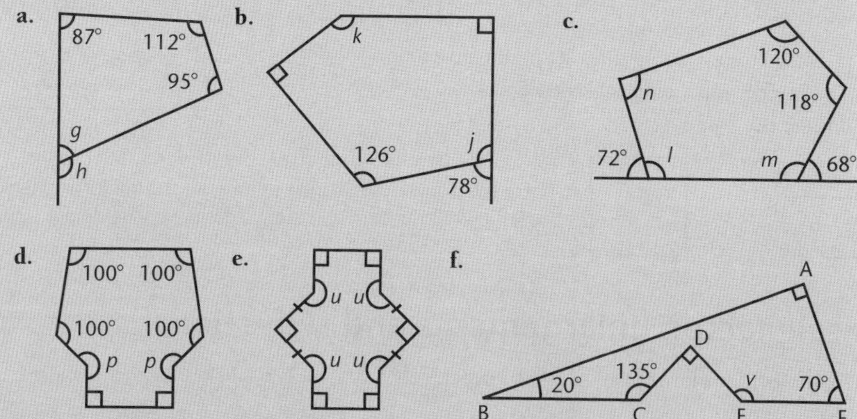

2. Find the size of each exterior angle of a regular nonagon (nine-sided figure) and use this to find the size of each interior angle.

3. Find the size of each interior angle of a regular 12-sided figure.

4. Each interior angle of a regular polygon is 150°. Calculate the size of each exterior angle and use this result to determine the number of sides in the polygon.

5. The size of each exterior angle of a regular polygon is 24°.

a. How many sides are there in the polygon? **b.** What is the size of each interior angle?

c. Find the sum of the interior angles.

6. If the size of each exterior angle of a regular polygon is 12°, find the sum of the interior angles.

7. Jason has designed a hexagonal paving brick, as shown in Diagram A. The 'long' sides are parallel and the 'large' angles are equal to 135°.

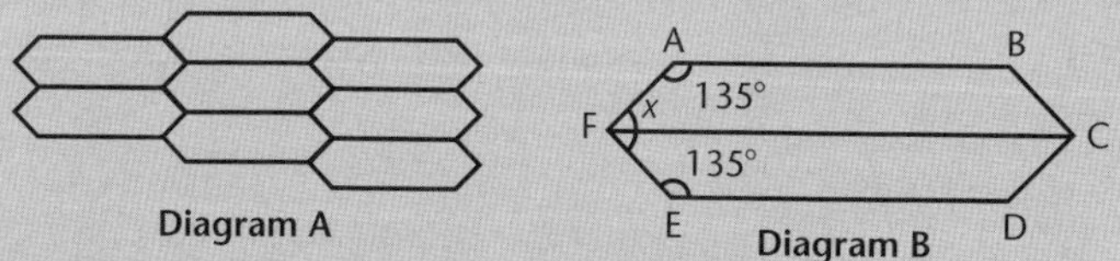

Diagram A

Diagram B

a. Using Diagram B, calculate the value of x (∠AFE).

b. Using your answer from **a.**, explain why Jason's paving bricks 'work', (ie meet without 'gaps').

8. Peter made some paving stones in the shape of a regular octagon.

a. Calculate the size of the exterior angle of a regular octagon.

b. What is the size of an interior angle of a regular octagon?

c. Explain why Peter's paving stones did 'not work' to pave an area (ie there were always gaps between some of the paving stones).

Similar figures

Similar figures have the same shape but may differ in size.

This means:

- Their corresponding sides are in the same ratio.
- Their corresponding angles are equal.

Consider this pair of squares:

All squares have the same shape but they can differ in size. These two squares are similar figures; all the angles are right angles and all the corresponding sides are in the ratio 1 : 2.

Scale factor, enlargements and reductions

The **scale factor** for the pair of similar squares above is $\frac{2}{1} = 2$; a side in the second figure is twice the length of the corresponding side in the first figure.

The second figure is an *enlargement* of the first figure and hence *the scale factor is greater than one*.

Consider these two regular hexagons:

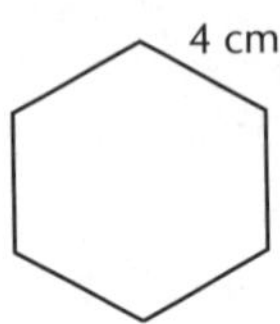

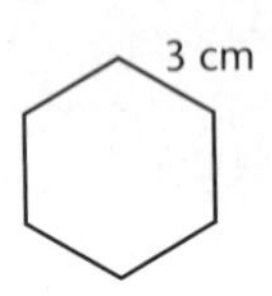

All regular polygons with the same number of sides are similar figures; their corresponding sides are in the same ratio and their interior angles are equal.

The scale factor in this case is $\frac{3}{4}$; a side in the second figure is $\frac{3}{4}$ of the length of the corresponding side in the first figure.

The second figure is a **reduction** of the first figure and hence the scale factor is less than one.

$$\text{Scale factor} = \frac{\textbf{Length of side in the second figure}}{\textbf{Length of corresponding side in the first figure}}$$

Example J

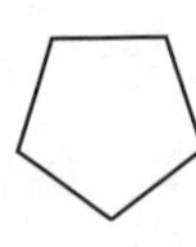

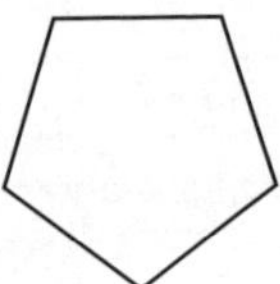

Q. The corresponding sides of these regular pentagons are in the ratio 2 : 3.

a. What is the scale factor for these pentagons?

b. If the length of one of the sides of the smaller pentagon is 15 centimetres, what is the length of one of the sides of the larger pentagon?

A. **a.** The pentagons are regular pentagons so they are similar figures.

$$\text{Scale factor} = \frac{\text{Length of side in the second figure}}{\text{Length of corresponding side in the first figure}} = \frac{3}{2}$$

b. Ratio of sides: 15 : x = 2 : 3 where x cm is the length of the side of the larger pentagon.

$$\frac{x}{15} = \frac{3}{2}$$

$$x = \frac{3 \times 15}{2}$$

$$x = 22.5$$

The length of one of the sides in the larger pentagon is 22.5 cm.

Enlarging and reducing figures

A figure can be enlarged or reduced by measuring distances from an arbitrary point, either within the figure or outside the figure.

Example K

Q. a. Enlarge Figure B by a factor of 3 (scale factor 3) using a point outside the figure as the 'centre'.

b. Reduce figure B by a factor of 2 (scale factor ½) using a point inside the figure.

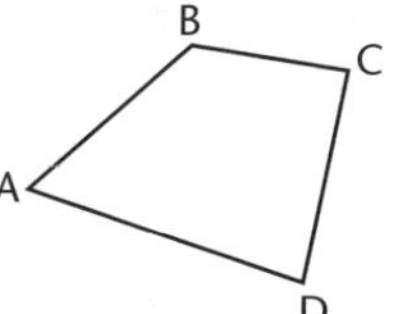

A. a. From a point O (this can be any point) outside the figure ABCD the distance to each of the points A, B, C and D is measured then multipied by 3 so that the distance from O to the image points **A′, B′, C′ and D′**, is three times the distance from O to A, B, C and D, respectively. The image points are then connected to form the enlarged image (shaded) **A′, B′, C′, D′**.

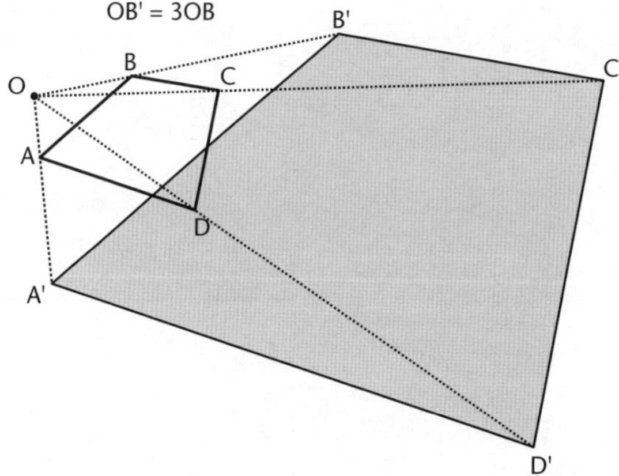

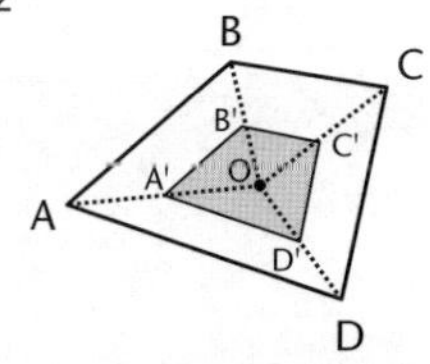

b. From a point O (this can be any point) *inside* the figure the distance to each of the points A, B, C and D is measured then halved to give the image points **A′, B′, C′ and D′**. These are then connected to form the reduce image **A′, B′, C′, D′**.

Enlargements and reductions can also be made by placing a grid over the figure. The size of the grid is then reduced (Figure C) or enlarged (Figure D) and the figure copied using the gridlines as a guide.

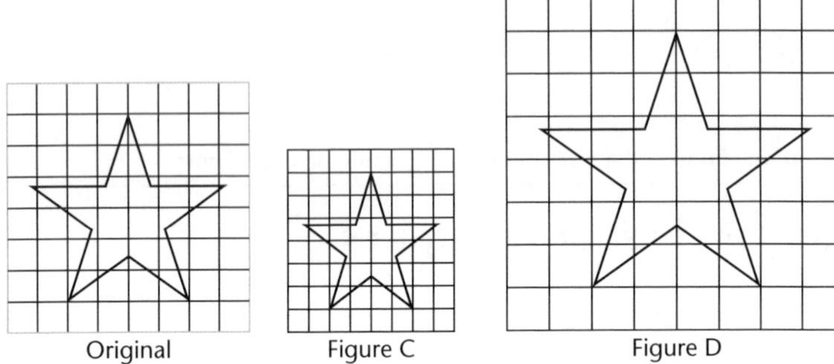

Original Figure C Figure D

Figure C is a reduction of the original, scale factor ¾, and Figure D is an enlargement of the original, scale factor 5/4.

The original, the reduction and the enlargement are all similar figures; their corresponding angles are equal and their corresponding sides are in the same ratio :- 1 : ¾ : 5/4 = 4 : 3 : 5

Unit 11.4 Activity 2D: Scale factor and enlargements

1. Find the scale factor for each of the following pairs of similar figures.

a.

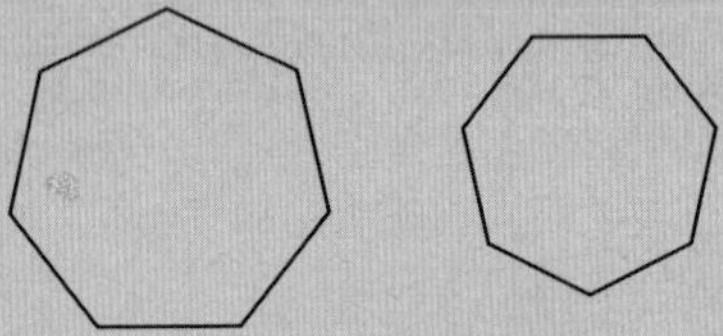

b.

c.

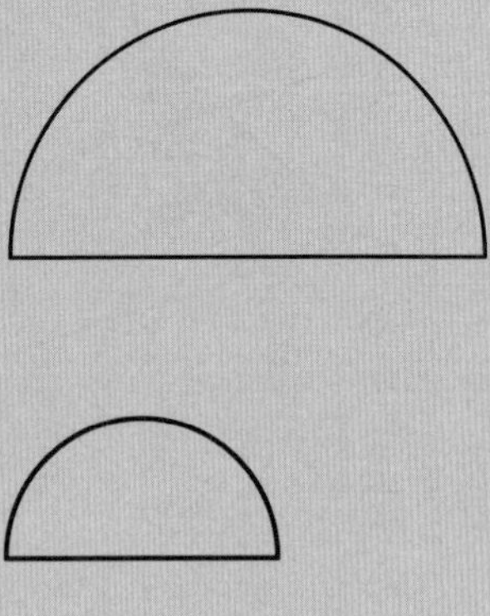

2. For each of the following pairs of similar figures:

i. find the scale factor by measuring a corresponding length.

ii. find the value of x

a.

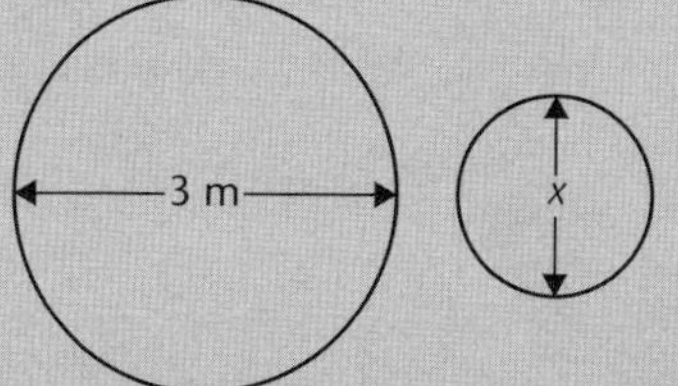

b.

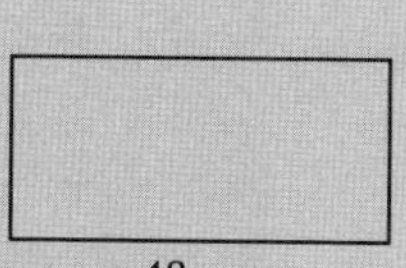

c.

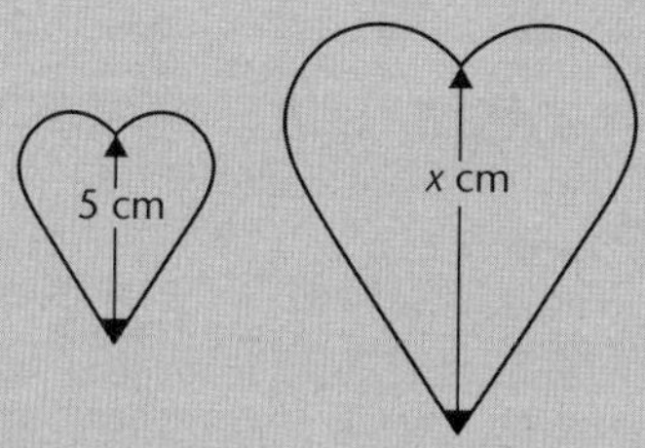

d.

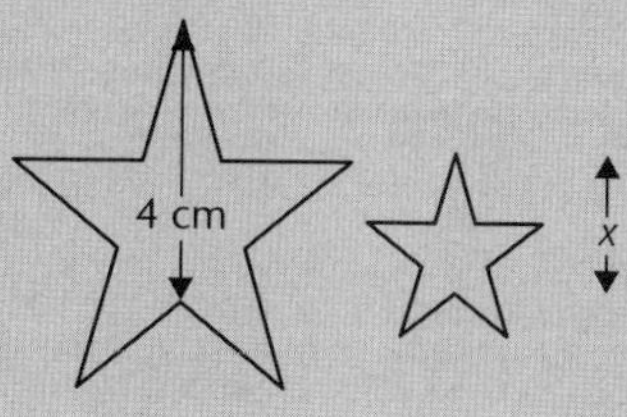

3. Determine whether the following pairs of figures are similar.

a.

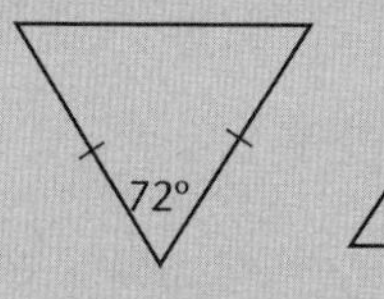

b.

c.

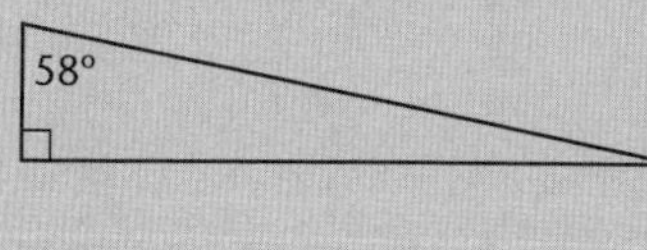

d.

4. Trace the figure below and enlarge, using the centre at point O, by a factor of 3.

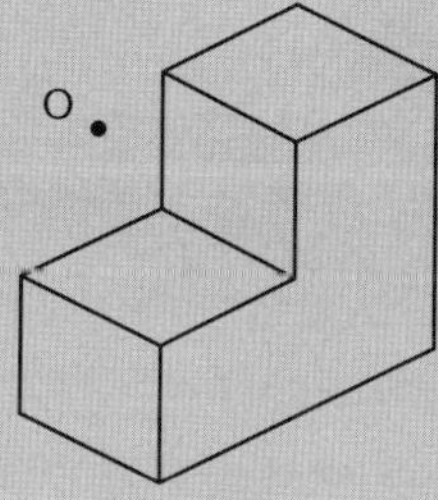

5. Use a grid to enlarge the following figure by a scale factor of 2.

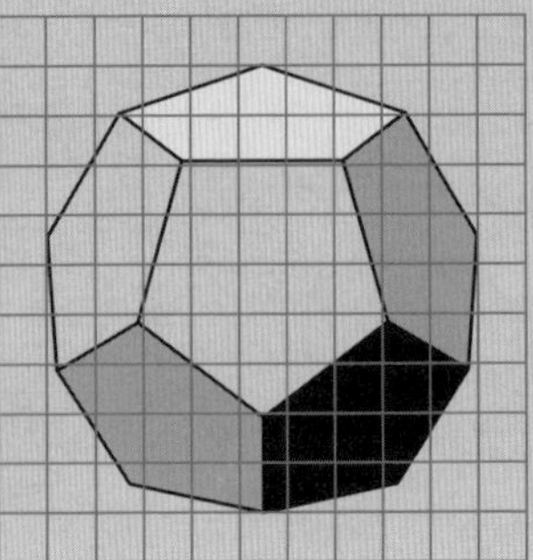

6. Draw a scalene triangle then:
   a. Enlarge it by a factor of 1.5 using a point inside the figure as 'centre'.
   b. Reduce it by a factor of 3 using a point outside the figure as 'centre'.
7. A map of a town has a ratio scale of 1 : 10 000. What is the scale factor of the map to the actual town?
8. Choose a figure or shape that you would not normally be able to construct using a ruler, compass or protractor. Enlarge this figure, stating the scale factor.

Applications of similar figures

Often we are told that a pair of figures are similar and we can use the fact that their corresponding lengths are in the same ratio and that their corresponding angles are equal.

Example L

Q. Find the value of x in the diagram below.

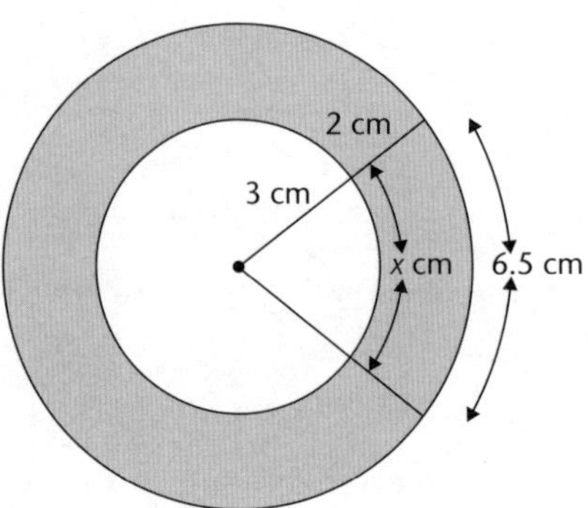

A. The two sectors from the diagram, separated and shown below, are similar:

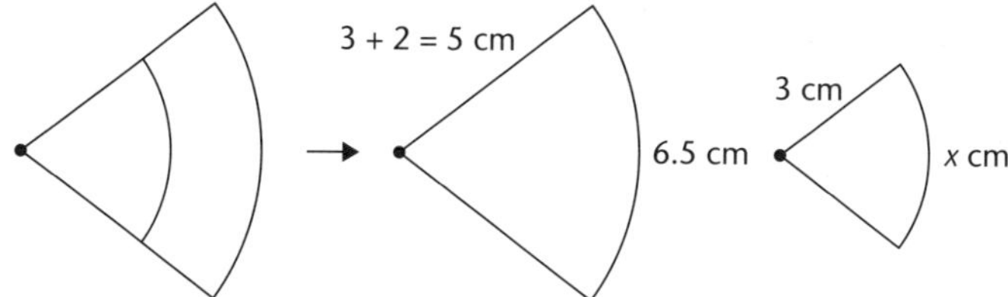

The radii are in the ratio 5 : 3 and therefore the arc lengths are in the ratio 5 : 3.

The scale factor is $\frac{3}{5}$

So $x = \frac{3}{5} \times 6.5 = 3.9$

Similar triangles

A special case of similar figures is similar triangles. Many construction principles are based on similar triangles and recognising that two triangles are similar enables us to use ratio, or scale factor, to calculate lengths.

There are three ways that we can establish that a pair of triangles are similar.

Two triangles are similar if:

i. all their corresponding sides are in the same ratio

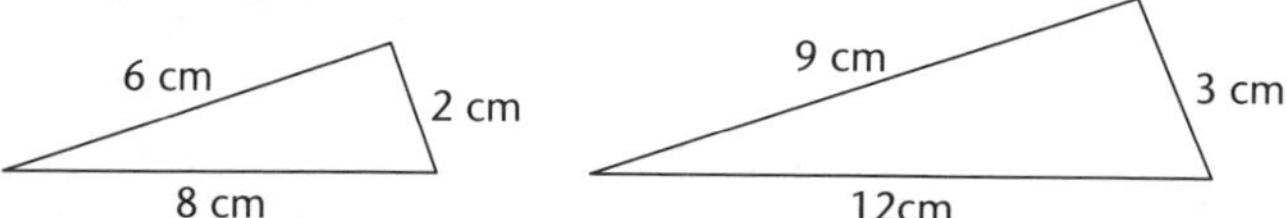

The corresponding sides in this pair of triangles are in the ratio 2 : 3
The scale factor is 3/2.

ii. all the corresponding angles are of equal size.

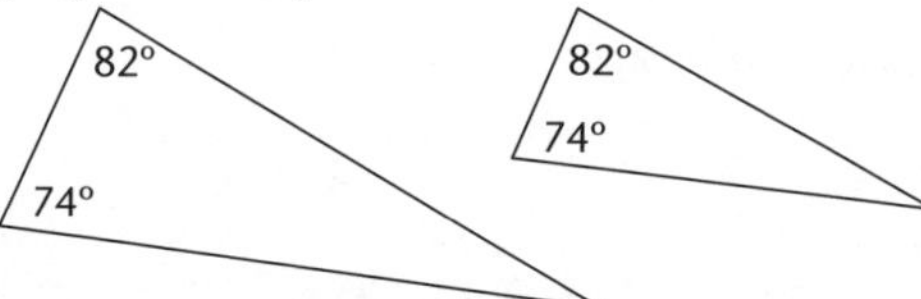

Note: If triangles show two angles that are the same size then the third angles must also be the same size because we can calculate the size of the third angles using the rule:
The sum of the three angles of a triangle is 180°.
For both of the triangles above the unknown angle is 180 – 82° – 74° = 24°.

iii. A corresponding pair of adjacent sides in the triangles are in the same ratio and the angle between these adjacent sides (the **included angle**) is the same size. These triangles have sides in the ratio 4:3; a scale factor of $\frac{3}{4}$.

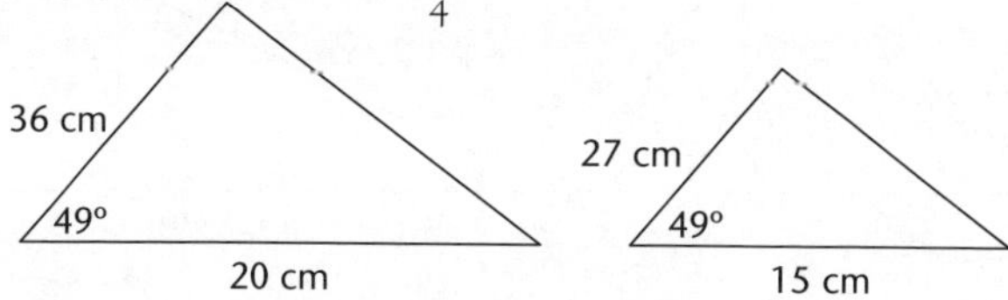

Often we are told, or it is obvious, that a pair of triangles are similar and in these cases we can use the fact that the corresponding sides are in the same ratio to solve problems.

Example M

Q. Triangles ABC and PQR are similar triangles.

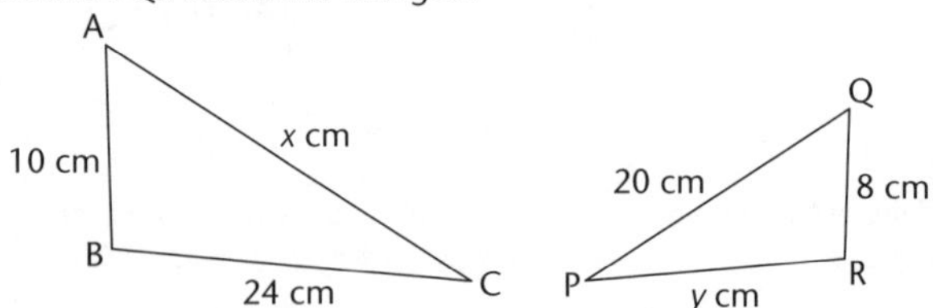

a. Which side in triangle PQR corresponds to side AB in triangle ABC?

b. Find the ratio of the sides of these triangles.

c. Find the value of x.

d. Find the value of y.

A. **a.** AB is the shortest side in triangle ABC. So AB corresponds to QR, the shortest side in triangle PQR.

b. The ratio of side lengths is AB : QR = 10 : 8 = 5 : 4 (scale factor $\frac{4}{5}$).

c. $x : 20 = 5 : 4$

$$\frac{x}{20} = \frac{5}{4}$$

$$x = \frac{5 \times 20}{4} = 25$$

d. $24 : y = 5 : 4$

$$\frac{y}{24} = \frac{4}{5}$$

$$y = \frac{4 \times 24}{5} = 19.2$$

Q. Find x in the following diagram:-

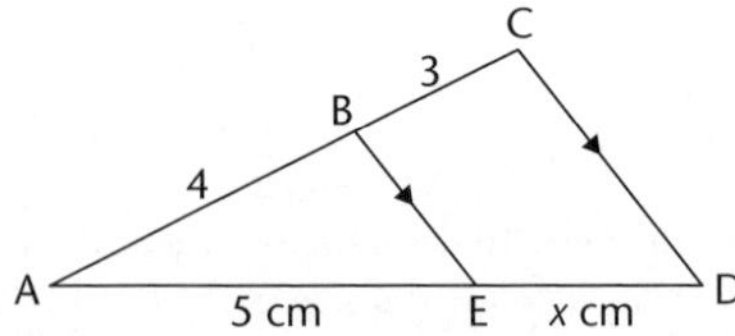

A. The triangles ABE and ACD are similar because their angles are the same; they have an angle at A in common and ∠ABE and ∠ACD are a pair of corresponding angles so are of equal size. Drawing the triangles separately:

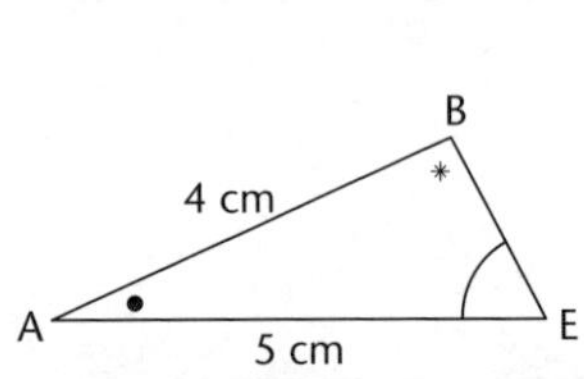

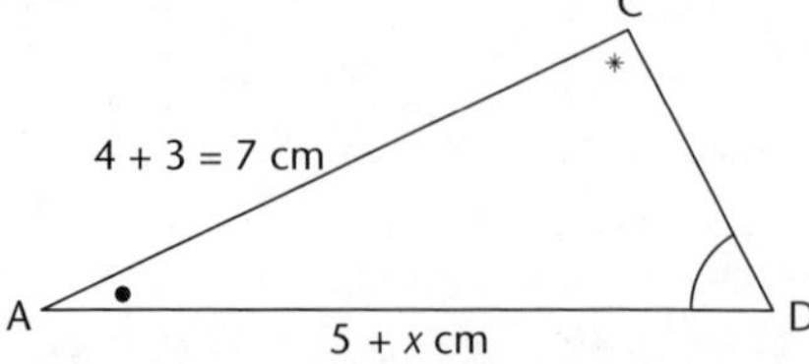

The value of x can be found using the fact that corresponding sides are in the same ratio.

AC and AB are corresponding sides: AC : AB = 7 : 4

AD and AE are corresponding sides : AD : AE = (5 + x): 5

The sides are in the same ratio hence:

$(5 + x): 5 = 7:4$

$$\frac{5 + x}{5} = \frac{7}{4}$$

$4\ (5 + x) = 5 \times 7$

$20 + 4x = 35$

$4x = 15$

$x = 3.75$

Unit 11.4 Activity 2E: Triangles and quadrilaterals

1. Find the value of x in each of the following:

a.

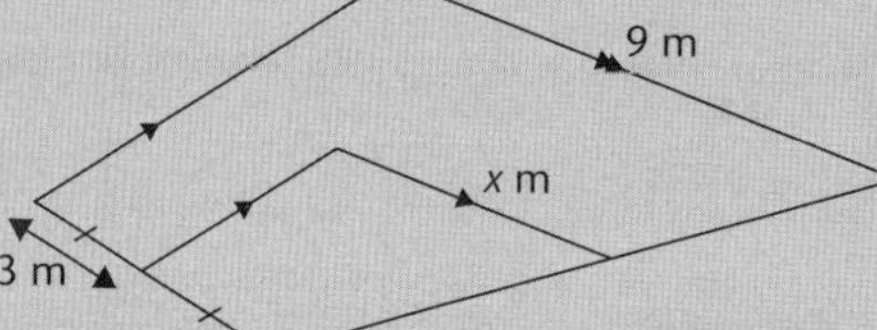

b.

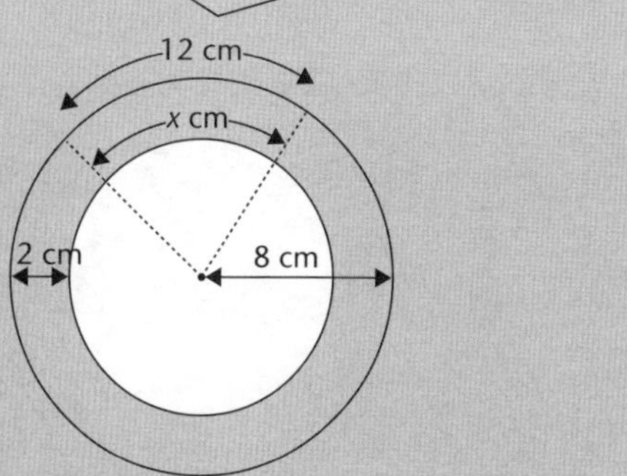

2. There is a pair of similar triangles in each of the following sets of three triangles (labelled A, B and C). State which pair is similar. The triangles are not drawn to scale.

a.

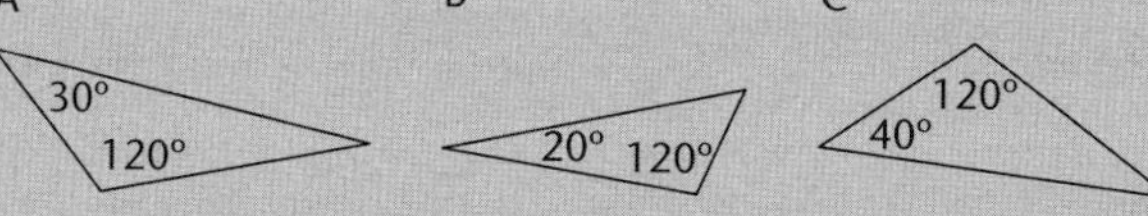

b.

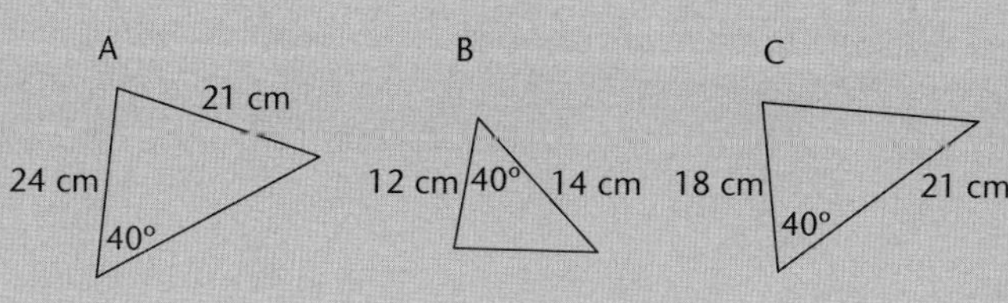

c.

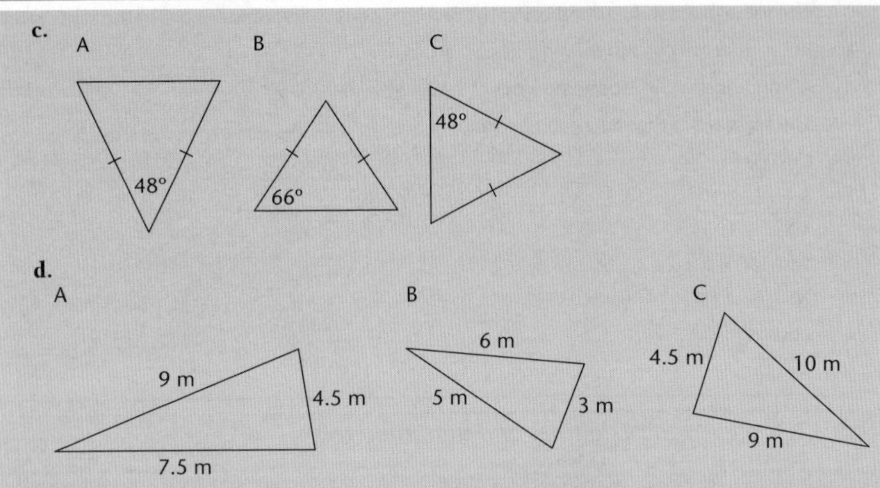

3. For each of the following triangles:
   i. Construct a triangle similar to the given triangles. Use a ruler, compass or protractor if necessary.
   ii. Mark on your triangle the angles or lengths that show that it is similar to the given triangle.
   iii. State the scale factor of your similar triangle.

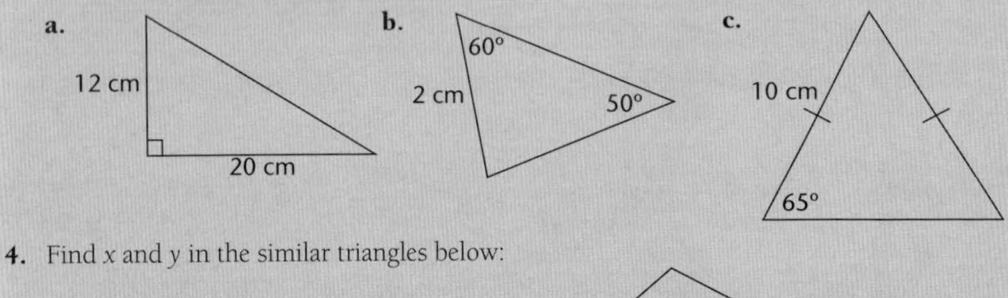

4. Find x and y in the similar triangles below:

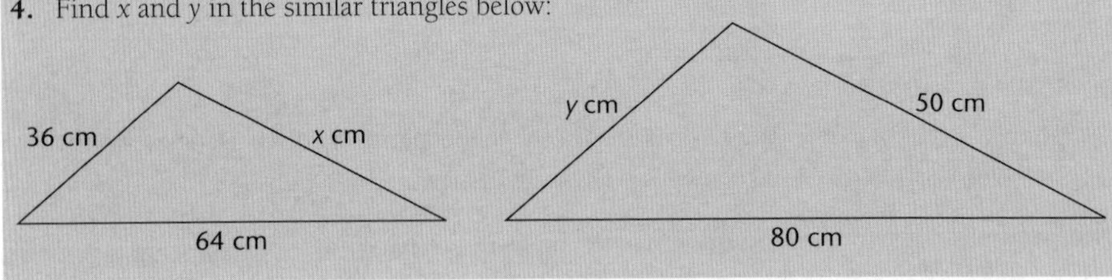

5. The figures below are similar and have a scale factor of 0.75. Find x and y.

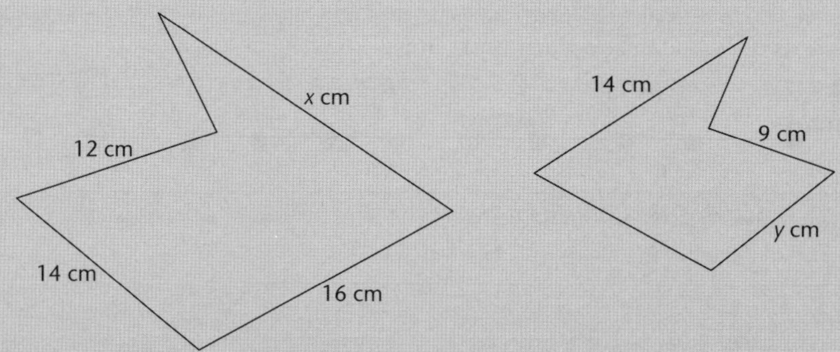

6. A 1 : 5 miniature scale model of a sculpture is to be made. If the original sculpture is 1.8 m high what will be the height, in centimetres, of the model?

7. Find the length of the sides marked with pronumerals in the following pairs of similar triangles. All measurements are in centimetres.

a.

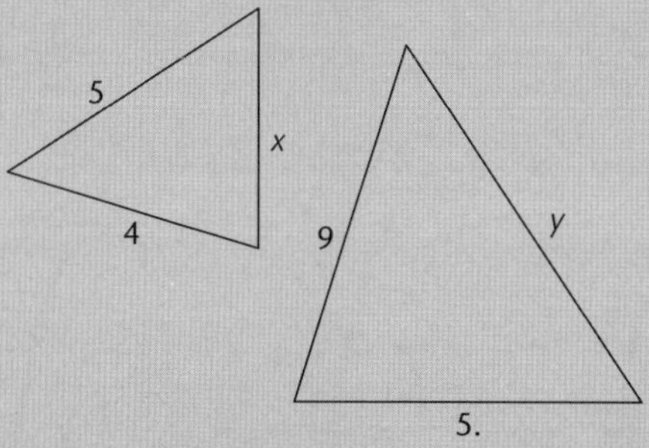

b.

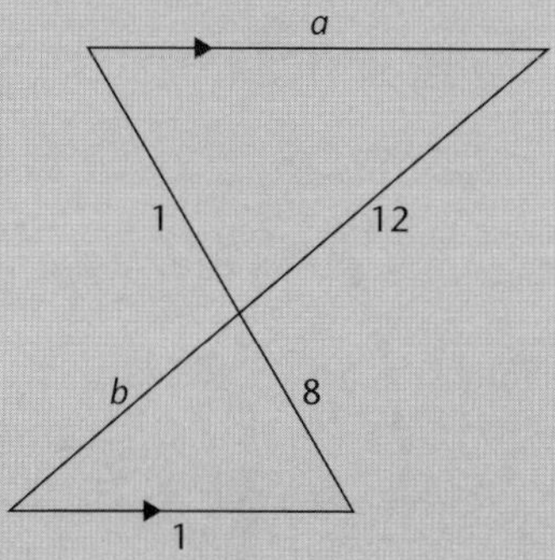

c.

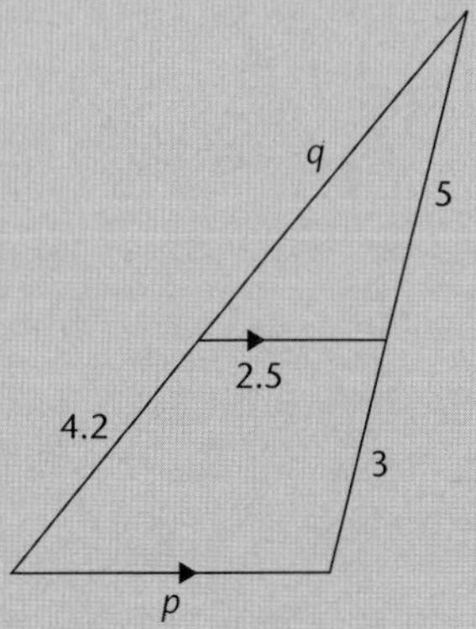

d.

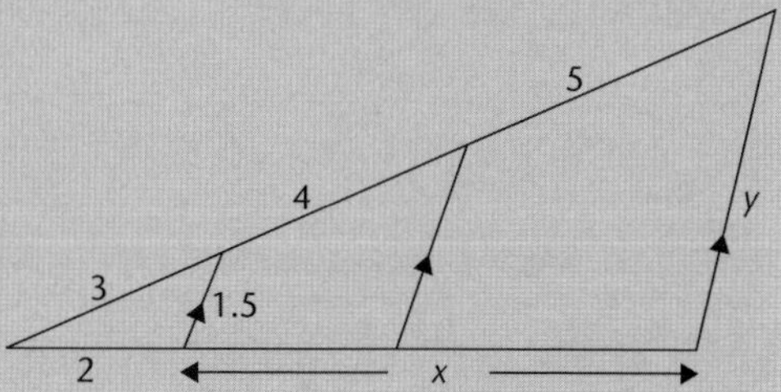

Congruent figures

Objects are said to be congruent if they are identical in size, shape and features. For example, out of the six trucks shown below, only trucks B and D are congruent figures.

Unit 11.4 Activity 2F: Congruent figures

1. Which of the following figures are congruent?

a. **b.** 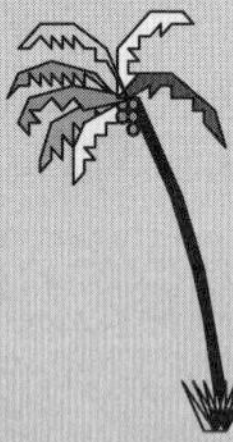 **c.** 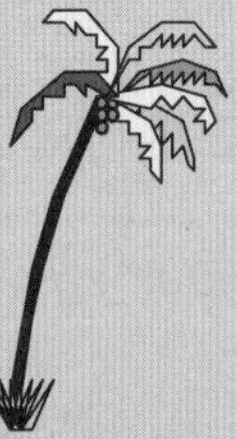 **d.**

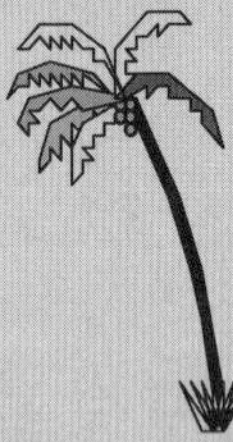

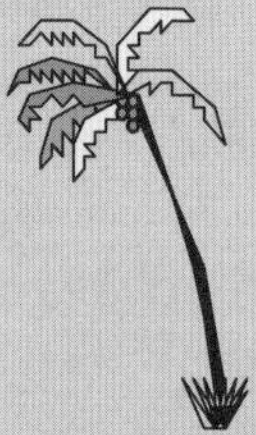

e. 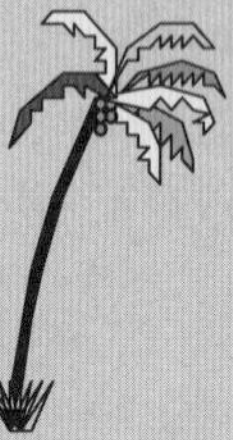 **f.** 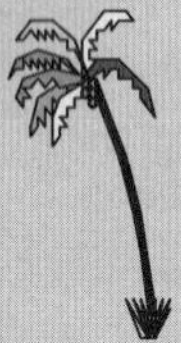 **g.** 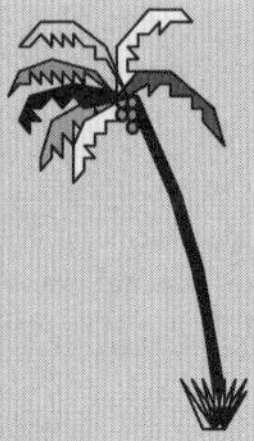 **h.**

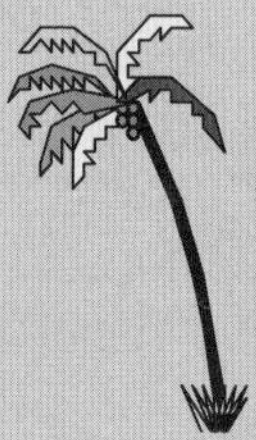

2. Which of the following geometric figures are congruent?

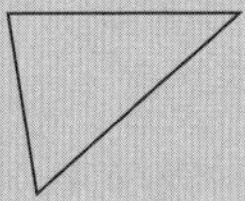

a **b** 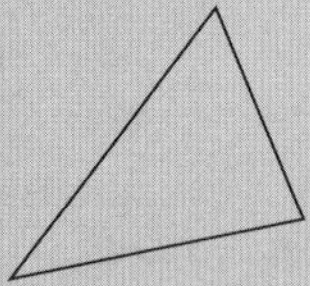 **c** 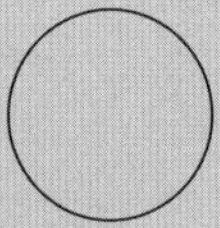 **d**

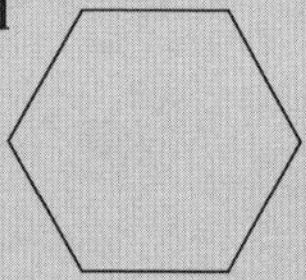

e 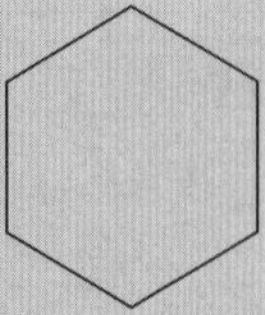 **f** 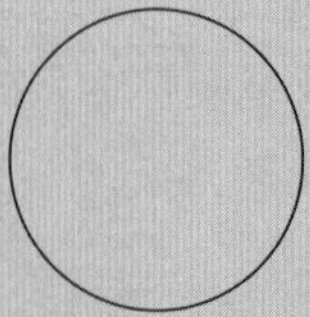 **g** 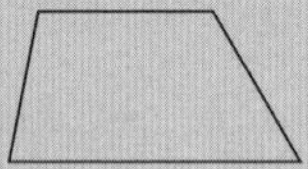 **h**

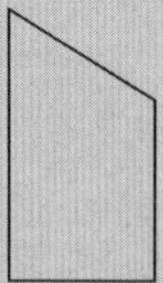

i 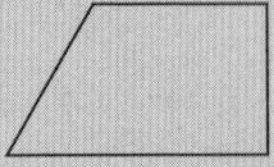 **j** 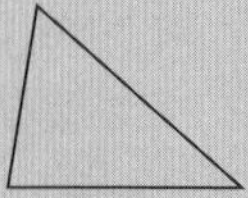 **k** 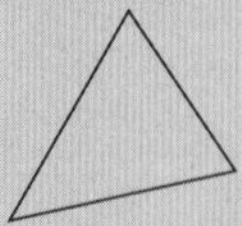 **l**

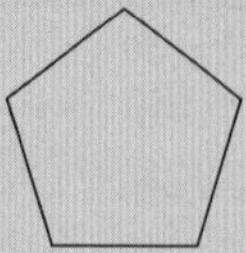

3. The quilting pattern, shown below, gives the impression of curved lines. It is made by joining identical square blocks, shown at right, that are made-up of triangles and quadrilaterals. Copy the square block and shade the congruent figures in the block in the same way (or use different colours).

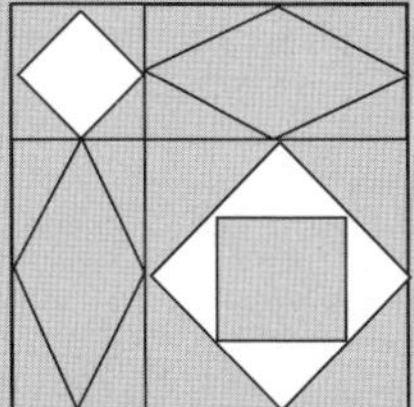

Congruent triangles

Two triangles are congruent if:

- All three corresponding sides are equal; this is called the side-side-side rule (SSS).

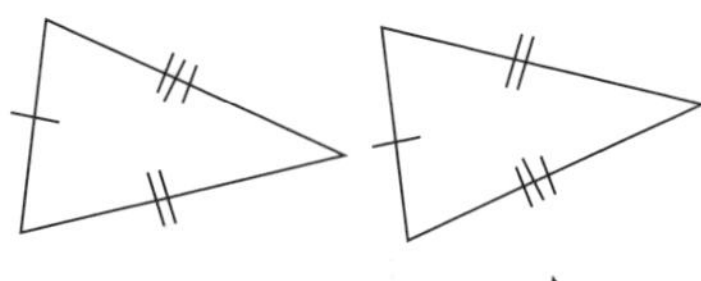

- Two sides and the included angle are equal; this is called the side-angle-side rule (SAS).

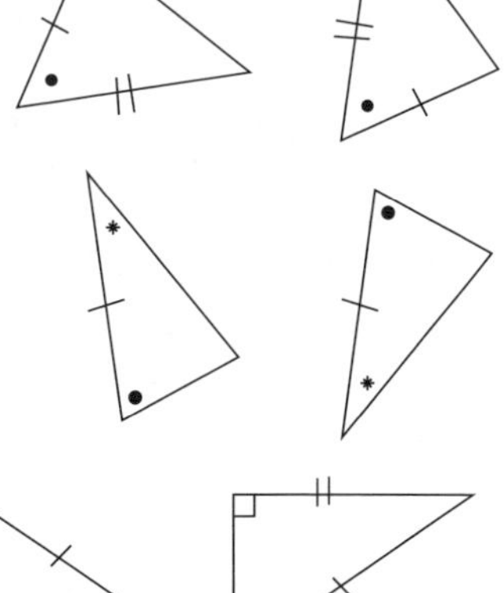

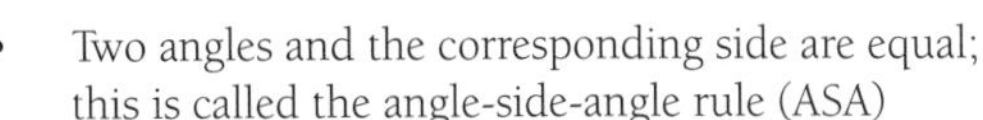

- Two angles and the corresponding side are equal; this is called the angle-side-angle rule (ASA)

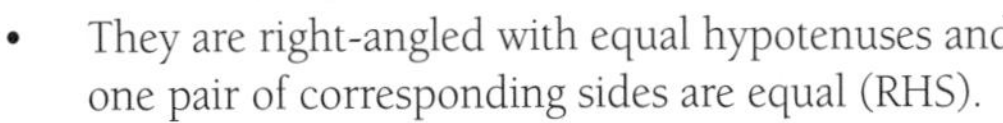

- They are right-angled with equal hypotenuses and one pair of corresponding sides are equal (RHS).

Constructing a triangle that is congruent to a given triangle

1. Where the lengths of the three sides are known:

i. Measure the length of one of the sides (AB) and rule a line of this length.

ii. Use a compass to measure and mark an arc for each of the other two sides.

iii. Join the ends of the drawn side to the point where the arcs meet.

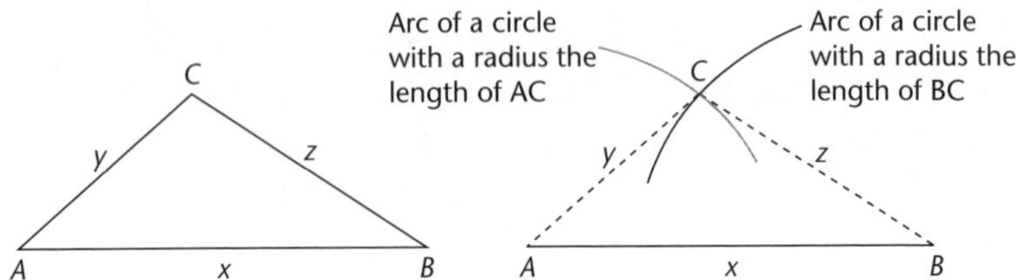

2. Where two angles and one side are known:

i. Measure the length of the known side (AB) and rule a line of this length.

ii. Use a protractor to find the size of one of the angles (∠CAB). Draw a straight line at this angle to the drawn line (AB).

iii. Repeat **ii** for the other angle (∠CBA).

iv. The vertex at C is the intersection of the lines from **ii** and **iii**.

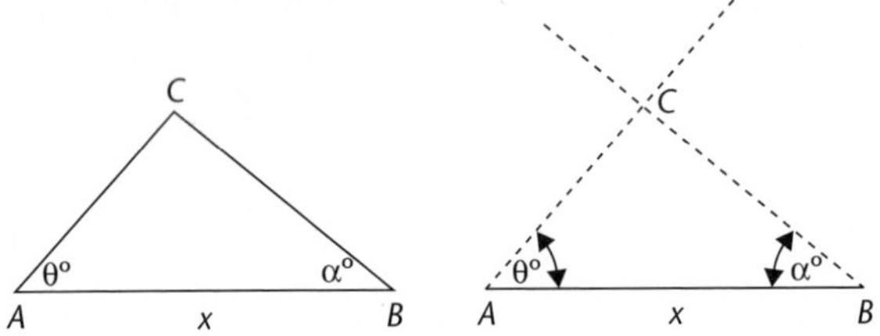

3. Where two sides and the included angle are given:

i. Measure the length of one of the known sides (AB) and rule a line of this length (x).

ii. Use a protractor to find the size of the included angle (∠CAB). Draw a straight line at this angle to the drawn line (AB). Mark the point that is the length of AC along this line.

iii. Join vertex B to the marked point.

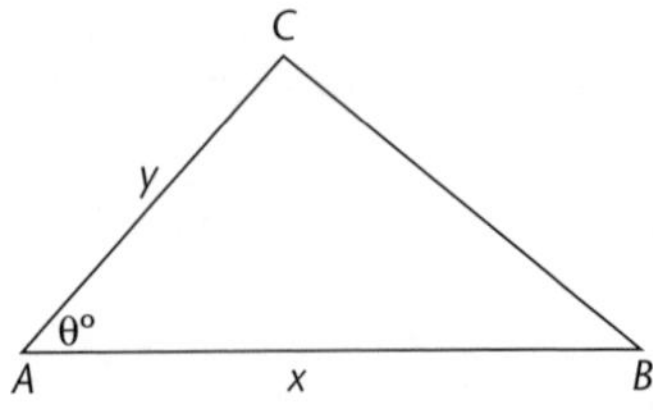

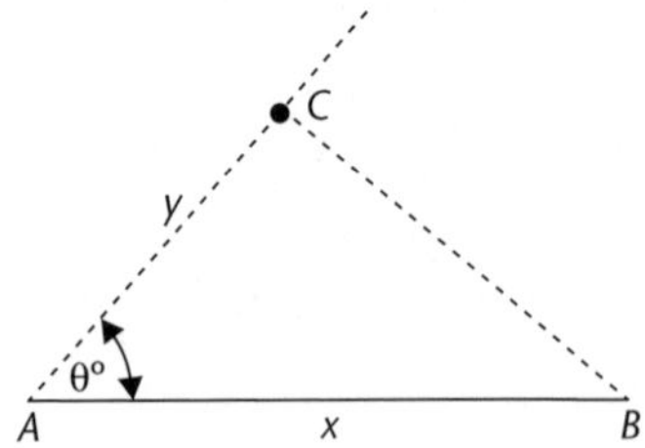

4. Where the hypotenuse and one other side in a right-angled triangle are known:
   - **i.** Measure the length of the known side (AB) and rule a line of this length (x).
   - **ii.** Measure an angle of 90° at B. Rule a line from B at this angle.
   - **iii.** Use a compass (point at A) to draw an arc of a circle with a radius the same length as the hypotenuse.
   - **iv.** The point where this arc cuts the line drawn from B is the vertex C.

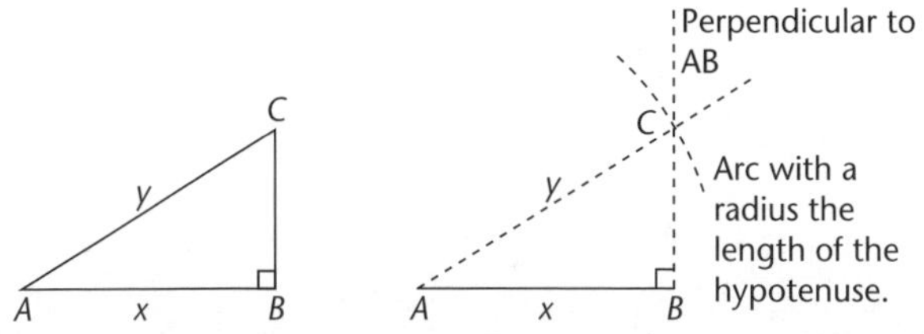

Unit 11.4 Activity 2G: Congruent triangles

1. Which of the triangles given below is congruent to triangle ABC? (Triangles are not drawn to scale.)

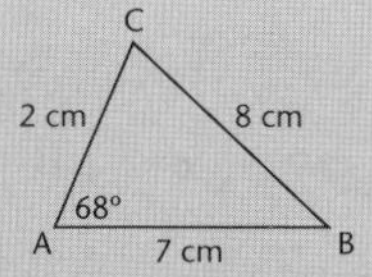

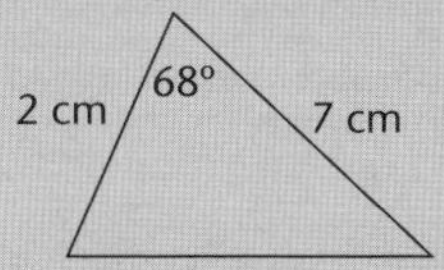

C
2 cm
8 cm
68°
A
7 cm
B

a. 68°
2 cm
7 cm

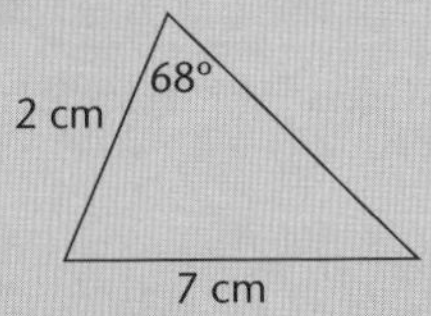

b. 68°
2 cm
7 cm

c. 22°
7 cm
68°
8 cm

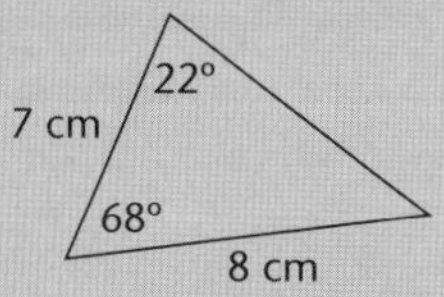

2. Determine the pairs of congruent triangles from those drawn below.

a.

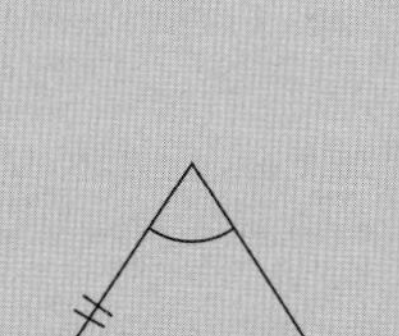

b.

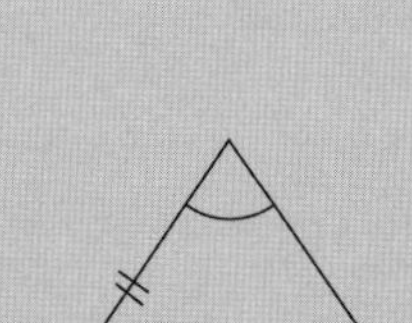

c.

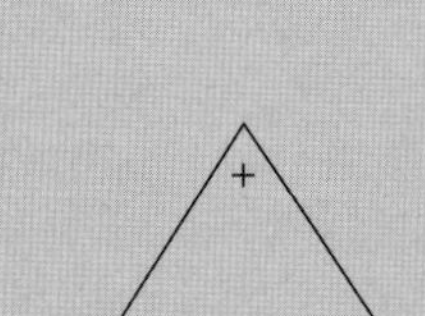

d.

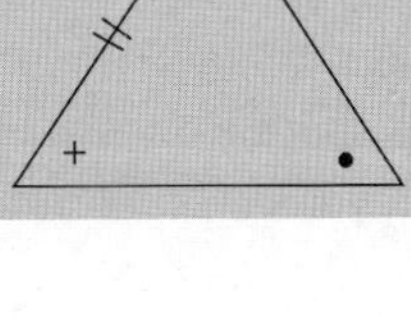

e.

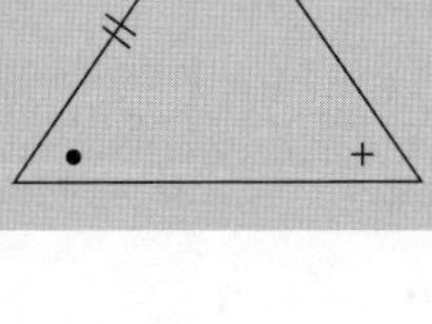

f.

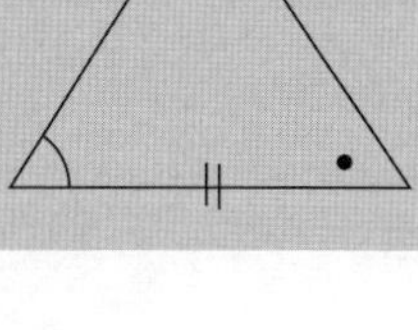

3. Use the construction methods given to construct a triangle that is congruent to each of the following triangles.

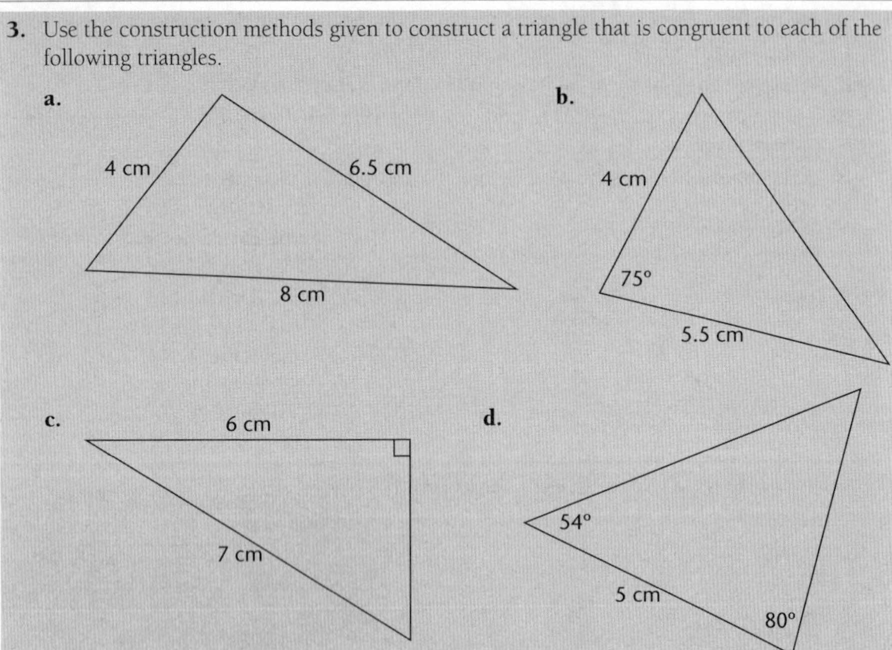

Unit 11.4 Geometry
Topic 3: Geometric constructions

A particular focus of Unit 11.4 is the development and assessment of mathematical skills in an applied context. Topic 3 deals with the use of geometric techniques to produce a pattern or object and covers:
- Constructions.
- Perpendiculars and bisectors.
- Angles.
- Parallel lines.
- Three-dimensional shapes.
- Truncated solids.

Introduction

The local environment can be used as the context for most of the application problems that students undertake. Students can examine, for example, the application of traditional patterns and measurement to expand their geometrical construction skills. Tasks will be assessed using performance criteria, for example:

- At the 'low achievement level' students rarely use appropriate mathematical techniques and their work contains many mathematical errors.
- At the 'satisfactory achievement level' students sometimes use appropriate and effective mathematical techniques, and their work may contain some mathematical errors.
- At the 'high achievement level' students usually use appropriate and effective mathematical techniques, and their work contains few mathematical errors.
- At the 'very high achievement level' students use appropriate and effective mathematical techniques at all times, and their work contains no mathematical errors, or almost none.

Construction

Construction requires the use of a pencil, compass, ruler, and sometimes a protractor or set-square to produce an *accurate* drawing.

Evidence of construction (lines indicating how the construction was done) should be left on the drawing, particularly if the drawings are being done for an assessment.

It is often useful to make a **sketch** (a freehand drawing) of a pattern first, to help decide how the construction should be done.

Perpendicular lines

A **perpendicular line** is a line at right angles (90°) to a given line.
For example a **vertical** line is perpendicular to a **horizontal** line.

The following three constructions involve perpendicular lines.

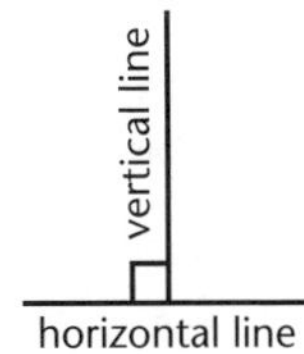

The perpendicular bisector

A **perpendicular bisector** (**mediator**) of a **line segment** cuts the line segment at right angles into two equal parts.

Construct the perpendicular bisector for a line segment AB

Set a compass to draw four arcs of the same radius. The radius of the arcs should be a little larger than half the distance between A and B.

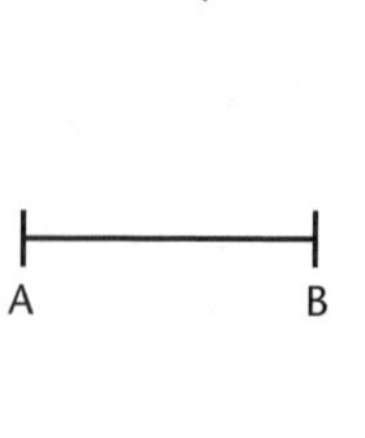

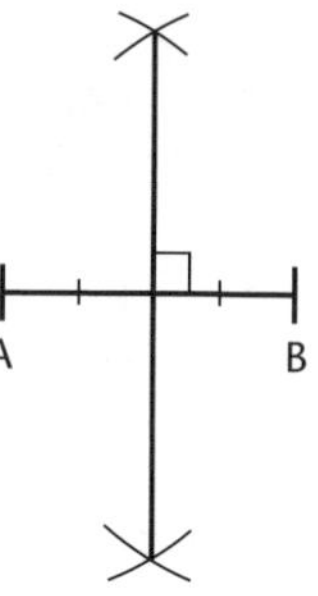

Two **arcs** are drawn from point A, one above the line segment and one below the line segment.	Two arcs are drawn from point B to intersect the arcs from A. **Note:** Don't change the radius.	Joining the points of intersection of the arcs forms the perpendicular bisector of AB.

Example A

Q. The **median** of a triangle is a line joining a vertex to the midpoint (middle) of the opposite side. In the triangle ABC shown, M is the midpoint of BC and AM is a median of the triangle. Describe how to construct the median AM of a triangle ABC.

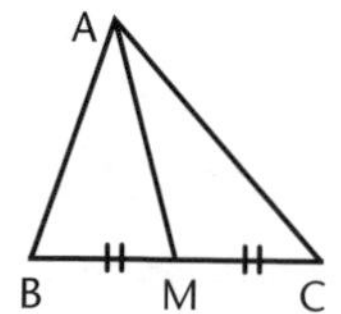

A. To find M, the midpoint of BC, the perpendicular bisector of BC is constructed. [see construction marks]

The perpendicular bisector of BC cuts BC at its midpoint, M.

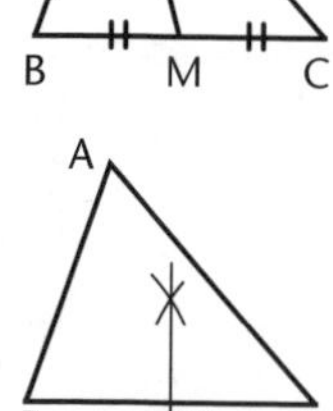

A is joined to M to make the median AM.

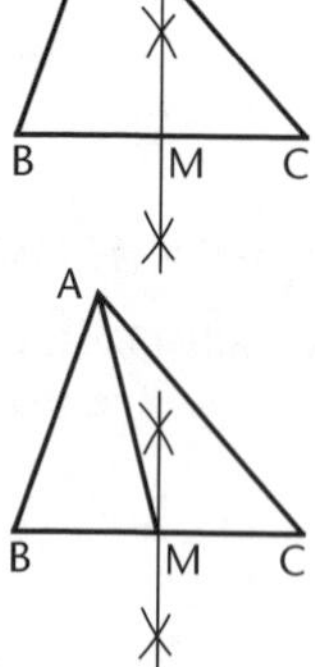

The perpendicular at a point on a line

A perpendicular line may be required through a given point on a line.

Construct the perpendicular to the line segment AB at the point C on the line segment, using a compass

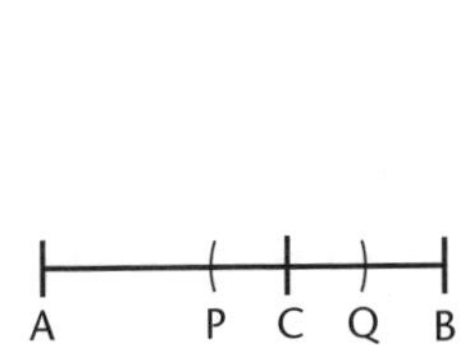

From C, a compass (set to any radius) is used to draw two arcs of equal radius on either side of C, intersecting the line segment AB. (Call these points of intersection P and Q.)

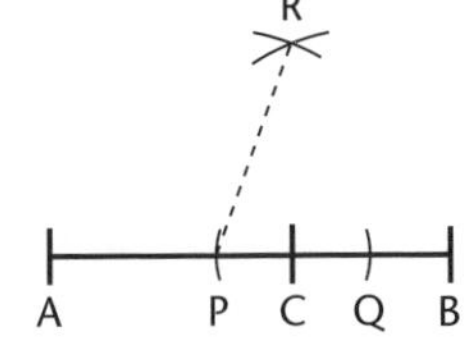

With the compass set to a *larger* radius, two more intersecting arcs of equal radius are drawn from points P and Q. Call this point of intersection R.

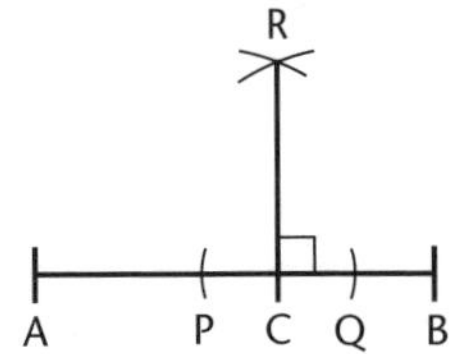

The required perpendicular is the line drawn from R to the point C.
Note: If C is near A or B, the line segment AB may be extended before beginning the construction.

Example B

Q. For a given side AB construct an isosceles triangle ABC which is right-angled at B, in which AC is the base.

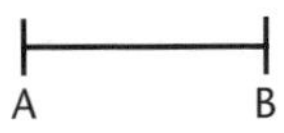

A. **1.** Extend AB and construct a perpendicular at B.

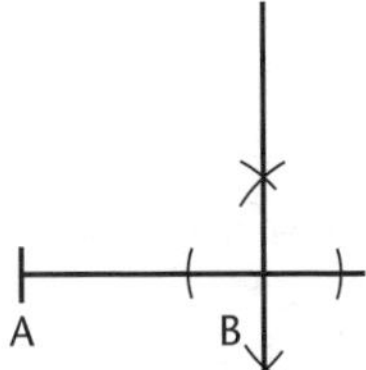

2. Open the compass to the length AB. With the point of the compass at B, mark an arc of length AB cutting the perpendicular through B at the point C.

3. Join ABC to form the required triangle ABC.

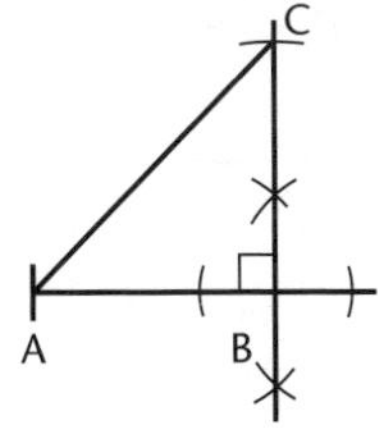

The perpendicular to a line from a point

The following method shows how to draw a perpendicular to a line from a point which is not on the line.

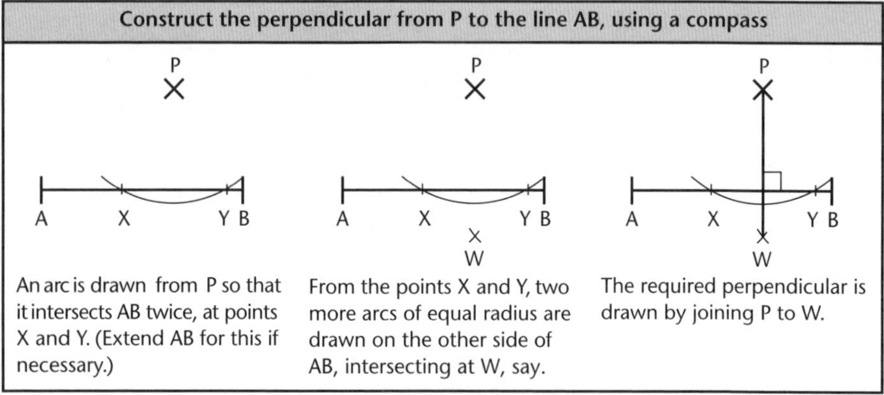

Note: If there is no space for W below the line AB, then W can be constructed above the line AB (using a different arc length).

Example C

Q. An **altitude** of a triangle ABC is a line from a vertex of the triangle which meets the opposite side at right angles. In the triangle ABC shown, AH is an altitude. Describe how to construct the altitude AH of triangle ABC.

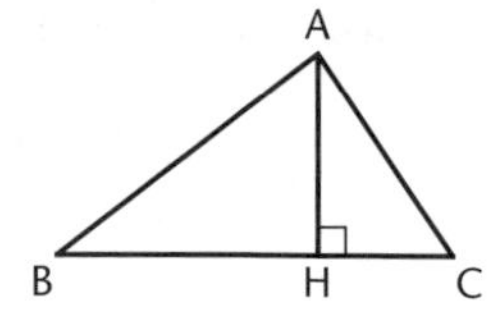

A. The perpendicular from the point A to the line AB is constructed, as shown.

This perpendicular cuts the line BC at the point labelled H.

AH is the required altitude.

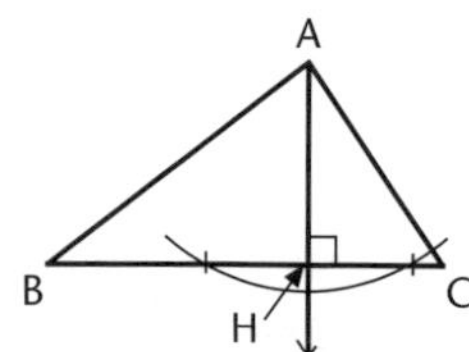

Angles

An **angle** is a turning. Some useful constructions involving angles follow.

Angle bisector

An **angle bisector** is a line which divides an angle into two angles of equal size.

To construct the angle bisector of angle ABC, using a compass

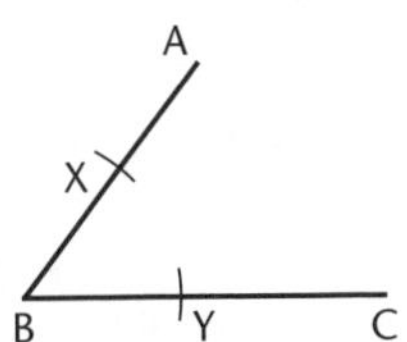

With the compass point at B, draw two arcs of the same radius, one intersecting AB (point X) and the other intersecting BC (point Y).

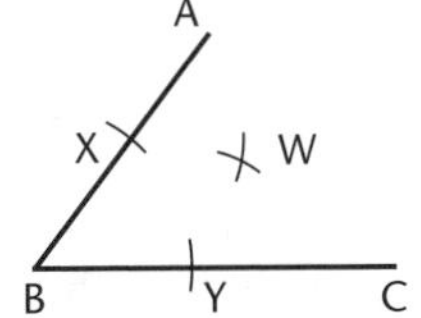

From these two points (X and Y), draw two more arcs of equal radius 'inside' the arms of the angle so that they intersect each other, at W, say.

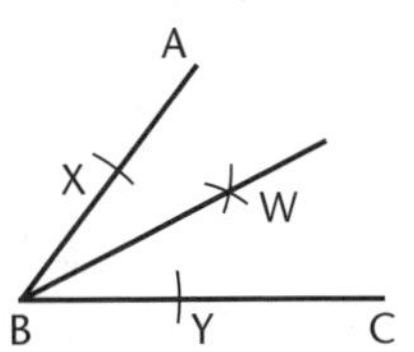

The line joining W and B is the required angle bisector.
Note: If extended, the line WB also bisects the reflex angle ABC.

Example D

Q. In a triangle, the angle bisectors are **concurrent** (they pass through the same point). This point is called the **incentre** of the triangle, and is the centre of the circle which is inside the triangle and touches all three sides of the triangle.

In triangle ABC shown, I is the incentre. Carry out the constructions needed to locate the position of I.

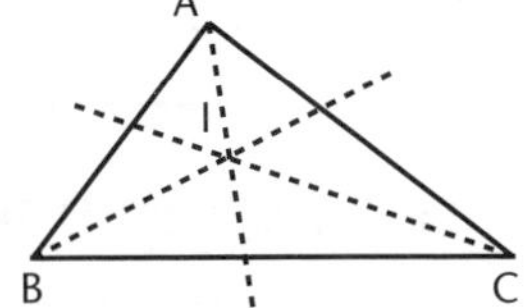

A. 1. Construct the angle bisector of C. [construction marks shown]

2. Construct the angle bisector of B. [construction marks shown]

3. The point of intersection of these two angle bisectors is the required point I (constructing the angle bisector of A will confirm this).

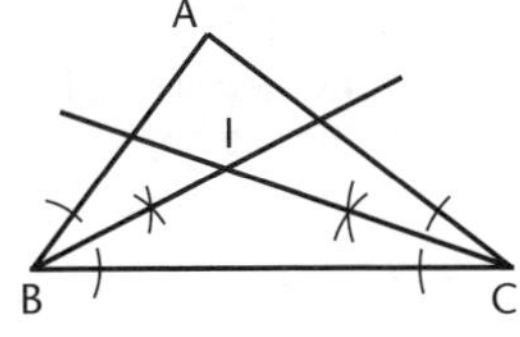

Note: Only two angle bisectors need to be constructed to locate I.

Constructing 30°, 45°, 60°, and 90° angles

30°, 45°, 60°, and 90° angles are often required, and can be drawn accurately *without* a protractor.

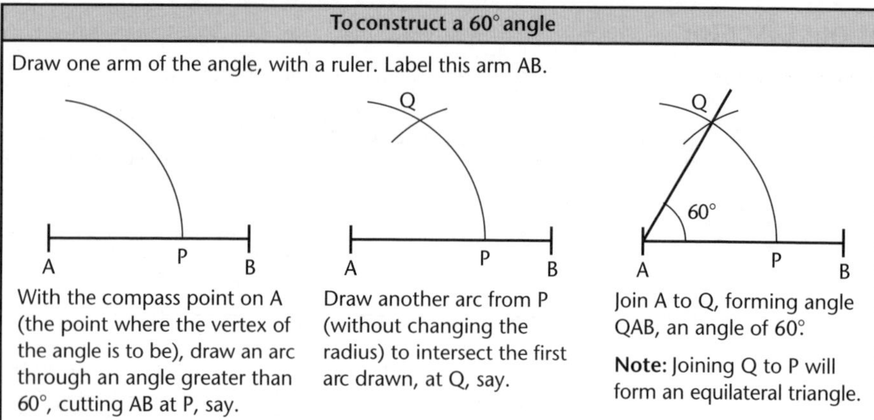

To construct a 60° angle

Draw one arm of the angle, with a ruler. Label this arm AB.

With the compass point on A (the point where the vertex of the angle is to be), draw an arc through an angle greater than 60°, cutting AB at P, say.

Draw another arc from P (without changing the radius) to intersect the first arc drawn, at Q, say.

Join A to Q, forming angle QAB, an angle of 60°.

Note: Joining Q to P will form an equilateral triangle.

The methods used to draw the angles 30°, 45°, and 90° are shown in the following table:

Angle	Procedure
30°	Construct a 60° angle then bisect the angle.
90°	Construct a perpendicular at a point on a line.
45°	Construct a 90° angle then bisect the angle.

Constructing a line parallel to another line

It may be required to construct a line **parallel** to a given line through a given point. The following example illustrates one possible method.

Example E

Q. Construct a line passing through a point P and parallel to the line AB.

A. Set the compass length to AP.

1. With point of compass at A mark point Q on line AB.
2. With point of compass at Q (same radius) mark an arc to the right of P.
3. With point of compass at P (same radius) mark an arc intersecting the arc in step **2** at R.

The line PR is the required parallel line.

Note: APRQ is a rhombus, therefore opposite sides are parallel.

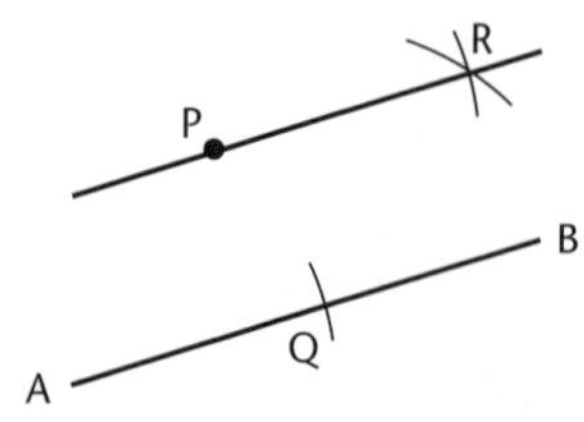

To construct a line parallel to, and a set distance from a given line, two possible techniques are shown.

To construct a line which is parallel to the line AB and 4 cm from AB

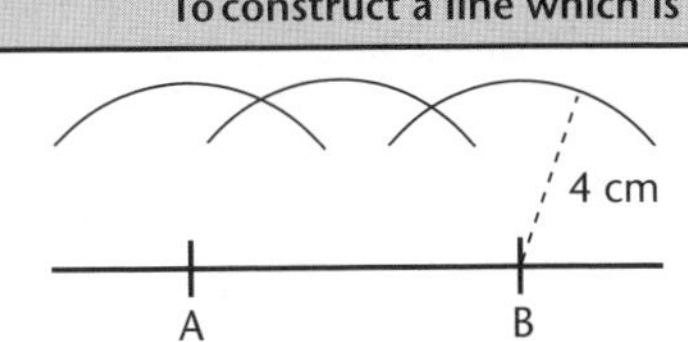

Set the radius of the compass at 4 cm and draw a series of arcs with the compass point on line AB.

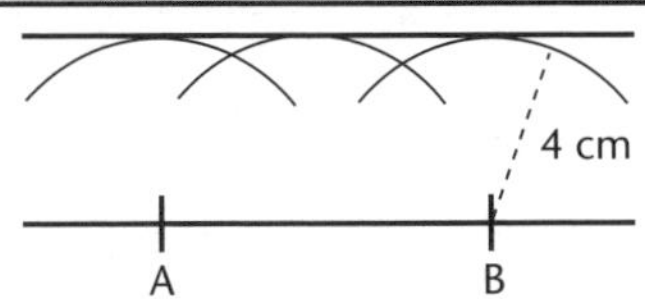

Draw the tangent to these arcs using a ruler.

Alternative solution

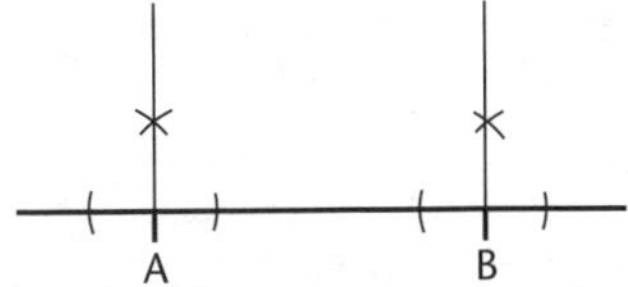

Extend AB on both sides. Construct a perpendicular at A and a perpendicular at B using the method already described.

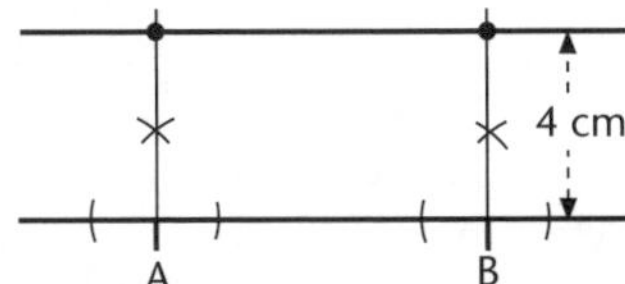

Measure a distance of 4 cm on each line. Join these points to form a line parallel to AB.

Unit 11.4 Activity 3A: Perpendiculars and bisectors

For each construction do a quick sketch first so that enough paper is allowed for the actual construction.

1. A seven-sided name badge is designed for the chain of SUPERSEVEN convenience stores. Follow the instructions to construct the shape.

- **a.** Rule a line of length 6 cm. Label this line AB.
- **b.** Construct the perpendicular bisector of AB.
- **c.** Locate the mid-point of AB and label it C.
- **d.** With centre C, construct an arc AB (the arc is a semicircle) above the line AB.
- **e.** Locate the point where the perpendicular bisector of line AB meets the semicircle AB and label this point D.
- **f.** Construct the bisector of angle DCB. Locate the point where the bisector meets the arc ADB and label this point E.
- **g.** Construct the bisector of angle ECB. Locate the point where the bisector meets the arc ADB and label this point F.
- **h.** Construct the bisector of angle DCA. Locate the point where the bisector meets the arc ADB and label this point G.
- **i.** Construct the bisector of angle GCA. Locate the point where the bisector meets the arc ADB and label this point H.
- **j.** Join points to create the required shape AHGDEFB.

2. A triangle ABC is to be constructed, following instructions **a.** to **d**.
   a. Draw a horizontal line segment AK approximately 5 cm long and construct a line at right angles to AK which passes through A.
   b. Mark and label the point B on AK, so that the length of AB is 4 cm.
   c. C is a point on the perpendicular drawn in **a.** above so that the length of BC is 6 cm. Locate and label point C.
   d. Complete the triangle ABC.
3. A shape is constructed following instructions **a.** to **g.** below.
   a. Draw a line of 5.5 cm. Label this line AB.
   b. Find, by construction, the mid-point of AB. Label this point C.
   c. Construct an equilateral triangle, CBD, so that D is 'above' the line ACB.
   d. Construct an equilateral triangle, CAE, so that E is 'above' the line ACB.
   e. Join DE.
   f. Repeat parts **c.** and **d.** so that the equilateral triangles CBF and CAG are 'below' the line ACB.
   g. Join FG.
   What is the name given to the shape AEDBFG?
4. In the sketch shown (not to scale), O is the centre of a circle of radius 12 mm. The length of OP is 5 cm.

   A quadrilateral is to be constructed as follows.
   a. Construct the diagram described above.
   b. Construct the perpendicular bisector of OP.
   c. Construct the circle using OP as a diameter. Label the points at which the two circles intersect X and Y.
   d. Join OX, OY, XP and YP.
   What type of quadrilateral is OXPY?
5. Nikau Housing Development Ltd created a simple logo for their company. The following steps will allow you to draw their logo.
   a. Rule a horizontal line 8 cm long. Label this line AB.
   b. Construct a line 3 cm above and parallel to AB. Label this new line CD.
   c. At A, construct the perpendicular line to AB. Locate the point where the perpendicular meets the line CD and label this point E.
   d. Bisect angle AED. Locate the point where this bisector meets line AB and label this point F.
   e. At F, construct a line perpendicular to AB. Locate the point where the perpendicular meets line CD and label this point G.
   f. Mark and label point H on the line AB, 3 cm to the right of F.
   g. At H, construct the perpendicular to line AB. Locate the point where the perpendicular meets line CD and label this point I.

h. Find, by construction, the mid-point of FG. Label this point J.

i. Join JI and JH.

j. Find, by construction, the mid-point of HI. Label this point K.

k. With K as the centre, construct the arc HI to the right of the line HI.

Three-dimensional shapes

Being able to represent a three-dimensional figure on a two-dimensional page is a skill that will enable you to see more clearly, and solve, three-dimensional problems.

Drawing three-dimensional shapes

Below are methods for drawing common three-dimensional figures.

1. **Cuboids (or cubes)**

   In a cuboid (or cube) all the faces are rectangles (or squares).

   i. Draw two identical rectangles (or squares).

   The second should be to the right and higher than the other.

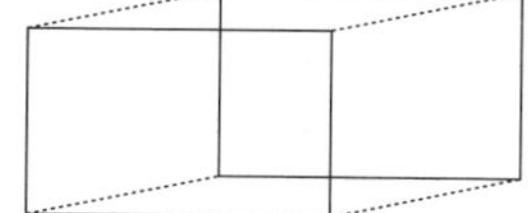

   ii. Join the corners as shown.

   Lines and shading can be added to the cuboid (or cube) to highlight different aspects of the figure:

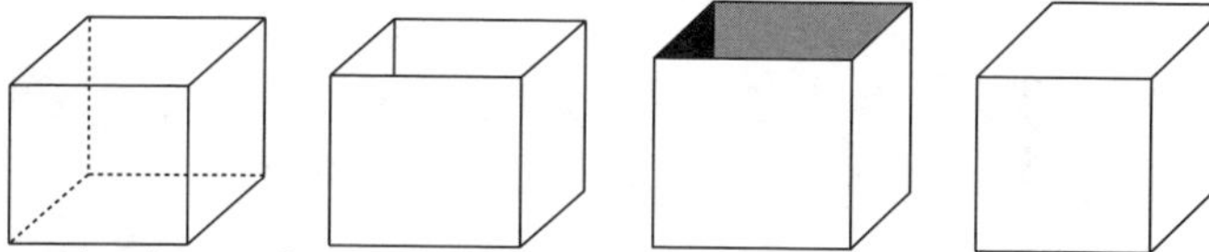

A see-through cube, used for illustrating inside measurements and angles. An open box. A shaded open box. A solid cube.

2. **Pyramids**

   a. **A right, rectangular pyramid**. The base is a rectangle, or square, and the vertex is directly above the 'centre' of the base.

   i. Draw a parallelogram to represent the base.

   ii. Find the centre of the base.

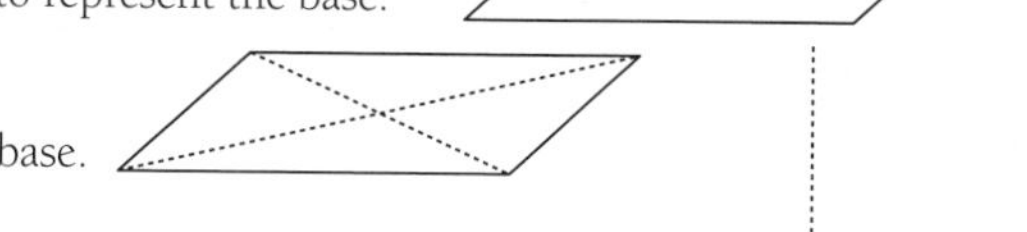

   iii. Draw a vertical line to give the pyramid height .

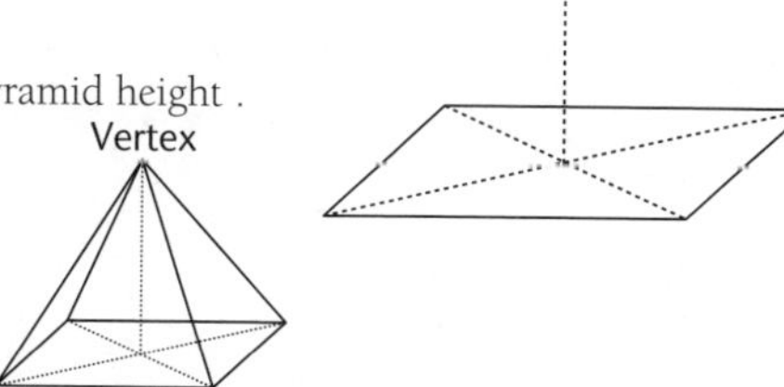

   iv. The highest point is called the vertex. Join the vertex to the corners of the base.

Lines or shading can be added to highlight different aspects of the figure:

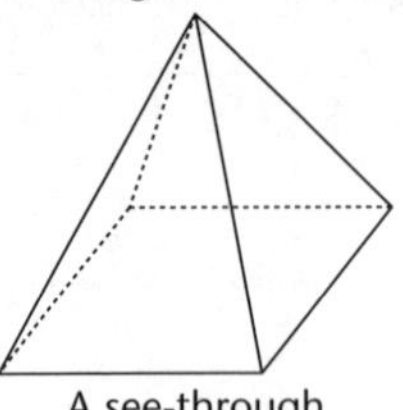
A see-through pyramid.

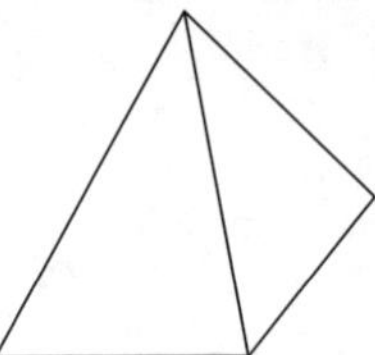
A solid pyramid.

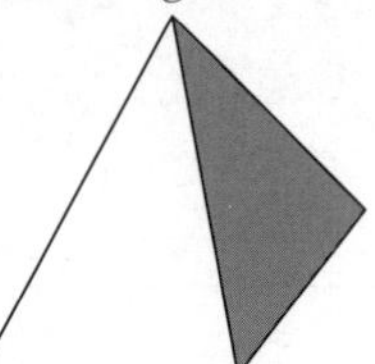
A shaded solid pyramid.

b. **Other right pyramids with a regular polygon as a base**.

Find the 'centre' of the base, locate the vertex directly over this 'centre' and connect the edges to the vertex.

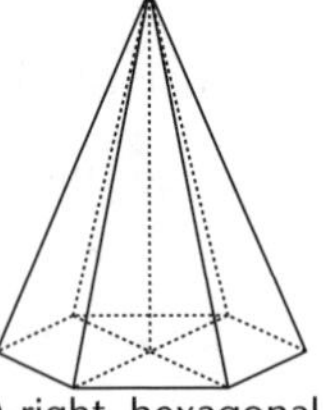
A right, hexagonal pyramid.

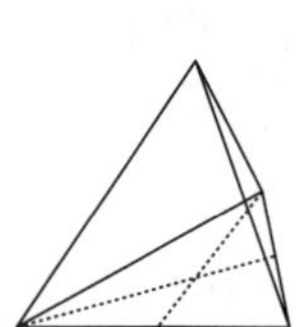
A right, triangular pyramid.

c. **Non-right pyramids**. These pyramids do not have the vertex above the 'centre' of the base.

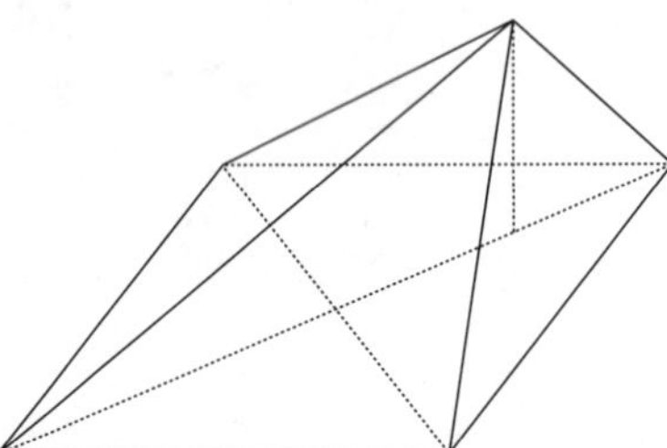

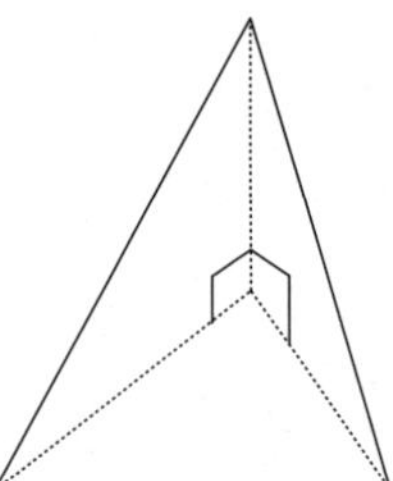

3. **An inclined plane**. One side of the plane is raised.

i. Draw a parallelogram for the base.

ii. Draw two vertical lines to give height at the back.

iii. Draw in the other three sides of the inclined plane.

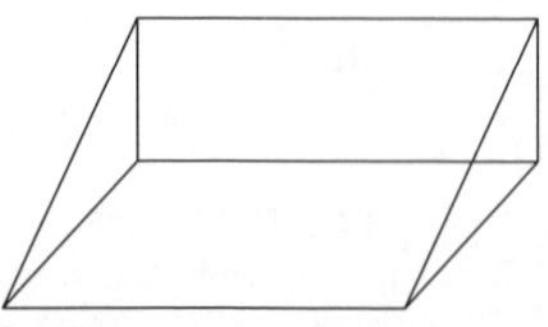

Lines or shading can be added to highlight different aspects of the figure:

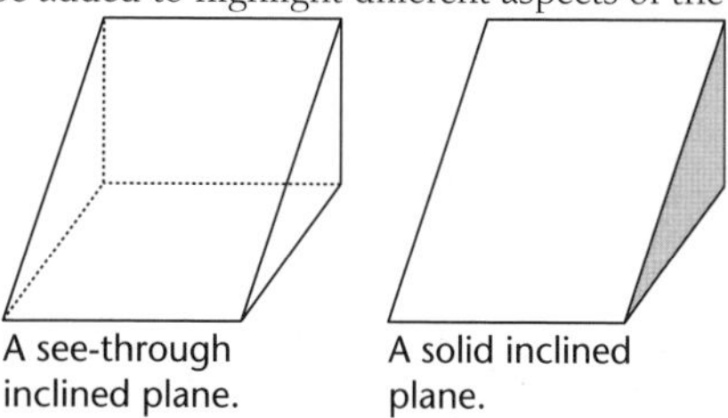

A see-through inclined plane. A solid inclined plane.

4. **Other prisms**

Prisms have a constant cross-section and rectangular sides.

Start by drawing two identical versions of the cross-section; one to the left and above the other. Join the vertices of the cross-sections. All of the following prisms have a constant vertical cross-section.

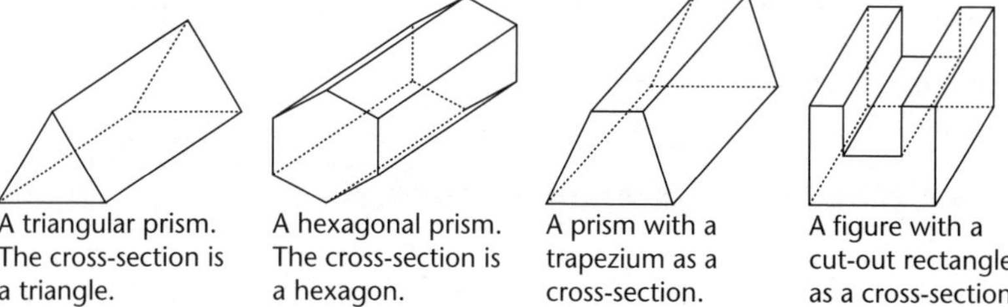

A triangular prism. The cross-section is a triangle. A hexagonal prism. The cross-section is a hexagon. A prism with a trapezium as a cross-section. A figure with a cut-out rectangle as a cross-section.

5. **Spheres and hemispheres**

i. Draw a circle.

ii. Connect two points on the circumference of the circle with an ellipse.

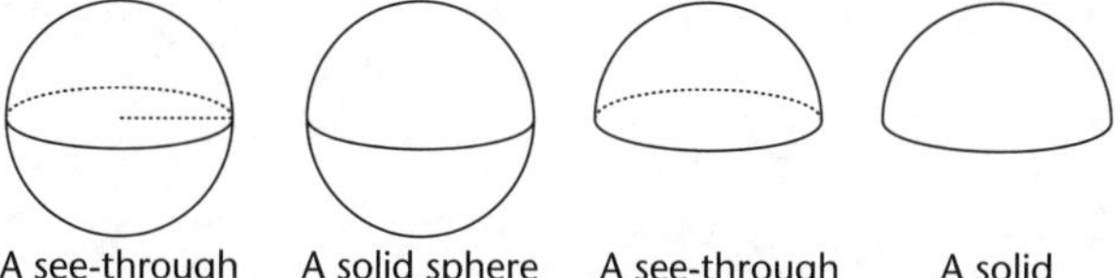

A see-through sphere. A solid sphere A see-through hemisphere. A solid hemisphere

6. **Cones**

i. Start by drawing an ellipse to represent the base.

ii. For a right cone, locate the centre of the base and draw a vertical line to the vertex.

iii. Join the outer points of the ellipse to the vertex.

Some variations of the cone:

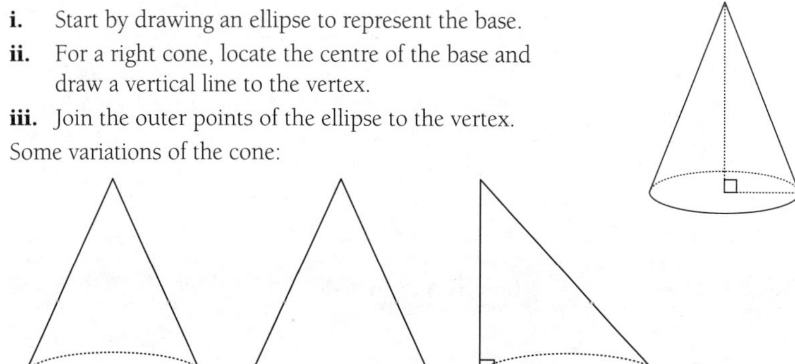

A see-through cone. A solid cone. A non-right cone.

7. **Cylinders**
    - **a.** An upright cylinder, resting on the circular base.
        - **i.** Draw two identical ellipses, one above the other, for the top and base of the cylinder.
        - **ii.** Connect the outer points of the ellipses with a straight line.
    - **b.** A cylinder resting on the curved surface.
        - **i.** Start by drawing two circles, one above and to the left of the other.
        - **ii.** Draw in two straight lines, tangents to the circles, to represent the curved surface.

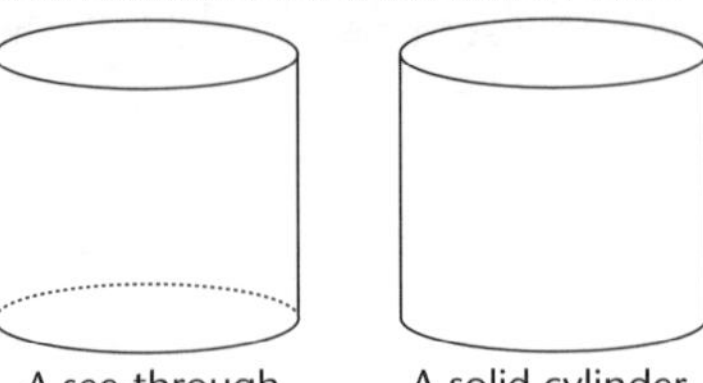
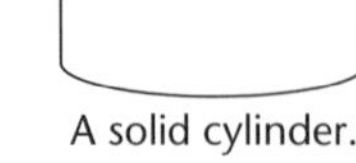

A see-through cylinder. A solid cylinder.

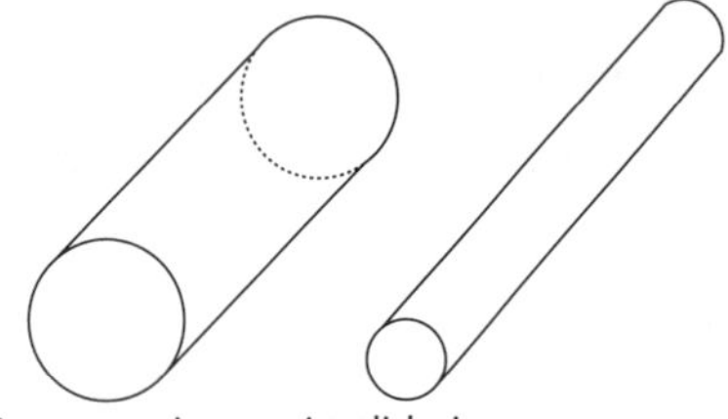

An open pipe A solid pipe.

If you have access to a template that enables you to easily draw circles, ellipses and parallelograms then constructing these three-dimensional shapes will be simplified. Isometric paper is also an aid to drawing prisms.

Unit 11.4 Activity 3B: Three-dimensional shapes

1. Construct an open cuboid that:
    - **i.** Resembles the large (bottom) part of a shoe box.
    - **ii.** Resembles the lid of a shoe box.
2. Construct a solid right pyramid with a square base. Use a pencil so that you can erase the construction lines.
3. Construct a non-right pyramid, with a square base, whose vertex is directly over one of the corners of the base.
4. Construct a see-through right pyramid whose base is a regular pentagon.
5. Construct this figure, which is a combination of an inclined plane and a cuboid.

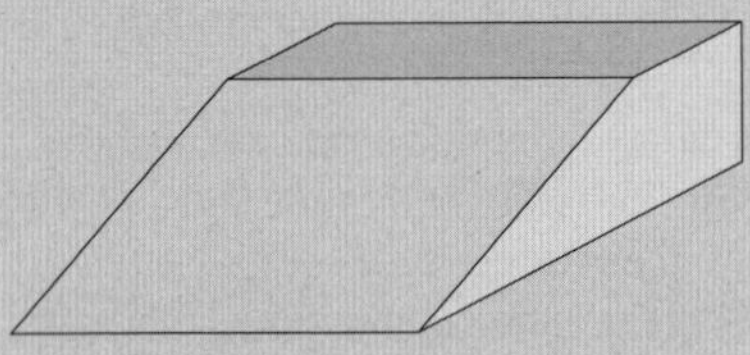

6. Construct a cylinder whose diameter is larger than its height.
7. For each of the following draw a prism with the given (shaded) cross-section:
    - **i.** Leave in all the construction lines.

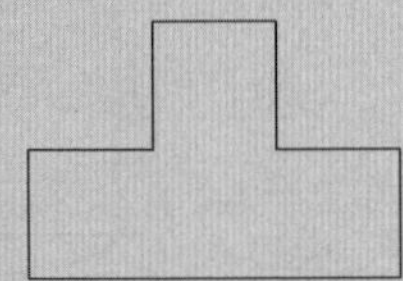

ii. Remove the construction lines so that it appears solid.

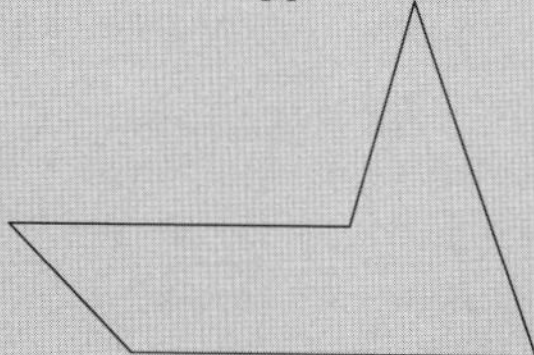

8. Construct this solid figure:

9. Draw a cone that appears to have a height that is:

i. twice the diameter. **ii.** half the diameter.

10. Construct a see-through hemisphere sitting on top of a cylinder of the same radius.

11. Construct a see-through figure with the following (shaded) cross-section:

a.

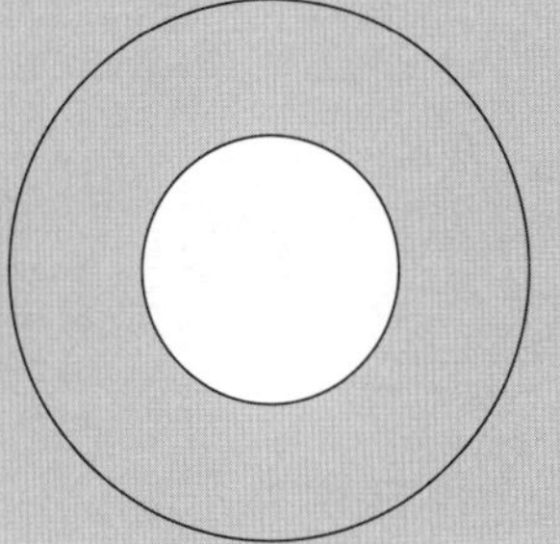

b.

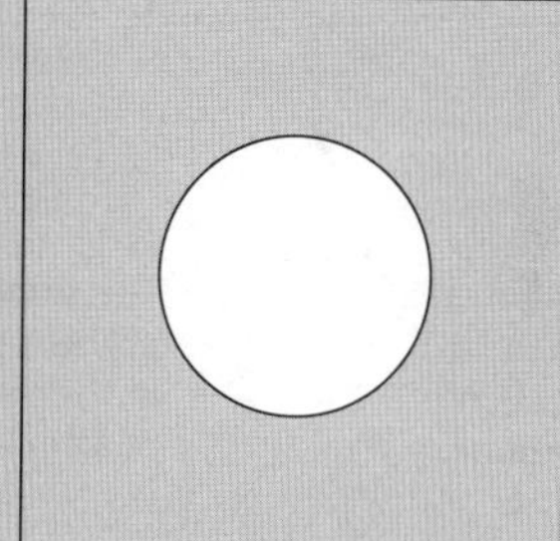

Truncated solids

Truncated means 'cut off'.

A truncated cone could look like:

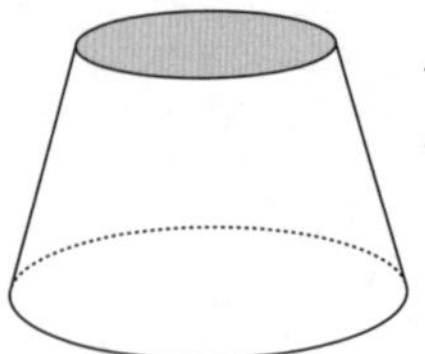

The cut surface is in the shape of a circle.

A truncated cube could look like:

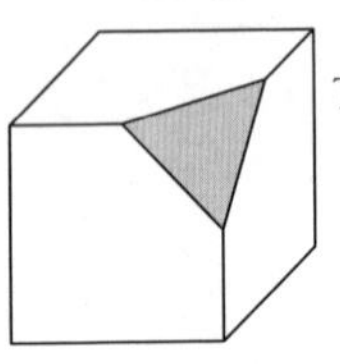

The cut surface is in the shape of a triangle.

A truncated sphere could look like:

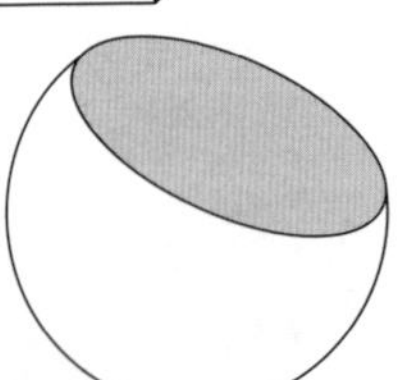

The cut surface is in the shape of a circle.

We can consider these truncated figures as being 'cut off' by a plane surface, with the orientation of the plane surface dictating the shape of the cut surface.

Below are three different shapes that can be produced by three different orientations of a plane surface intersecting with a right cone:

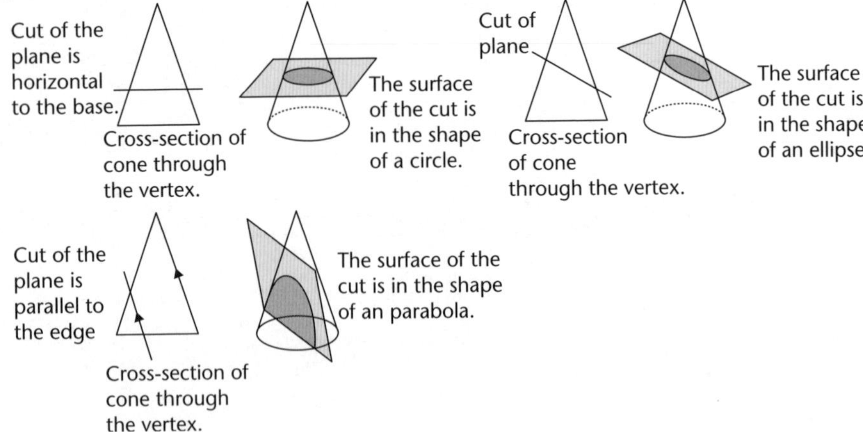

Example F

Q. A right, square pyramid is shown at right.

If the square base of this pyramid is horizontal, describe the shape of the surface produced by the intersection of the pyramid with:

a. A horizontal plane.

b. A plane that is at an angle less than 90° with the base.

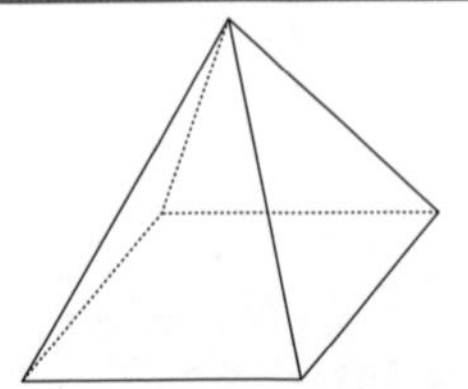

A. a. The horizontal plane is parallel to the base so the shape of the cut will be a square.

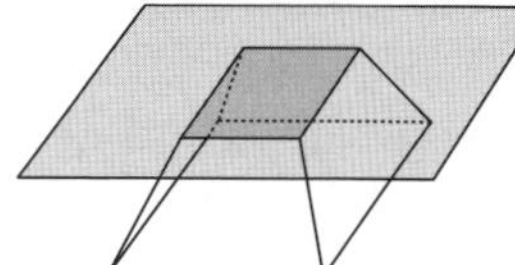

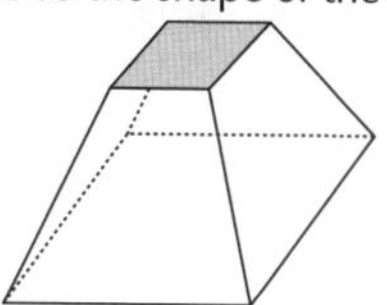

b. The plane is at an angle to the base so the shape of the cut is a trapezium if it cuts off two vertices.

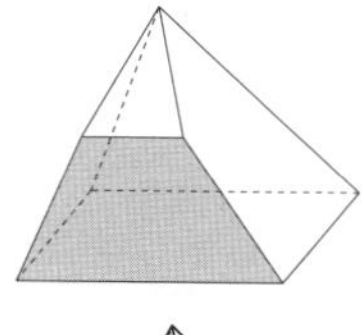

A triangle can be produced if the plane cuts off one of the corners of the pyramid.

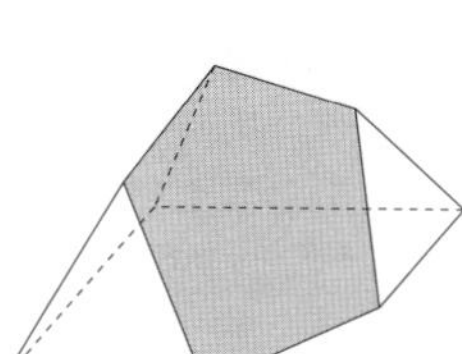

A pentagon is produced if the vertex and one corner of the pyramid is cut off by the plane.

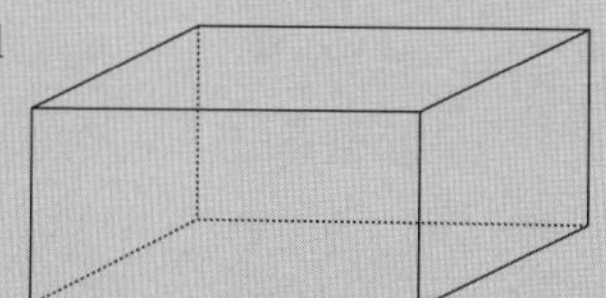

Unit 11.4 Activity 3C: Truncated solids

1. Investigate the different cut surfaces that can be produced on the cuboid, given at right, by planes at different orientations. Show, with diagrams, at least three different cut shapes, and describe all the possible cut shapes.

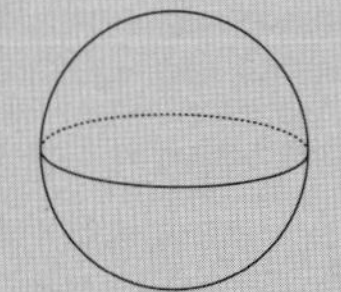

2. Investigate the different cut surfaces that can be produced on the sphere, given at right, by planes at different orientations. Show, with diagrams, at least three different cut positions, and describe all the possible cut shapes.

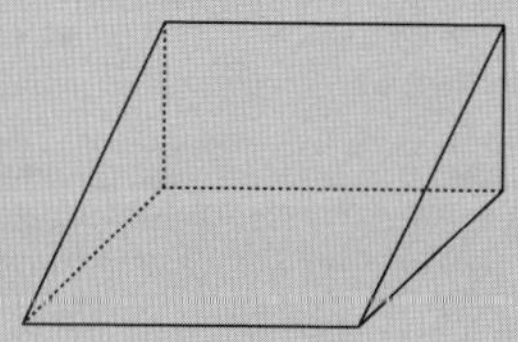

3. Investigate the different cut surfaces that can be produced on the inclined plane, given at right, by cut planes at different orientations. Show, with diagrams, at least three different cut shapes, and describe all the possible cut shapes.

4. Investigate the different cut surfaces that can be produced on the figure, given at right, by planes at different orientations. Show, with diagrams, at least three different cut shapes, and describe all the possible cut shapes.

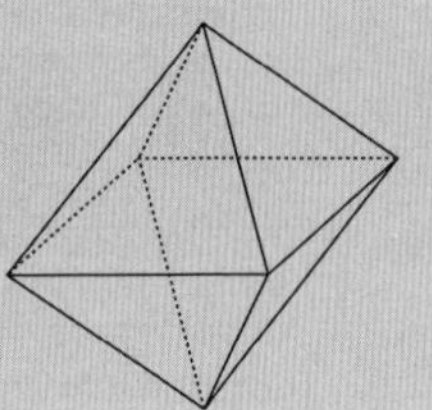

Unit 11.4 Geometry
Topic 4: Circle geometry

This Topic deals with the content described under the sub-heading 'Circles' in the Grade 11 syllabus (see Syllabus p. 18). It covers:

- Circle geometry (limited to angle geometry).
- Finding unknowns in circles using two-step processes.
- Investigating a conjecture, presenting a proof or solving a problem using at least three steps of reasoning.

The circle

A **circle** is the set of all points in a plane a fixed distance (the length of the **radius**) from a particular point (the **centre**).

Lines and areas relating to circles are described in the following diagrams.

Centre	Circumference	Diameter	Radius	Tangent
The point about which the circle is formed. 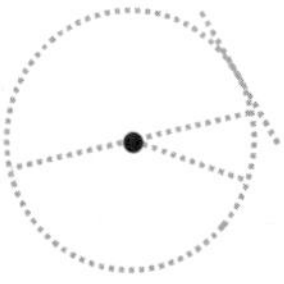	The line that forms the circle. 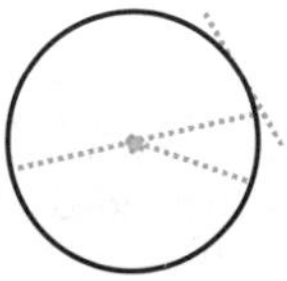	A straight line segment passing through the centre of a circle and joining two opposite points on the circumference. 	A straight line segment from the centre of a circle to a point on the circumference. 	A straight line touching one point on the circumference of a circle. 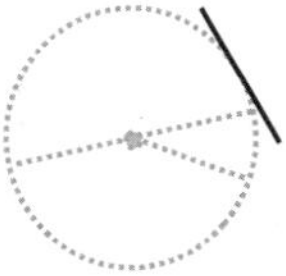

An **arc** is part of the circumference. There are always two arcs between any two points P, Q on the circumference, a **minor arc** and a **major arc**. When the length of the minor arc is the same as the length of the major arc, the arcs are called **semi-circles**. Arcs can be drawn using a compass, in the same manner as for drawing a full circle.

Minor Arc	Major Arc	Semi-circle
The shorter arc between any two points on the circumference of a circle.	The longer arc between any two points on the circumference of a circle.	Occurs when the length of the major arc and minor arc are the same.
 P, minor arc, Q	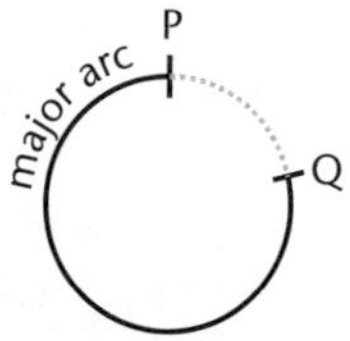 P, major arc, Q	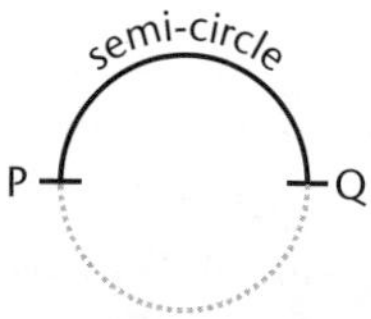 semi-circle, P, Q

In general, the arc PQ refers to the minor arc PQ (of length less than the semicircle).

Sector	Chord	Segment
A region inside a circle enclosed by an arc and two radii.	A straight line segment joining any two points on the circumference.	A region inside a circle enclosed by a chord and an arc.

The diagrams show angles which the arc AB **subtends** (lies under).

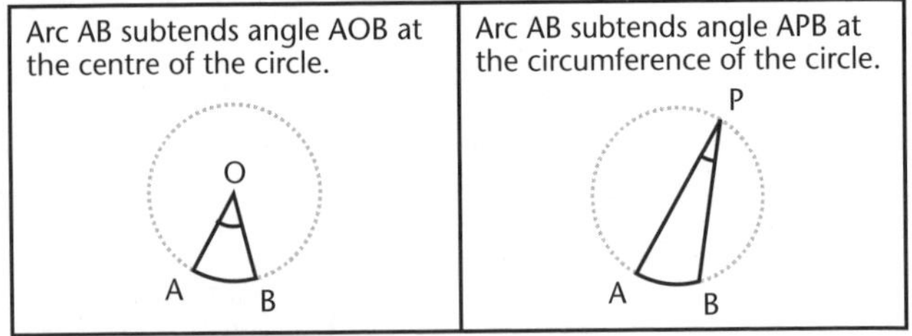

Angles on the same arc

Angles on the circumference of a circle which are subtended by the *same* arc (or chord) are equal in size. (Abbreviated to: ∠s same arc or ∠s same chord.)

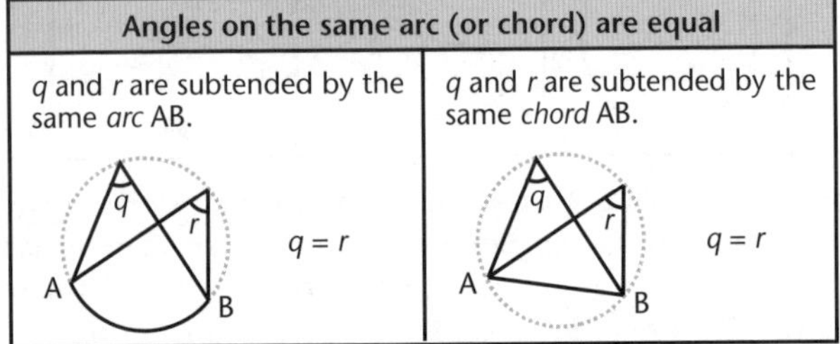

Note: In the diagram shown opposite:

$x \neq y$ because x and y are subtended by different arcs.

$x \neq z$ because, although these two angles are subtended by the same arc, z is not an angle at the circumference.

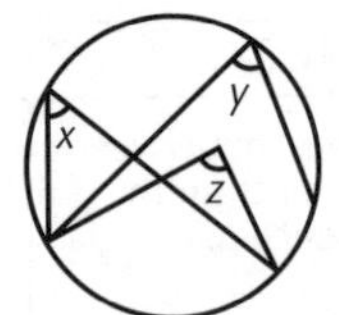

Example A

In the diagram:

$x = 35°$ (∠s same arc)

$y = 42°$ (∠s same arc)

$y + z + 35 = 180$ (sum ∠s triangle)

$42 + z + 35 = 180$ [$y = 42°$]

$\therefore\ z = 103°$ [simplifying]

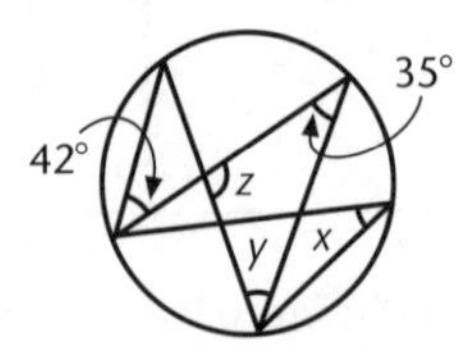

Angles at the centre and circumference

If an angle at the circumference of a circle and an angle at the centre are subtended by the same arc (or chord), then the angle at the centre is twice the angle at the circumference. (Abbreviation: ∠ at centre.)

Angle at the centre is twice angle at circumference

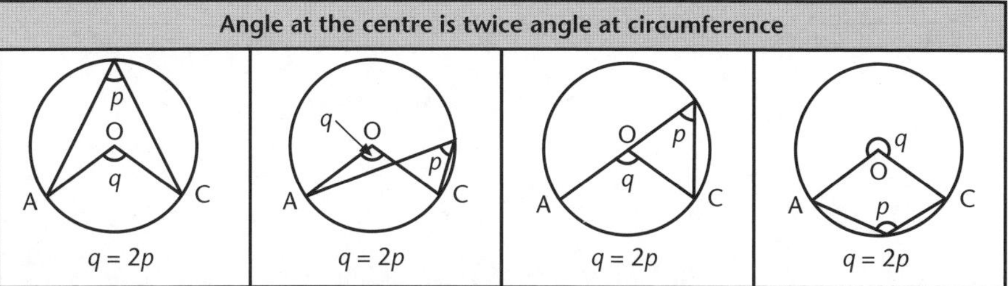

Note: In the diagram shown opposite, $q \neq 2r$, because the angle q is subtended by the arc ABC and the angle r is subtended by the arc ADC.

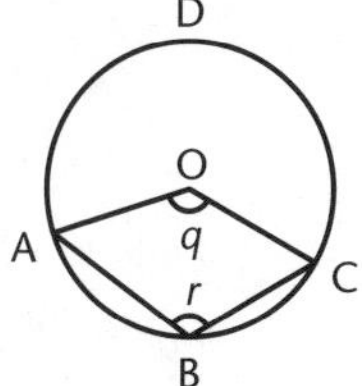

Example B

In the diagram:

$$a = \frac{100}{2} \quad (\angle \text{ at centre})$$

$$= 50°$$

Reflex ∠AOC = 260° (∠s at a point)

$$\angle b = \frac{1}{2} \times 260 \quad (\angle \text{ at centre})$$

$$= 130°$$

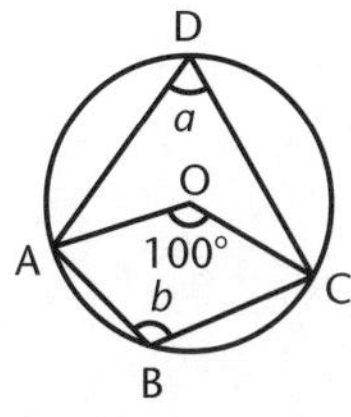

Angle in a semi-circle

The following result is a special case of the rule for the angle at the centre.

When the angle at the centre is 180°, the angle at the circumference is a right angle. (Abbreviation: ∠ in semi.)

The angle in a semi-circle is a right angle

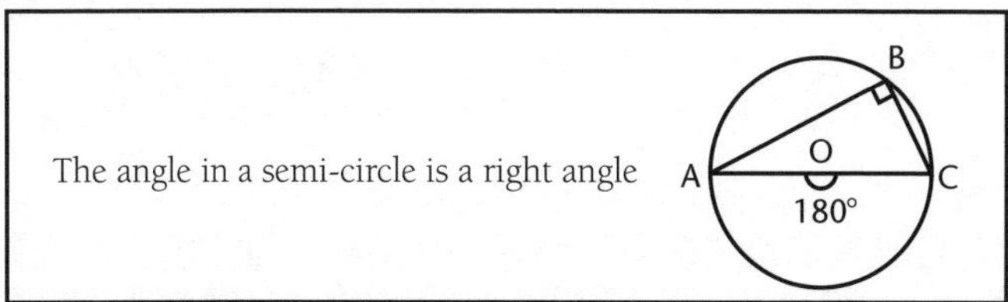

Note: When solving problems involving triangles inside circles, be on the lookout for isosceles triangles, formed when two of the sides of the triangle are radii of the circle.

Unit 11.4 Activity 4A: Circle geometry

1. Find the size of the angles labelled with letters. Give a brief reason for each answer.

a.

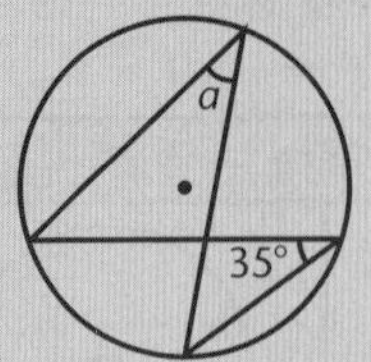

b.

c.

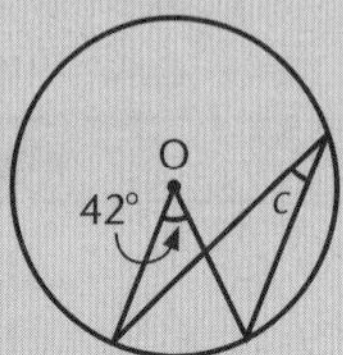

d.

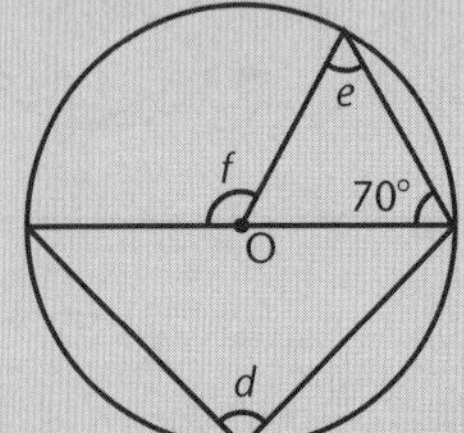

e.

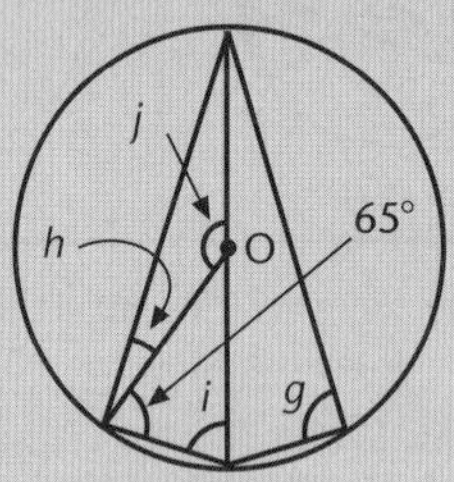

f.

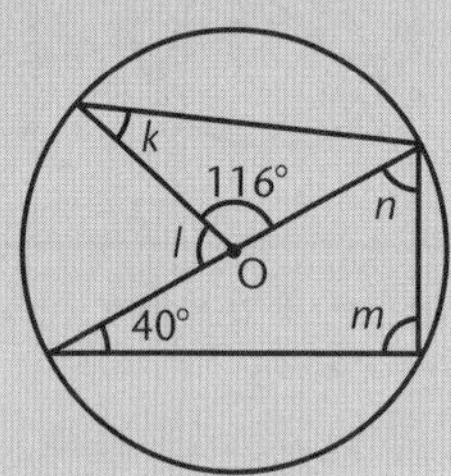

2. Find, with reasons, the size of the angles marked.

a.

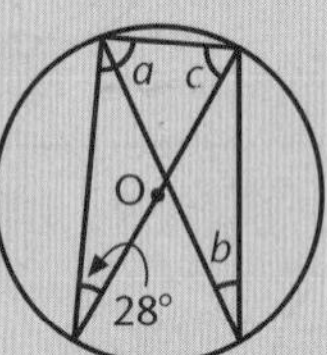

b.

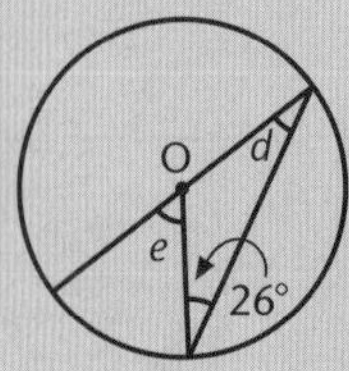

c.

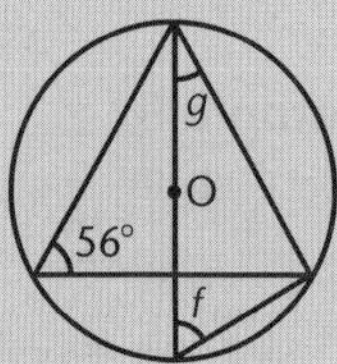

d.

e.

f.

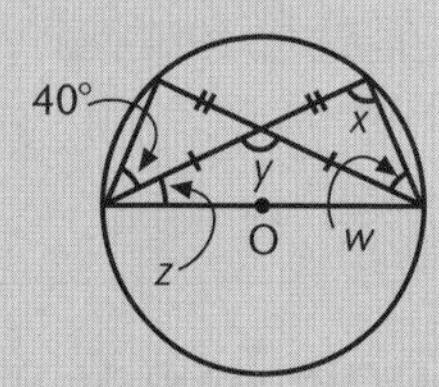

3. In the semi-circle shown O is the centre, ABCD is an isosceles trapezium with ∠BAD = 64°. Find, with reasons, the size of ∠DOC.

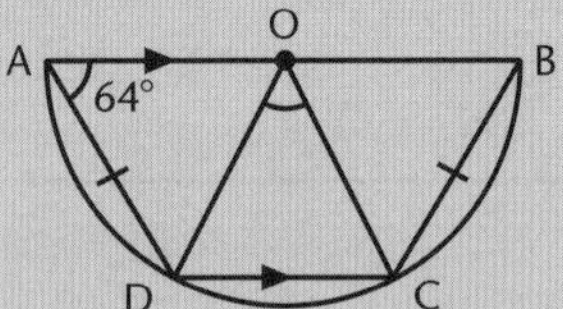

4. A logo for a sail-maker is shown, where O is the centre of the circle and PQS is an isosceles triangle. If ∠OSQ = 35°, find, with reasons, the size of QPS.

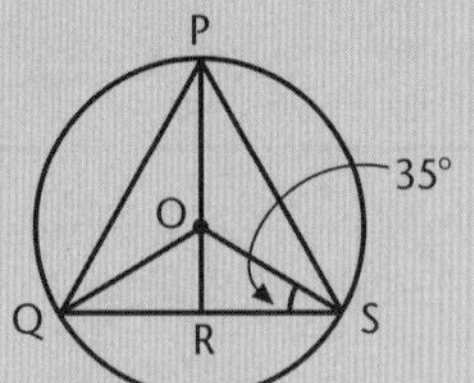

5. A circle has a centre O, and the points A, B, C, D lie on its circumference. If ∠AOB = 54° find:

a. ∠OBC

b. ∠ODA

c. Why is AD parallel to BC?

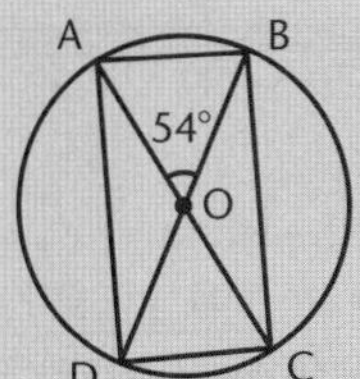

6. O is the centre of the circle. Use the angle at the centre rule to prove that $a = b$.
Hint: Consider the relationship between x and a and x and b.

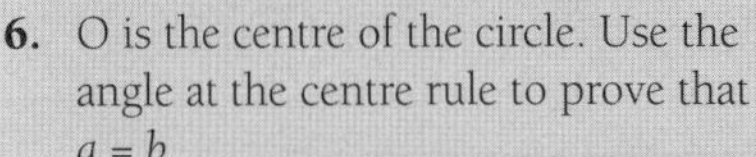

7. O is the centre of the circle ABCD. Prove that ∠AOC = 2 ∠ABC.
Hint: Find the size of ∠BAO in terms of a, then find the size of ∠AOD in terms of a.

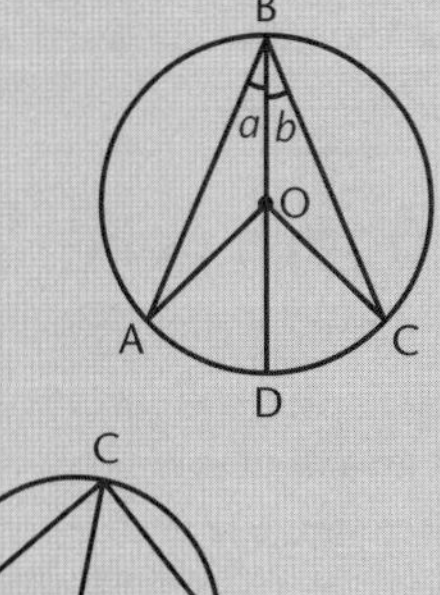

8. AB is the diameter of circle centre O. Prove that ∠BCA = 90°.

Hint: Extend CO to D on the circumference and let ∠AOD = x, ∠BOD = y.]

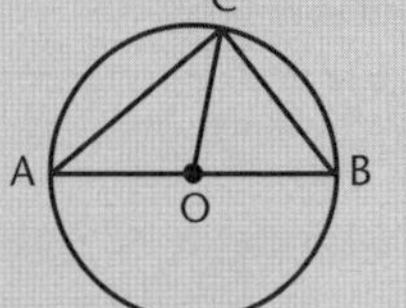

Further properties of the circle

Radius and tangent

A right angle is formed at the point where the radius of a circle meets the tangent to the circle at that point. (Abbreviation: rad ⊥ tan.)

The radius of a circle is perpendicular to the tangent

Example C

Q. A circle centre O has a tangent which touches it at T.

Find the sizes of the angles *a, b, c, d*. Give reasons.

A. $a = 90°$ (rad ⊥ tan)

OT and OR are radii, so ΔOTR is isosceles.

$\angle OTR = 35°$ (base ∠ isos Δ)

$b + 35 = 90$ (rad ⊥ tan)

$b = 55°$

$c + 35 = 90$ ($\angle TRU = 90°$, angle in semicircle)

$c = 55°$

ΔORU is also isosceles, with base angles *c* and ∠OUR. [since OR and OU are radii]

So $d = 180 - 55 - 55$ (sum ∠s Δ)

$= 70°$

Tangents from a point

From a point T outside a circle, two tangents to a circle, TA and TB, can be drawn.

The tangents are equal in length. The line, TO, drawn from the point outside, T, to the centre of the circle, O, is an axis of symmetry.

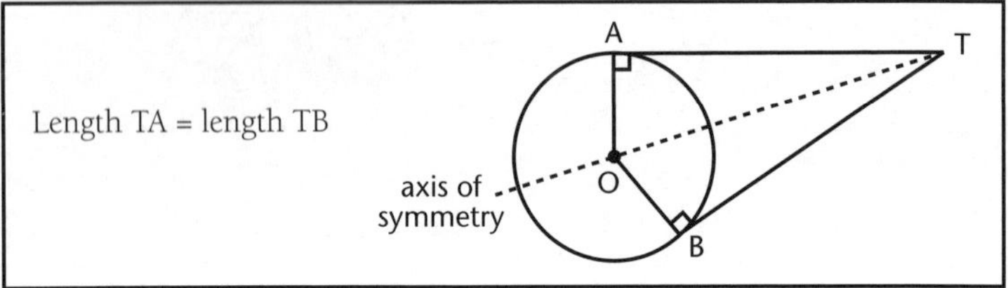

Chords and mediators

The **mediator** (perpendicular bisector) of a chord passes through the centre of the circle, O.

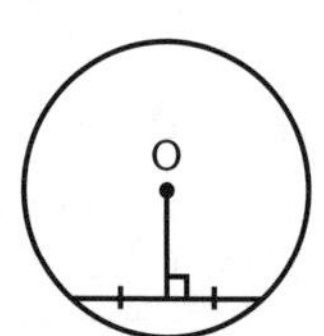

Likewise, the perpendicular from the centre of a circle to a chord will bisect the chord.

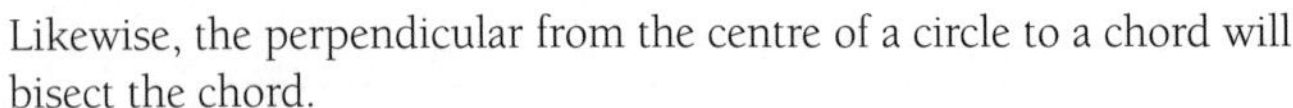

Cyclic quadrilaterals

A **cyclic quadrilateral** is a quadrilateral with all four vertices (corners) on the circumference of the same circle.

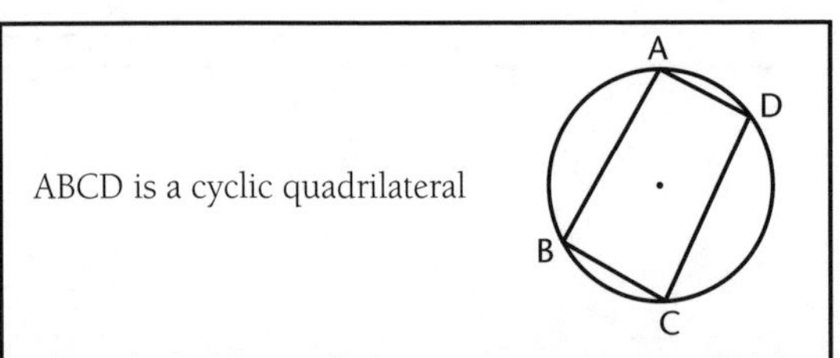

Note: OPQR is not a cyclic quadrilateral, because point O does not lie on the circumference.

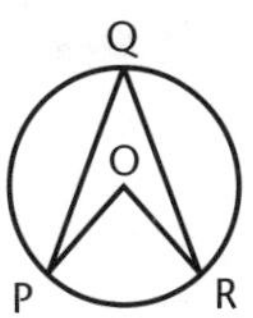

Properties of cyclic quadrilaterals

The *opposite* angles of a cyclic quadrilateral are **supplementary** (add to 180°). (Abbreviation: opp ∠s cyc quad.)

$a + b = 180°$
$c + d = 180°$

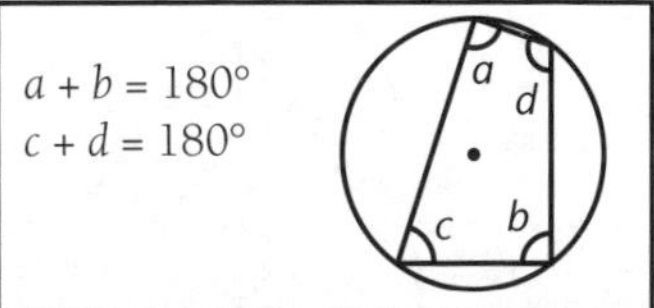

The *exterior* angle of a cyclic quadrilateral is equal to the interior opposite angle. (Abbreviation: ext ∠ cyc quad.)

$x = y$

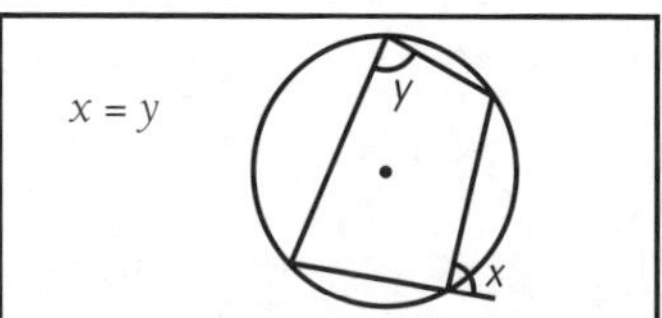

Example D

In the diagram:

$x + 68 = 180$ (opp ∠s cyc. quad)

$x = 112°$

$a = 80°$ (∠s on line)

$y = 80°$ (ext ∠ cyc quad)

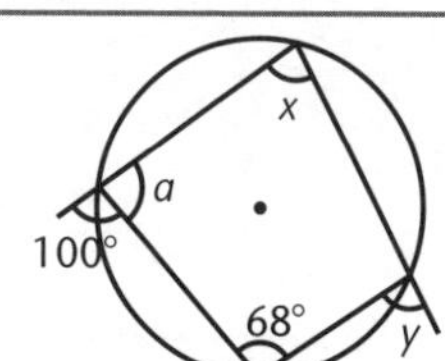

Concyclic points

Concyclic points are points which lie on the circumference of the same circle.

Four points will be concyclic if, in the quadrilateral they form:

1. the opposite angles add to 180°, *or*
2. an exterior angle equals the interior opposite angle, *or*
3. the same straight line subtends two equal angles.

If four points are concyclic, then they form a cyclic quadrilateral.

Example E

Q. Test the following quadrilaterals to see if they are concyclic:

1.

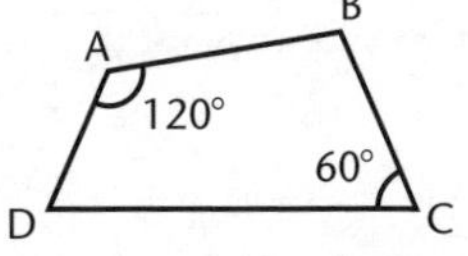

2.

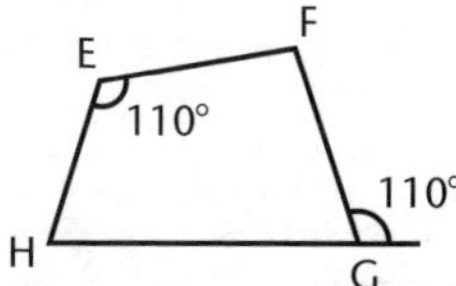

3.

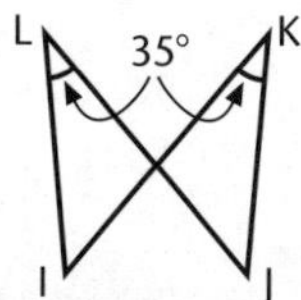

A. **1.** ∠A + ∠C = 180°, so ABCD is a cyclic quadrilateral since opposite angles add to 180°. (A, B, C, and D are concyclic points.)

2. EFGH is a cyclic quadrilateral since the exterior angle (at G) is equal to the interior opposite angle (at E). (E, F, G, and H are concyclic points.)

3. I, J, K, and L are concyclic points since IJ subtends equal angles at K and L.

Unit 11.4 Activity 4B: Cyclic quadrilaterals

1. Find the size of the angles labelled with letters. In each case O is the centre of the circle. Give a brief reason for each answer.

a.

b.

c.

d.

e.

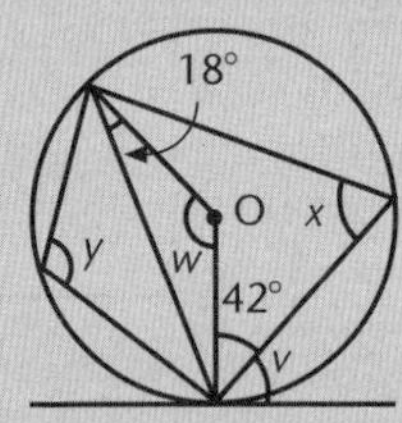

f.

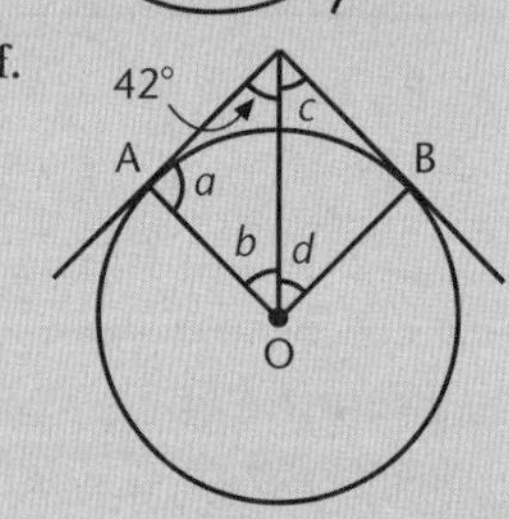

2. Find, with reasons, the sizes of the angles marked. In each case O is the centre of the circle.

a. Circle has tangent at D.

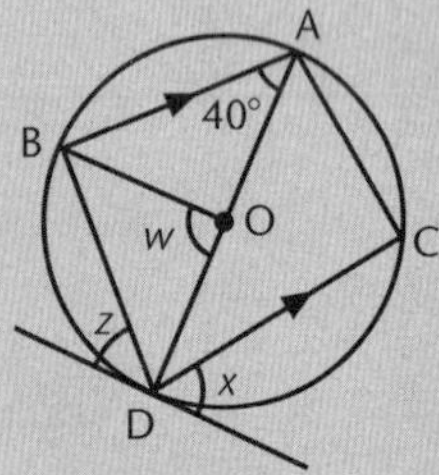

b. AB and AC are tangents to the circle.

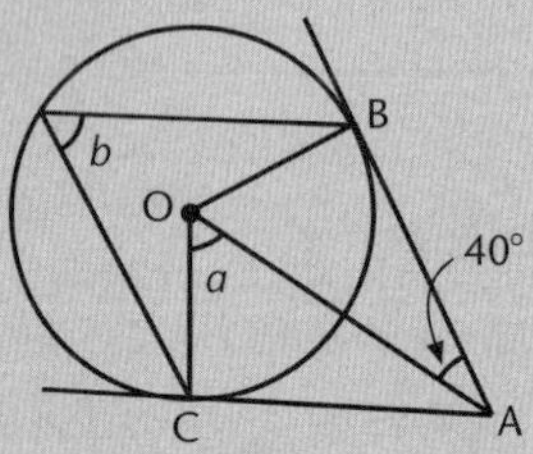

c. TA and TB are tangents to the circle.

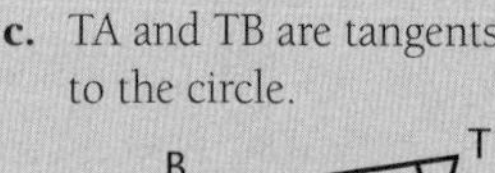

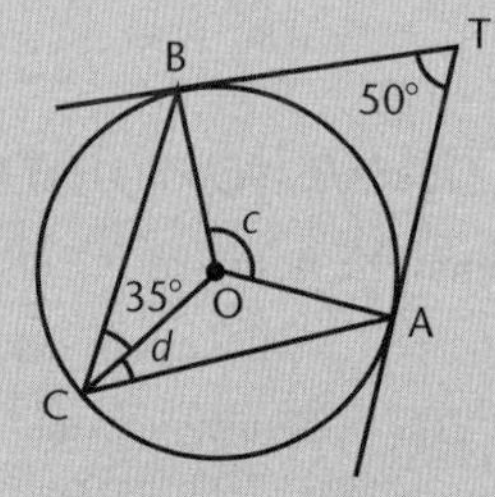

d.

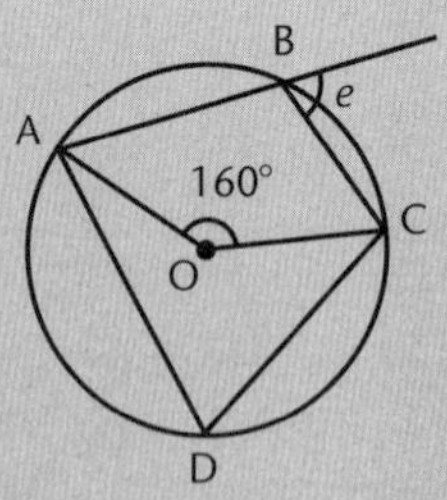

e.

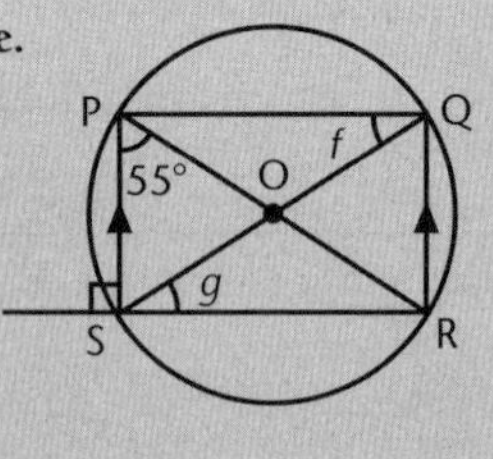

f. Circle has a tangent at T.

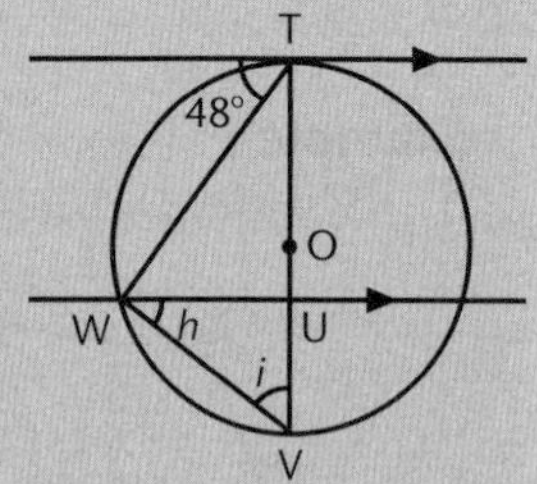

3. In the diagram shown, find the sizes of angles a and b. The circle has centre O and OB = AB. Give reasons.

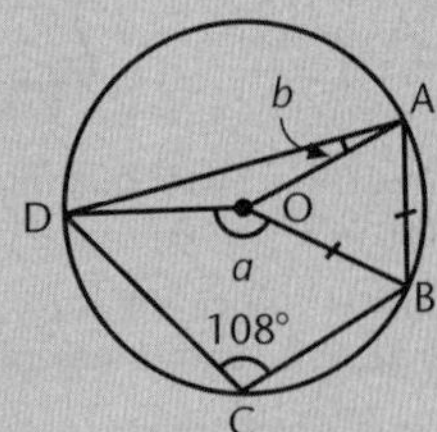

4. Name a set of four concyclic points on each diagram.

a.

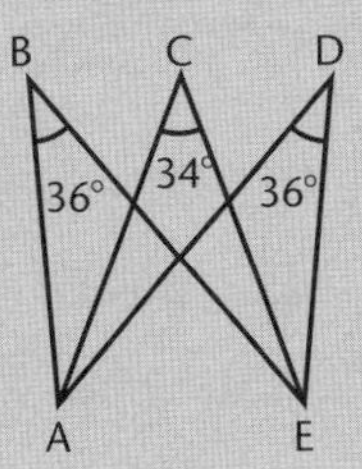

b.

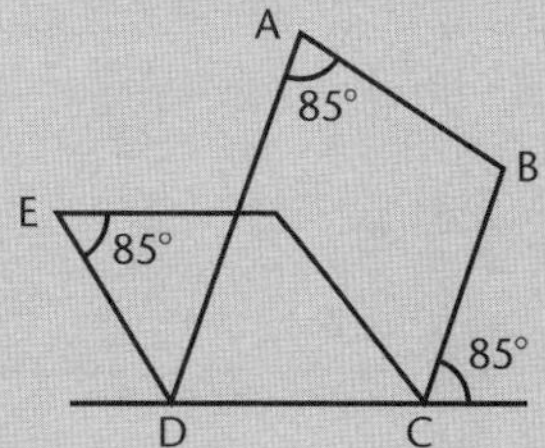

c.

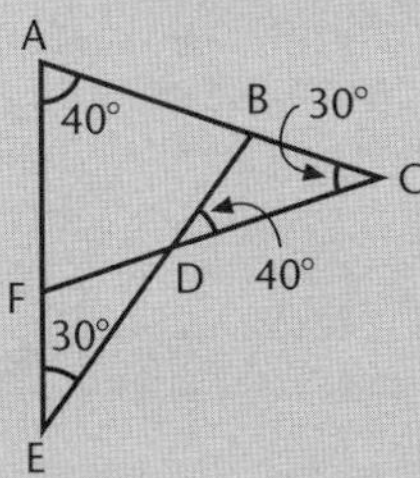

d.

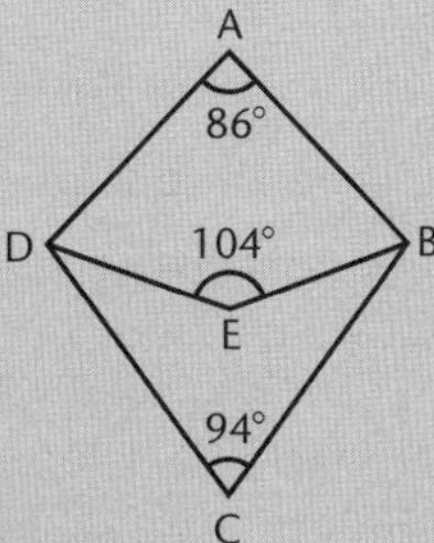

e.

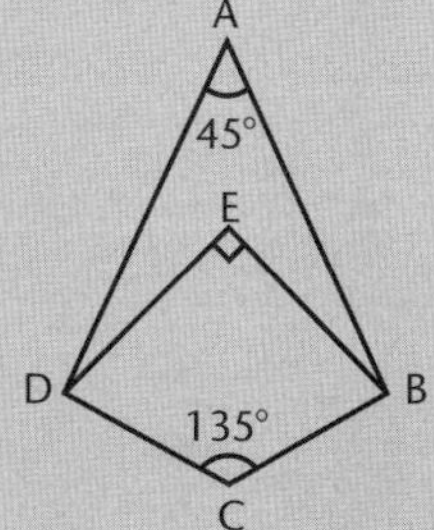

f.

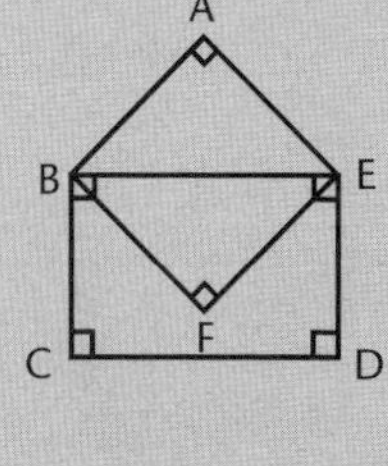

5. A semicircle, centre O, is drawn with ∠BAO = 60°.

Copy the figure and:

a. Mark all sides of equal length on the diagram.

b. Draw the tangent at B. Mark a point on the tangent and name it P. Join P to O.

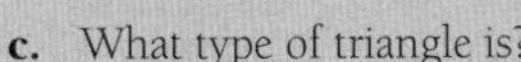

c. What type of triangle is?

i. ABC **ii.** OBC **iii.** OBP **iv.** AOB

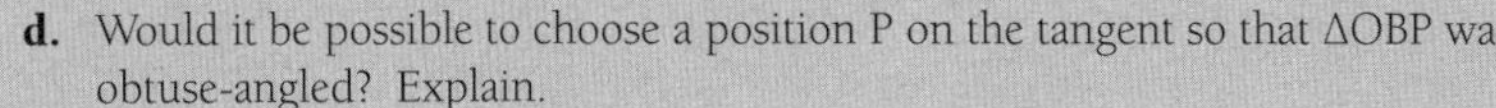

d. Would it be possible to choose a position P on the tangent so that ΔOBP was obtuse-angled? Explain.

6. ABC and PQR are straight lines. Prove that AP is parallel to RC.

Hint: Consider ∠s RCB, BQR and PAB.

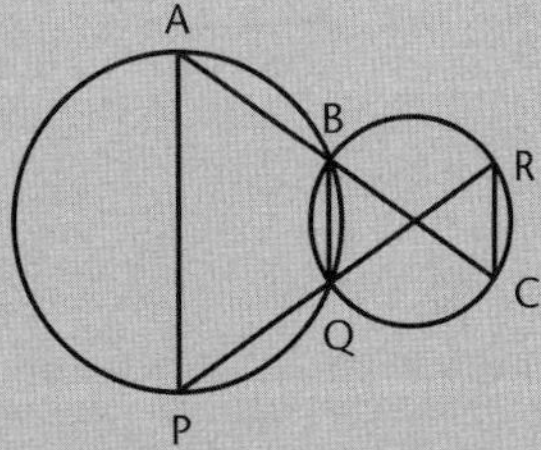

7. TA is a tangent to the circle centre O. Prove that $x = y$.

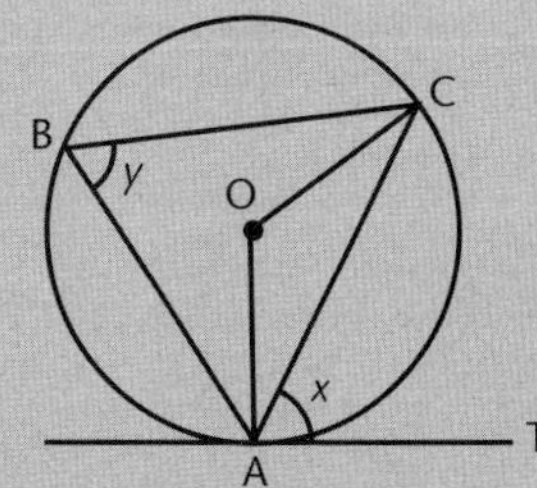

8. Prove that the opposite angles of a cyclic quadrilateral add to 180° (ie A + C = 180° or B + D = 180°).

Hint: Join BO and OD and consider ∠ at centre rule.

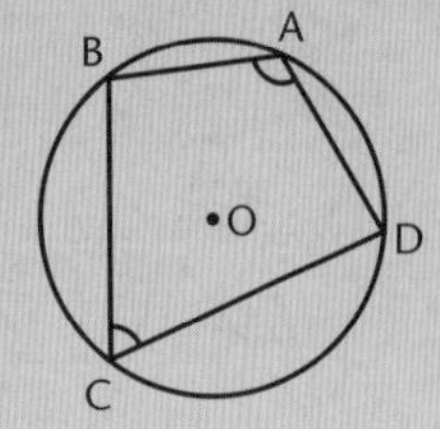

Unit 11.5 Trigonometry

Topic 1: Pythagoras' Theorem—revision

Pythagoras' Theorem is mentioned under Trigonometry on p. 30 of the General Mathematics Teacher Guide, so we have included it as the first Topic in Unit 11.5. This Topic covers:

- The use of Pythagoras' Theorem to find unknown sides in right-angled triangles.
- Finding unknown sides in practical situations involving right-angled triangles (the right-angled triangle will either be given or easily identified and extracted from a diagram).
- Finding unknown sides in word problems (student to extract right-angled triangles implicit in a context and use them to solve the problem in a well-reasoned and logical solution).
- Using appropriate rounding, units and mathematical statements.

Introduction

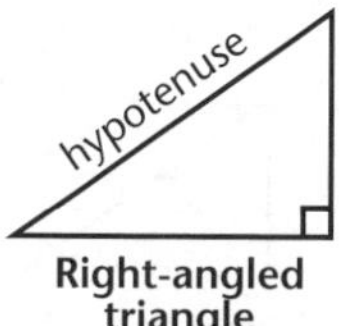

Right-angled triangle

In a **right-angled triangle**, one of the angles of the triangle is 90° (on a diagram a small square is used to mark the right angle). The side opposite the right angle is called the **hypotenuse** and is always the longest side of the triangle.

This Achievement Standard focuses on finding unknowns in right-angled triangles.

Pythagoras' Theorem

There is an important relationship between the lengths of the three sides of a right-angled triangle. This relationship is called **Pythagoras' Theorem**, and states:

> For any right-angled triangle, the square on the hypotenuse is equal to the sum of the squares on the other two sides.

Pythagoras' Theorem is more useful when written as a mathematical formula:

$$a^2 = b^2 + c^2$$

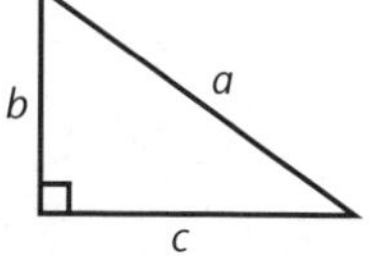

where a is the length of the hypotenuse, and b and c are the lengths of the other two sides.

The following example illustrates Pythagoras' Theorem.

Example A

A right-angled triangle with sides of length 3, 4, and 5 is shown alongside.

A square is drawn on each side of the triangle. Each of these squares is divided into square units labelled 1–9 and a–p. The 'square on each side' is the number of square units on that side. The number of square units on the hypotenuse is compared with the total number of square units on the other two sides.

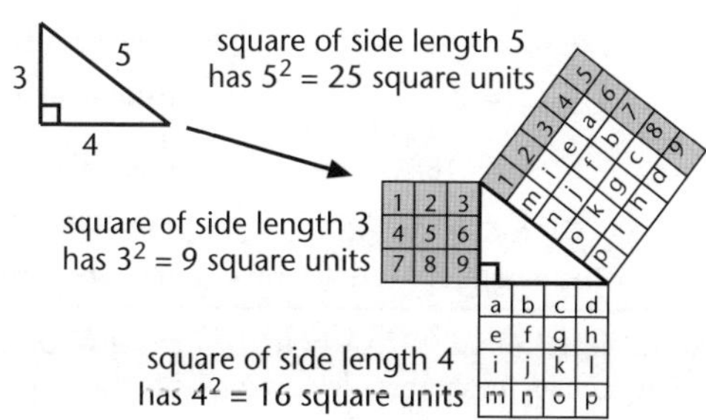

The number of square units on the hypotenuse (25) is equal to the sum of the numbers of square units on the other two sides (9 + 16).

Conversely, if the three sides of a triangle obey Pythagoras' Theorem, then the triangle is right-angled. For example, if the three sides of a triangle are 5 cm, 12 cm and 13 cm, then the relationship $a^2 = b^2 + c^2$ holds , where $a = 13$, $b = 5$ and $c = 12$ [since $13^2 = 5^2 + 12^2$] and the triangle is right-angled.

Finding unknown sides using Pythagoras' Theorem

Pythagoras' Theorem is used to find an unknown side of a right-angled triangle when two sides of the triangle are known. Be sure to set out each step of the working carefully and use appropriate rounding, including units with your answer where required.

Example B

Q. **1.** Find x in the triangle shown.

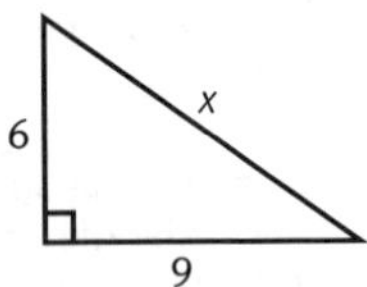

2. Find y in the triangle shown.

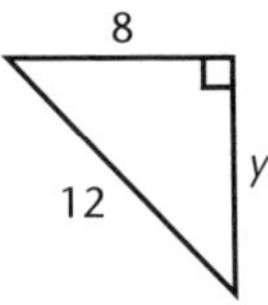

A. **1.** $x^2 = 6^2 + 9^2$ [Pythagoras' Theorem]

$= 36 + 81$

$= 117$

$x = \sqrt{117}$ [taking square roots]

$= 10.8$ (1 dp)

2. $12^2 = y^2 + 8^2$ [Pythagoras' Theorem]

$144 = y^2 + 64$

$80 = y^2$ [subtracting 64]

$y = \sqrt{80}$ [taking square roots]

$= 8.9$ (1 dp)

Note: Both parts of Example B begin with 'hypotenuse squared ='. This is the 'safest' way of writing an equation involving Pythagoras' Theorem.

In a word problem, no diagram may be provided. A right-angled triangle must be drawn from the information given.

Example C

Q. In a triangle ABC, angle A = 90°, the length of AB is 2.7 cm and the length of BC is 4.9 cm. Find the length of AC.

A. Draw a triangle, right-angled at A, hypotenuse BC, using the information given.

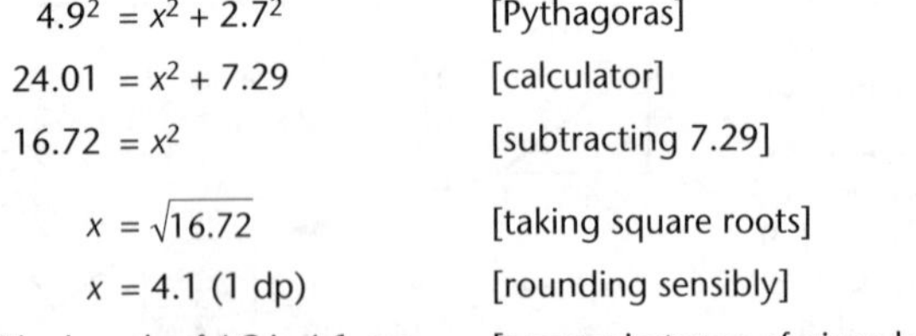

$4.9^2 = x^2 + 2.7^2$ [Pythagoras]

$24.01 = x^2 + 7.29$ [calculator]

$16.72 = x^2$ [subtracting 7.29]

$x = \sqrt{16.72}$ [taking square roots]

$x = 4.1$ (1 dp) [rounding sensibly]

The length of AC is 4.1 cm [answer in terms of given labels, not x]

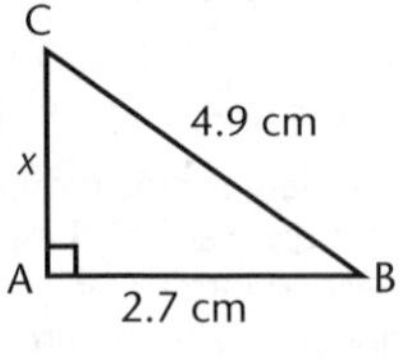

Note: Never round an answer to the nearest whole number when using Pythagoras' Theorem. Doing so may imply that all the lengths of the sides are whole numbers, and this is seldom so.

Unit 11.5 Activity 1A: Pythagoras' Theorem

1. Find the lengths of the marked sides, rounding sensibly when required.

a. 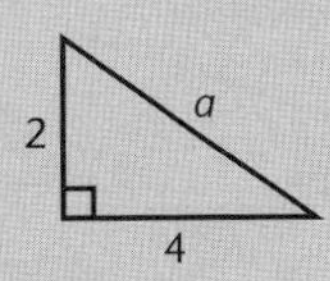

b. 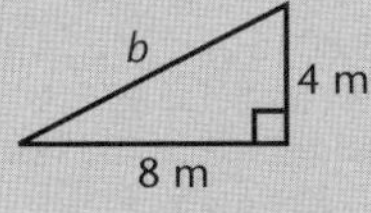

c. 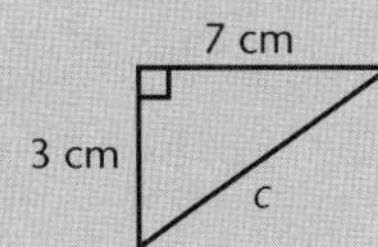

d. 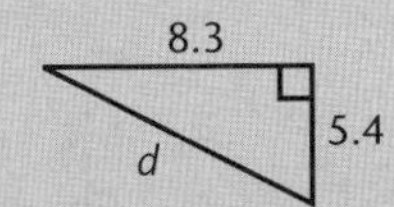

e. 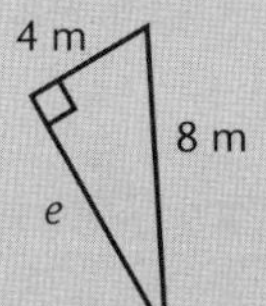

f. 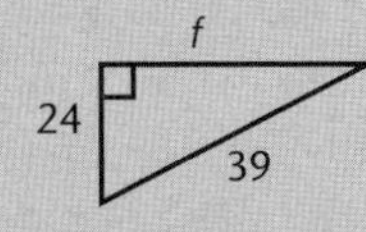

g. 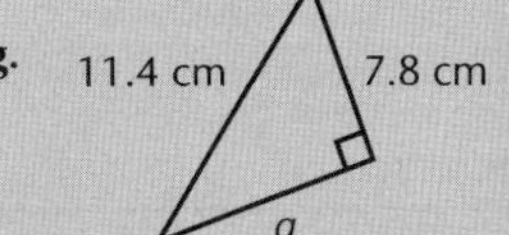

h. 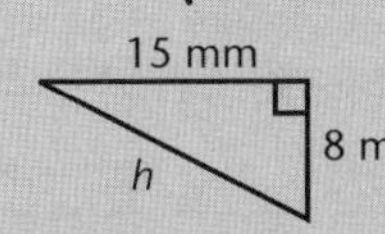

i. 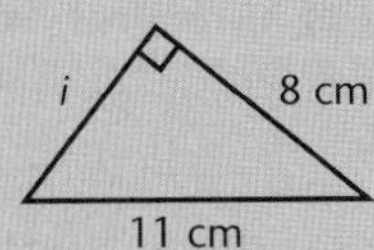

2. A right-angled triangle has a hypotenuse of unknown length. Its other two sides are of length 4.7 m and 6.3 m. Find the length of the hypotenuse.

3. A right-angled triangle has a hypotenuse of length 154 cm and another side of length 65 cm. Find the length of the third side.

4. The longest side of a right-angled triangle is 16.5 mm. If another side is 9.8 mm find the length of the third side.

5. If a triangle is right-angled then its sides obey Pythagoras' Theorem. Which of the following triangles, whose side lengths are given, are right-angled? (Remember the hypotenuse is always the longest side.)

a. 2.4 cm, 3.2 cm, 4 cm **b.** 6 cm, 9 cm, 12 cm **c.** 3.5 cm, 12 cm, 12.5 cm

Solving problems using Pythagoras' Theorem

Many practical problems involve finding side lengths of right-angled triangles. A clear diagram, including all information given, is the starting point for solving such problems.

Example D

Q. A 4 m ladder is leaning against a wall. The foot of the ladder is 1 m out from the wall, as the diagram shows. How far up the wall does the ladder reach?

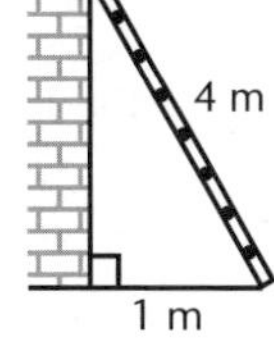

A. A right-angled triangle is drawn with h being the height the ladder reaches up the wall.

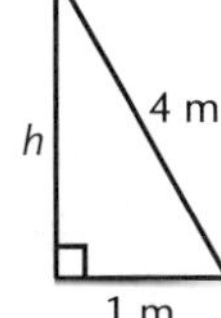

Using Pythagoras' Theorem:

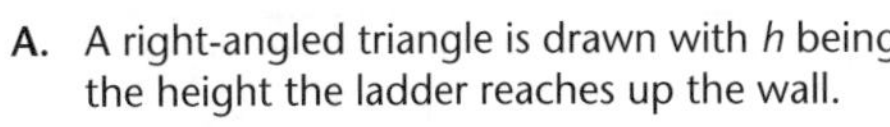

$4^2 = h^2 + 1^2$

$16 = h^2 + 1$

$h^2 = 15$ [subtracting 1 and rearranging]

$h = 3.87$ m (2 dp) [taking square roots]

Include extra lines where necessary to complete a right-angled triangle in your diagram.

Example E

Q. Two parallel sides of a field are 80 m and 140 m long, as shown in the diagram. If the parallel sides are 68 m apart, how long is the fourth side of the field?

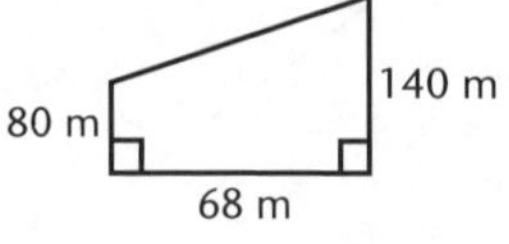

A. Label the corners of the field A, B, C, D as shown.

Draw a line DE parallel to CB so that AED is a right-angled triangle.

AE = 140 – 80 = 60 m [subtracting CD from AB]

DE = 68 m

By Pythagoras' Theorem $x^2 = 68^2 + 60^2$

$= 8\,224$

$x = 90.7$ m (1 dp) [taking square root]

Drawing the axis of symmetry of an isosceles triangle creates two right-angled triangles of the same size and shape. Pythagoras' Theorem can then be applied.

Example F

Q. A flagpole is held upright by two wires both of length 25 m. The poles are 16 m apart. How tall is the flagpole?

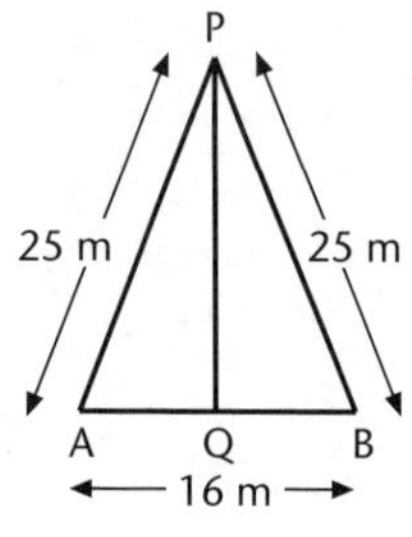

A. Draw a flagpole PQ and wires PA and PB as shown.

PQ is the axis of symmetry in the triangle APB.

Let the height of the pole be h.

In triangle APQ, AQ = 8 m [half of AB by symmetry]

$25^2 = h^2 + 8^2$ [Pythagoras' Theorem]

$625 = h^2 + 64$

$h^2 = 561$ [subtracting 8^2 and swapping sides]

$h = 23.7$ m [taking square root]

The flagpole is 23.7 m tall.

Unit 11.5 Activity 1B: Solving problems using Pythagoras' Theorem

1. In an orienteering event, Simon ran from point A to point B along two paths at right angles to each other. He ran 460 m along one path and 372 m along the other.

   Mary ran from point A to point B in a direct line.

   How much shorter was Mary's course, to the nearest metre?

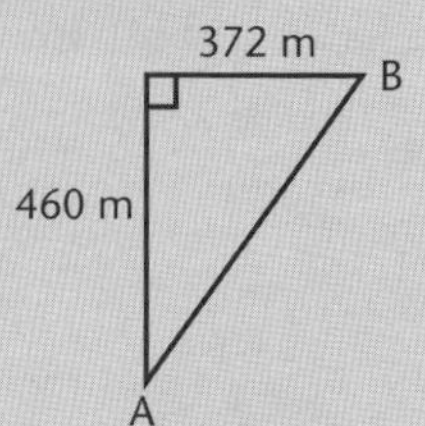

2. A flagpole is held upright by four wires. Each wire is fastened to the top of the pole, 24 m above ground, and to pegs in the ground 16 m from the base of the pole. How long is each wire?

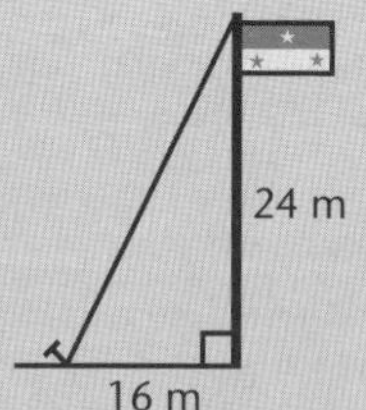

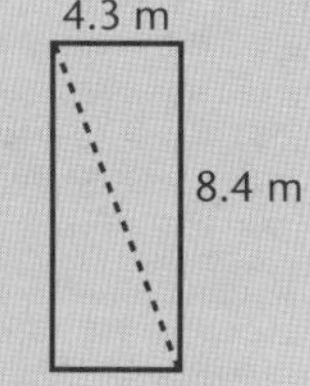

3. A builder laying the foundations of a house wants to be sure she has the walls of the house at right angles.

   She knows the length of the house should be 8.4 m and the width 4.3 m. To check this, she will measure the length of the diagonal.

   a. Explain how measuring the length of the diagonal will check that the walls of the house meet at right angles.

   b. How long should the diagonal be so that the walls are at right angles?

4. In the diagram, AB is the sloping roof of a shed. The front A is 3 m high, the back B is 6 m high. The width of the shed is 6.5 m.

   What length (AB) of roofing iron, to the nearest centimetre, is needed for the roof?

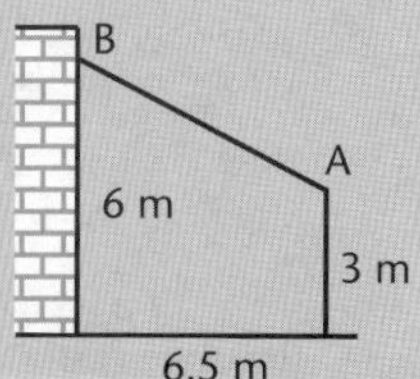

5. The dimensions of the roof and ceiling in a house are shown. Calculate the distance between the top of the roof and the ceiling, h.

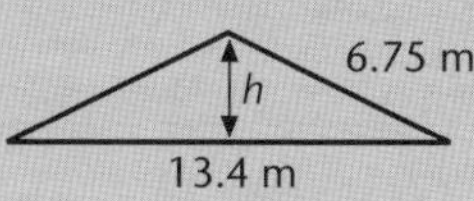

6. Susan has a suitcase with dimensions 90 cm by 60 cm by 30 cm. She has an umbrella 1.05 m long.

   Can she fit her umbrella in the bottom of her suitcase?

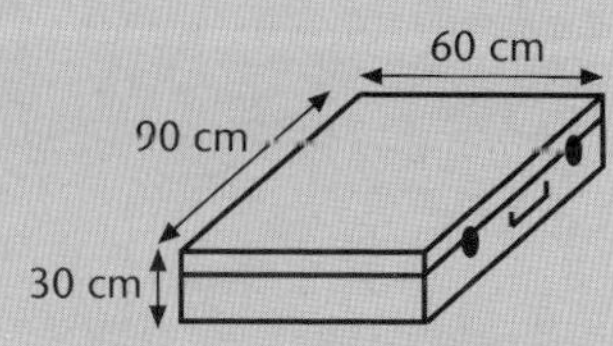

7. A rugby goal kicker places the ball, 18 m to the right of the centre of the goal posts and 34 m out from the goal-line.

How far is it from the centre of the goal posts to the ball?

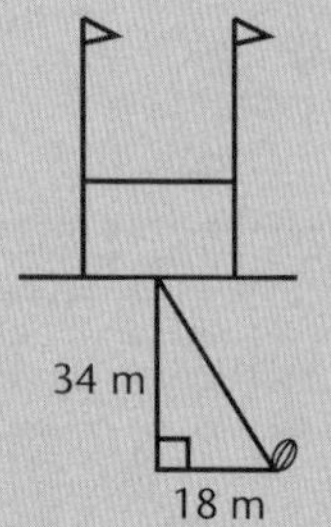

8. James and Rachel were laying a footpath for their new home, and they wanted to make sure that they had a right angle at point A.

James placed a peg in the ground at B, 4 m from A. He tied a 10 m length of twine to the peg. Rachel held a tape at point A and James held the free end of the twine.

At what point on the tape should the twine meet the tape to get a right angle at A?

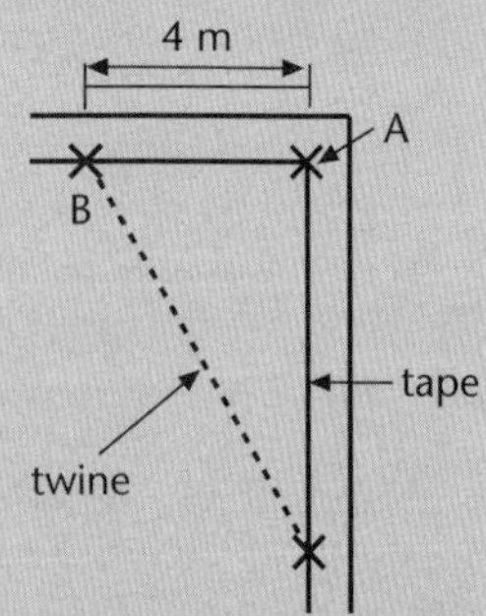

9. A triangle ABC has side lengths AB = 31 m, BC = 42 m and the distance between the vertex B and the side AC is 15 m. Find the length of AC.

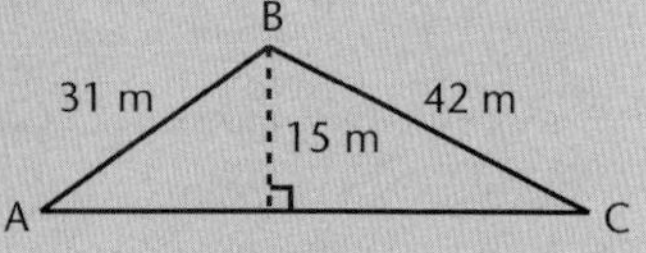

10. Ming is designing a 3 cm wide headband. She draws lines 4 cm long making a zig-zag pattern.

a. How far apart are the points P and Q marked on Ming's headband?

b. Ripeka wants to make a similar headband. It is to be the same width as Ming's band, but Ripeka wants the points P and Q to be exactly 4 cm apart. How long should each of the zig-zag lines be?

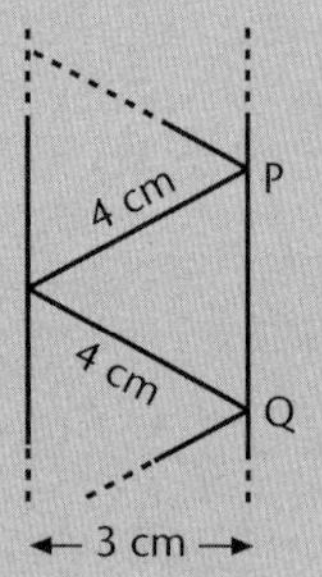

Unit 11.5 Trigonometry

Topic 2: Trigonometric ratios

The material in this Topic relates to the Grade 11 syllabus (see Syllabus p. 19): 'solve application problems on right-angled triangles'. It covers:

- The use of trigonometric ratios to find unknowns in right-angled triangles.
- Using trigonometric ratios to find and interpret unknowns in practical situations involving right-angled triangles (the right-angled triangle will either be given or easily identified and extracted from a diagram).
- Using trigonometric ratios to find unknowns in word problems (student to extract right-angled triangles implicit in a context and use them to solve the problem in a well-reasoned and logical solution).
- Using appropriate rounding, units and mathematical statements.

Introduction

Trigonometric ratios provide a relationship between angle size and side lengths in right-angled triangles. In this chapter trigonometric ratios are used with right-angled triangles:

- To find an unknown side length when an angle and side are known, or
- To find the size of an unknown angle when two side lengths of the triangle are known.

Ratios of side lengths in right-angled triangles

Any angle between 0° and 90° can be drawn as part of a right-angled triangle. No matter what *size* right-angled triangle is drawn for a particular angle size, the *ratio* of one side of the triangle compared with another side of the same triangle *will always be equal*, as shown below.

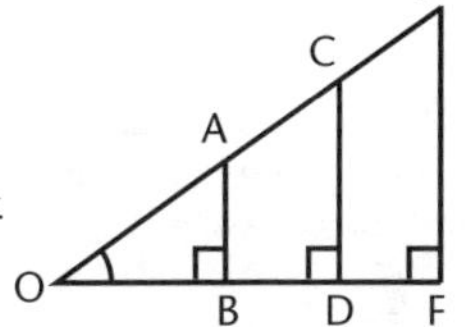

$$\frac{\text{length AB}}{\text{length OB}} = \frac{\text{length CD}}{\text{length OD}} = \frac{\text{length EF}}{\text{length OF}}$$

This relationship is the basis of the **trigonometric ratios** called the **tangent**.

Similar equalities give rise to other trigonometric ratios called the **sine** and **cosine**.

Labelling the sides of right-angled triangles

The three differe.nt sides of a right-angled triangle are named the **hypotenuse**, **opposite side** and **adjacent side**. The labelling of the adjacent and opposite side is relative to the angle chosen in the triangle.

h	**h**ypotenuse	The side *opposite the right angle*, as shown by the arrow pointing from the right angle to *h* in the diagram.
o	**o**pposite side	The side *opposite* the angle referred to (in this case θ), as shown by the arrow pointing from 0 to *o* in the diagram.
a	**a**djacent side	The side *adjacent* (next) to the angle referred to (in this case θ).

The trigonometric ratios

In a right-angled triangle each trigonometric ratio compares the length of one side with another. These rules are shown in full and in abbreviated form in the box below.

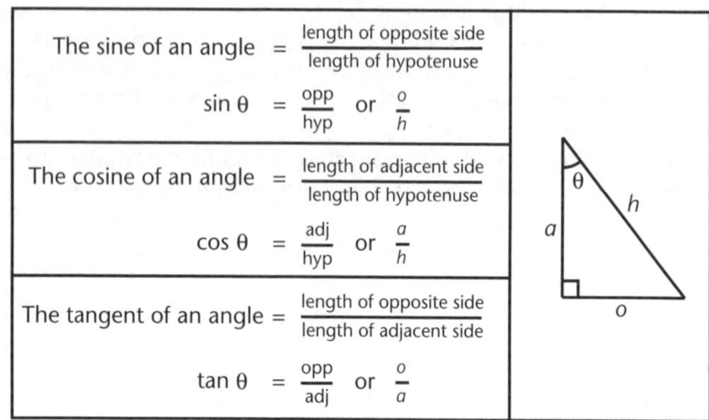

The sine of an angle $= \dfrac{\text{length of opposite side}}{\text{length of hypotenuse}}$

$\sin\theta = \dfrac{\text{opp}}{\text{hyp}}$ or $\dfrac{o}{h}$

The cosine of an angle $= \dfrac{\text{length of adjacent side}}{\text{length of hypotenuse}}$

$\cos\theta = \dfrac{\text{adj}}{\text{hyp}}$ or $\dfrac{a}{h}$

The tangent of an angle $= \dfrac{\text{length of opposite side}}{\text{length of adjacent side}}$

$\tan\theta = \dfrac{\text{opp}}{\text{adj}}$ or $\dfrac{o}{a}$

These three trigonometric ratios must be learnt. **SOH CAH TOA** is a useful mnemonic for doing this:

S	Sine of angle	C	Cosine of angle	T	Tangent of angle
O	Opposite over	A	Adjacent over	O	Opposite over
H	Hypotenuse	H	Hypotenuse	A	Adjacent

Example A

Using ΔUVW shown, the hypotenuse is UV (5.8 cm). Relative to angle V, the adjacent is VW (5 cm) and the opposite is UW (3 cm).

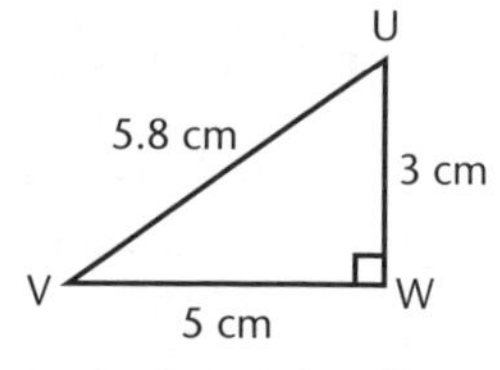

$\sin V = \dfrac{3}{5.8} = 0.52$ (2 dp) $\quad [\sin = \dfrac{\text{opp}}{\text{hyp}}]$

$\cos V = \dfrac{5}{5.8} = 0.86$ (2 dp) $\quad [\cos = \dfrac{\text{adj}}{\text{hyp}}]$

$\tan V = \dfrac{3}{5} = 0.6$ $\quad [\tan = \dfrac{\text{opp}}{\text{adj}}]$

Relative to angle U, the opposite is VW (5 cm) and the adjacent is UW (3 cm).

$\tan U = \dfrac{5}{3} = 1.67$ (2 dp) $\quad [\tan = \dfrac{\text{opp}}{\text{adj}}]$

Unit 11.5 Activity 2A: Trigonometric ratios

1. Use the diagram of ΔABC to state:

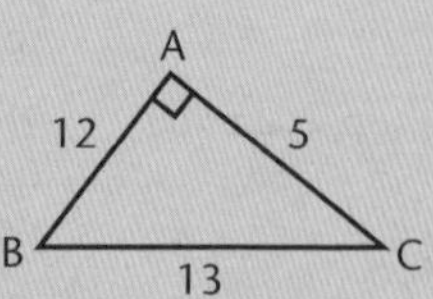

a. The length of the hypotenuse.

b. The length of the side adjacent to angle B.

c. The length of the side opposite angle C.

d. sin C **e.** tan B **f.** cos B **g.** sin (90° – C)

2. In ΔPQR, state the values of:

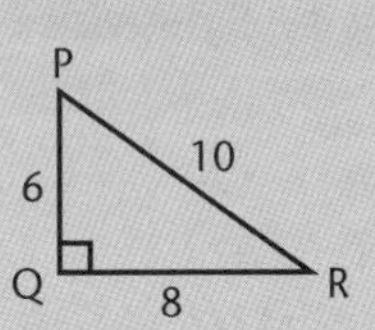

a. The length of the hypotenuse.

b. The length of the side opposite angle P.

c. The length of the side adjacent to angle P.

d. tan P **e.** sin R **f.** cos (90° – P) **g.** sin (90° – R)

Finding the value of a trigonometric ratio

The procedure for finding the value of a trigonometric ratio with a calculator is shown in the following example (in this book, access to a calculator with trigonometric function keys is assumed).

Example B

All answers are rounded to four decimal places.

sin 60° = 0.8660

tan 49.2° = 1.159

Make sure your calculator is in the correct MODE, ie DEG (degrees).

Finding angle sizes from the value of a trigonometric ratio

The inverse [inv] or second function key [Shift] or [2nd F] is used to find an angle size when a trigonometric ratio is known.

Example C

Q. Find θ if **1.** sin θ = 0.766 **2.** cos θ = 0.3420 **3.** tan θ = 0.7266

A. **1.** If sin θ = 0.7660, θ = 50.0° [Shift] [sin] [•] [7] [6] [6] [=]

2. If cos θ = 0.3420, θ = 70.0° [Shift] [cos] [•] [3] [4] [2] [=]

3. If tan θ = 0.7266, θ = 36.0° [Shift] [tan] [•] [7] [2] [6] [6] [=]

Note: It is common practice to round angle sizes to 1 dp, unless an alternative accuracy (rounding) is suggested.

The notation '$\mathbf{cos^{-1}}$' means 'the angle whose cosine is'. For example, '$\cos^{-1}$ (0.3420)' means 'the angle whose cosine is 0.3420'. From Example C it can be seen $\cos^{-1}$ (0.3420) = 70.0°, ie cos 70.0° = 0.3420.

Similarly, the notation '$\sin^{-1}$' means 'the angle whose sine is' and the notation '$\tan^{-1}$' means 'the angle whose tangent is'.

Unit 11.5 Activity 2B: Using your calculator

1. Use your calculator to find the value of each of the following to 4 decimal places:

   a. sin 20° **b.** cos 10° **c.** tan 70° **d.** cos 38° **e.** tan 48°

   f. cos 64° **g.** sin 56.3° **h.** tan 78.4° **i.** sin 12.4°

2. Find the value of θ in each of the following, giving answers to 1 decimal place.

   a. $\sin\theta = 0.6428$ **b.** $\cos\theta = 0.7660$ **c.** $\tan\theta = 1.7324$

   d. $\cos\theta = 0.6428$ **e.** $\tan\theta = 0.7931$ **f.** $\sin^{-1}(0.4532) = \theta$

   g. $\theta = \cos^{-1}(0.8751)$ **h.** $\theta = \tan^{-1}(1.234)$ **i.** $\theta = \sin^{-1}(0.1159)$

Finding unknown side lengths (unknown in numerator)

For any right-angled triangle, the length of an unknown side can be found if the length of another side *and* the size of another angle (other than the right angle) are known.

Example D

Q. Find the length of the unknown side (*a*, *b* or *c*) in each of the following triangles:

1.

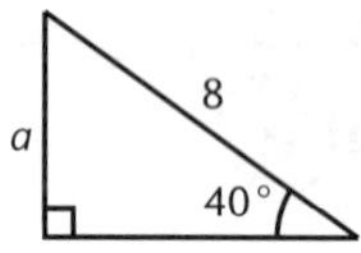

2.

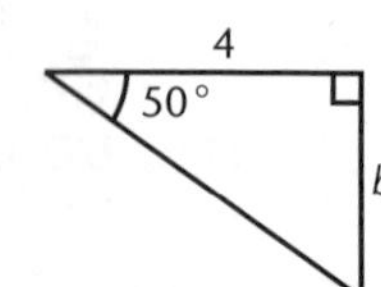

3.

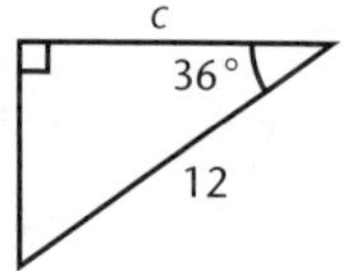

A. **1.** In the triangle, the hypotenuse is 8 and (relative to the given angle) the opposite side is *a*. The rule linking opposite and hypotenuse is sine [SOH].

$\sin 40° = \frac{a}{8}$ $[\sin = \frac{o}{h}]$

$8 \times \sin 40° = a$ [multiplying by 8]

$a = 5.14$ (2 dp) [using a calculator]

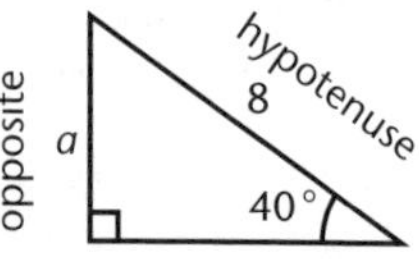

To calculate 8 sin 40° using a scientific calculator:

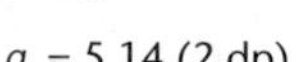

Note: Since the horizontal side is neither given, nor required, it is left unlabelled (this makes the identification of the correct trigonometric ratio easier).

2. $\tan 50° = \frac{b}{4}$ $[\tan = \frac{o}{a}]$

$b = 4 \times \tan 50°$ [multiplying by 4]

$= 4.77$ (2 dp) [using a calculator]

3. $\cos 36° = \frac{c}{12}$

$c = 12 \cos 36°$

$= 9.71$ (2 dp)

adjacent 4 50° b opposite

adjacent c 36° 12 hypotenuse

Finding unknown side lengths (unknown in denominator)

Sometimes, the unknown side appears in the denominator of the trigonometric ratio. An extra line of working will be required to rearrange the equation in these types of problem.

Example E

Q. Find d in the right-angled triangle shown.

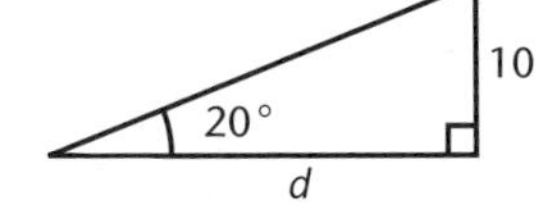

A. Relative to the angle 20°, d is 'adjacent' and 10 is 'opposite', so tangent is used (TOA).

$\tan 20° = \frac{10}{d}$ [$\tan = \frac{o}{a}$]

$10 = d \tan 20°$ [multiplying by d (to bring d into the numerator)]

$\frac{10}{\tan 20°} = d$ [dividing by tan 20° (the coefficient of d)]

$d = 27.47$ (2 dp) [using a calculator]

To calculate $\frac{10}{\tan 20°}$ using a calculator:

[1] [0] [÷] [tan] [2] [0] [=]

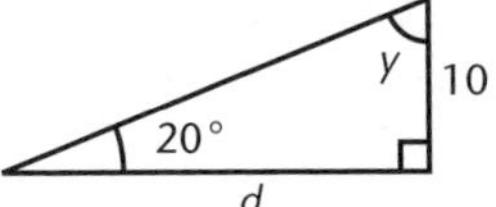

Note: d can be found using the other angle in the triangle, labelled y. Investigate this for yourself.

Example F

Q. Find x in the right-angled triangle shown.

A. $\sin 24° = \frac{8}{x}$ [using SOH]

$x \sin 24° = 8$ [multiplying by x]

$x = \frac{8}{\sin 24°}$ [dividing by sin 24°]

$x = 19.67$ (2 dp) [using a calculator]

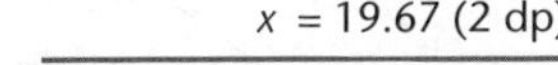

Unit 11.5 Activity 2C: Finding unknown sides

1. Find the lengths of the lettered sides for each of the triangles. When required, round your answers sensibly.

a.

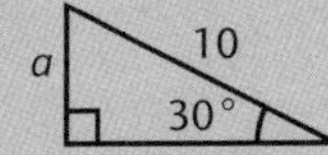

b.

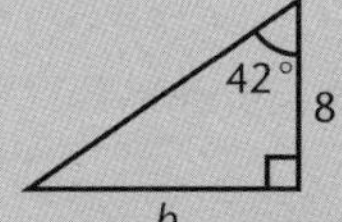

c.

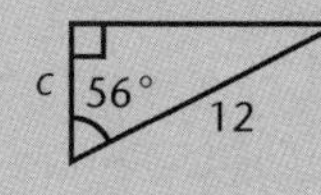

d.

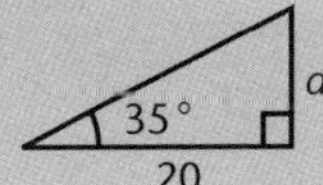

e.

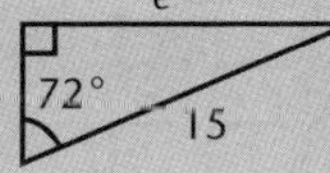

f. 15, 26°, f

g.

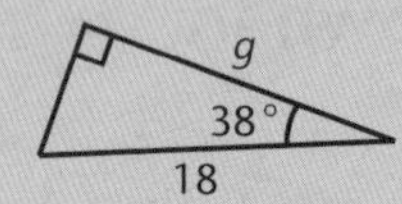

h.

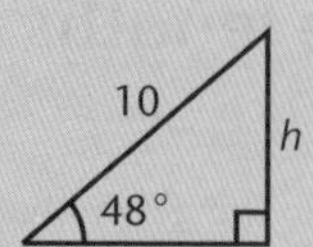

i.

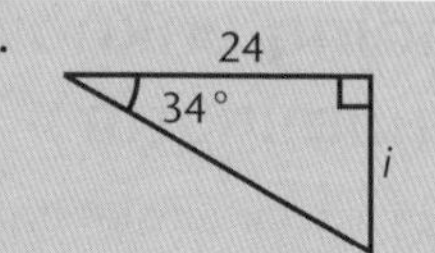

2. Find the values of the unknown sides, as marked.

a.

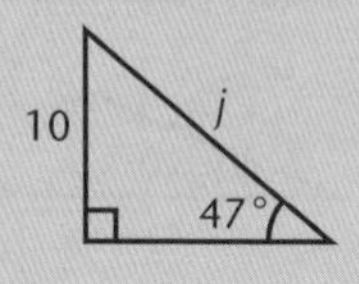

b.

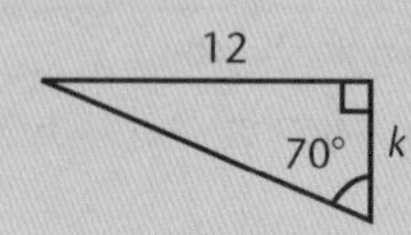

c.

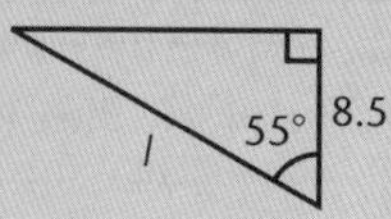

d.

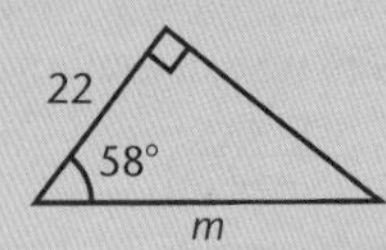

e.

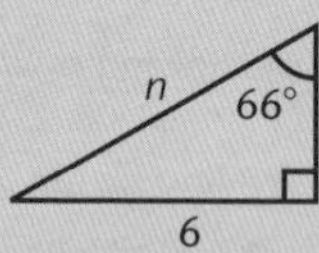

f.

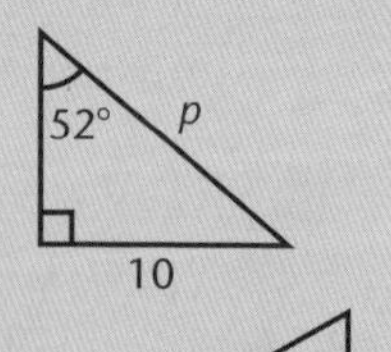

g.

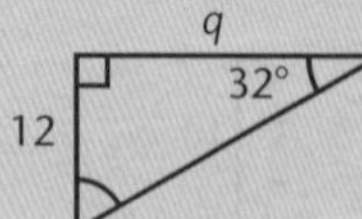

h.

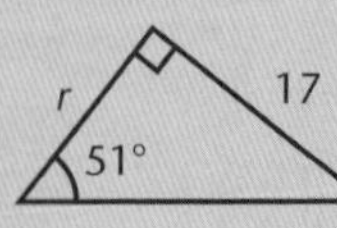

i.

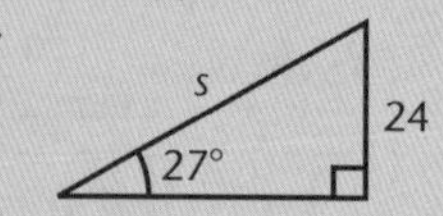

Finding unknown angles

If two sides of a right-angled triangle are known, the size of an unknown angle can be found.

Label the known sides relative to the angle required and choose the relevant trigonometric ratio for this angle.

Example G

Q. Find the size of the angles marked in each of the following:

1.

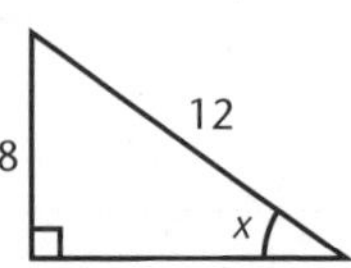

2.

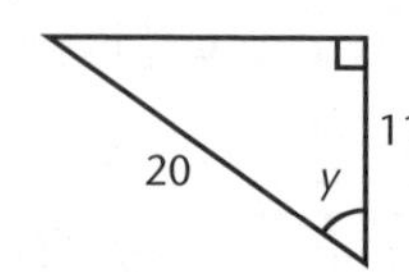

3.

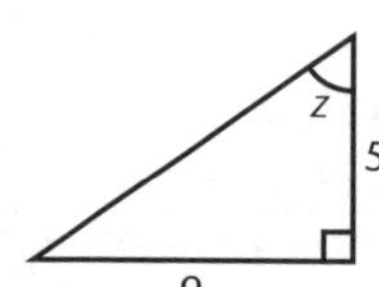

A. **1.** $\sin x = \frac{8}{12}$ $\quad$ $[\sin = \frac{o}{h}]$

$x = \sin^{-1}\left(\frac{8}{12}\right)$

$x = 41.8°$ (1 dp)

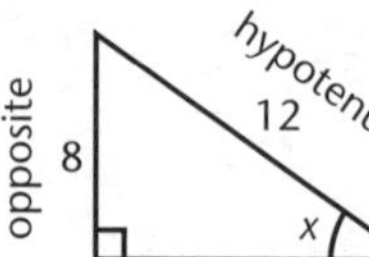

To find the value of x using a calculator:

 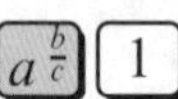 

or

Note: No information is given about the adjacent side, which means that the trigonometric ratios cos and tan cannot be used directly to solve this problem.

2. $\cos y = \frac{11}{20}$ $\quad [\cos = \frac{a}{h}]$

$y = \cos^{-1}\left(\frac{11}{20}\right)$

$y = 56.6°$ (1 dp)

3. $\tan z = \frac{9}{5}$ $\quad [\tan = \frac{o}{a}]$

$z = \tan^{-1}\left(\frac{9}{5}\right)$

$z = 60.9°$ [rounding to one decimal place]

Note: Once one unknown angle in a right-angled triangle is found, the third angle can be found using the angle sum of a triangle rule. For example, in part **3** of Example G, the third angle in the triangle is $90° - z = 90° - 60.9° = 29.1°$.

Unit 11.5 Activity 2D: Finding unknown angles

1. For each of the triangles find the size (to one decimal place) of the angle labelled.

a.

b.

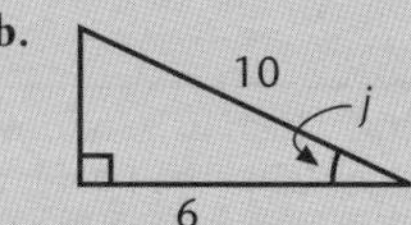

c.

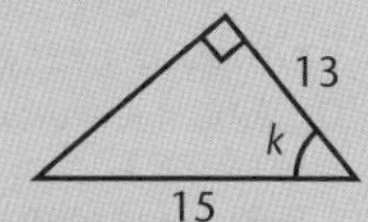

d.

e.

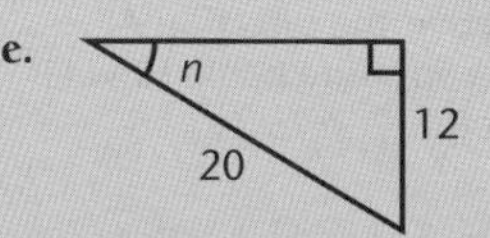

f.

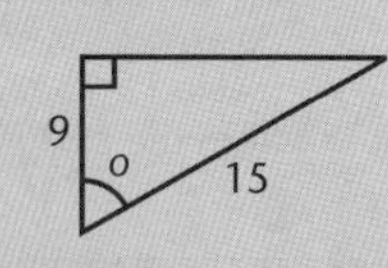

g.

h.

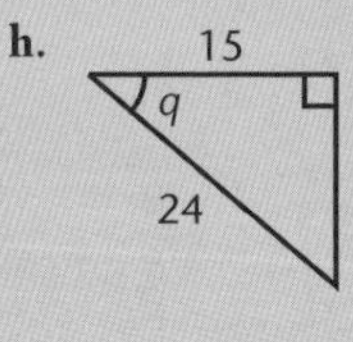

i.

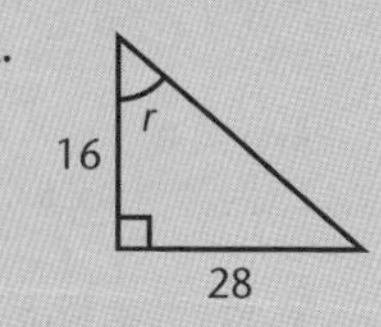

2. Find the size (to one decimal place) of the angles labelled.

a.

b.

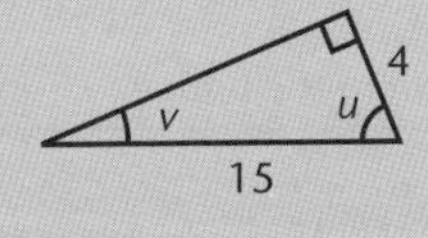

c.

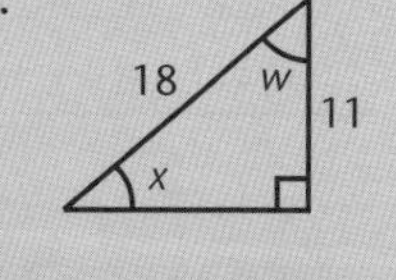

d.

e.

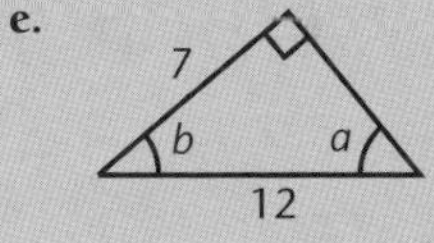

f.

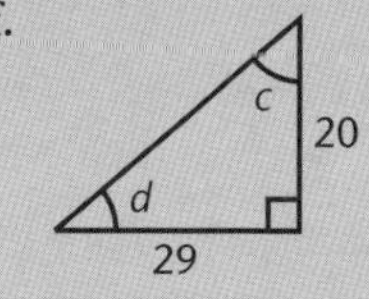

Solving problems using trigonometry

To solve practical problems involving right-angled triangles, draw a clear diagram from the given information.

Example H

Q. A ladder 3.5 m long is leaning against a wall. If the foot of the ladder is 1 m out from the base of the wall, what angle does the ladder make with the wall?

A. $\sin x = \dfrac{1}{3.5}$ $\quad$ [$\sin = \dfrac{o}{h}$]

$x = 16.6°$ $\quad$ [taking $\sin^{-1}\left(\dfrac{1}{3.5}\right)$]

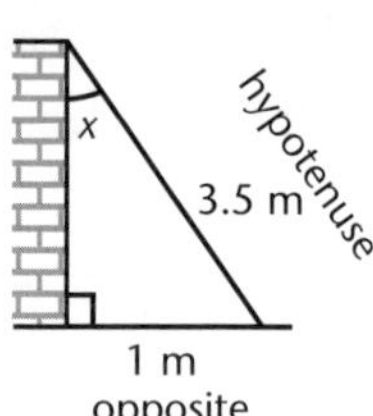

Note: Once the equation has been established to find the missing angle, the entire calculation may be done on the calculator:

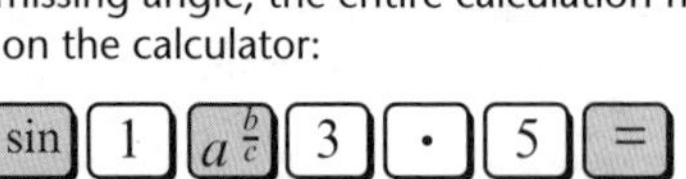

If your diagram of a practical problem does not contain a right-angled triangle, you may need to add at least one extra line perpendicular to an existing line of your diagram.

Example I

Q. A boat leaves from point A which is 10 m downstream from a 12.5 m long bridge (CD). The boat is aimed directly at the opposite side of the river, but is pushed downstream, so that it travels in the direction AB, making a 73° angle with the opposite bank. How far from the bridge is the point B?

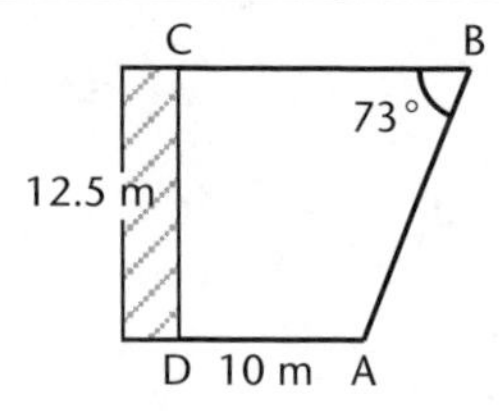

A. In the diagram, construct AP, so that AP is perpendicular to BC and ΔAPB is right-angled, AP = 12.5 m $\quad$ [same as CD]

Require to find x (the length PB).

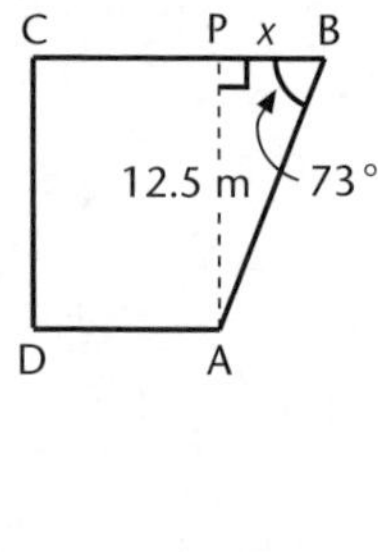

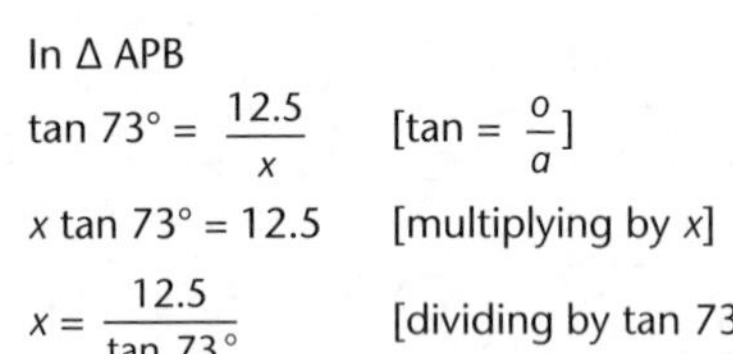

In Δ APB

$\tan 73° = \dfrac{12.5}{x}$ $\quad$ [$\tan = \dfrac{o}{a}$]

$x \tan 73° = 12.5$ $\quad$ [multiplying by x]

$x = \dfrac{12.5}{\tan 73°}$ $\quad$ [dividing by tan 73°]

$= 3.8$ m

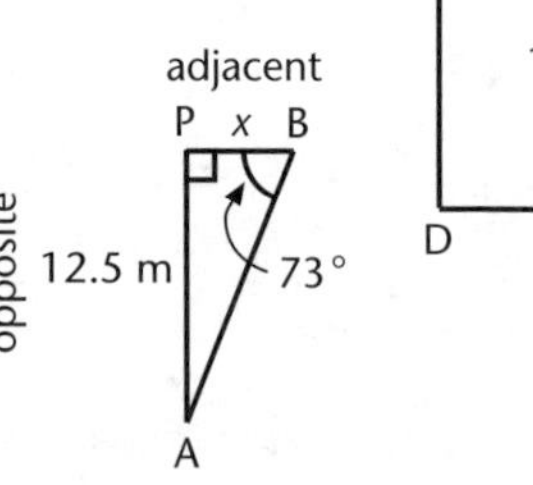

Thus B is 10 + 3.8 = 13.8 m from the bridge $\quad$ [since CP = DA = 10 m]

Angles of elevation and depression

Two expressions in common usage when describing the position of an object are **angle of elevation** and **angle of depression**.

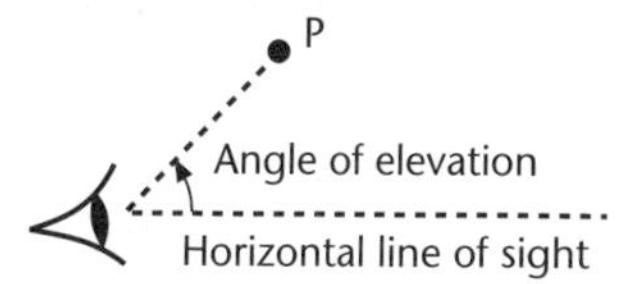 If you look up to a position P, ie P is above the horizontal line of sight, then the angle of elevation is measured up from the horizontal.	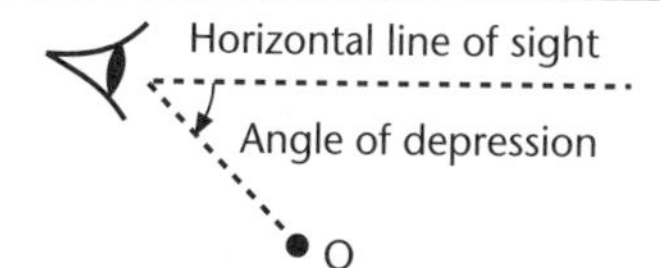 If you look *down* to a position Q, ie Q is below the horizontal line of sight, then the angle of depression is measured down from the horizontal.

Example J

Q. A kite is flying at the end of a 60 m length of string. How high above the ground is the kite if the angle of elevation to the kite is 55°?

A. The information can be shown on a diagram. K is the position of the kite and S is the other end of the string. The height of the kite above the ground is *d*.

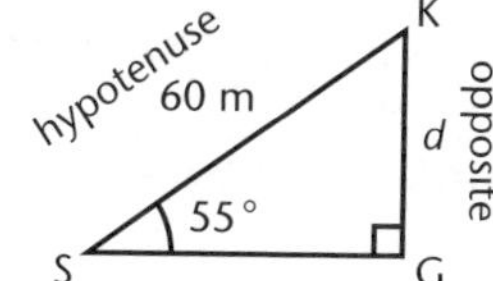

$\sin 55° = \frac{d}{60}$ $\quad$ $[\sin = \frac{o}{h}]$

$d = 60 \sin 55°$ $\quad$ [multiplying by 60 and swapping sides]

$= 49.1$ m (1 dp)

Unit 11.5 Activity 2E: Problem solving using trigonometry

1. A ladder 2.4 m long is leaning against a wall.

If the ladder reaches 2.0 m up the wall, what angle does the ladder make with the ground?

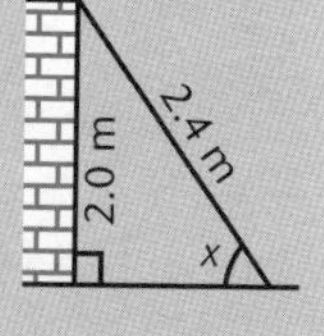

2. A farmer has to build a ramp to load cattle onto a truck.

The ramp is to make an angle of 30° with the ground. The horizontal distance between the beginning and end of the ramp is 2.3 m.

Calculate the length *h* to find the height of the end of the ramp.

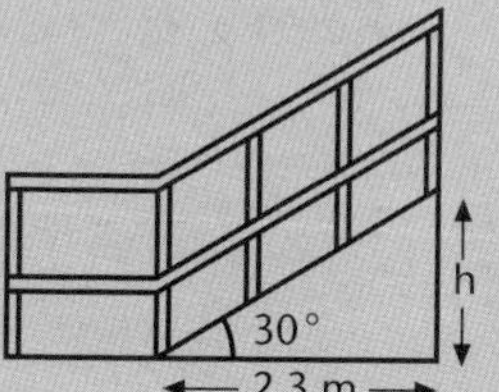

3. In a calm sea, a sailor measures the angle of elevation of a ship's telescope to the top of a lighthouse on a cliff to be 21°. The telescope is 28.4 metres above sea level, and radar places the ship 1 250 metres from the foot of the cliff.

Calculate the height of the top of the lighthouse above sea level, to the nearest metre.

4. A sailor, 1 km from a vertical cliff, notes that the angle of elevation of the top of the cliff is 10°.

a. How high is the cliff? Give answer to the nearest metre.

b. An observer on the top of the cliff later notices that a boat is passing a marker 500 m from the base of the cliff. Calculate the angle of depression of the observer's line of vision.

5. A tent has a cross-section which is an isosceles triangle. The base of the triangle is of length 3.5 m and the base angles are of size 51°.

What is the height of the tent?

6. A kite is flown on a cord 125 m long. The cord, held taut with its end at a height of 1.5 m above the ground, makes an angle of 70° with the horizontal.

How high above the ground is the kite?

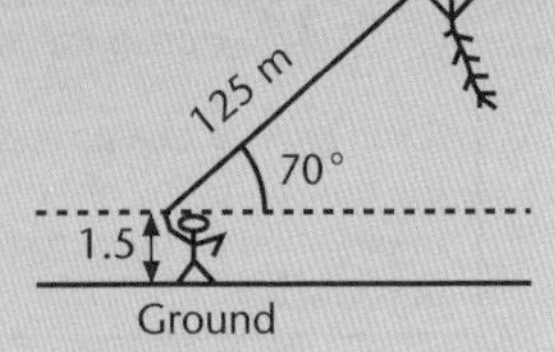

7. Anna was given the task of calculating the height of the school hall. She stood 6 metres from the wall and, using a clinometer, measured the angle of elevation to the topmost point of the hall.

The angle of elevation was measured to be 56° and Anna's eye level was 1.5 metres above the ground.

a. Calculate the height, h, of the school hall. Give your answer to 1 dp.

b. Anna's answer to the height of the school hall was wrong by several metres. Give a possible source for this error.

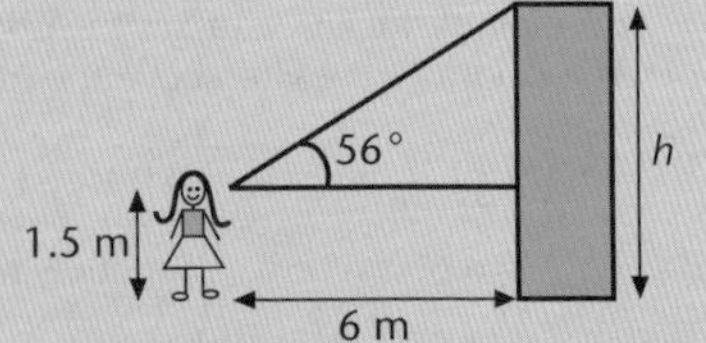

8. From a position 3.5 m above ground level in a building, an an observer measures the angle of elevation of the top of a flagpole to be 48°, and the angle of depression of the foot of the flagpole to be 35°.

a. How far away from the flagpole is the building?

b. How tall is the flagpole?

48°
35°
3.5 m

9. A boat sets off from a straight shore at an angle of 67° to the shore and travels in a straight line for 14.5 km. The boat then turns and travels in a straight line for home as shown in the diagram.
If the homeward journey is 17.6 km, find **θ**, the acute angle the line of the homeward journey makes with the shore.

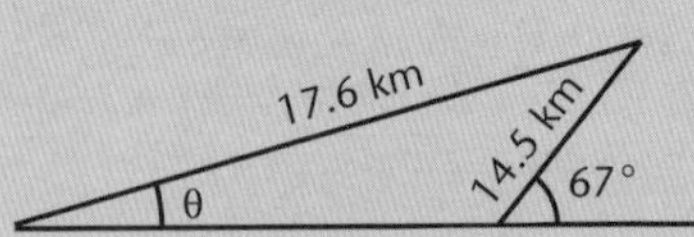

10. A beam is supported by two ropes both attached to a horizontal rod.

One rope is 3 m long and makes an angle of 120° with the beam. The other rope is 4 m long.

What angle does the 4 m rope make with the beam?

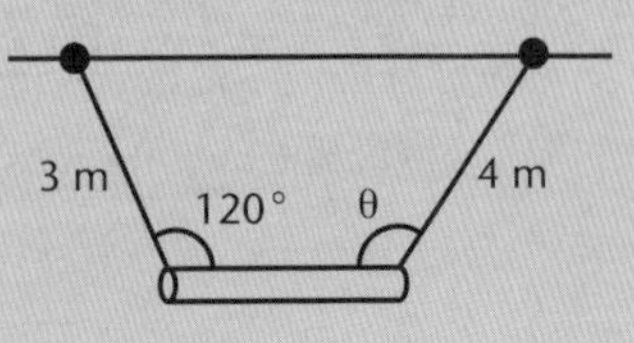

Unit 11.5 Trigonometry

Topic 3: Three-dimensional trigonometry

The material in this Topic continues the coverage of solving right-angled triangle problems. It covers:

- The use of trigonometric ratios to find and interpret unknowns in straightforward 3-D situations.
- Using trigonometric ratios to find unknowns in word problems (student to extract right-angled triangles implicit in a context and use them to solve the problem in a well-reasoned and logical solution).
- Using appropriate rounding, units and mathematical statements.

(Note: 'Angle between a line and a plane' and 'angle between two plane', mentioned under Unit 11.4 in the syllabus, is covered in this Topic.)

Introduction

In order to find the sizes of angles and lengths of lines in three-dimensional figures, right-angled triangles must be identified within the figure. Trigonometric ratios and/or **Pythagoras' Theorem** may then be applied to find the unknowns.

Finding unknown lengths in 3-D figures

If a length is to be found in a 3-D figure, look for a right-angled triangle in which the unknown length is a side. If the other two side lengths are known, Pythagoras' Theorem may then be applied to find the unknown side. (It is best to redraw these right-angled triangles.)

Example A

Q. Rebecca's bag is 75 cm by 40 cm by 16 cm. Rebecca wants to buy an umbrella which will fit in her bag. What is the longest umbrella she could buy?

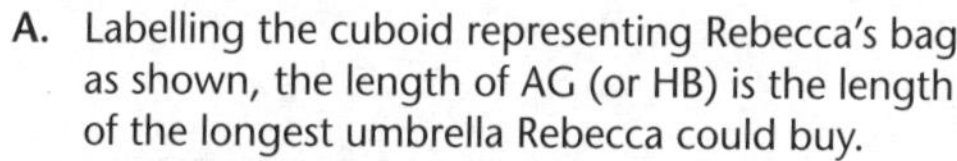

A. Labelling the cuboid representing Rebecca's bag as shown, the length of AG (or HB) is the length of the longest umbrella Rebecca could buy.

AG is part of the right-angled triangle AEG drawn alongside.

The length EG is the hypotenuse of the right-angled triangle EHG (as shown). Therefore,

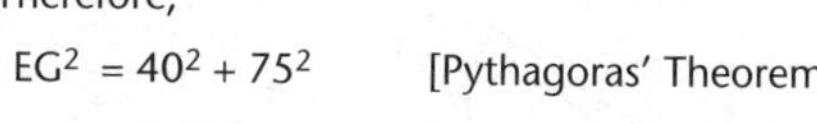

$EG^2 = 40^2 + 75^2$ [Pythagoras' Theorem]

$= 7\,225$

$\therefore$ EG = 85 cm [taking square root]

In triangle AEG

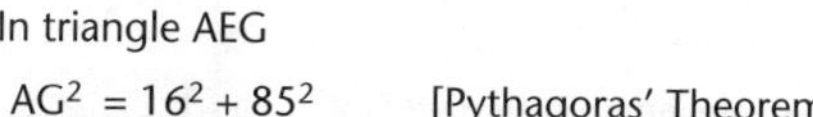

$AG^2 = 16^2 + 85^2$ [Pythagoras' Theorem]

$= 7\,481$

$\therefore$ AG = 86.5 cm (3 sf)

The longest umbrella Rebecca could buy is 86.5 cm.

Angle between a line and a plane

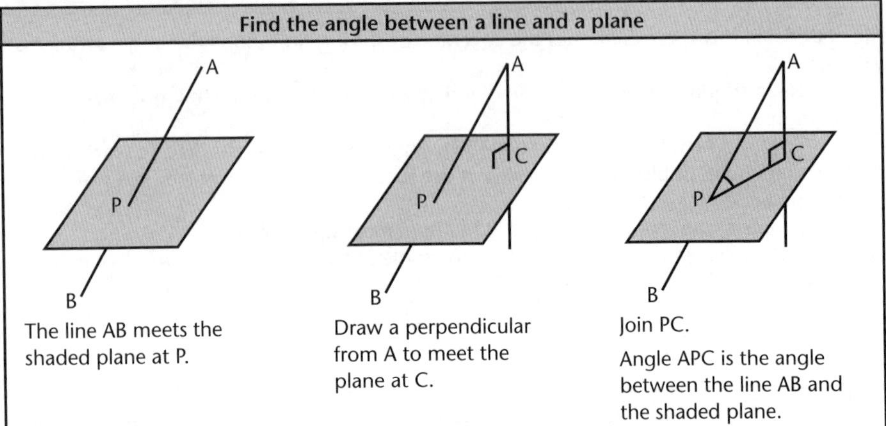

Example B

Q. Name the angle between the line EC and the plane EFGH in the cuboid shown.

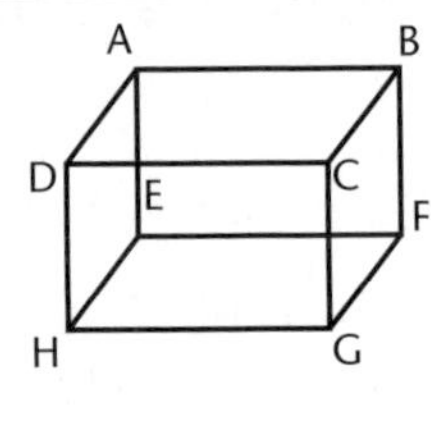

A. Join EC and shade the plane EFGH.

The line EC meets the plane at E.

The perpendicular (CG) from C meets the plane at G.

The required angle is ∠CEG as shown on the diagram.

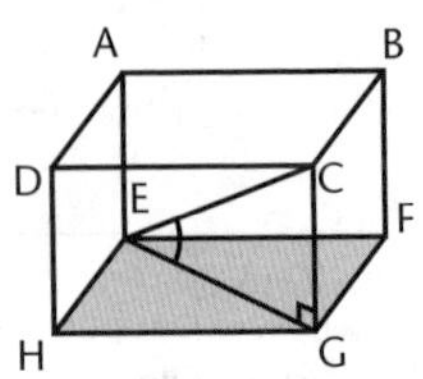

When measurements are given on the figure Pythagoras' Theorem and/or trigonometry can be used to calculate the sizes of angles.

Example C

Q. The cuboid shown has length 4, width 2 and height 3.

Calculate the size of the angle between the line HB and the plane ABFE.

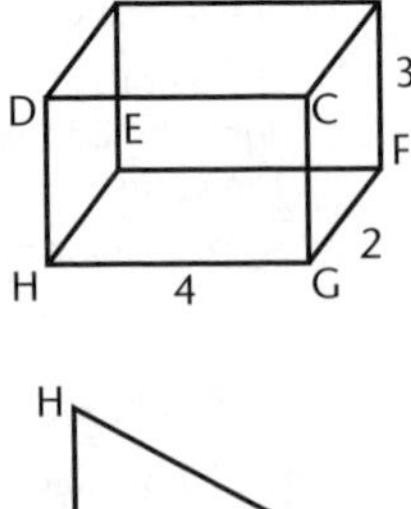

A. HB meets the plane (shown shaded) at B.

The perpendicular from H meets the plane at E.

The required angle is ∠HBE.

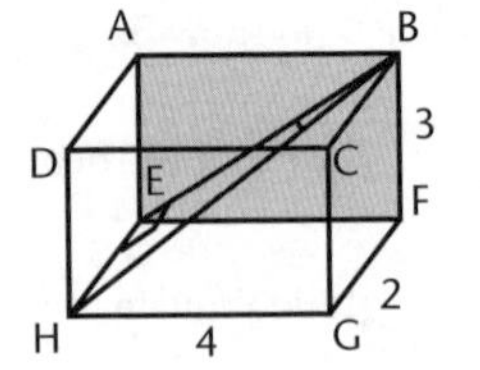

Triangle HBE is drawn alongside.

The length EB is the hypotenuse of the right-angled triangle EFB.

Thus $EB^2 = EF^2 + FB^2$ [Pythagoras' Theorem]

$= 4^2 + 3^2$

$= 25$

$\therefore EB = 5$ [taking square root]

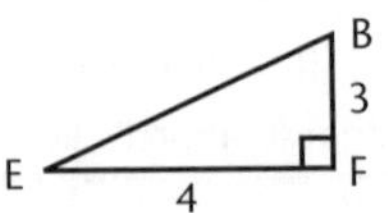

In triangle HEB, let $\angle HBE = \theta$

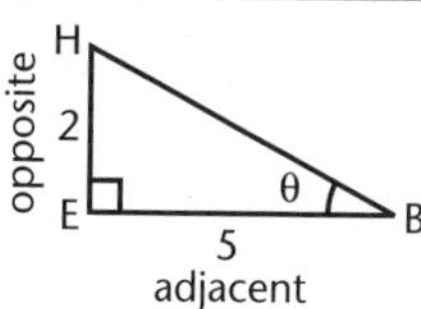

$$\tan\theta = \frac{2}{5}$$
$$\theta = \tan^{-1}\left(\frac{2}{5}\right)$$
$$= 21.8^\circ$$

The angle between the line HB and the plane ABFE is 21.8°.

When working with pyramids, use the symmetry of the figure to find unknown lengths. The apex (top point of the pyramid) sits over the 'centre' of the base. In a square base the 'centre' is where the diagonals intersect.

Example D

Q. In the pyramid shown, ABCD is a square with sides of 4 cm. The length of EB is 8 cm.

Calculate the size of the angle between line EB and the plane ABCD.

A. The line EB meets the plane ABCD (shown shaded) at B.

The perpendicular from E meets the plane at F, the 'centre' of the square (where its diagonals meet).

The required angle, labelled θ, is $\angle EBF$.

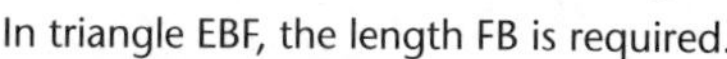

In triangle EBF, the length FB is required.

FB is half the length of the hypotenuse DB of the right-angled triangle DCB (shown alongside).

Thus $DB^2 = 4^2 + 4^2$ [Pythagoras' Theorem]
$= 32$

$\therefore DB = 5.6569$ [taking square root]

$FB = 2.8284$ [since FB is half DB]

Triangle EFB is drawn alongside.

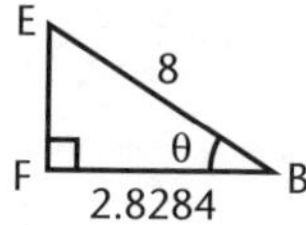

$$\cos\theta = \frac{2.8284}{8}$$
$$= 0.3536$$
$$\theta = \cos^{-1}(0.3536)$$
$$= 69.3^\circ \text{ (1 dp)}$$

The angle between EB and the plane ABCD is 69.3°.

Note: Work to several decimal places and round at the end to avoid rounding error.

Unit 11.5 Activity 3A: 3-D trigonometry

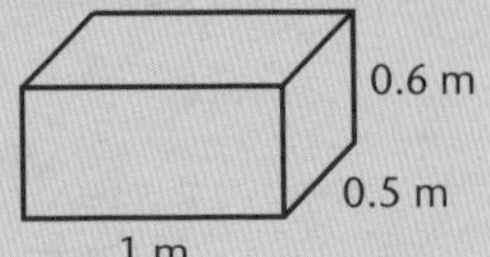

1. Giwi is packing tent poles in a box. He notices that the longest pole just fits in the box. If the box is 1 m by 0.5 m by 0.6 m, how long is the longest pole?

2. The diagram shows a cuboid. Name the following angles:

   a. The angle between line AH and plane EFGH.

   b. The angle between line CE and plane EFGH.

   c. The angle between line BH and plane ABCD.

   d. The angle between line DF and plane BCGF.

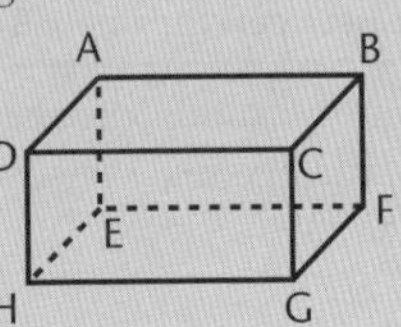

3. The diagram shows a wedge in which ABCD, AEFD and BCFE are rectangles.

   a. Calculate the length of CD.

   b. Name and calculate the size of the angle between the planes ABCD and AEFD.

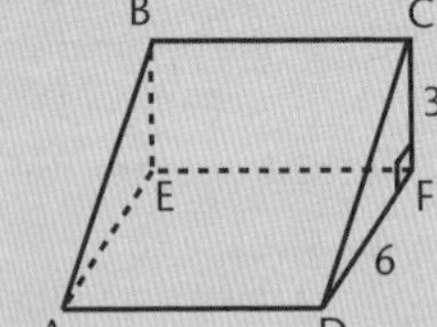

4. ABCDEFGH is a cuboid.

   a. Calculate the length of EG.

   b. Calculate the length of EC.

   c. Name and calculate the angle between CE and the plane EFGH.

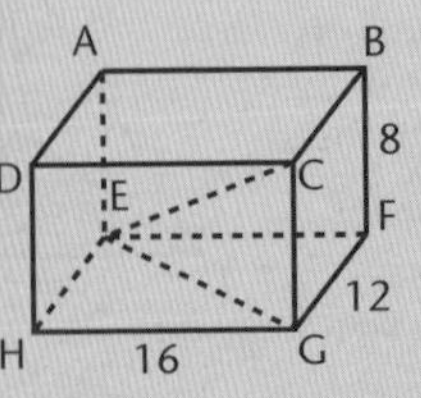

5. A pole QP is held upright by four equal ropes PA, PB, PC, PD. The ropes are fastened to the pole at 18 m above the ground and to pegs at A, B, C and D, each 12 m from the base of the pole Q.

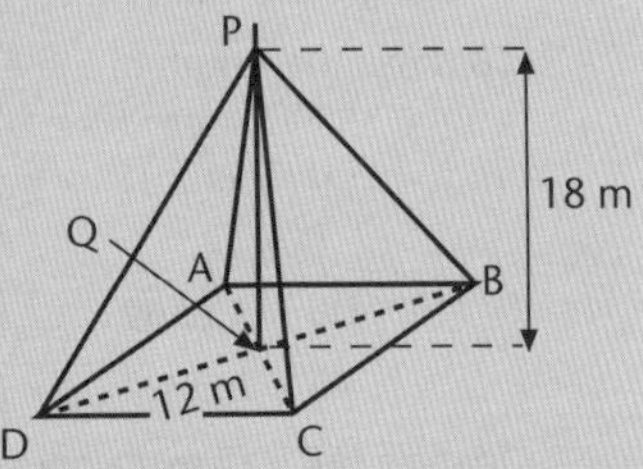

   a. Calculate the size of the angle each rope makes with the ground.

   b. The four pegs form a square ABCD. What is the length of each side of this square, to the nearest centimetre?

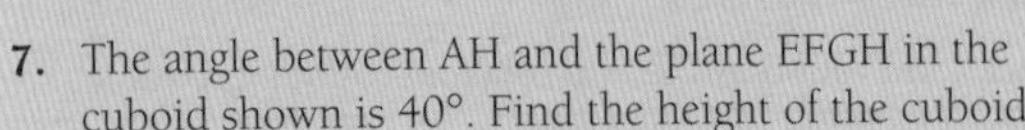

6. A pencil box in the shape of a cuboid is 16 cm long and 6 cm wide. If the longest pencil that fits in the box is 18 cm long, how deep is the box?

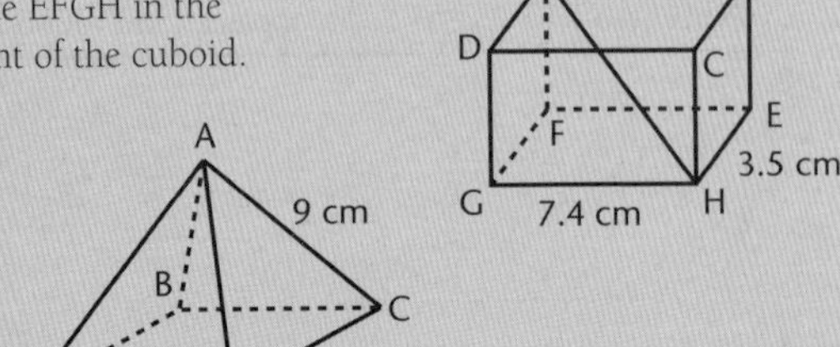

7. The angle between AH and the plane EFGH in the cuboid shown is 40°. Find the height of the cuboid.

8. A 12 cm by 10 cm rectangle-based pyramid ABCDE is shown. Its sides are isosceles triangles of sloping side length 9 cm.

   Find the angle between the line AE and the base plane BCDE.

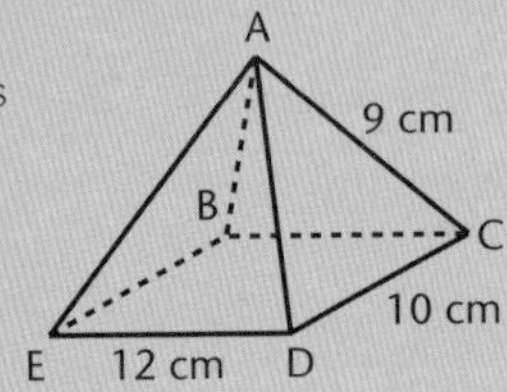

The angle between two planes

Find the angle between two planes

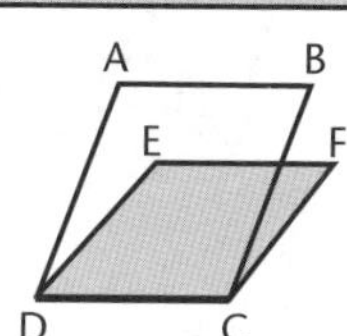

Find the line of intersection: planes ABCD and CDEF meet on the line CD.

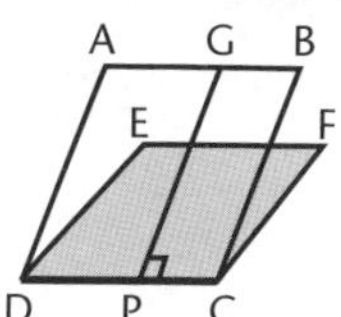

On the plane ABCD draw a line GP perpendicular to CD, meeting CD at P.

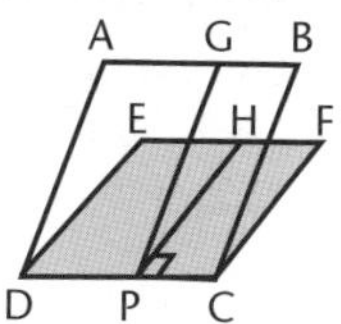

On plane CDEF draw a line HP perpendicular to CD meeting CD at P.

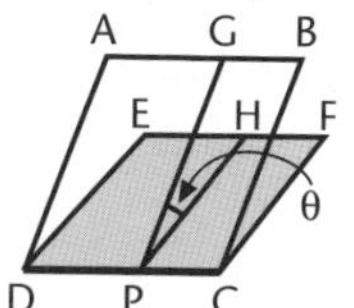

The angle GPH (marked θ) is the angle between the plane ABCD and the plane CDEF.

Example E

Q. In the cuboid shown, find the size of the angle between planes AFGD and EFGH.

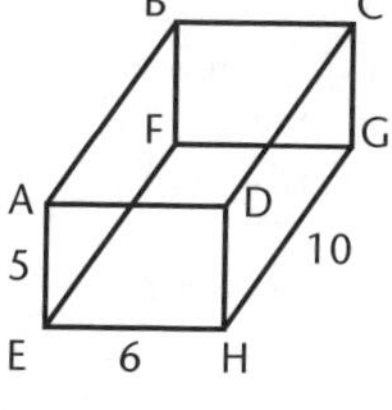

A. The two planes meet on line FG.

- On the plane AFGD, DG is perpendicular to FG.
- On the plane EFGH, HG is perpendicular to FG.

Since DG and HG are perpendicular to FG, the required angle is $\angle DGH$.

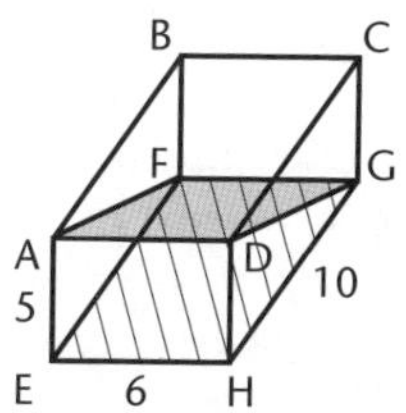

Triangle DGH is shown alongside.

$\tan \angle DGH = \frac{5}{10}$

$\angle DGH = \tan^{-1} 0.5$

$= 26.6°$

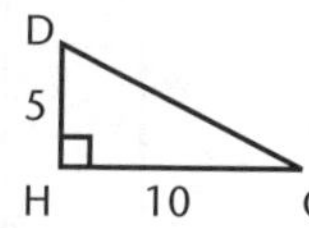

Note: $\angle AFE$ (= $\angle DGH$) is also the angle between the two planes.

In pyramids, take care to *identify perpendiculars correctly* when working with triangular planes.

Example F

Q. In the square-based pyramid shown, the sloping faces are isosceles triangles. X is the midpoint of BC and Y is the midpoint of AD. Name the angle between the planes ECB and ABCD.

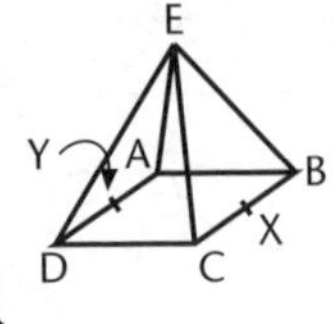

A. The diagrams below show how to identify the required angle, EXY.

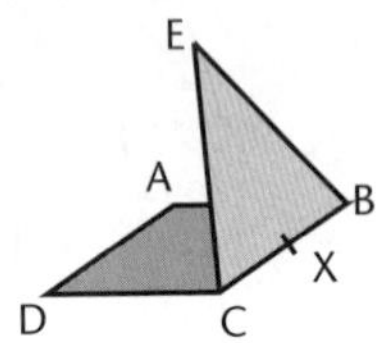

BC is the line of intersection of planes EBC and ABCD.

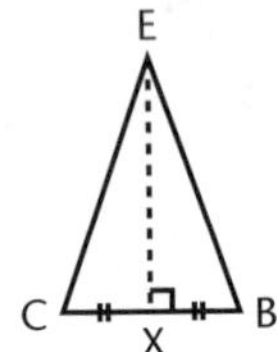

EX is at right angles to BC (ΔEBC isosceles).

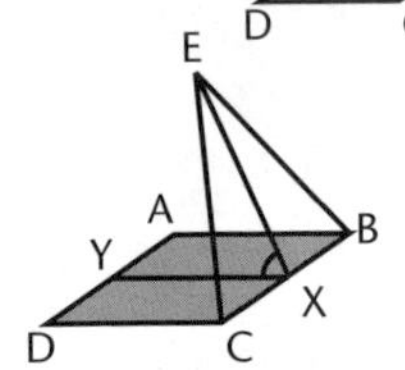

On the plane ABCD, the line XY is perpendicular to BC.

Shortest distance along the faces of a solid

You may be required to find the shortest distance between two points on a 3-D object when distances are measured along the *faces* of the object. To do this, draw a net of the object. Join the points concerned with a straight line and use trigonometry and/or **Pythagoras' Theorem** to find the length of the line joining the two points. There may be more than one diagram possible.

Example G

Q. Find the shortest distance between A and B over the faces of the cuboid shown.

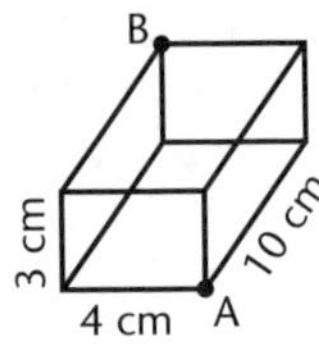

A. The relevant parts of the net of the cuboid are drawn, and the points A and B are joined with a straight line.

There are two cases.
A right-angled triangle is formed in each case.

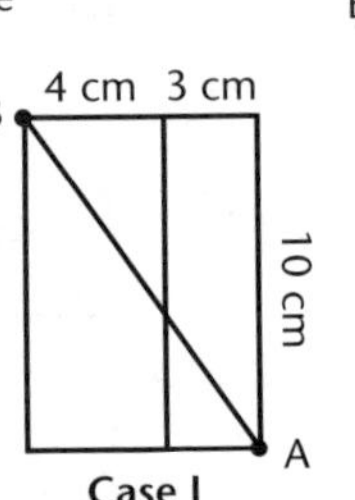

Case I

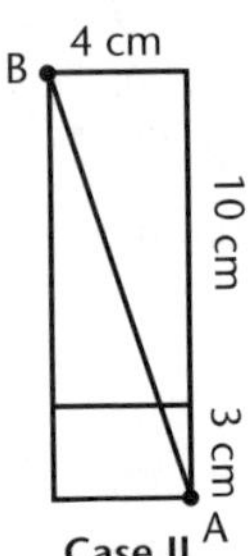

Case II

Case I

Using Pythagoras' theorem:

$AB^2 = 7^2 + 10^2$

$= 149$

$AB = \sqrt{149}$

$= 12.2$ cm (1 dp)

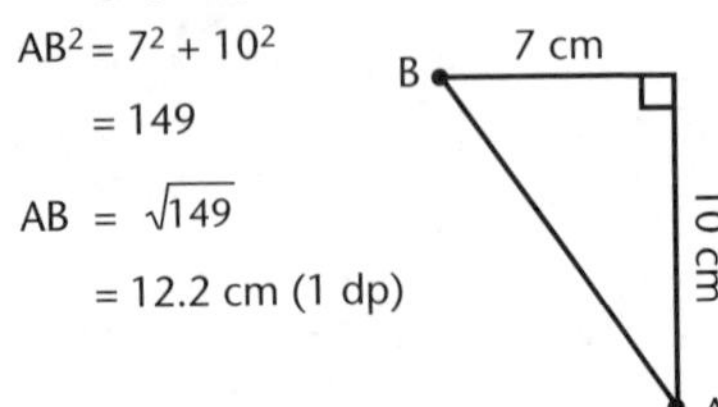

Thus the shortest distance from A to B is 12.2 cm.

Case II

Using Pythagoras' theorem:

$AB^2 = 4^2 + 13^2$

$= 185$

$AB = \sqrt{185}$

$= 13.6$ cm (1 dp)

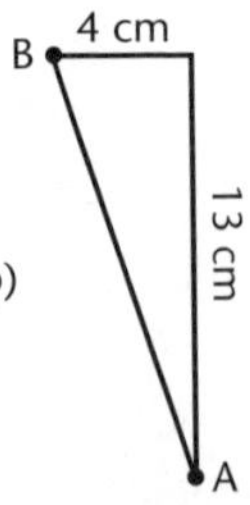

Note: The shortest distance between A and B is still through the interior of the object (check this distance is 11.18 cm) but here distances are being calculated along the faces.

There are many practical applications for this type of problem.

Example H

Q. Sam wants to run a cord from one end of the top of the roof of the village Haus Bung (D) to the bottom corner at the opposite end (C).
What is the shortest length of cord Sam will need?

A. Part of the *net* of the Haus Bung is drawn. The roof peak opposite D is labelled A. The length x is the length of the roof from the top to the side wall.

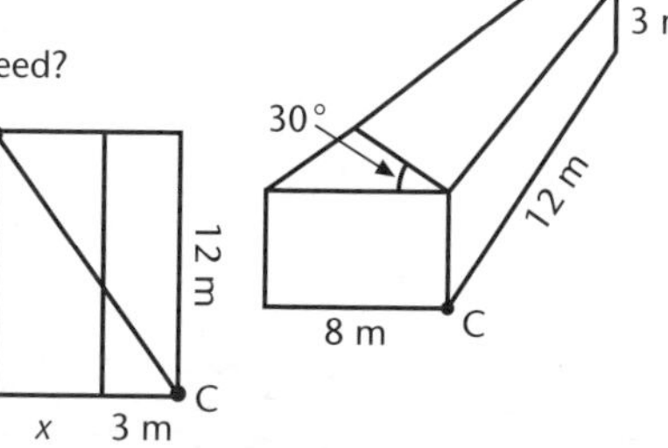

In the triangle of the cross-section of the roof

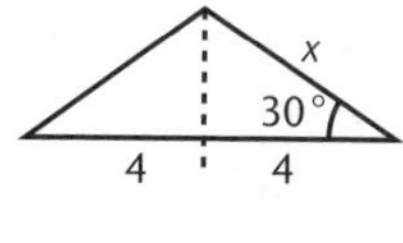

$$\cos 30° = \frac{4}{x}$$

$$\therefore x = \frac{4}{\cos 30°} \qquad \text{[rearranging]}$$

$$= 4.619 \text{ m}$$

In triangle DAC drawn alongside,

$$AC = x + 3$$

$$= 7.619 \text{ m}$$

$$CD^2 = 12^2 + 7.619^2 \qquad \text{[Pythagoras' Theorem]}$$

$$= 202.049$$

$$CD = \sqrt{202.049}$$

$$= 14.2 \text{ m (1 dp)}$$

The shortest length of cord Sam will need is 14.2 m.

Unit 11.5 Activity 3B: Further 3-D trigonometry

1. The diagram shows a cuboid. Name the following angles.
   - **a.** The angle between planes ABGH and EFGH.
   - **b.** The angle between planes CDEF and ABCD.
   - **c.** The angle between planes EFGH and DCFE.
   - **d.** The angle between planes ACGE and BCGF.

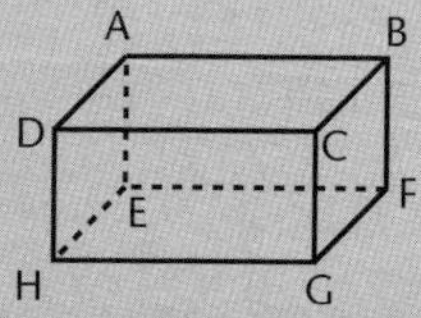

2. A cuboid ABCDEFGH has dimensions 12 cm by 5 cm by 8 cm.

   Find the size of the angle between the plane AFGD and the plane BCGF.

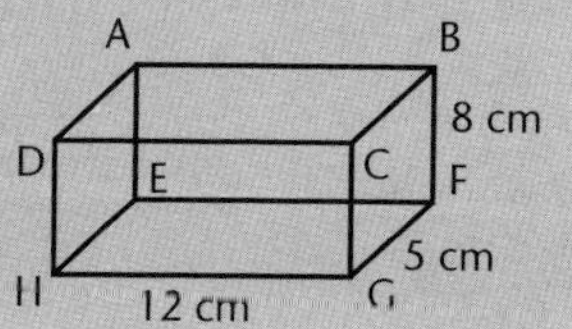

3. ABCDEFGH is a cube of side length 12 cm.

   Find the size of the angle between the planes ACGE and BDHF.

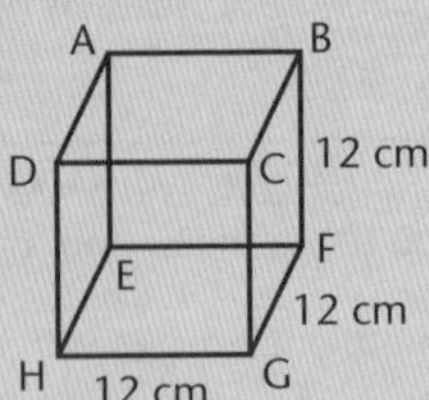

4. The diagram shows a triangular-based prism.

   Triangles ABC and DEF are right-angled, as shown.

   The length of AB is 5 cm. The length of BC is 8 cm and the length of CF is 18 cm.

   Calculate the size of the angle between planes BCFE and ADFC.

5. ABCDE is a right pyramid with a square base.

   Length of AD is 10 cm. Length of EH is 15 cm. X is the midpoint of CD.

   Name and calculate the angle between planes ECD and ABCD.

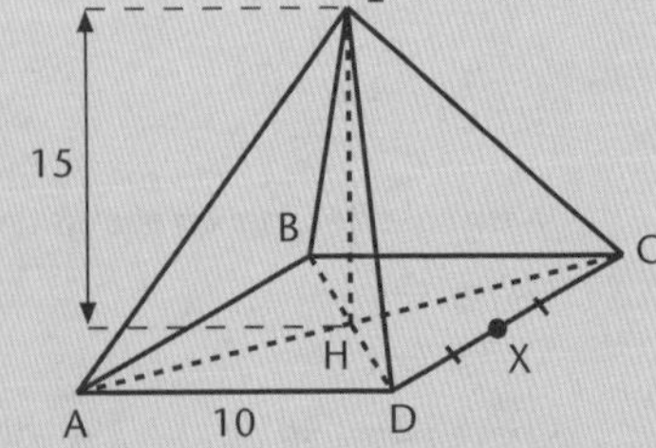

6. A rectangle-based pyramid is shown below.

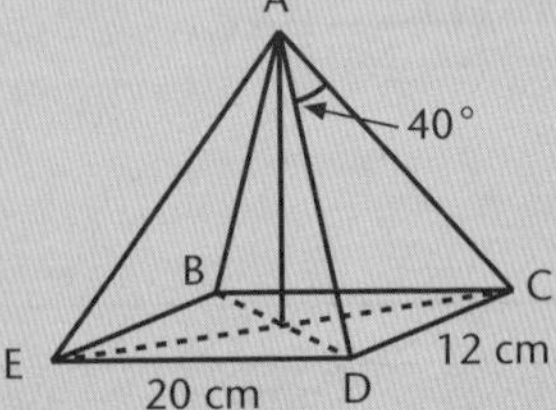

   Find the angle between the planes ADC and BCDE.

7. Find the shortest distance between A and B on the cuboid drawn. (Check both possible cases.)

8. Hivai wants to run a power cable from the top of the school hall (T) to a fuse box (F) on the edge where the side wall meets the front wall of the hall.

   The fuse box is 2 m above the ground, and the walls of the hall are 3.2 m high. The hall is 20 m long and 8.4 m wide. The pitch of the roof is 25°.

   The cable is to be attached to the roof and wall *at all times*.

   a. What is the shortest length of cable Hivai needs?

   b. Where along the roof/wall line will the cable cross?

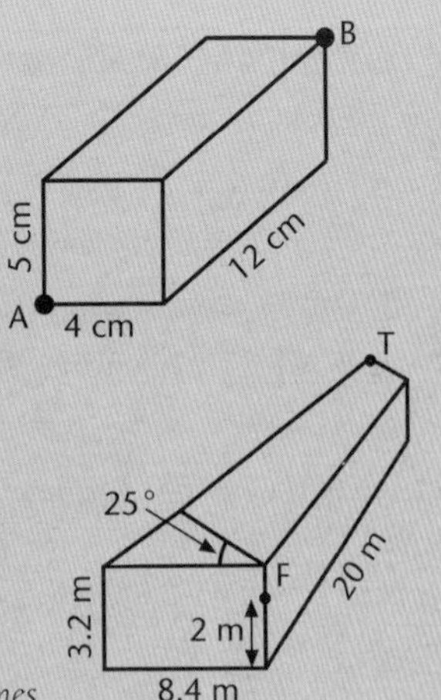

Unit 11.5 Trigonometry

Topic 4: The sine and cosine rules

The material in this Topic deals with the sine and cosine rule and the solving of practical problems using those rules (see Syllabus p. 19). This Topic covers:

- Trigonometric modelling of practical problems, using non right-angled triangles.
- Using the sine and cosine rules to find lengths and angles.
- Finding the area of a triangle.
- Solving multi-step trigonometry problems, including 2-D representations of a 3-D situation.

Introduction

Trigonometry involving right-angled triangles (and the mnemonics SOH, CAH, TOA) should by now be familiar to the reader.

In this chapter, trigonometry is extended to include triangles which are not right-angled.

Conventions

An important **convention** in this section is the method used to name the angles and sides of a triangle.

The diagram shows how capital letters are used to show the angles at the corners of the triangle. The length of the side *opposite* each corner is denoted by the same letter in the lower case (side a is opposite angle A, etc).

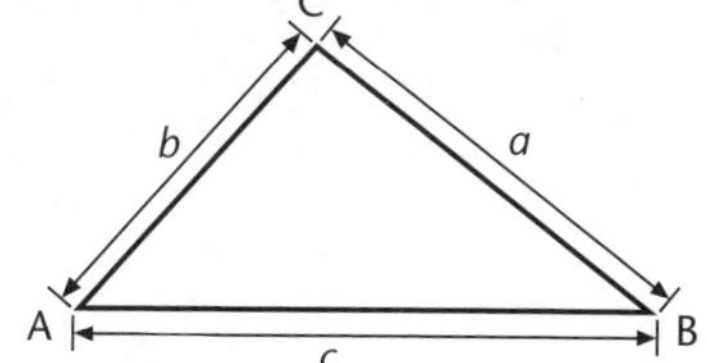

The cosine rule

For any triangle ABC, the **cosine rule** states:

$$a^2 = b^2 + c^2 - 2bc \cos A$$

Making cos A the subject of $a^2 = b^2 + c^2 - 2bc \cos A$ gives another form of the cosine rule, useful for calculating unknown angles:

$$\cos A = \frac{b^2 + c^2 - a^2}{2bc}$$

Note: Different letters can be made subject of these formulae.
By symmetry, $b^2 = a^2 + c^2 - 2ac \cos B$ and $c^2 = a^2 + b^2 - 2ab \cos C$.
Also $\cos B = \dfrac{a^2 + c^2 - b^2}{2ac}$ and $\cos C = \dfrac{a^2 + b^2 - c^2}{2ab}$

Example A

Q. A fishing vessel sets out to sea. It is 5.2 km from point P on the shore and 4.7 km from another point, Q, on the shore. The angle between the boat and these two points is 75°.

1. How far apart are the two points?
2. Failing to catch any fish, the captain sails the boat to a point which is now 5.4 km from point P and 6.3 km from Q. What is the angle between P, the boat, and Q?

A. 1. The diagram shows the points P and Q.

The length PQ is required. Using the cosine rule,

$PQ^2 = 5.2^2 + 4.7^2 - 2 \times 5.2 \times 4.7 \cos 75°$ [substituting into $a^2 = b^2 + c^2 - 2bc \cos A$]

$= 36.48$

$\therefore$ PQ = 6.0 km (1 dp) [taking square roots]

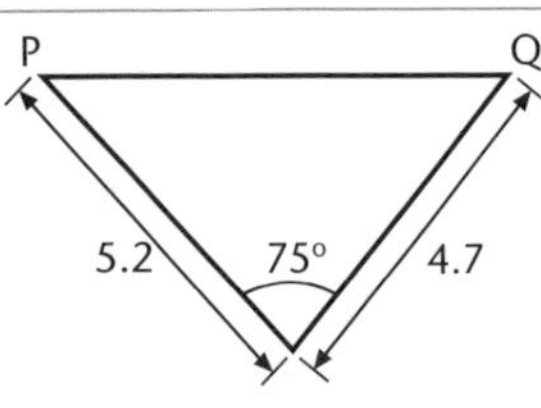

2. The required angle, x, is shown below.

$$\cos x = \frac{6.3^2 + 5.4^2 - 6.04^2}{2 \times 6.3 \times 5.4}$$

[substituting into $\cos A = \frac{b^2 + c^2 - a^2}{2bc}$]

$\therefore \cos x = 0.4757$

$\therefore x = 61.6°$ (1 dp) [taking $\cos^{-1}$]

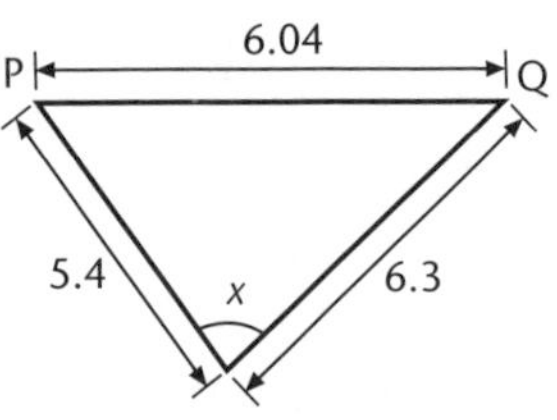

Unit 11.5 Activity 4A: The cosine rule and applications

1. In the diagram shown, find:

a. The length x.

b. The size of $A\hat{B}C$.

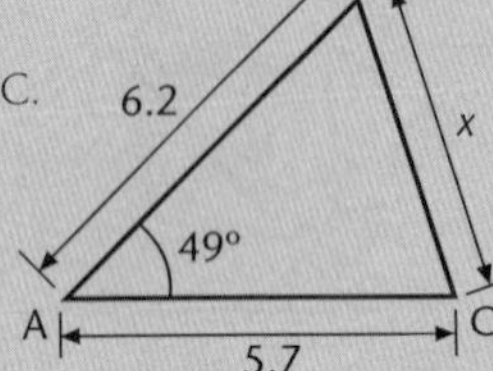

2. In this diagram find:

a. The length x.

b. The angle y.

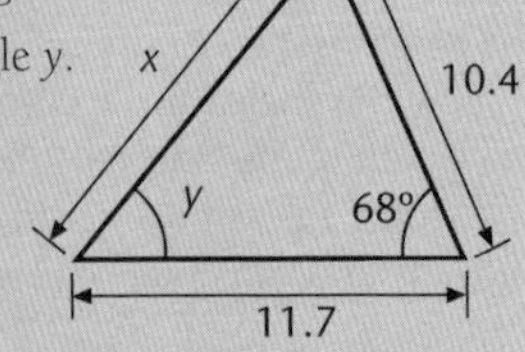

3. In this diagram find:

a. The length of QR.

b. The size of $P\hat{Q}R$.

Q
15.3
23°
P
R
13.4

4. In the triangle shown, calculate

a. The length of BC.

b. The size of angle ABC.

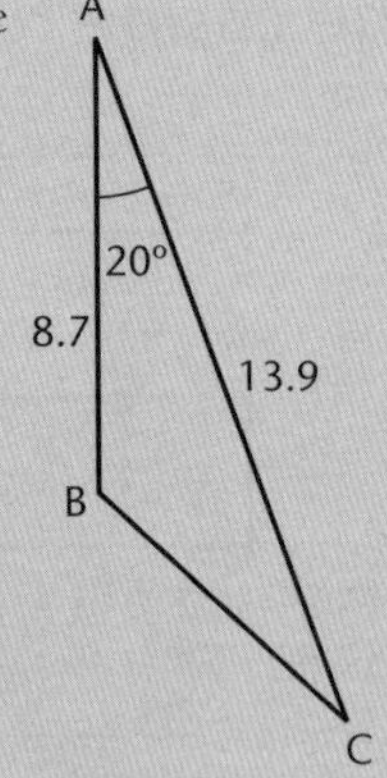

5. A bar of length 2 m is suspended by two strings of 1.3 m and 1.8 m long, attached to a nail.

a. Find the angle between the two pieces of string.

b. If the bar is now shortened so that the angle between the two strings is 75°, find the new length of the bar.

6. A triangular shaped field has one boundary extending 350 m from a gate in a direction N35°E. The other boundary extends N32°W from the gate for 200 m.

a. Find the length of the third side of the field to the nearest metre.

b. Find the largest angle of the field.

The sine and area rules

Using the same triangle labelling conventions as before, the **sine rule** states:

$$\frac{\sin A}{a} = \frac{\sin B}{b} = \frac{\sin C}{c} \quad \text{or} \quad \frac{a}{\sin A} = \frac{b}{\sin B} = \frac{c}{\sin C}$$

Note: The sine rule is a short way of expressing three separate relationships:

$\frac{a}{\sin A} = \frac{b}{\sin B}$, $\frac{a}{\sin A} = \frac{c}{\sin C}$ and $\frac{b}{\sin B} = \frac{c}{\sin C}$ (and their reciprocal relationships).

The **area rule** states that the area of a triangle is:

$$\text{Area} = \frac{1}{2}ab \sin C$$

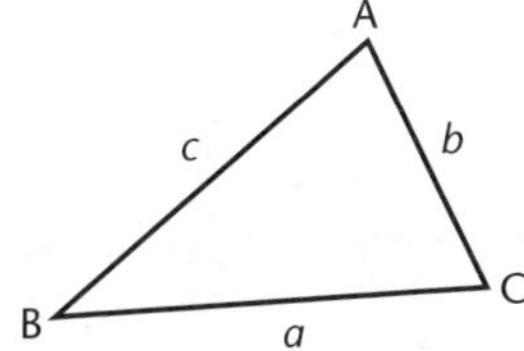

Note: Alternatively, the area is equal to $\frac{1}{2}bc \sin A$ or $\frac{1}{2}ac \sin B$.

Example B

Q. Peter and Moreni tie two ropes to the top of a pole. They stake the other end of each rope into the ground so they are tight and in the same plane with the pole. Peter's rope is 5.6 m long and makes an angle of 26° with the ground. Moreni's rope makes an angle of 32° with the ground.

1. Find the length of Moreni's rope.
2. Find the distance between the two stakes.
3. Find the area of the triangle formed by the two ropes and the line along the ground which joins the two stakes.

A. The following diagram shows Moreni's rope with length L, and the distance between the stakes d. Let the area required be A.

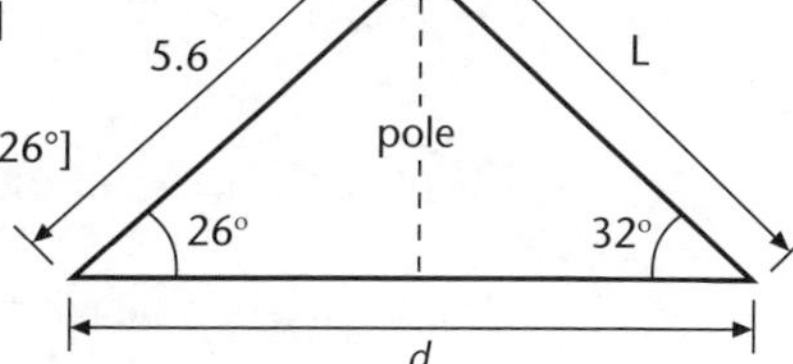

1. $\frac{L}{\sin 26°} = \frac{5.6}{\sin 32°}$ [using the sine rule]

$\therefore\ L = \frac{5.6}{\sin 32°} \sin 26°$ [multiplying by sin 26°]

$= 4.63255\ldots$

$\therefore$ Moreni's rope is 4.6 m (1 dp)

2. The third angle of the triangle is 180 – 26 – 32 = 122°

$\frac{d}{\sin 122°} = \frac{5.6}{\sin 32°}$ [using the sine rule]

$\therefore\ d = \frac{5.6 \times \sin 122°}{\sin 32°}$ [multiplying by sin 122°]

$= 8.96187\ldots$

$\therefore$ the distance between the stakes is 9.0 m (1 dp)

3. $A = \frac{1}{2} \times 5.6 \times 8.962 \times \sin 26°$ [using area = $\frac{1}{2}ab \sin C$]

$= 11.000\ldots$

$\therefore$ the area of the triangle is 11.0 m^2 (1 dp)

Unit 11.5 Activity 4B: The sine and area rules and applications

1. In this diagram find:
   a. The length AC.
   b. The length BC.
   c. The area of ΔABC.

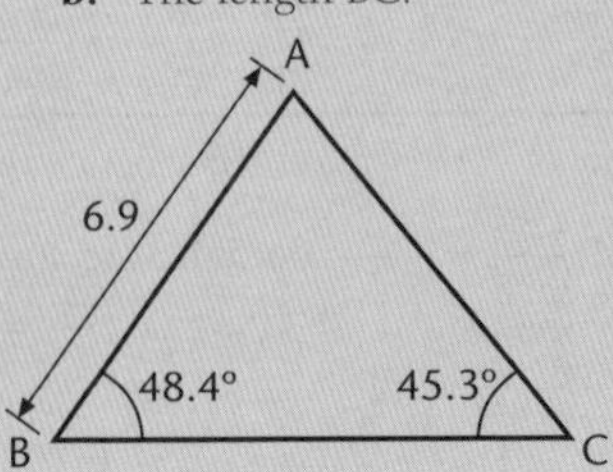

2. In this diagram find the length of:
   a. AB.
   b. AC.

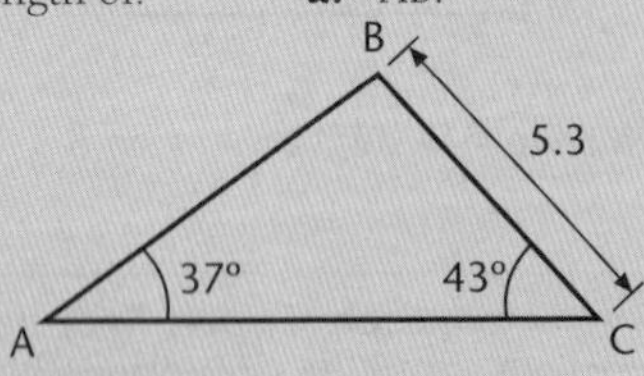

3. In this diagram find:
   a. The length of AC.
   b. The length of CD.
   c. The area of ABCD.

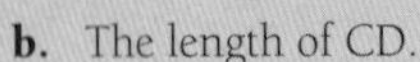

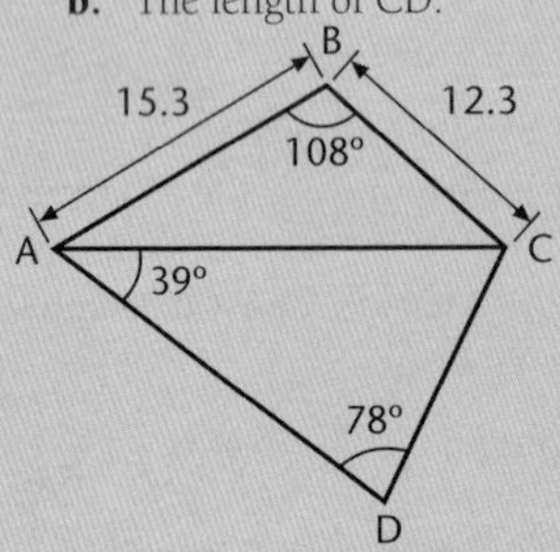

4. A ship sails for 105 km in a direction E32°S. It then sails for another 200 km in a direction N10°E.
   a. How far is it from its original position?
   b. What is its compass bearing from the original position?

5. A father looks down at his son's eyes. The man's eyes are tilted at 38° to the vertical. His son looks at his father's feet. The boy's eyes are tilted at 42° to the vertical. The son's eyes are 1.1 m from his father's eyes. Find the height of the father's eyes from the ground.

6. A triangular wedge (prism) has a base with an angle of 43° enclosed between sides of 4.3 cm and 5 cm. Find:
   a. the length of the third side of the triangle.
   b. the area of the cross-section.
   c. the volume of the wedge if the length is 51 cm.

7. A section of land is of triangular shape. One corner of the land has fences of lengths 52 m and 54 m enclosing an angle of 66°.

a. Find the cost to the nearest dollar of fencing the land at $6.80 a metre.

b. Find the area of the section.

8. A man walks for 5.6 km on a bearing of N33°E from O. He stops and finds that he is now at a bearing of N27°W from P which is due East of O.

a. How far is he from P?

b. How far is P from O?

9. A cyclist rides for 11 km in a direction of N38°E from O. At this time she dismounts and finds she is 12 km away from P which is due East of O.

a. What is her compass bearing from P?

b. What distance is P from O?

10. ABC is an equilateral triangle with sides of length 2. AE, BE, CE are of equal length. Find that length.

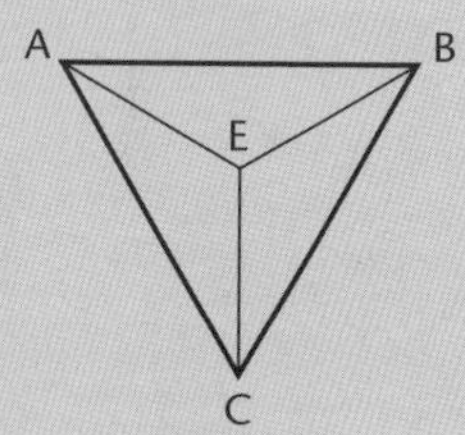

11. A student placed a piece of clean paper on a table in the middle of a field ABCD and by use of a device called an alidade was able to construct the following diagram (not to scale). The lines intersect at a point E on the paper. AE is 30 m long, BE is 40 m long, DE is 40 m long and CE is 35 m long.

a. Find angle BAE.

b. Find the perimeter of the field ABCD.

c. Find the area of the field.

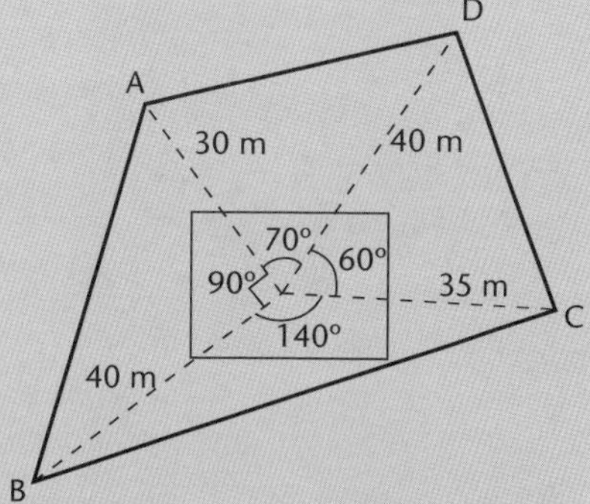

12. A small boat, which is experiencing difficulties, sets off an alarm. This alarm is detected by two aeroplanes which are 8 km apart in a North-South direction. The northernmost plane detects the alarm at a bearing of 120° and the other plane detects the alarm at a bearing of 040°. How far is the northernmost plane from the boat?

Unit 11.5 Trigonometry
Topic 5: Location and vectors

The material in this Topic deals with vector notation and position vector, and the use of scalar multiplication to explain and apply parallel vectors (see Syllabus p.19). This Topic covers:

- Interpretation of bearings and contour maps.
- The use of trigonometric ratios to find unknowns in straightforward practical situations involving vectors.
- Using trigonometric ratios to find unknowns in word problems (student to extract right-angled triangles implicit in a context and use them to solve the problem in a well-reasoned and logical solution).
- Use appropriate rounding, units and mathematical statements.

Directions

The position of a place or feature on a map or diagram can be represented in a number of different ways.

Compass points

The basic method of describing direction is by referring to the eight main **compass points**. Eight more compass points lie between the main compass points.

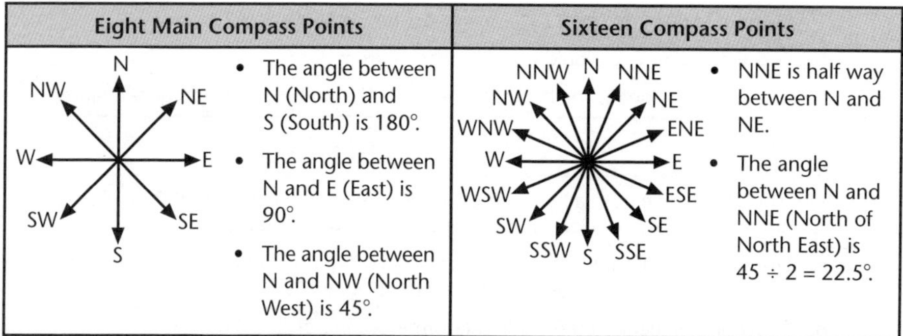

Eight Main Compass Points	Sixteen Compass Points
• The angle between N (North) and S (South) is 180°. • The angle between N and E (East) is 90°. • The angle between N and NW (North West) is 45°.	• NNE is half way between N and NE. • The angle between N and NNE (North of North East) is 45 ÷ 2 = 22.5°.

Compass points which lie between these 16 points are usually named starting from North or South. Starting from North or South, the angle must be less than 90°.

Example A

1. A direction N30°E means: from North turn 30° towards East.

N

N30° E

30°

2. A direction S52°W means: from South turn 52° towards West.

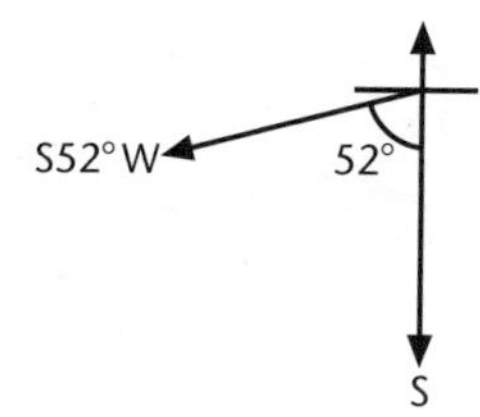

Bearings

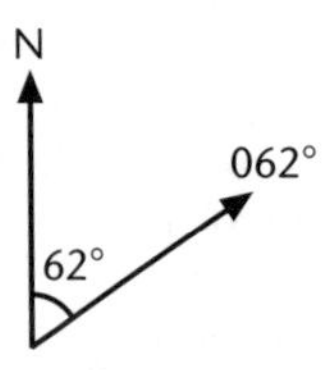

A **bearing** is a means of describing a direction.

- A bearing is an angle measured *clockwise* from North.
- A bearing is described *using three* digits.

For example, the direction shown alongside by an angle of 62° clockwise from North is N62°E, and is written 062° as a bearing.

Example B

Q. Find the bearing of B from A in the diagram alongside.

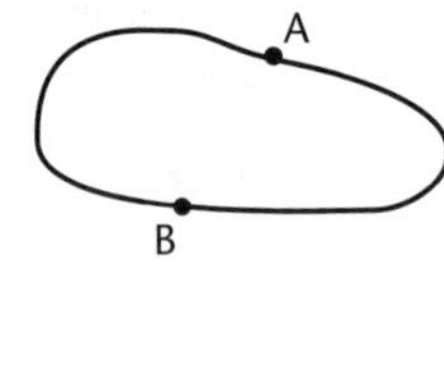

A. To find the bearing of B from A, draw a North line at A and join A to B.

The required bearing is the clockwise turning from A to B marked θ.

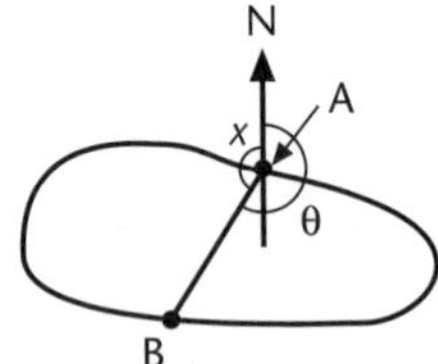

To find θ, use a protractor to measure the angle marked x on the diagram. (This is the angle measured anticlockwise from North.)

$x = 147°$

Bearing = 360° – 147° = 213° [a full turn is 360°]

This is the direction one would travel to reach B from A.

Note: To reach A from B, the bearing is 180° – 147° = 033° [property of co-interior ∠s]

Example C

Q. Two ships are visible from a lighthouse. Ship A is 5 km, on a bearing of 056°, from the lighthouse and ship B is 6 km, on a bearing of 125°, from the lighthouse.

a. Draw a scale diagram to represent the positions of the lighthouse and the two ships. Use the scale: 1 cm represents 1 km on the diagram.

b. Show on the diagram the angle that is the bearing of ship B from ship A.

c. Use the scale diagram to find the distance between ship A and ship B.

d. Use the cosine rule to find the actual distance between ship A and ship B.

e. Use the sine rule to find the angle in the lighthouse–ship A–ship B triangle at ship A.

f. Find the bearing of ship B from ship A, to the nearest degree.

A.

a.

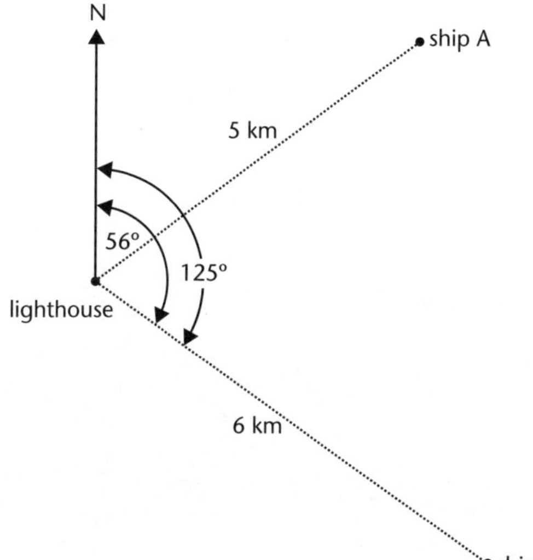

b.

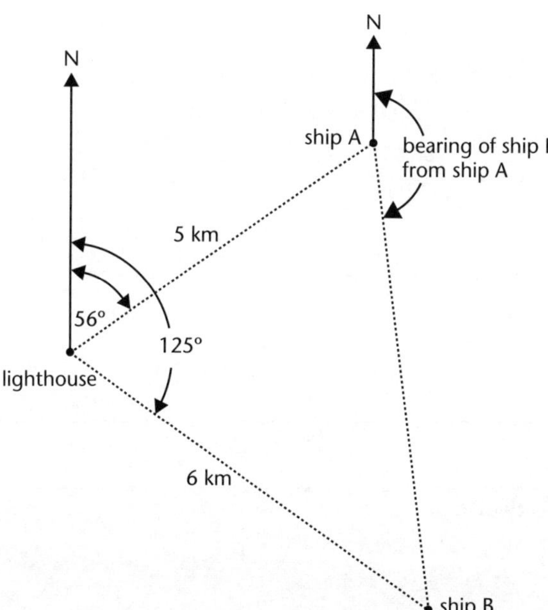

c. The line joining ship A and ship B measures 6.3 cm, so the distance is 6.3 km.

d. The angle θ is equal to 125° – 56° = 69°.

Using the cosine rule for triangle LAB:

$x^2 = 5^2 + 6^2 - 2 \times 5 \times 6 \cos 69$

$- 39.4979 \ldots$

$x = 6.2847 \ldots$

The actual distance between ship A and ship B is 6.285 km correct to three decimal places.

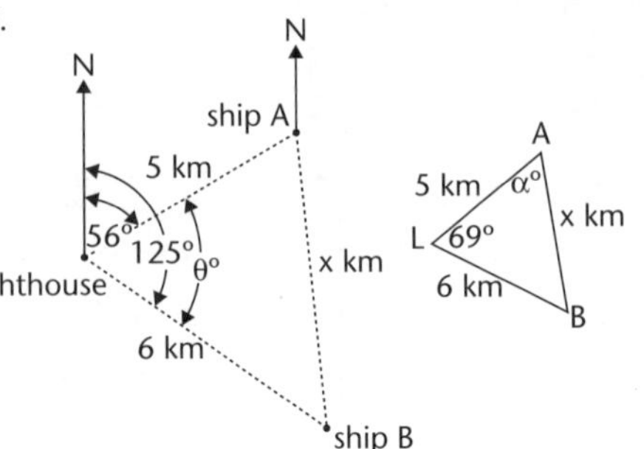

e. Using the sine rule to find the size of angle α:

$$\frac{\sin \alpha}{6} = \frac{\sin 69}{x}$$

$$\sin \alpha = \frac{6 \sin 69}{6.2847}$$

$$= 0.8913$$

$$\alpha = 63.03°$$

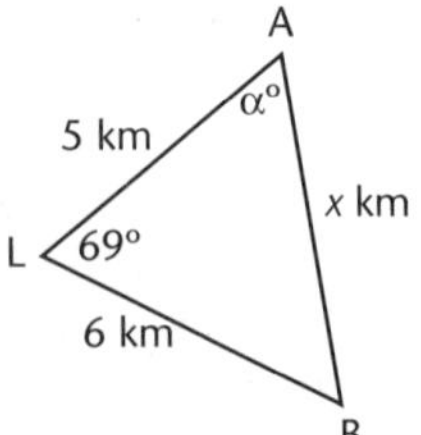

f. The bearing of ship B from ship A is equal to 360° – 124° – 63.03° = 173° (to the nearest degree).

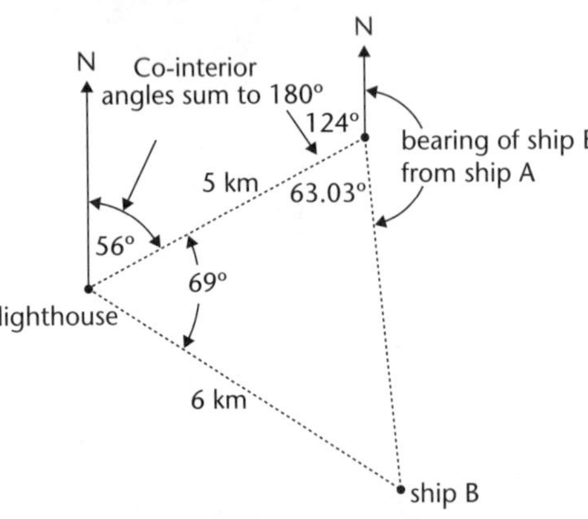

Unit 11.5 Activity 5A: Bearings

1. Convert the following compass bearings to true bearings:

a. N35°W

b. S63°E

c. ESE

d. S72°W

2. Write the following directions as true bearings.

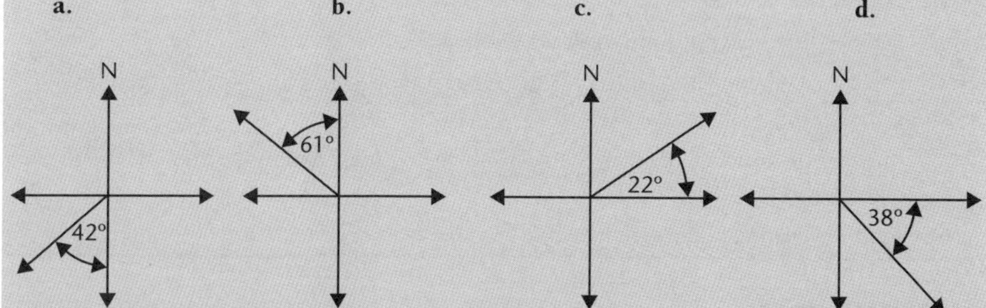

3. The relative position of three towns, P, Q and R, is given on the diagram below. Find the bearing of:

a. Q from P

b. R from P

c. R from Q

d. P from Q

e. P from R

f. Q from R

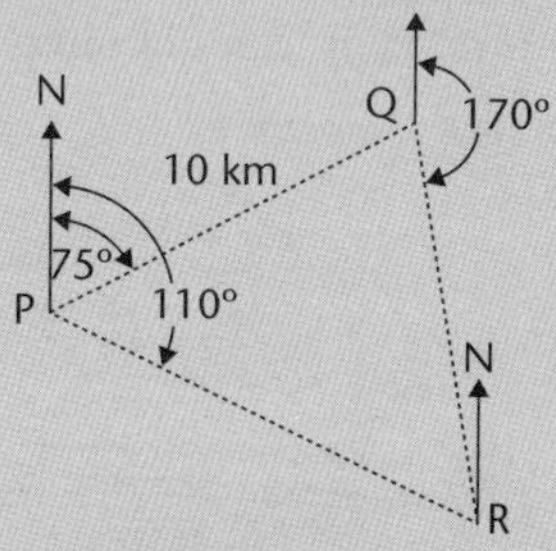

4. For each of the following, construct a scale diagram and state the scale that you have used. Use a protractor for the angles and choose a suitable scale for the distances.

a. Peter walks 3.5 km due East (bearing 090°) and then 6 km on a bearing of 330°. Calculate the distance of Peter's finishing point from his starting point and check this with your measurement of this distance.

b. Town B is 20 km due East of town A. From town A, town C is on a bearing of 055° and from town B, town C is on a bearing of 300°. Locate the position of town C and calculate its distance from town A.

c. Manni travels 56 nautical miles on a bearing of 265° then 38 nautical miles on a bearing of 150°. Calculate the distance of Manni's finishing point from his starting point and check this with your measurement of this distance.

d. Lae is 190 km on a bearing of 045° from Kerema and Goroka is 190 km on a bearing of 295° from Lae. Find the distance and bearing of Kerema from Goroka.

5. Use the map at right to describe the relative positions of the mine and village Q from town P. Use bearings (measured with a protractor) and distances in kilometres.

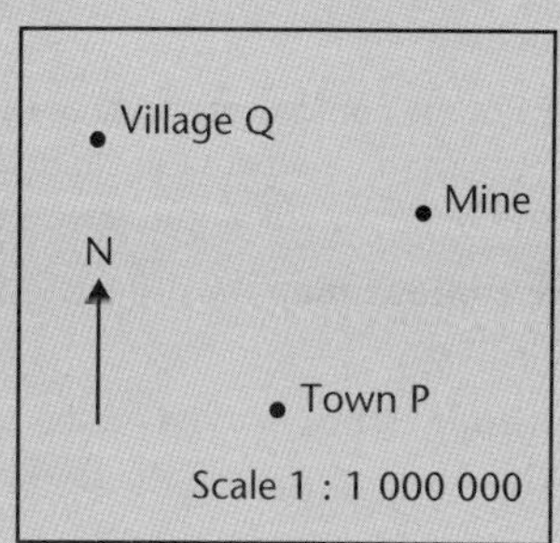

6. A villager walks 4 km on a bearing of 35° and then walks a further 5 km on a bearing of 320°. At this point he stops to rest.
   a. Draw a scale drawing of the villager's journey using a scale of 1 : 10 000.
   b. Use the cosine rule to calculate the distance of the villager's resting point from his starting point.
   c. Measure the distance between the villager's resting point and his starting point and convert this to kilometres.
   d. Find the bearing of the starting point from the resting point, to the nearest degree.

Contour maps

A contour map gives a view *from above* of the terrain of an area. Contour lines join all points that have the same altitude in the area. The distance between contour lines is usually constant; this distance is called the **contour interval**.

For example:

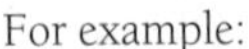

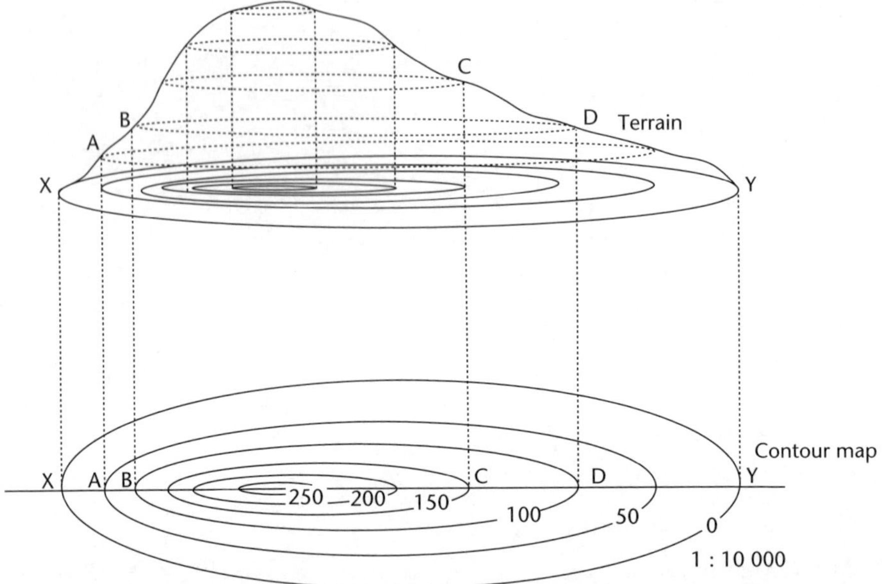

A contour map for the terrain given is depicted in the diagram beneath it. The **contour interval** is 50 metres.

Sections of the contour map where the contour lines are close together will have a steeper gradient than those where the contour lines are further apart. For example: if you were travelling from A to B the terrain would have a steeper slope than if you were travelling from C to D.

On the contour map the distance AB represents the horizontal distance between the points A and B.

The **altitude** can be marked on the contour lines as in the diagram or quoted as a contour interval.

A **ratio scale** on a contour map refers to the horizontal distance between points.

For the contour map above, the horizontal distance between points C and D measures 1.5 centimetres on the contour map. The ratio scale is 1 : 100 000 which means that the horizontal distance between C and D is 150 000 centimetres = 1500 metres = 1.5 kilometres.

The **average slope** of the hill between points C and D can be found using

$$\text{The average slope between two points} = \frac{\text{the difference in height of the points}}{\text{the horizontal distance between the points}}$$

$$\text{The average slope between points C and D} = \frac{50}{1\,500} = \frac{1}{30} \approx 0.033$$

Example D

Q. A contour map of a local hill is shown at left. Points B and C are at sea level and the top of the hill has an altitude of 210 m.

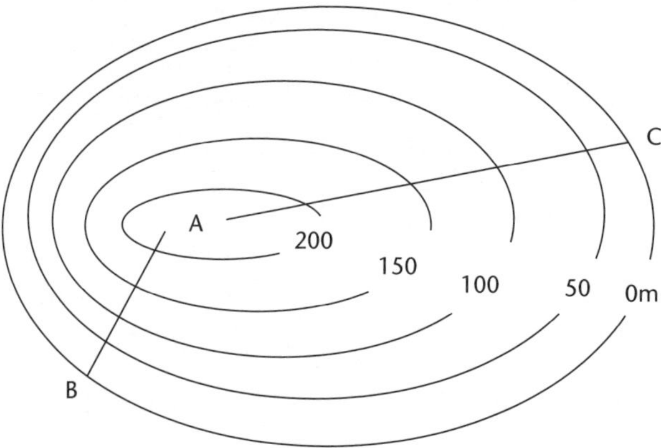

1 : 40 000

To reach the top of the hill, A, Tom has the choice of two paths: BA or CA.

a. Which path would provide the steepest walk to the top of the hill?

b. If AC measures 5 cm on the contour map find the average slope from point C to point A.

A. **a.** The steepest path will be the one where the contour lines are closest together. Hence path BA will provide the steepest walk to the summit.

b. Using the ratio scale the horizontal distance between points A and C is 5 × 40 000 = 200 000 cm = 2000 m.

Point A is at a height of 210 m. The average slope

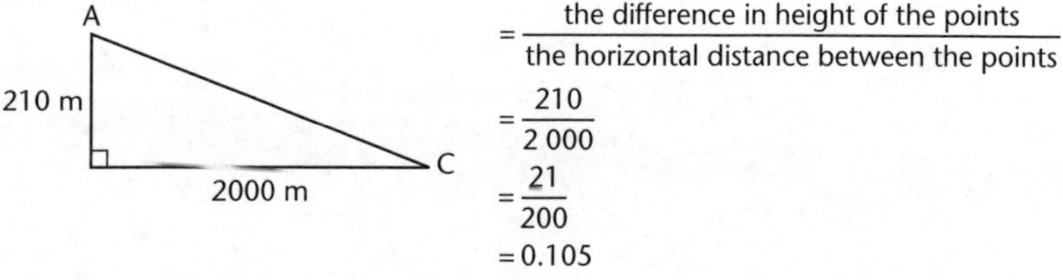

$$= \frac{\text{the difference in height of the points}}{\text{the horizontal distance between the points}}$$

$$= \frac{210}{2\,000}$$

$$= \frac{21}{200}$$

$$= 0.105$$

Drawing cross-sections from contour maps

We can use a contour map of terrain to draw its cross-section. For example: Consider the following contour map where the cross-section axis is the line XY.

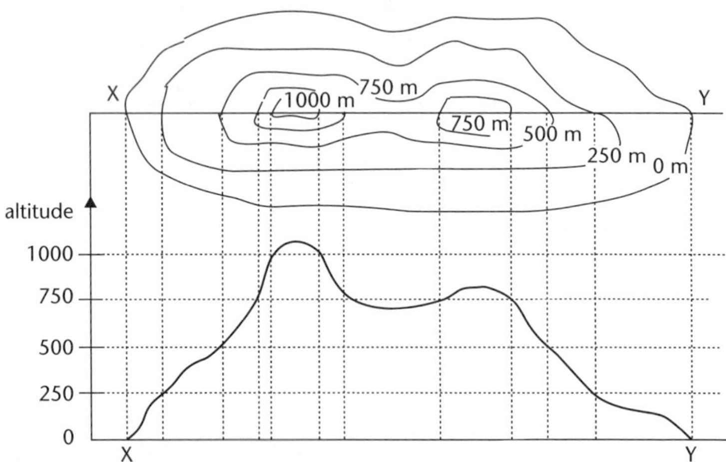

The highest part of the hill would be more than 1000 m high (but less than 1250 m) so the vertical scale is marked from 0 to 1250. The horizontal scale should be as long as the line XY.

Points along the line XY are transferred directly from the contour map onto the set of axes below so that it will have the same position on the horizontal axis, but its altitude is shown on the vertical axis. This produces a cross-section of the hill along the axis XY.

A different axis line drawn on the contour map could produce a different cross-section.

Unit 11.5 Activity 5B: Contour maps

Answers should be given correct to two decimal places.

1. A contour map for a geographical feature is given below as well as four cross-section diagrams. There are four cross-section axes drawn on the contour map. Match these axes to the cross-section diagrams.

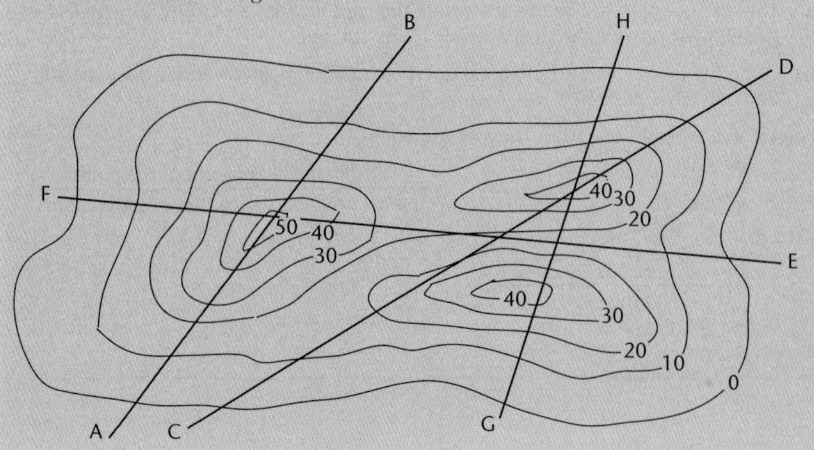

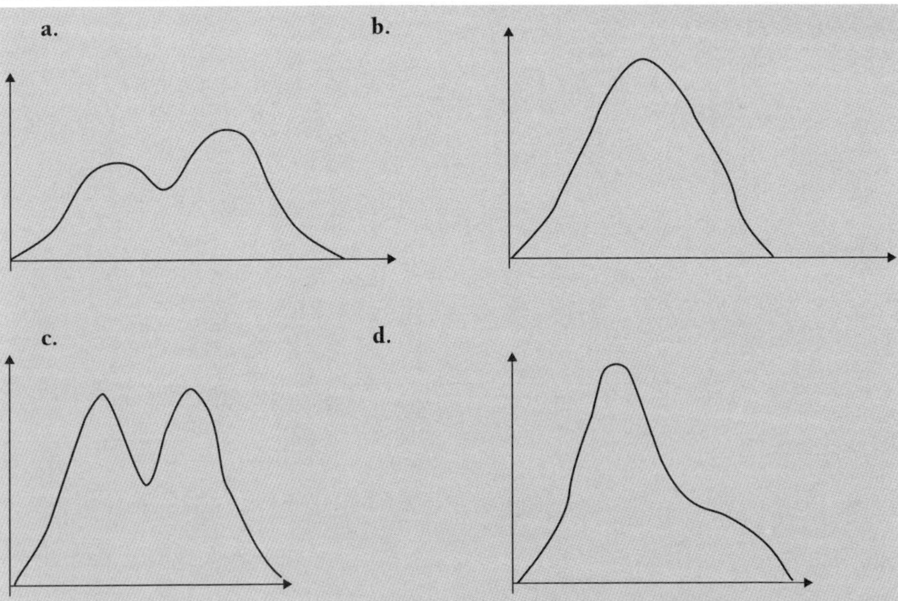

2. The ratio scale for the contour map below is 1 : 100 000 and the contour interval is 50 metres. An axis XY has been drawn on the contour map.

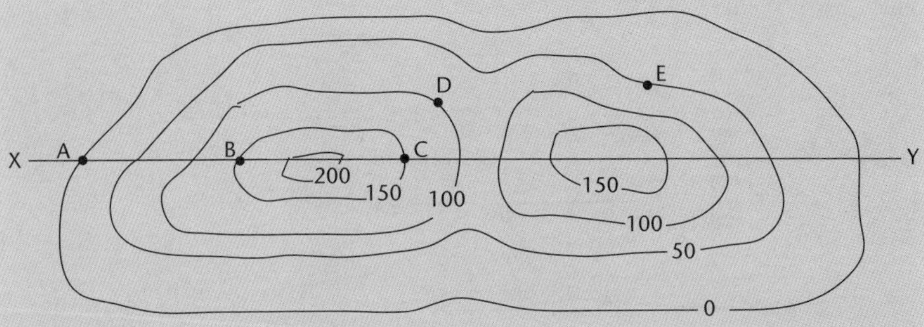

a. Draw a cross-section of the terrain along the axis XY. (Carefully measure the distances between the contour lines or place your workbook below the contour map.)

b. Find the average slope of the terrain between the points:

   i. A and B.

   ii. C and D.

c. Find an estimate of the actual distance covered when travelling from:

   i. A to B.

   ii. C to E.

3. A cross-section of a geographical feature is given below.

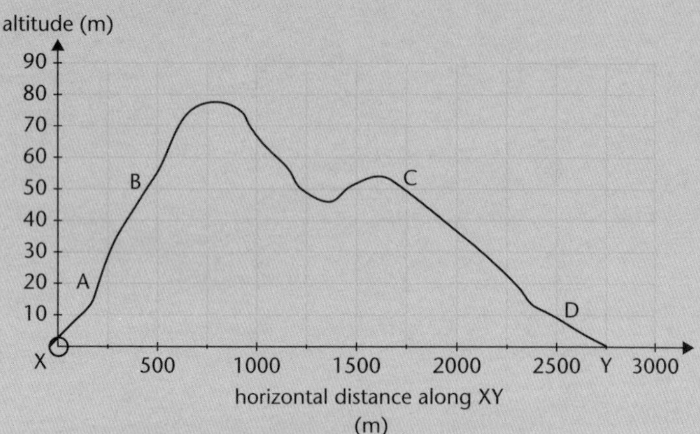

a. Draw a possible contour map for the cross-section shown above using a contour interval of:

i. 10 metres. **ii.** 20 metres.

b. Explain why it is preferable to use a contour interval of 10 metres for the contour map.

c. Find the average slope of the terrain between points:

i. A and B. **ii.** ii. C and D.

4.

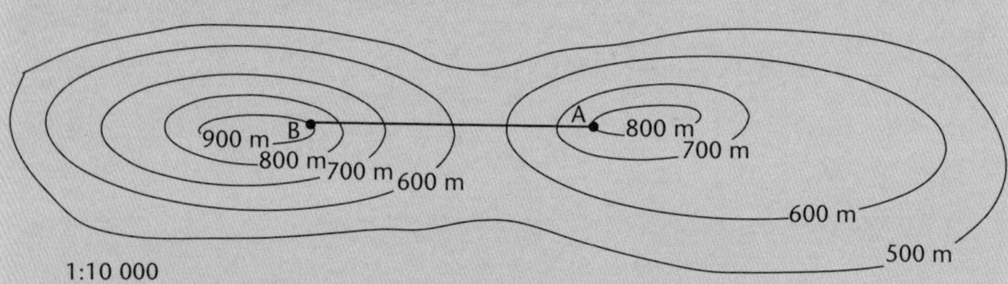

The ratio scale on the contour map above is 1 : 10 000 and the contour interval is 100 metres.

a. Construct a cross-section of the terrain along the axis AB.

b. If the line joining points A and B measures 4.5 centimetres on the contour map:

i. Find the average slope from point A to point B.

ii. Find the straight line distance from point A to point B.

Vectors

Vectors are quantities which have **magnitude** (size) *and* **direction**. Vectors are represented by directed line segments, as shown in the **vector diagram** alongside.

- The length of the line segment represents the magnitude of the vector.
- The arrow on the line segment shows the direction of the vector.

Examples of vector quantities include velocity, acceleration and force. Quantities such as time, volume and length are *not* vector quantities, because there is no direction associated with them. Such quantities are called **scalar** quantities.

Example E

Q. Draw a vector diagram representing a velocity of

1. 30 km h^{-1} south **2.** 45 km h^{-1} west

A. Using a scale of 1 cm = 10 km h^{-1}

1. The vector representing 30 km h^{-1} south is as shown alongside.
[length is 3 cm which represents 3 x 10 = 30 km h^{-1};
arrow indicates direction is due south]

2. The vector representing 45 km h^{-1} west is as shown below.
[length is 4.5 cm which represents 4.5 x 10 = 45 km h^{-1};
arrow indicates direction is due west]

Vector notation

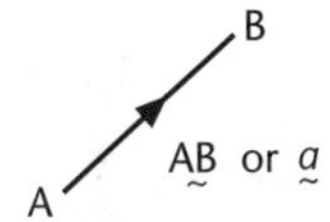

A vector joining two points A and B can be written $\underset{\sim}{AB}$, or it can be represented by a single lower case letter, such as $\underset{\sim}{a}$.

A vector is often written in **matrix** form $\begin{pmatrix} x \\ y \end{pmatrix}$.

- x is the size of the **horizontal component** (negative for left, or positive for right) of the vector and represents the shift parallel to the x-axis.
- y is the **vertical component** (positive for up, or negative for down) of the vector and represents the shift parallel to the y-axis.

Example F

1. $\underset{\sim}{AB} = \begin{pmatrix} 3 \\ 2 \end{pmatrix}$ is a vector beginning at A and ending at B.

B is 3 units to the *right* and 2 units *up* from A.

2. $\underset{\sim}{CD} = \begin{pmatrix} 3 \\ -2 \end{pmatrix}$ is a vector beginning at C and ending at D.

D is 3 units to the *right* and 2 units *down* from C.

3. $\underset{\sim}{EF} = \begin{pmatrix} -2 \\ 3 \end{pmatrix}$ is a vector beginning at E and ending at F.

F is 2 units to the *left* and 3 units *down* from E.

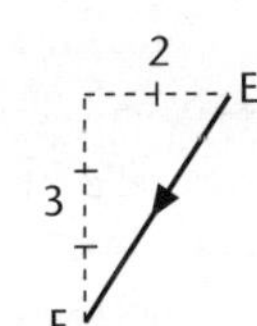

Equal and opposite vectors

Two vectors that have the same length and direction are said to be **equal vectors** regardless of their starting point.

The **opposite** of a vector is another vector of the *same* magnitude but *opposite* direction.

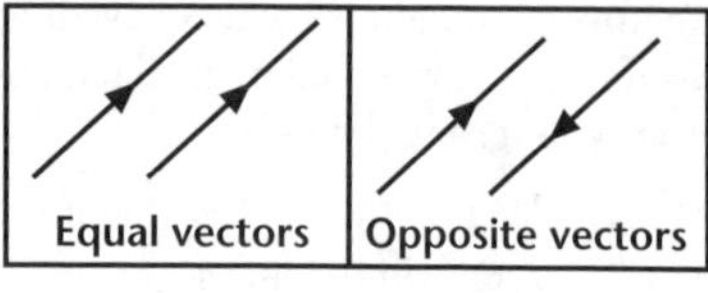

The magnitude of a vector

The **magnitude** of a vector is the length of the vector. The length of a vector can be found by using Pythagoras' theorem.

Example G

If $\underset{\sim}{a} = \begin{pmatrix} 3 \\ 2 \end{pmatrix}$, the magnitude of $\underset{\sim}{a}$ (symbol $|\underset{\sim}{a}|$), is given by:

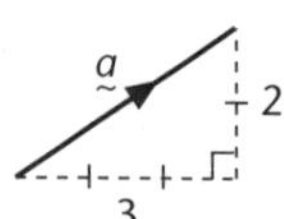

$|\underset{\sim}{a}| = \sqrt{3^2 + 2^2} = \sqrt{9 + 4} = \sqrt{13}$

$= 3.6$ units (to 1 dp)

The direction of a vector

The **direction** of a vector can be found using trigonometry.

Example H

If $\underset{\sim}{a} = \begin{pmatrix} 3 \\ 2 \end{pmatrix}$

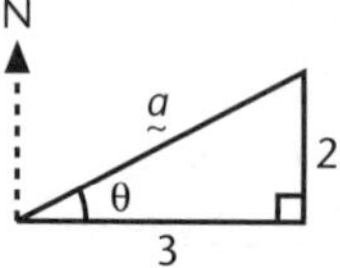

then $\tan\theta = \frac{2}{3}$

$\theta = \tan^{-1}\begin{pmatrix} 3 \\ 2 \end{pmatrix} = 33.7°$

$\underset{\sim}{a}$ is in the direction N56.3°E (or E33.7°N).

Multiplication of a vector by a scalar

Vectors can be multiplied by a **constant** (scalar). This is done by multiplying each component of the vector by the constant. The length of a vector changes, but the direction does not.

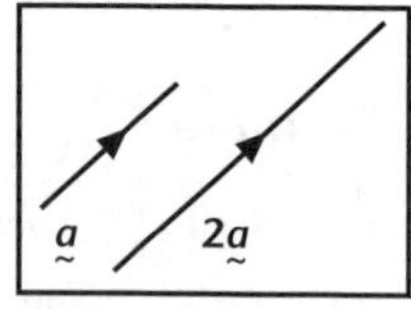

Addition of vectors

Vectors can be added together by using either a vector diagram, or by adding the matrices.

When adding two vectors together using a vector diagram.

- The 'tail' (beginning) of the second vector is joined to the 'head' (end) of the first vector.
- The resulting vector (called the **resultant**) is found by joining the 'tail' of the first vector to the 'head' of the second vector.

The resultant vector is often shown by a two-headed arrow.

Example I

If $\underset{\sim}{a} = \begin{pmatrix} 3 \\ 2 \end{pmatrix}$ and $\underset{\sim}{b} = \begin{pmatrix} 1 \\ -3 \end{pmatrix}$ then the vector diagram for the addition of $\underset{\sim}{a} + \underset{\sim}{b}$ is:

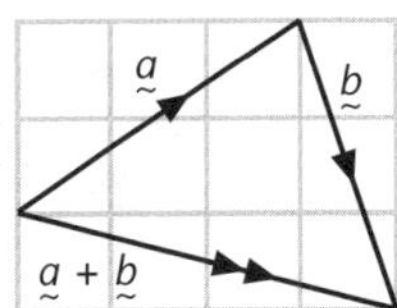

[vector $\underset{\sim}{b}$ is drawn starting at the end of vector $\underset{\sim}{a}$]

The resultant vector starts at the beginning of $\underset{\sim}{a}$ and finishes at the end of $\underset{\sim}{b}$.

From the diagram, $\underset{\sim}{a} + \underset{\sim}{b} = \begin{pmatrix} 4 \\ -1 \end{pmatrix}$

Note: A 'move' described by $\underset{\sim}{a}$ followed by a 'move' described by $\underset{\sim}{b}$ is the same as a single move described by the resultant $\underset{\sim}{a} + \underset{\sim}{b}$.

In the matrix method, the corresponding elements of the vectors are added together.

Example J

From Example I where $\underset{\sim}{a} = \begin{pmatrix} 3 \\ 2 \end{pmatrix}$ and $\underset{\sim}{b} = \begin{pmatrix} 1 \\ -3 \end{pmatrix}$:

$$\underset{\sim}{a} + \underset{\sim}{b} = \begin{pmatrix} 3 \\ 2 \end{pmatrix} + \begin{pmatrix} 1 \\ -3 \end{pmatrix}$$

$$= \begin{pmatrix} 3 + 1 \\ 2 + -3 \end{pmatrix} \quad \text{[adding horizontal and vertical components]}$$

$$= \begin{pmatrix} 4 \\ -1 \end{pmatrix}$$

More than two vectors can be added together, as the diagram shows.

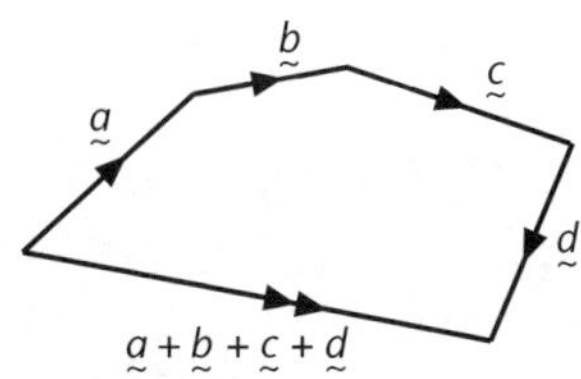

A resultant vector is always formed by joining the 'tail' of the first vector to the 'head' of the last vector.

The arrows on the resultant vector go in the direction of the *first vector to the last vector*.

Note: A vector is subtracted by 'adding the opposite vector'. Thus $\underset{\sim}{a} - \underset{\sim}{b}$ becomes $\underset{\sim}{a} + (-\underset{\sim}{b})$.

Components of vectors

Finding the horizontal shift and the vertical shift is called **resolving** a vector into its **components**. Resolving a vector into its components is done using trigonometry.

Example K

Q. Resolve the vector shown into its horizontal and vertical components.

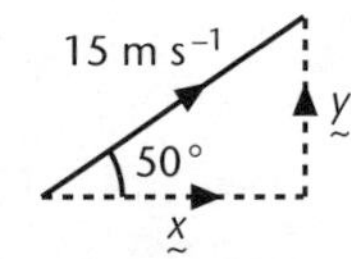

A. Let the magnitude of the horizontal component be x, and the magnitude of the vertical component be y.

Then, using trigonometry:

$\cos 50° = \frac{x}{15}$ $\qquad$ $\sin 50° = \frac{y}{15}$

$x = 15 \cos 50°$ $\qquad$ $y = 15 \sin 50°$

$= 9.6 \text{ m s}^{-1}$ (1 dp) $\qquad$ $= 11.5 \text{ m s}^{-1}$ (1 dp)

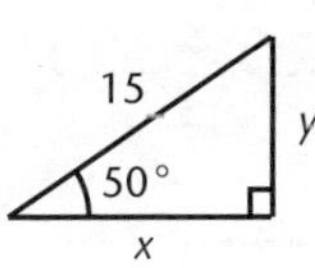

The horizontal component is 9.6 ms^{-1} and the vertical component is 11.5 ms^{-1} in the directions shown.

Vectors with negative components are treated in the same way.

Example L

Q. Resolve the vector shown into its horizontal and vertical components.

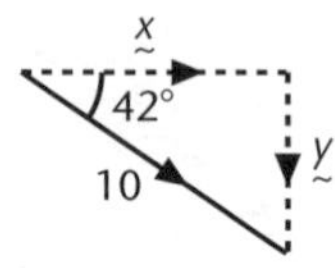

A. Let the magnitude of the horizontal component be x, and the magnitude of the vertical component be y.

Using trigonometry,

$\cos 42° = \frac{x}{10}$ $\sin 42° = \frac{y}{10}$

$x = 10 \cos 42°$ $y = 10 \sin 42°$

$= 7.4$ (2 sf) $= 6.7$ (2 sf)

The vector shown has a positive (right) horizontal component of 7.4 and a negative (down) vertical component of 6.7, ie it can be represented by the vector $\begin{pmatrix} 7.4 \\ -6.7 \end{pmatrix}$ to 2 sf.

Unit 11.5 Activity 5B: Vectors

1. Find the magnitude and direction of the following vectors (use compass directions, eg S14°E). Round answers to 1 dp where appropriate.

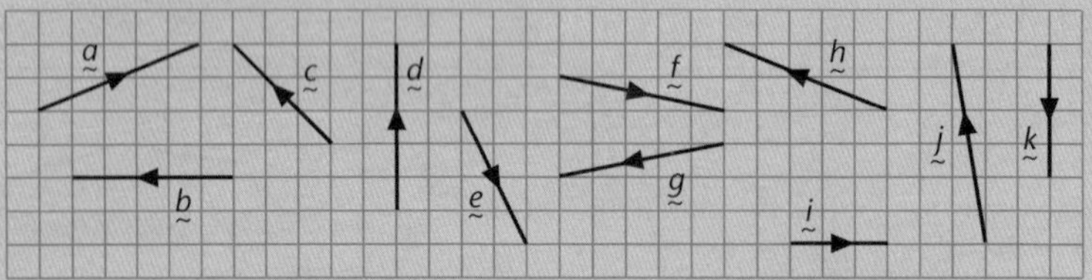

2. Draw each of the following vectors. Find the magnitude and direction of each vector (use compass directions eg N24.2°W). Round to 1 dp where appropriate.

a. $\begin{pmatrix} 2 \\ 1 \end{pmatrix}$ **b.** $\begin{pmatrix} 4 \\ 0 \end{pmatrix}$ **c.** $\begin{pmatrix} -2 \\ 3 \end{pmatrix}$ **d.** $\begin{pmatrix} -4 \\ -2 \end{pmatrix}$ **e.** $\begin{pmatrix} 5 \\ -3 \end{pmatrix}$

f. $\begin{pmatrix} 0 \\ 5 \end{pmatrix}$ **g.** $\begin{pmatrix} -3 \\ 0 \end{pmatrix}$ **h.** $\begin{pmatrix} 4 \\ 5 \end{pmatrix}$ **i.** $\begin{pmatrix} 0 \\ -4 \end{pmatrix}$ **j.** $\begin{pmatrix} -3 \\ 5 \end{pmatrix}$

3. Vector $\underset{\sim}{a}$ is 4 units long in the direction East. Vector $\underset{\sim}{b}$ is 3 units long in the direction North. Find the magnitude and direction of the vector

a. $2\underset{\sim}{a}$ **b.** $-\underset{\sim}{b}$ **c.** $\underset{\sim}{a} + \underset{\sim}{b}$

4. Resolve the following vectors into their horizontal and vertical components, $\underset{\sim}{x}$ and $\underset{\sim}{y}$. Round answers to 2 dp.

a.

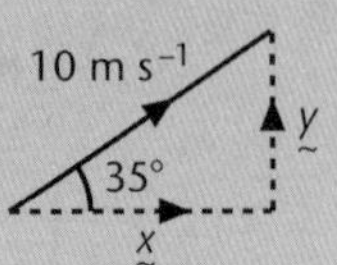

b.

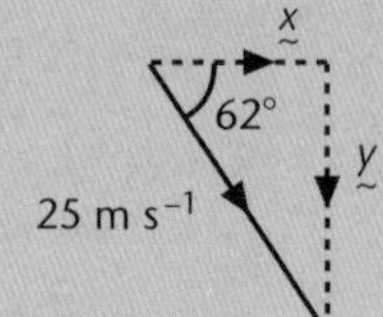

c.

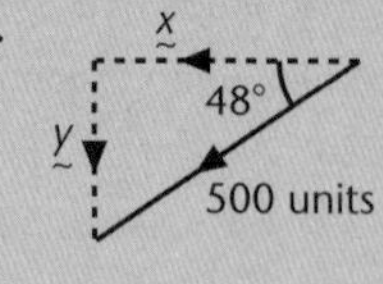

Applications of vectors

Vectors are used in navigation. The position of a ship or an aeroplane can be given as a bearing and a distance from a particular point. While this type of problem can also be solved by using a scale drawing, using trigonometry/Pythagoras' Theorem gives more accurate results.

Example L

Q. Mary is going to paddle her canoe across a river which is flowing at 3 m s^{-1}. She sets out from point A on one bank, heading directly across the river and paddling at 4 m s^{-1}.

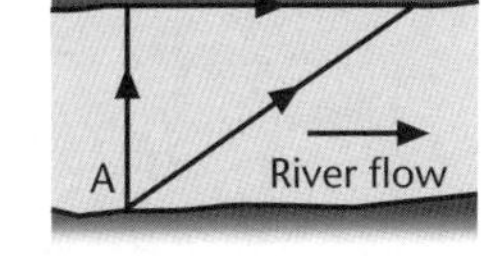

1. Label the vector diagram with the appropriate values.
2. Calculate Mary's speed with respect to the bank.
3. Calculate the direction Mary ends up travelling.

A. 1.

3 m s^{-1}

4 m s^{-1}

$\underset{\sim}{v}$

A

where $\underset{\sim}{v}$ is the resultant vector

2. Mary's resultant speed v is given by Pythagoras' theorem.

$$v^2 = 4^2 + 3^2$$
$$= 16 + 9$$
$$= 25$$
$$v = \sqrt{25}$$
$$= 5 \text{ m s}^{-1}$$

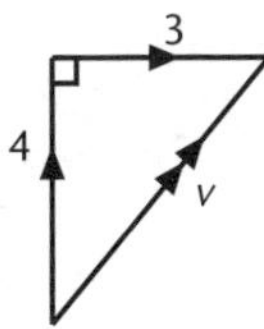

3. Mary's direction is given by $\tan \theta = \dfrac{3}{4}$

$\theta = 36.9°$ [taking $\tan^{-1}$]

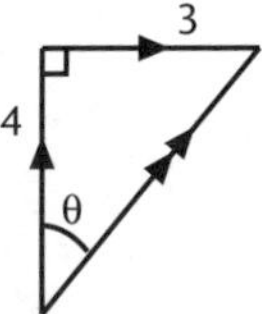

Mary is thus travelling at a speed of 5 m s^{-1} at 36.9° downstream, or at an angle of 90 – 36.9 = 53.1° from the bank.

Note:
1. Mary is travelling faster than the 4 m s^{-1} she is rowing. The 'extra' speed comes from the current.
2. The magnitude of velocity is speed.
3. Diagrams of vectors do not need to be drawn to scale when using trigonometry/Pythagoras' Theorem to determine the sizes of lengths or angles.

Unit 11.5 Activity 5C: Problem solving using vectors

1. A tractor pulls a log, as shown in the diagram.

The angle between the tow rope and the horizontal is 20°.

The pull of the rope is a force of 1 000 units.

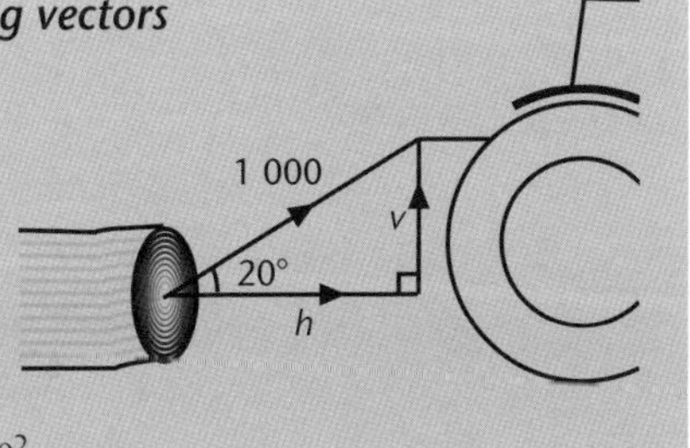

a. What is the horizontal component h of this force on the log?

b. What is the vertical component v of the force?

2. A plane leaves from town A and flies 400 km on a bearing of 210°. Calculate how far south of A the plane has travelled.

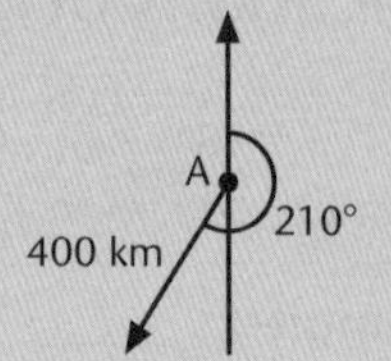
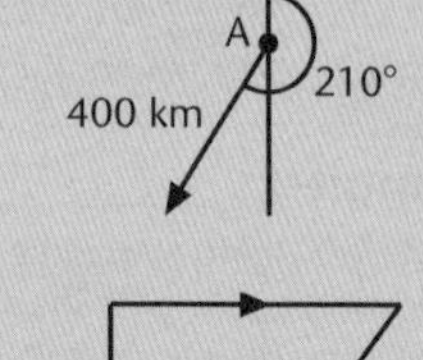

3. An aeroplane heads due north at an airspeed of 200 km h^{-1}. It is carried off course by a steady wind of 40 km h^{-1} blowing from the west.

a. Copy the vector diagram and label each vector.

b. Calculate the actual speed of the aircraft.

c. Calculate the direction in which the aircraft is actually heading. Give your answer as a bearing.

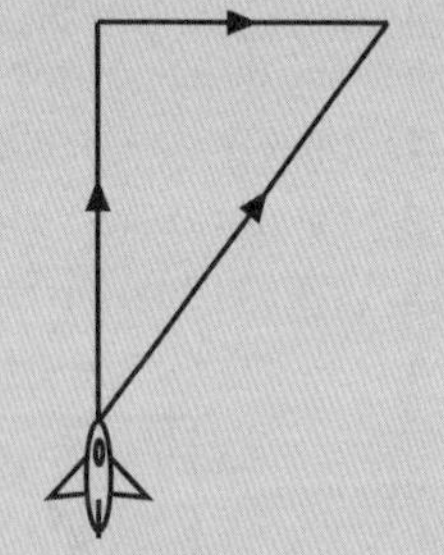

4. A tow truck pulls a car, as shown in the diagram. The angle between the tow rope and the horizontal is 75°. The pull of the rope is a force of 2 000 units.

What is the vertical component of this force?

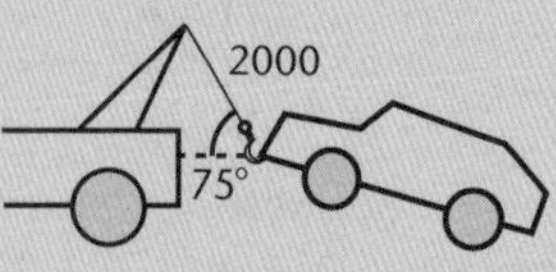

5. A ship leaves from A and sails 20 km on a course of 300°. Calculate how far west of A the ship has travelled.

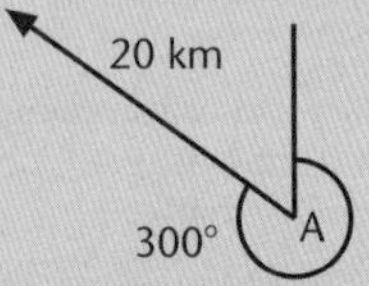

6. A fisherman travels at a speed of 25 km h^{-1} on a bearing of 140° from his mooring, He arrives at his fishing spot after 36 minutes. Later the fisherman travels from the fishing spot at the same speed on a bearing of 047° to an island due East of his mooring.

Draw a vector diagram for this information and use it to determine how long the fisherman's second journey took, and how far the island is from his mooring.

Appendix

Mathematical symbols

Sets

$b \in A$	b is an element of A
$c \notin A$	c is not an element of A
U or ε	the universal set
ϕ or $\{\ \}$	the empty or null set
A′	the complement of A
$A \subset B$	A is a subset of B
$A \not\subset B$	A is not a subset of B
$A \cup B$	the union of A and B, ie, the set containing elements in A or B (or in both)
$A \cap B$	the intersection of A and B, ie, the set containing elements in both A and B
$n(A)$	the number of elements in set A

Number sets

N	Natural numbers
W	Whole numbers
I	Integer numbers
Q	Rational numbers
Q′	Irrational numbers
R	Real numbers
∞	infinity
$-\infty$	negative infinity

Language

:	such that
…	and so on
$\therefore$	therefore
$=$	is equal to
$\neq$	is not equal to
$\approx$	is approximately equal to
$<$	is less than
$\leq$	is less than or equal to
$>$	is greater than
$\geq$	is greater than or equal to
$\Rightarrow$	implies
Σ	sum
$\sqrt{x}$	the positive square root of x
$\pm x$	plus or minus x
$\lvert x \rvert$	the absolute value of x

Geometry

$\angle$	angle
$\angle ABC$ or $A\hat{B}C$	the angle at B defined by the points A, B and C.
$\overline{AB}$	the line segment A to B.
//	parallel to
$\perp$	perpendicular to
Δ	triangle
$^\circ$	degree
π	Pi = 3.14159 (5 dp)

Statistics

$\bar{x}$	the mean of a sample
s	the standard deviation of a sample
μ	population mean
σ	population standard deviation

Mathematical formulae

Area

$\frac{1}{2}bh$	triangle	a^2	square (side a)
bh	rectangle or parallelogram	$\frac{1}{2}(a+b)h$	trapezium
$\frac{1}{2}d_1d_2$	rhombus or kite	πr^2	circle
$\frac{\theta}{360}\times\pi r^2$	sector (θ in degrees)	$4\pi r^2$	sphere (surface area)
$\frac{1}{2}r^2\theta$	sector (θ in radians)	$2\pi r(r+h)$	closed cylinder (surface area)

Volume

l^3	cube (side l)	Ah	prism (A = base area)
lbh	cuboid	$\frac{1}{3}Ah$	pyramid (A = base area)
$\frac{1}{2}bhl$	triangular prism	$\frac{4}{3}\pi r^3$	sphere
πr^2h	cylinder	$\frac{1}{3}\pi r^2h$	cone

Logarithms and indices

Logarithms

If $y = b^x$ then $\log_b y = x$

$\log_b x + \log_b y = \log_b xy$

$\log_b x - \log_b y = \log_b \frac{x}{y}$

$\log_b x^n = n\log_b x$

$\log_b x = \frac{\log_a x}{\log_a b}$

Indices

$a^m \times a^n = a^{m+n}$ $\quad$ $a^m \div a^n = a^{m-n}$

$(a^m)^n = a^{mn}$ $\quad$ $(ab)^m = a^mb^m$

$a^0 = 1$ (if $a \neq 0$) $\quad$ $\sqrt{a^m} = a^{\frac{m}{2}}$

$\frac{1}{a^m} = a^{-m}$ (if $a \neq 0$) $\quad$ $a^{\frac{m}{n}} = \sqrt[n]{a^m} = \left(\sqrt[n]{a}\right)^m$

Co-ordinate geometry

Distance

$d = \sqrt{(x_1 - x_2)^2 + (y_1 - y_2)^2}$

Mid-point of a line

mid-point $\left(\frac{x_1 + x_2}{2}, \frac{y_1 + y_2}{2}\right)$

Gradient

$m = \dfrac{y_2 - y_1}{x_2 - x_1}$

Lines

$y = mx + c$

$y - y_1 = m(x - x_1)$

Perpendicular and parallel lines

$y = m_1x + c_1$ and $y = m_2x + c_2$ are parallel if $m_1 = m_2$ and perpendicular if $m_1 \times m_2 = -1$

Circles

$x^2 + y^2 = r^2$ circle with centre (0, 0) and radius r

$(x - a)^2 + (y - b)^2 = r^2$ circle with centre (a, b) and radius r

Perimeter

$C = \pi d$ or $2\pi r$ circumference of a circle

$\dfrac{\theta}{360} \times 2\pi r$ arc length of a sector (θ in degrees)

$r\theta$ arc length of a sector (θ in radians)

Trigonometry

$\sin\theta = \dfrac{o}{h}$ o = opposite

$\cos\theta = \dfrac{a}{h}$ h = hypotenuse

$\tan\theta = \dfrac{o}{a}$ a = adjacent side

$h^2 = a^2 + o^2$ Pythagoras' theorem

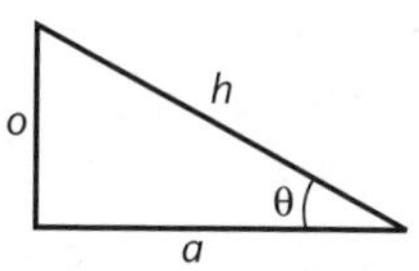

Special angles

θ	0°	30°	45°	60°	90°
$\sin\theta$	0	$\frac{1}{2}$	$\frac{1}{\sqrt{2}}$	$\frac{\sqrt{3}}{2}$	1
$\cos\theta$	1	$\frac{\sqrt{3}}{2}$	$\frac{1}{\sqrt{2}}$	$\frac{1}{2}$	0
$\tan\theta$	0	$\frac{1}{\sqrt{3}}$	1	$\sqrt{3}$	

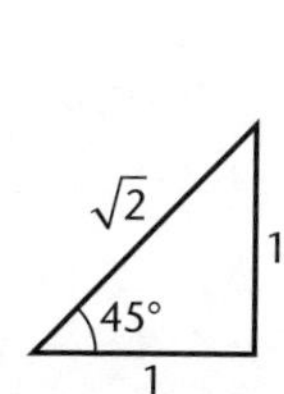

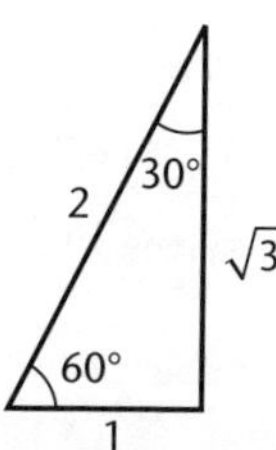

Trigonometric Identities

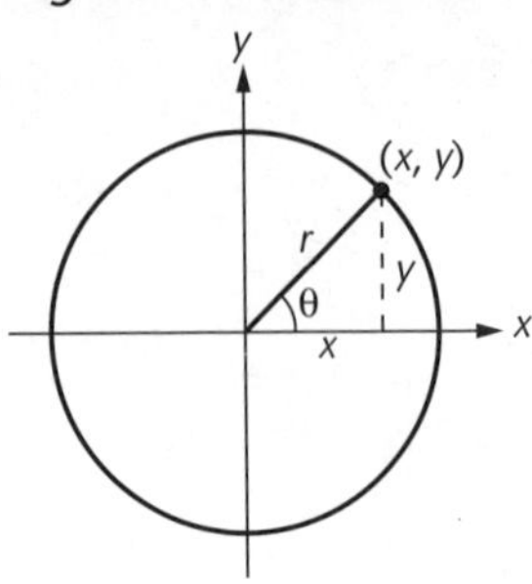

$\sin(90° \pm \theta) = \cos\theta$

$\cos(90° \pm \theta) = \mp\sin\theta$

$\sin(180° \pm \theta) = \mp\sin\theta$

$\cos(180° \pm \theta) = -\cos\theta$

$\sin(360° - \theta) = -\sin\theta$

$\cos(360° - \theta) = \cos\theta$

Triangles

Sine rule: $\dfrac{a}{\sin A} = \dfrac{b}{\sin B} = \dfrac{c}{\sin C}$ or $\dfrac{\sin A}{a} = \dfrac{\sin B}{b} = \dfrac{\sin C}{c}$

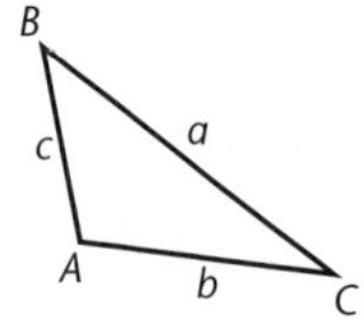

Cosine rule: $a^2 = b^2 + c^2 - 2bc\cos A$ or $\cos A = \dfrac{b^2 + c^2 - a^2}{2bc}$

Area of triangle: Area $= \dfrac{1}{2}ab\sin C$

Sequences and series

Arithmetic series

$a + (a + d) + (a + 2d) + (a + 3d) + \ldots$

$t_n = a + (n - 1)d$

$S_n = \dfrac{n}{2}(2a + (n - 1)d)$

Geometric series

$a + ar + ar^2 + ar^3 + \ldots$

$t_n = ar^{n-1}$

$S_n = \dfrac{a(1 - r^n)}{(1 - r)}, r \neq 1$

$S_\infty = \dfrac{a}{1 - r}$ for $|r| < 1$

Probability

$P(A) = 1 - P(A')$

$P(A \text{ or } B) = P(A \cup B) = P(A) + P(B) - P(A \cap B)$

A and B are independent events $\Leftrightarrow P(A \cap B) = P(A) \times P(B)$

Mean, variance and standard deviation

Statistic	Sample of Ungrouped Data	Samples of Grouped Data
Mean	$\overline{x} = \dfrac{\Sigma x_i}{n}$	$\overline{x} = \dfrac{\Sigma x_i f_i}{\Sigma f_i}$
Variance	$s^2 = \dfrac{\Sigma (x_i - \overline{x})^2}{n-1}$	$s^2 = \dfrac{\Sigma (x - \overline{x})^2 f_i}{\Sigma f_i - 1}$
Standard Deviation	$s = \sqrt{s^2}$	$s = \sqrt{s^2}$

Answers

Answers for many questions include a 'Marking Guide':
A. ('Achievement', meaning 'satisfactory achievement').
M. ('Merit', meaning 'high achievement').
E. ('Excellence', meaning 'very high achievement').
The Marking Guide has been made by the authors and the publishers and is not an official guide but we hope it will be a help to students who are striving for the best possible results.

Unit 11.1 Activity 1A: The real number system (page 4)

1. **a.** $\frac{15}{1}$ **b.** $\frac{-4}{1}$ **c.** $\frac{0}{1}$ **d.** $\frac{125}{10}$ **e.** $\frac{45}{1\,000}$

f. $\frac{19}{8}$ **g.** $\frac{23224}{1000}$ **h.** $\frac{-62}{9}$ **i.** $\frac{-5}{1\ 00000}$ **j.** $\frac{98}{25}$

2. **a.** $\frac{1}{3}$ **b.** $\frac{7}{9}$ **c.** $\frac{56}{99}$ **d.** $\frac{67}{99}$ **e.** $\frac{23}{9}$

f. $\frac{137}{11}$ **g.** $\frac{20920}{333}$

3. There exists a real number, 0, that for every real number, n, $n + 0 = n$

Example $5 + 0 = 5$

The additive identity is 0

4. There exists a real number, 1, that for every real number, n, $n \times 1 = n$

Example $7 \times 1 = 7$

The multiplicative identity is 1

5. The associative axiom of multiplication.

Unit 11.1 Activity 2A: Order of operations and integers (page 9)

Note: *These problems practise essential skills but only problems in context are given a Marking Guide.*

1. **a.** 10 **b.** 14 **c.** 32 **d.** 18 **e.** 10
f. 2 **g.** 28 **h.** 36 **i.** 9 **j.** 9
k. 4 **l.** 15 **m.** 6 **n.** 5 **o.** 2
p. 2 **q.** 21 **r.** 0 **s.** Undefined. **t.** Undefined.

2. K130 **(A)** **3.** 55 kg **(A)**

4. **a.** –4 **b.** 7 **c.** 12 **d.** –3 **e.** –20
f. 12 **g.** –24 **h.** 12 **i.** 0 **j.** 3
k. –7 **l.** –21 **m.** 5 **n.** –1 **o.** 12
p. 1 **q.** 25 **r.** –7 **s.** –3 **t.** 1

5. **a.** 2°C **(A)** **b.** –10°C **(A)** **c.** 12°C **(A)**

6. **a.** **i.** 7°C **(A)** **ii.** 7°C **(A)** **b.** 5°C **(A)** **c.** –2°C **(A)**

7. –K14.90, ie K14.90 overdrawn. **(A)**

8. **a.** 4 months **(E)** **b.** K418 **(E)**

Unit 11.1 Activity 2B: Powers and roots of numbers (page 13)

1. **a.** 25 **b.** 81 **c.** 625 **d.** 1 849 **e.** 6
f. 13 **g.** 56 **h.** 98 **i.** 18.601 **j.** 35.128
k. 29.155 **l.** 7.746 **m.** 3 **n.** 4 **o.** 5
p. 9 **q.** –11 **r.** –20 **s.** –24.495 **t.** 5

2. **a.** –36 **b.** 1 **c.** 36 **d.** –75 **e.** 7 **f.** 0

3. **a.** 11–12 **b.** 9–10 **c.** 5–6 **d.** 20–21

4. **a.** 37 cm **b.** 15.326 cm longer

5. **a.** 81 m^2 **b.** 9 m

6. **a.** 512 **b.** 6 561 **c.** 243 **d.** 4 096 **e.** 3
f. 11 **g.** 4 **h.** 2 **i.** 1.348 **j.** 1
k. 1.479 **l.** 7.990 **m.** 2 **n.** 5 **o.** 7
p. 4 **q.** –6 **r.** –6 **s.** Undefined. **t.** –1.369

7. **a.** 6 **b.** 81 **c.** 1.968 **d.** 1 **e.** 2 **f.** –1

8. **a.** 125 cm^3 **b.** 6.3 cm

9. **a.** 5.848 cm **b.** 7.368 cm **c.** $\frac{\text{long}}{\text{short}} = \frac{\sqrt[3]{400}}{\sqrt[3]{200}} = \sqrt[3]{2}$ so $k = 2$ **(E)**

10. 6 cm **(E)**

Unit 11.1 Activity 3A: Equivalent fractions and mixed numbers (page 17)

Note: *Only problems in context are given A, M, E ratings.*

1. **a.** $\frac{7}{15}$ **b.** $\frac{3}{8}$ **c.** $\frac{9}{16}$

2. **a.** $\frac{8}{17}$ **b.** $\frac{5}{17}$ **c.** $\frac{12}{17}$

3. **a.** $\frac{25}{50} = \frac{1}{2}$ **b.** $\frac{10}{50} = \frac{1}{5}$ **c.** $\frac{15}{50} = \frac{3}{10}$

4. **a.** $\frac{1}{2}$ **b.** $\frac{1}{4}$ **c.** $\frac{7}{15}$ **d.** $\frac{1}{8}$
e. $\frac{6}{11}$ **f.** $\frac{1}{2}$ **g.** $\frac{3}{4}$ **h.** $\frac{3}{5}$
i. $\frac{7}{10}$ **j.** $\frac{5}{8}$

5. **a.** $9\frac{1}{2}$ **b.** $9\frac{3}{5}$ **c.** $13\frac{1}{6}$ **d.** $8\frac{7}{10}$
e. $8\frac{1}{3}$ **f.** $5\frac{1}{2}$ **g.** $5\frac{2}{3}$ **h.** $5\frac{3}{5}$
i. $7\frac{5}{6}$ **j.** $9\frac{1}{2}$

6. a. $\frac{5}{2}$ b. $\frac{14}{3}$ c. $\frac{23}{4}$ d. $\frac{47}{8}$
e. $\frac{247}{20}$ f. $\frac{7}{3}$ g. $\frac{19}{5}$ h. $\frac{51}{20}$
i. $\frac{93}{8}$ j. $\frac{191}{15}$

7. a. $\frac{2}{7}, \frac{1}{3}, \frac{3}{8}$ b. $\frac{2}{15}, \frac{3}{20}, \frac{1}{6}$ c. $\frac{7}{8}, 1\frac{1}{5}, \frac{4}{3}$ d. $\frac{12}{5}, \frac{8}{3}, 2\frac{3}{4}$

8. a. Bob ***(A)*** b. Anne ***(A)***

9. 60 ***(A)***

10. a. $\frac{1}{3}$ ***(A)*** b. $\frac{4}{5}$ ***(A)*** c. $\frac{5}{8}$ ***(A)***

Unit 11.1 Activity 3B: Operations with fractions (page 22)

Note: *Only problems in context are given A, M, E ratings.*

1. a. $\frac{4}{5}$ b. $\frac{7}{10}$ c. $\frac{6}{8} = \frac{3}{4}$ d. $\frac{5}{6}$
e. $\frac{13}{15}$ f. $\frac{41}{40} = 1\frac{1}{40}$ g. $\frac{9}{20}$ h. $\frac{17}{30}$
i. $\frac{31}{20} = 1\frac{11}{20}$ j. $1\frac{2}{5}$ k. $3\frac{3}{4}$ l. $4\frac{7}{12}$

2. a. $\frac{3}{8}$ b. $\frac{3}{10}$ c. $\frac{7}{12}$ d. $\frac{1}{6}$
e. $\frac{5}{12}$ f. $\frac{13}{72}$ g. $\frac{1}{20}$ h. $\frac{2}{48} = \frac{1}{24}$
i. $\frac{4}{96} = \frac{1}{24}$ j. $\frac{2}{3}$ k. $1\frac{2}{15}$ l. $1\frac{7}{8}$

3. a. $\frac{1}{6}$ b. $\frac{1}{6}$ c. $\frac{7}{30}$ d. $\frac{1}{20}$
e. $\frac{2}{3}$ f. $\frac{10}{27}$ g. $\frac{32}{3} = 10\frac{2}{3}$ h. $\frac{247}{10} = 24\frac{7}{10}$
i. $\frac{49}{4} = 12\frac{1}{4}$ j. 17 k. $\frac{20}{3} = 6\frac{2}{3}$ l. $\frac{172}{3} = 57\frac{1}{3}$

4. a. $\frac{3}{2} = 1\frac{1}{2}$ b. $\frac{3}{2} = 1\frac{1}{2}$ c. $\frac{15}{16}$ d. $\frac{5}{4} = 1\frac{1}{4}$
e. $\frac{11}{15}$ f. $2\frac{13}{18}$ g. 2 h. $\frac{1}{5}$
i. $\frac{5}{4} = 1\frac{1}{4}$ j. $\frac{2}{9}$ k. $\frac{26}{35}$ l. $\frac{98}{75} = 1\frac{23}{75}$

5. a. $\frac{4}{9}$ b. $\frac{125}{729}$ c. 16 d. $22\frac{1}{2}$
e. $2\frac{1}{4}$ f. $5\frac{4}{9}$ g. $54\frac{109}{125}$ h. $7\frac{58}{81}$

6. **a.** $14\frac{1}{16}$ m^2 **b.** $11\frac{25}{64}$ cm^2

7. 40 **(A)** **8.** $\frac{1}{15}$ **(A)**

9. $\frac{17}{20}$ **(A)** **10.** $\frac{3}{8}$ **(A)**

11. 16 **(A)** **12.** $1\frac{1}{6}$ cups **(A)**

Unit 11.1 Activity 3C: Problem solving involving fractions (page 25)

Note: *Only problems in context are given A, M, E ratings.*

1. **a.** $\frac{9}{14}$ **b.** $-16\frac{7}{10}$ **c.** $-12\frac{19}{40}$ **d.** $\frac{1}{7}$ **e.** $\frac{3}{25}$ **f.** $3\frac{3}{8}$ **g.** $5\frac{1}{16}$

h. $\frac{9}{22}$ **i.** $4\frac{3}{4}$ **j.** $1\frac{329}{400}$ **k.** $1\frac{5}{8}$ **l.** $\frac{3}{4}$

2. 3 kg **(A)** **3.** $\frac{1}{4}$ **(A)** **4.** $\frac{9}{16}$ **(A)** **5.** $\frac{11}{15}$ **(A)** **6.** $\frac{1}{6}$ **(A)**

7. $\frac{21}{24} = \frac{7}{8}$ **(A)** **8.** K800 **(M)** **9.** 2 000 L **(M)** **10.** 45 **(M)**

Unit 11.1 Activity 3D: Multiple choice – fractions (page 26)

1. D **2.** B **3.** B **4.** A **5.** D

6. C **7.** A **8.** C **9.** B

Unit 11.1 Activity 4A: Calculations involving decimals (page 29)

Note: *Only problems in context are given A, M, E ratings.*

1. **a.** 0.8 **b.** 3.76 **c.** $0.8\dot{3}$ **d.** $18.58\dot{3}$ **e.** $0.\dot{1}\dot{3}$

f. $\frac{273}{500}$ **g.** $3\frac{1}{50}$ **h.** $4\frac{1}{3}$ **i.** $\frac{4}{9}$ **j.** $-11\frac{1}{1\,000}$

2. **a.** 15.2 **b.** 1.82 **c.** 1.88 **d.** 0.02

e. 4 **f.** 9.8 **g.** 30.29 **h.** 0.624

3. **a.** 54 **b.** 70 **c.** 420 **d.** 0.238

e. 57 630 **f.** 0.002958 **g.** 13 581

4. 10.5 km **(A)** **5.** K6.95 **(A)** **6.** 3.52 **(A)** **7.** Short answer. **(A)**

8. **a.** 3.2 m **(A)** **b.** Less. **(A)** **9.** 77 **(A)** **10.** K55.27 **(A)**

Unit 11.1 Activity 4B: Significant figures and rounding (page 33)

Note: *Only problems in context are given A, M, E ratings. Approximate answers will vary.*

1. **a.** 120 **b.** 5 910 **c.** 31 800 **d.** 14.7 **e.** 89.9 **f.** 0.049
2. **a.** 1.54 **b.** 11.7 **c.** 0.414 **d.** 89.86 **e.** 0.0041 **f.** 10.0
3. **a.** 13 500 **b.** 32 000 **c.** 280 **d.** 30 **e.** 20 **f.** 45 or 50
   g. 200 **h.** 1 100 **i.** 4 000
4. **a.** 720 **b.** 665.25 **(A)**
5. **a.** The radius was measured to 2 significant figures and the area should then be given to 2 significant figures. 8 significant figures implies too much accuracy. **(A)**
   b. 120.8 cm^2 **(A)**
6. 10 (9 packets would not be enough since she needs at least 184 tiles). **(A)**
7. **a.** 80 cm^3 **b.** 87.1 cm^3 **(A)**
8. **a.** 20 m^2 **b.** 18.5 m^2 (1 dp) **(A)**

Unit 11.1 Activity 4C: Standard form (page 37)

Note: *Only problems in context are given A, M, E ratings.*

1. **a.** 4.87×10^2 **b.** 9.12×10^3 **c.** 7.63×10^5 **d.** 7.43×10^{-2}
   e. 5.18×10^{-3} **f.** 4.51×10^{-6} **g.** 6.95×10^7 **h.** 4.958×10^2
   i. 1.6401×10^1 **j.** 1.265×10^4 **k.** 3.2×10 **l.** 4.859×10^{-2}
2. **a.** 251 **b.** 6 390 **c.** 84 120 **d.** 0.357
   e. 0.0017 **f.** 0.000975 **g.** 34.8 **h.** 46 000
   i. 0.0783 **j.** 18 450 **k.** 1.264 **l.** 0.00057
3. **a.** 3.65×10^2, 3.7×10^2, 2.4×10^3, 1.901×10^4
   b. 0.0014, 0.002, 4.1×10^{-2}, 2.6×10^{-1}
   c. 9.9×10^3, 12.6×10^3, 12 965, 1.3×10^4
4. **a.** 4.8×10^3 **b.** 5.665×10^{-3} **c.** 9.0×10^4
   d. 7.8×10^{-2} **e.** 4×10^5 **f.** 1.0×10^{-2}
5. **a.** 2.7×10^6, 2.21×10^6 **b.** 6.5×10^5, 7.01×10^5
   c. 4.9×10^9, 4.79×10^9 **d.** 700, 7.92×10^2
6. **a.** 4.07×10^6 **b.** 3.02×10^6 **(M)** **c.** 8.80×10^5 **(M)**
7. 1 837 **(M)**
8. 1.415×10^9 km **(M)**

9. a. 1.286×10^9. Answer to 3 sf would be 1.29×10^6, the original figure for the population of China. **(M)**

b. 327 **(M)**

10. a. Sweets by 1.40×10^7 **(M)** **b.** 3.52×10^5 **(M)** **c.** 1.61×10^5 **(M)**

Unit 11.1 Activity 4D: Multiple choice – decimals (page 38)

1. C **2.** B **3.** C **4.** D **5.** A

Unit 11.1 Activity 5A: The numbers π and e (page 40)

1. a. 3.140845 … correct to 3 d.p.

b. 3.142857 … correct to 2 d.p.

c. 3.14626 … correct to 2 d.p.

d. 3.141592920 … correct to 6 d.p.

e. 3.01707 … correct to 0 d.p. (including $-\frac{1}{31}$ correct to 1 d.p.)

f. 3.04936 … correct to 0 d.p.(including $\frac{1}{20^2}$; correct to 1 d.p.)

2. a. 2.593742 … correct to 0 d.p.

b. 2.704813 … correct to 1 d.p.

c. 2.716923 … correct to 2 d.p.

d. 2.716666 … 2 d.p.

e. Including $\frac{1}{7!}$: 2.7182539· ...correct to 4 d.p.

Unit 11.1 Activity 5B: Surds (page 41)

1. $\sqrt{101}, \sqrt{17}, \sqrt{8}, \sqrt{83.4}$

2. a. 8, 0.5, 4.33, 20, 63.25, –0.6, –0.6, 0.04, –0.2, 500

b. $\sqrt{64}, \sqrt{0.25}, \sqrt{400}, \sqrt{-0.36}, \sqrt{0.0016}$

3. a. Q **b.** Q' **c.** Q **d.** Q'

e. Q **f.** Q' **g.** Q **h.** Q

4. a. $5\sqrt{2}$ **b.** $4\sqrt{3}$ **c.** $7\sqrt{2}$ **d.** $8\sqrt{5}$

e. $15\sqrt{3}$ **f.** $6\sqrt{3}$ **g.** 18 **h.** $2\sqrt{2}$

5. a. $\sqrt{75}$ **b.** $-\sqrt{28}$ **c.** $\sqrt{1000}$ **d.** $\sqrt{150}$

e. $-\sqrt{45}$ **f.** $\sqrt{32}$ **g.** $\sqrt{44}$ **h.** $\sqrt{0.56}$

Unit 11.1 Activity 5C: Simplifying surds (page 42)

1. **a.** $10\sqrt{7}$ **b.** $2\sqrt{3}+3\sqrt{2}$ **c.** $\sqrt{6}+2\sqrt{3}+3$ **d.** Already simplified
2. **a.** $\sqrt{6}$ **b.** $13\sqrt{5}$ **c.** $-2\sqrt{3}$ **d.** $12\sqrt{3}$
   e. $\sqrt{3}$ **f.** $\sqrt{5}$ **g.** $11\sqrt{2}+3\sqrt{11}$ **h.** $6\sqrt{2}$
3. **a.** $2\sqrt{3}+2\sqrt{3}-1$ **b.** $3\sqrt{5}-5\sqrt{3}$ **c.** $16\sqrt{3}-6\sqrt{6}-8$
   d. $12\sqrt{2}-\sqrt{6}-8\sqrt{3}$ **e.** $14+14\sqrt{2}$ **f.** $13\sqrt{3}-39$

Unit 11.1 Activity 5D: Simplifying and expanding surds (page 44)

1. **a.** 5 **b.** $5\sqrt{3}$ **c.** $3\sqrt{6}$ **d.** $15\sqrt{2}$
   e. $6\sqrt{6}$ **f.** $30\sqrt{2}$
2. **a.** $9+4\sqrt{5}$ **b.** $30-12\sqrt{6}$ **c.** $49-6\sqrt{10}$
   d. $41+24\sqrt{2}$ **e.** $219-120\sqrt{3}$
3. **a.** $6-\sqrt{3}$ **b.** $10+2\sqrt{21}$ **c.** $\sqrt{6}$ **d.** $6+2\sqrt{3}-2\sqrt{6}-2\sqrt{2}$
   e. $8-15\sqrt{6}$ **f.** $15\sqrt{3}-9$
4. **a.** $2\sqrt{3}+4$ **b.** $2-2\sqrt{6}$ **c.** $5+3\sqrt{5}$ **d.** $5\sqrt{3}-8$
5. **a.** −1 **b.** 3 **c.** 4 **d.** 13
   e. 23 **f.** 13 **g.** 94

Unit 11.1 Activity 5E: Simplest form (page 46)

1. **a.** $\frac{1}{5}$ **b.** $\frac{4}{3}$ **c.** $\frac{6}{5}$ **d.** $\sqrt{7}$ **e.** $\frac{\sqrt{6}}{2}$
2. **a.** $\frac{\sqrt{5}}{5}$ **b.** $\frac{\sqrt{2}}{4}$ **c.** $\sqrt{3}$ **d.** $\frac{\sqrt{6}}{12}$ **e.** $\frac{5\sqrt{2}+4}{2}$
3. **a.** $\sqrt{3}+\sqrt{2}$ **b.** $5-2\sqrt{5}$ **c.** $3+\sqrt{3}$ **d.** $\frac{11-4\sqrt{7}}{3}$ **e.** $\frac{2\sqrt{5}+20}{19}$
   f. $\frac{29-6\sqrt{6}}{7}$

4. a. $17+12\sqrt{2}$ b. $\frac{12+6\sqrt{2}+2\sqrt{3}-\sqrt{6}}{11}$ c. $\frac{10-15\sqrt{2}-4\sqrt{3}+6\sqrt{6}}{-14}$

d. $\frac{61+7\sqrt{5}}{44}$ e. $\frac{2+\sqrt{2}}{2}$ f. $\frac{41+4\sqrt{10}}{39}$ g. $\frac{\sqrt{15}-3}{2}$

h. $\frac{8\sqrt{21}-28}{5}$ i. $\frac{2\sqrt{11}+2\sqrt{6}+\sqrt{66}+6}{5}$ j. $\frac{57+\sqrt{15}}{42}$

Unit 11.1 Activity 6A: Negative indices (page 49)

1. a. 0.0625 b. 0.4019 c. 7.3046 d. 5.0027 e. 4.1890 f. 9
   g. 27.648 h. 0.0865 i. 1.06 j. −1.2022 k. 0.2712

2. a. $\frac{1}{9}$ b. $\frac{1}{64}$ c. $\frac{1}{16}$ d. $\frac{27}{64}$ e. $\frac{8}{25}$ f. 9
   g. 16 h. $\frac{25}{9}$ i. $\frac{125}{64}$ j. $\frac{27}{4}$ k. $\frac{25}{27}$ l. $\frac{1}{24}$
   m. $\frac{1}{10}$ n. $\frac{7}{5}$ o. $\frac{8}{3}$ p. $\frac{1}{6}$ q. $\frac{1}{15}$ r. $\frac{1}{12}$
   s. 10 t. 12 u. $\frac{1}{36}$ v. $\frac{1}{175}$ w. 3 x. 5
   y. $\frac{1}{14}$ z. $\frac{4}{3}$

3. a. $\frac{3}{2}$ b. $\frac{1}{36}$ c. 144 d. $\frac{1}{576}$ e. $\frac{5}{6}$ f. $\frac{5}{18}$
   g. $\frac{33}{10}$ h. $\frac{1}{6}$ i. $\frac{1}{2}$ j. $\frac{17}{12}$ k. $-\frac{73}{18}$ l. $\frac{6}{5}$
   m. $\frac{5}{36}$ n. $\frac{11}{16}$ o. $\frac{11}{4}$ p. $\frac{4}{5}$

4. a. $\frac{1}{2}$ b. $\frac{1}{6}$ c. 2 d. $\frac{5}{9}$ e. $\frac{3}{2}$ f. $\frac{2}{3}$
   g. 9 h. $\frac{3}{2}$ i. $1\frac{3}{5}$

5. a. 3×2^2 b. $2^3\times 5^2$ c. 5×3^2 d. $5^2\times 2^{-4}$ e. 2×3^{-1} f. 22×3^{-3}
   g. $2^4\times 3^{-4}$ h. $2^6\times 3^{-4}$ i. $3^2\times 2^{-2}$

6. a. 2^4 b. 5^3 c. 3^4 d. 2^{-4} e. 2^6
   f. 5^{-4} g. 3^{-2} h. 3^{-5}

7. a. 25 b. − 25 c. $\frac{1}{25}$ d. $\frac{5}{6}$ e. $\frac{1}{36}$
   f. $-\frac{6}{7}$ g. − 1

8. **a.** 2^{C-2} **b.** 2^{2m} **c.** 2^{3+t} **d.** 2^{2x+2} **e.** 2^{n-1} **f.** 2^1
g. 2^{1+x} **h.** 3^{2+x} **i.** 3^{y-3} **j.** 3^{5a} **k.** 3^3 **l.** 3^{3a-1}

Unit 11.1 Activity 6B: Simplifying expressions with indices (page 51)

1. **a.** $\frac{1}{x^3}$ **b.** $\frac{1}{y^2}$ **c.** $\frac{1}{a^5}$ **d.** $\frac{1}{b^4}$ **e.** $\frac{1}{z^7}$ **f.** $\frac{1}{y^6}$
g. $\frac{1}{z^{11}}$ **h.** $\frac{1}{xy}$ **i.** $\frac{1}{abc}$ **j.** $\frac{1}{x^2y^2}$ **k.** $\frac{1}{2x}$ **l.** $\frac{1}{4x^2}$
m. $\frac{1}{9a^2}$ **n.** $\frac{1}{8x^3}$ **o.** $\frac{3}{x}$ **p.** $\frac{a}{b^3}$ **q.** $\frac{b^2}{a^3}$ **r.** $\frac{1}{a}$
s. $\frac{1}{2x}$ **t.** $\frac{1}{4x^3}$ **u.** $\frac{3}{5x^2}$ **v.** $\frac{2}{7a^3}$ **w.** $\frac{4}{9a^2}$ **x.** $\frac{3}{5x^4}$
y. $\frac{4}{5x^5}$ **(A)**

2. **a.** x^{-6} **b.** y^{-5} **c.** z^{-2} **d.** $4x^{-1}$ **e.** $5y^{-2}$ **f.** $7z^{-5}$
g. $\frac{x^{-2}}{2}$ **h.** $\frac{a^{-4}}{3}$ **i.** $\frac{y^{-4}}{5}$ **j.** $\frac{2x^{-2}}{5}$ **k.** $\frac{3y^{-3}}{4}$ **l.** $\frac{3x^{-2}}{4}$
m. $\frac{4y^{-2}}{25}$ **n.** $\frac{x^{-1}}{8}$ **o.** $\frac{x^{-3}}{8y^{-2}}$ **(A)**

3. **a.** $\frac{y^3}{2x^2}$ **b.** $\frac{2b^4}{3a^3}$ **c.** $\frac{x}{5}$ **d.** $\frac{2y^4}{x^2}$ **e.** $\frac{x^5}{y}$ **f.** $\frac{2y^3}{x}$
g. $\frac{8}{x^7y^2}$ **h.** $\frac{1}{25x^3}$ **i.** $\frac{1}{6xy}$ **j.** $\frac{1}{27x^4}$ **k.** $\frac{x}{2y^2}$ **l.** $\frac{x^2}{9y^3}$
m. $\frac{2}{3x^2}$ **n.** $\frac{4y^2}{5x^3}$ **o.** $\frac{x+y}{xy}$ **p.** $\frac{x+1}{x^2}$ **q.** $\frac{xy^2+2}{2y}$ **(A)**

Unit 11.1 Activity 6C: Fractional indices (page 53)

1. **a.** $a^{\frac{1}{2}}$ **b.** $c^{\frac{1}{2}}$ **c.** $x^{\frac{1}{3}}$ **d.** $a^{\frac{1}{5}}$ **e.** $u^{\frac{1}{6}}$ **f.** $x^{\frac{2}{5}}$
g. $x^{\frac{3}{7}}$ **h.** $a^{\frac{3}{8}}$ **i.** $x^{-\frac{1}{2}}$ **j.** $y^{-\frac{1}{2}}$ **k.** $a^{-\frac{1}{2}}$ **l.** $x^{-\frac{1}{5}}$
m. $x^{-\frac{2}{7}}$ **n.** $x^{-\frac{5}{9}}$ **o.** $a^{-\frac{3}{7}}$ **p.** $a^{-\frac{8}{9}}$ **q.** $4a^{\frac{1}{3}}$ **r.** $6y^{-\frac{1}{3}}$
s. $8x^{-\frac{7}{3}}$ **t.** $\frac{x^{-\frac{2}{3}}}{7}$ **u.** $\frac{2x^{-\frac{3}{4}}}{5}$ **v.** x^{-1} **w.** $x^{\frac{1}{6}}$ **x.** $a^{-\frac{5}{3}}$
y. $x^{-\frac{1}{12}}$ **(A)**

2. a. $\sqrt[7]{a^3}$ b. $\sqrt[4]{b^3}$ c. $\sqrt[5]{x^2}$ d. $\frac{1}{\sqrt[3]{x^2}}$ e. $\frac{1}{\sqrt[4]{y^3}}$ **(A)**

3. a. $(xy)^{\frac{1}{2}}$ or $(x)^{\frac{1}{2}}(y)^{\frac{1}{2}}$ b. $(x)^{\frac{1}{2}}y$ c. $x(y)^{-\frac{1}{2}}$ d. $x^2(y)^{-\frac{1}{3}}$ e. $2x(y)^{-\frac{3}{5}}$

f. $(x)^{-\frac{5}{4}}y$ g. $(x)^{\frac{1}{2}}(a)^{\frac{1}{3}}$ h. $(a)^{-\frac{1}{2}}(c)^{\frac{1}{2}}$ i. b **(A)**

4. a. $\frac{1}{8x^3}$ b. $\frac{1}{9x^6}$ c. $\frac{1}{4y^4}$ d. $27A^3$ e. $\frac{1}{32B^{10}}$

f. $\frac{1}{4x^4}$ g. $4A^2B^2$ h. $\frac{1}{25A^2B^4}$ i. $3x^{\frac{3}{2}}$ j. $256x^{10}$ **(A)**

Unit 11.1 Activity 6D: Calculating with indices (page 55)

1. 16 **2.** 0.11 **3.** 82.19 **4.** –32 **5.** 2.25 **6.** 1.32
7. –2.11 **8.** No sol. **9.** 3.51 **10.** 1720.51 **11.** 2.73 **12.** 1.83
13. 74.30 **14.** –0.5109 **15.** 1.25 **16.** 2.1715 **17.** 14.64 **18.** 0.0833
19. No sol. **20.** 0.0638

Unit 11.1 Activity 6E: Equations with indices (page 57)

1. a. 1.26 **b.** 8 **c.** 81 **d.** 25 **e.** $\frac{32}{243}$ **f.** 0.2 or $\frac{1}{5}$

g. $\frac{1}{4}$ or 0.25 **h.** 81 **i.** 32 **j.** 8 **k.** $\frac{1}{6}$ **l.** $\frac{1}{6}$

m. 4 **n.** 8 **o.** 5 **p.** 27 **q.** $\frac{1}{27}$ **r.** $\frac{1}{4}$

s. $\frac{1}{9}$ **t.** $\frac{1}{243}$ **(A)**

2. a. 64 **b.** 2 381 **c.** 0.20 **d.** 3.04 **e.** No Sol. **f.** –10.08

g. 6.55 **h.** 270.8 **i.** 1.67 **(A)**

3. a. $x = 3$ **b.** $x = -\frac{1}{2}$ **c.** $x = \frac{5}{2}$ **d.** $x = 0$ **e.** No solution

f. $x = 2$ **g.** $x = 5$ **h.** $x = -3$ **i.** $x = 3$ **j.** $x = 1\frac{1}{2}$

k. $x = -3$ **l.** $x = \frac{1}{2}$ **m.** $x = -\frac{1}{2}$ **n.** $x = -\frac{1}{4}$

4. a. $x = 3$ **b.** $x = 3$ **c.** $x = 2$ **d.** $x = -2$ **e.** $x = 2$ **f.** $x = -2$

5. a. K51 **b.** 81.51 m **(A)**

6. a. 14 336 **b.** 4 **c.** 1.8 **(M)**

7. a. *T* gets smaller **b.** 0.01094 **(M)**

Unit 11.1 Activity 7A: Other logarithmic functions (page 59)

1. 2.10 **2.** 199.53 **3.** 2 508.52 **4.** 0.04 **5.** 3.68 **6.** 2.5

Unit 11.1 Activity 7B: Using logarithmic properties (page 63)

1. **a.** $\log_2 8 = 3$ **b.** $\log_3 81 = 4$ **c.** $\log_5 25 = 2$ **d.** $\log_6 216 = 3$ **e.** $\log_x b = a$

2. **a.** $3^2 = 9$ **b.** $(9)^{\frac{1}{2}} = 3$ **c.** $25^{1.5} = 125$ **d.** $(1000)^{\frac{1}{3}} = 10$ **e.** $3^{-1} = \frac{1}{3}$

3. **a.** 6 **b.** –1 **c.** 6 **d.** 7 **e.** –1 **f.** –3
g. $\frac{1}{2}$ **h.** 3 **i.** $\frac{1}{2}$ **j.** $\frac{1}{4}$ **k.** $\frac{1}{2}$ **l.** $\frac{3}{2}$
m. 2 **n.** $\frac{5}{6}$ **o.** $-\frac{1}{2}$ **p.** 1 **q.** –1 **r.** –3
s. $\frac{5}{3}$ **t.** $\frac{3}{4}$ **(A)**

4. **a.** 0.954 **b.** –2.117 **c.** 2.510 **d.** 2.885 **e.** 0.972 **f.** 1.262
g. 1.390 **h.** 1.710 **i.** 2.767 **j.** 2.484 **k.** 31 **(A)**

5. **a.** 1 **b.** 1 **c.** 3 **d.** $\frac{1}{2}$ **e.** $\frac{1}{4}$ **f.** $\frac{5}{4}$
g. $\frac{3}{4}$ **h.** $\frac{1}{2}$ **i.** $\frac{4}{5}$ **j.** –1

6. **a.** 64 **b.** 81 **c.** 2 **d.** 216 **e.** 1.73 **f.** 64
g. 81 **h.** 3 **i.** 2 **j.** 36 **k.** 5 **l.** 1
m. –3 **n.** $\frac{1}{3}$ **o.** $\frac{2}{3}$ **(A)**

7. **a.** $\log xyz$ **b.** $\log x^2y$ **c.** $\log xy^2$ **d.** $\log \frac{xy}{z}$ **e.** $\log \frac{ab^3}{c^2}$ **f.** $\log \frac{(p)^{\frac{1}{2}}}{(q)^{\frac{3}{2}}}$
g. $\log xyz$ **h.** $\log y$ **i.** $\log xy^4$ **j.** $\log y$ **k.** $\log \frac{1}{y^2}$ **l.** $\log y^2$
m. $\log xy$ **n.** $\log \frac{1}{z^2}$ **o.** $\log xy^3$ **p.** $\log_a x^2a^5$ **(A)**

8. **a.** log 60 **b.** log 12 **c.** log 75 **d.** log 12 **e.** log 24 **f.** $\log \frac{6}{7}$
g. log 1 **h.** $\log \frac{25}{3}$ **i.** log 96 **j.** log 12 **(A)**

9. **a.** 3 **b.** $\frac{2}{3}$ **c.** $\frac{2}{3}$ **d.** $\frac{4}{3}$ **e.** $\frac{6}{5}$ **f.** 2
g. $\frac{11}{5}$ **h.** 1 **i.** 1 **j.** 2 **(A)**

10. **a.** 1.285 **b.** 0.715 **c.** 0.717 **d.** 1.781 **(M)**

Unit 11.1 Activity 7C: Further logarithmic problems (page 65)

1. **a.** 571 429 **(M)** **b.** 2 711 864 **(M)** **c.** 1.63 years **(M)** **d.** 3 million **(E)**

2. **a.** 1 000 **b.** 1 911 **c.** 2000 **d.** 1 year **(M)**

3. **a.** K4 500 **b.** K1 824 **c.** 3.65 years **d.** 2.3 years **(M)**

4. **a.** $A = 100\ 000$, $k = 0.0395$ **b.** K148 043 **c.** 58.3 years later **(M)**

5. 10 m **(E)**

Unit 11.1 Activity 8A: Units of measurement (page 70)

1. **a.** x **b.** vii **c.** xix **d.** v **e.** xi **f.** i
g. xv **h.** xviii **i.** ix **j.** viii **k.** ii **l.** iii
m. xii **n.** iv **o.** xvi

2. **a.** 800 cm **b.** 40 mm **c.** 4 kg **d.** 3 500 mm **e.** 30 000 cm
f. 0.6 L **g.** 4 800 m **h.** 8.3 t **i.** 370 mg **j.** 40 cm
k. 750 mL **l.** 0.945 g **m.** 0.14 km **n.** 0.0137 t **o.** 0.045 km
p. 4 700 kg **q.** 0.12 kL **r.** 0.64 km **s.** 4 000 mm **t.** 7 300 000 g
u. 1.86 m **v.** 63 kg **w.** 3 800 g **x.** 9 370 mL

3. **a.** 18.50 m **b.** 240 cm or 2.4 m **c.** 38.5 cm

4. **a.** 1.34 kg **b.** 4 g **c.** 140 g

5. **a.** 1:22 pm **(A)** **b.** 2:12 pm **(A)** **c.** 0430 **(A)**
d. April 4 **(A)** **e.** 1 hr 45 min **(A)**

6. **a.** 3:35 pm **(A)** **b.** 5 hr 32 min **(A)** **c.** 18 hr 10 min **(A)**
d. 2 678 400 seconds **e.** 2325 **(A)**

Unit 11.1 Activity 8B: Reading measuring instruments (page 73)

This exercise gives practice in reading scales and other measuring equipment. While these skills are important, no A, M, E rating is given as these skills are usually part of a problem-solving exercise.

1. **a.** 0.8 kg **b.** 800 g **c.**

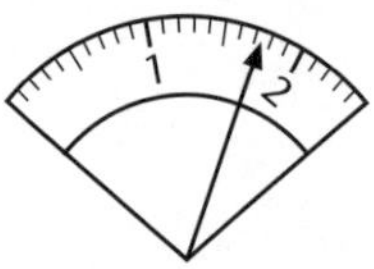

2. **a.** 116° **b.** 64°

3. **a.** **i.** 380 mL **ii.** 230 mL **iii.** 150 mL
b. **i.** 75 mL **ii.** 225 mL **iii.** 450 mL
c. **i.** 410 mm **i.** 425 mm **iii.** 437 mm
d. **i.** 2 m 30 cm (2.3 m) **ii.** 2 m 32 cm (2.32 m) **iii.** 2 m 27.2 cm (2.272 m)
e. **i.** 85 cm **ii.** 86.7 cm **iii.** 89.3 cm
f. **i.** 500 g **ii.** 1 075 g **iii.** 1 250 g
g. **i.** 41°C **ii.** 37.8°C **iii.** 36.6°C
h. **i.** 0.3 Amp **ii.** 4 Amp **iii.** 6 Amp
i. **i.** 4 **ii.** 30 **iii.** 400
j. **i.** 540 kHz **ii.** 675 kHz **iii.** 850 kHz
k. **i.** 52 **ii.** 54 **iii.** 58

Unit 11.1 Activity 8C: Limits of accuracy (page 75)

1. **a.** 5.5 m $\le x <$ 6.5 m **(M)**
b. 4.25 cm $\le x <$ 4.35 cm **(M)**
c. 5.235 m $\le x <$ 5.245 m **(M)**
d. 1.75 cm $\le x <$ 1.85 cm **(M)**
e. 23.5 mm $\le x <$ 24.5 mm **(M)**
f. 757.5 km $\le x <$ 758.5 km **(M)**
g. 1.035 L $\le x <$ 1.045 L **(M)**
h. 0.55mL $\le x <$ 0.65mL **(M)**
i. 6.95 mg $\le x <$ 7.05 mg **(M)**
j. 8.995 m $\le x <$ 9.005 m **(M)**

2. 15.415 s ≤ time < 15.425 s **(M)**

3. 3.395 m ≤ length < 3.405 **(M)**

4. 10.4 m ≤ perimeter < 10.8 m **(M)**

5. 47 500 L ≤ capacity < 48 500 L **(M)**

6. The limits of accuracy for the old record are 12.35 seconds ≤ time taken < 12.45 seconds. The limits of accuracy for the recent runner's time are

12.405 seconds ≤ time taken < 12.415 seconds.

Within the limits of accuracy for the old record, there are times such as 12.42 seconds which are slower than the times within the recent runner's limit of accuracy. Therefore the recent runner could well have run faster that the record holder. However, 12.41 seconds as a single time would not break the record. ***(E)***

Unit 11.1 Activity 8D: Conversion of units (page 78)

1. **a.** 2.7432 yd **b.** 321.860 km **c.** 1.0160 m
d. 5468.32 yds **e.** 1853 m **f.** 3.937 in.

2. **a.** 1196 yd^2 **b.** 7.7229 sq miles **c.** 247.104 acres
d. 0.1672 ha **e.** 1935.48 mm^2 **f.** 1416.415 m^2

3. **a.** 141.343 ft^3 **b.** 7.2083 ft^3 **c.** 1.7047 litres
d. 1136.6 cm^3 **e.** 0.5368 pints **f.** 26.3197 fl oz.

Unit 11.1 Activity 8E: Multiple choice – measurement (page 78)

1. A **2.** C **3.** A **4.** C

5. B **6.** D **7.** A **8.** B

Unit 11.1 Activity 9A: Ratios (page 81)

Note: *Only problems in context are given A, M, E ratings.*

1. **a.** 1 : 2 **b.** 1 : 3 **c.** 5 : 3 **d.** 1 : 8 **e.** 3 : 2 **f.** 4 : 1
g. 10 : 1 **h.** 1 : 3 **i.** 20 : 1 **j.** 1 : 250 **k.** 1 : 500 **l.** 14 : 1

2. **a.** 15 : 25 **b.** 40 : 16 **c.** 9 : 36 : 45 **d.** K3.20 : K4.80
e. K13.50 : K18.90 **f.** 36 : 54 : 90

3. **a.** K3 000 **b.** K112.50 **(*A*)**

4. **a.** K20 000 **b.** 2 : 5 **c.** Mr Liklik, K50 000; Mr Bik, K125 000 **(*A*)**

5. **a.** 1 : 3 **b.** 6 L **c.** 4.5 L **(*A*)** **d.** 16 **(*M*)**

6. **a.** 900 g **(*A*)** **b.** 2.25 kg **(*M*)**

Unit 11.1 Activity 9B: Map scales (page 84)

1. **a.** 100

b.
0 100 200 300 400 km

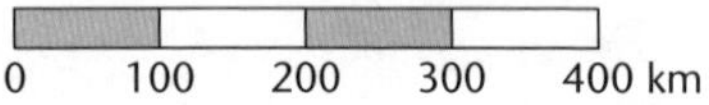

c. **i.** 3.3 cm **ii.** 5.5 cm

d. **i.** 330 km **ii.** 550 km

2. **a.** 1 : 10 000

b. 1 : 100 000

c. 1 : 20 000 000

3. **a.**
0 20 40 60 80 100 m

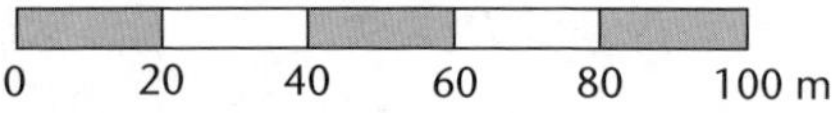

b.
0 3 6 km

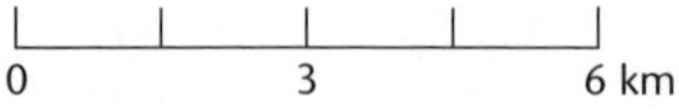

c.
0 1 km

4. **a.** **i.** 507 m **ii.** 780 m

b. 6.7 cm

c.
0 130 260 390 520 m

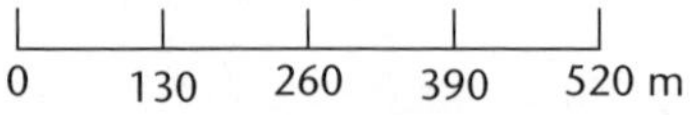

5. 15.4 cm

6. 11.2 cm

7. 1 : 5 000 000

Unit 11.1 Activity 9C: Variation (page 87)

1. a. $y \propto x$; $y=kx$ **b.** 20 **c.** 2.8

2. a. 7.5 **b.** 252

3. a. 270 **b.** 10

4. a. and **b.**

5. a. $A = 0.421362P$ **b.** AU$33.71 **c.** PGK237.33

6. a. 13 cm **b.** 3.846 kg

7. K381.25

8. 8.92 cm

9. 10.102 sec.

10. 12.247 km

11. a. 0.57 sec **b.** 24.5 cm

Unit 11.1 Activity 9D: Inverse variation (page 89)

1. a. 11 **b.** 6.6

2. a.

x	2	3	8	12
y	60	40	15	10

b.

p	100	50	25	7.5
q	0.75	1.5	3	10

3. a. $V = \frac{k}{P}$ **b.** 1480 **c.** 2.96 L

4. 4 hours 10 minutes.

5. a. 53.0 cm **b.** 35.4 cm

6. a. i. 212.5 units **ii.** 34 units

b. 3.54 m

Unit 11.1 Activity 10A: Algebraic expressions (page 93)

1. $9 + a$ **2.** $b - 4$ **3.** $5c$ **4.** $x + 6$ **5.** $y - 7$ **6.** $4(x + y)$

7. $\frac{x}{6}$ **8.** $\frac{x + y}{10}$ **9.** $12 - y$ **10.** $3(x + 4)$ **11.** $\frac{y - 7}{2}$ **12.** $\sqrt{x + 10}$

13. $n + n + 1$ or $2n + 1$ **14.** $\frac{a + 10}{7b}$ **15.** $2y - 3$ **16.** $(p - 2)^3$ **17.** $x - 5$

18. $t + 3l$ **19.** $10x + 15y$ **20.** $a + b + 4$

Unit 11.1 Activity 10B: Combining algebraic terms (page 95)

1. $8a$ **2.** $6a$ **3.** $3 + 2x$ **4.** $x^2 + x + 1$

5. $4x^2 - 3x + 5$ **6.** $4x^2y - 3xy^2$ **7.** $10x$ **8.** $3a$

9. $4 + 3x$ **10.** $x^2 + 8x + 2$ **11.** $3x^2 - x - 3$ **12.** $x^2 - 9x + 9$

13. $24ab$ **14.** $12a^2b$ **15.** $24pqr$ **16.** $10a^2b^2$

17. $3x^6 - x^4$ **18.** $3x$ **19.** $\frac{3bc}{2}$ **20.** $\frac{2}{3y}$

Unit 11.1 Activity 10C: Algebraic fractions (page 98)

1. **a.** x^8 **(A)** **b.** $3y^8$ **(A)** **c.** $4a^3$ **(A)** **d.** $4b$ **(A)** **e.** $3x$ **(A)**

f. $3a^4$ **(A)** **g.** $\frac{a^3b}{2}$ **(A)** **h.** $\frac{3x^2}{2y}$ **(A)** **i.** $\frac{2a^2b}{5}$ **(A)** **j.** $\frac{2x^5}{3z^2}$ **(A)**

2. **a.** $\frac{x+1}{3}$ **(M)** **b.** $\frac{y-2}{9}$ **(M)** **c.** $\frac{7x}{12}$ **(M)** **d.** $\frac{a}{10}$ **(M)** **e.** $\frac{7k}{15}$ **(M)**

f. $\frac{y}{6}$ **(M)** **g.** $\frac{7}{x+1}$ **(M)** **h.** $\frac{x^2+2}{2x}$ **(M)** **i.** $\frac{4b+3a}{ab}$ **(M)**

j. $\frac{4}{3x}$ **(M)** **k.** $\frac{1+2x}{x^2}$ **l.** $\frac{3-2y}{3y^2}$ **(M)**

3. **a.** $\frac{a^2}{12}$ **b.** 2 **c.** 1 **d.** $\frac{x^2}{10}$ **e.** $\frac{2}{5}$

f. $\frac{4}{3}$ **(A)** **g.** $\frac{1}{9}$ **(A)** **h.** $\frac{8a^3}{45}$ **(A)** **i.** $\frac{9x}{2}$ **(A)**

j. a^2b **(A)** **k.** $\frac{9xy^2}{10}$ **(A)** **l.** $\frac{2y^3}{3}$ **(A)**

4. **a.** $\frac{x+9}{20}$ **(M)** **b.** $\frac{9x+1}{20}$ **(M)** **c.** $\frac{3-x}{4}$ **(M)** **d.** $\frac{9x-5}{14}$ **(M)**

e. $\frac{8-5x}{12}$ **(M)** **f.** $\frac{4x+3}{12x}$ **(M)**

5. **a.** $\frac{7a}{12}$ **(M)** **b.** $\frac{-1}{x}$ **(M)** **c.** $\frac{a^2}{24}$ **(M)** **d.** $\frac{7y}{30}$ **(M)**

e. $\frac{w}{3}$ **(M)** **f.** $\frac{10a^2}{3}$ **(M)**

Unit 11.1 Activity 11A: Expanding brackets (page 100)

Many of these problems practise basic skills.

1. $3x + 3$ **2.** $4x + 8$ **3.** $ab - 2a$ **4.** $x^2 - 3x$

5. $-4x + 20$ **6.** $2x^3 + 6x^2 - 8x$ **7.** $2 - 6x$ **8.** $11 - x$

9. $2x + 10$ **10.** $11x + 6y$ **(A)** **11.** $10a + 21b$ **(A)** **12.** $2x - y$ **(A)**

13. $x^2 + 5x - 2$ **(A)** **14.** $3p^2 - 2pq + 2q^2$ **(A)** **15.** $-x^2 - x$ **(A)** **16.** $x^2 - 6x$ **(A)**

17. $3x^2 - 6x^3 - 3x^4$ **(A)** **18.** $-2x^2y + 2xy^2 + 6xyz$ **(A)** **19.** $x^2 + 3x$ **(A)**

20. a. K(y – 10) **(A)** **b.** K3(y – 10) or K(3y – 30) **(A)**

Unit 11.1 Activity 11B: Further expansions (page 102)

1. $x^2 + 9x + 20$ **(A)** **2.** $x^2 - x - 6$ **(A)** **3.** $x^2 - x - 20$ **(A)**

4. $x^2 - 2x - 15$ **(A)** **5.** $x^2 - 11x + 28$ **(A)** **6.** $x^2 - 11x + 30$ **(A)**

7. $2x^2 + 11x + 12$ **(A)** **8.** $12x^2 - 7x - 12$ **(A)** **9.** $2x^2 - 15x + 18$ **(A)**

10. $6 + 7x - 3x^2$ **(A)** **11.** $35 + 3x - 2x^2$ **(A)** **12.** $6x^2 - 23x + 20$ **(A)**

13. $3x^2 + 10x + 3$ **(A)** **14.** $15x^2 + 14x - 8$ **(A)** **15.** $12 + x - 6x^2$ **(A)**

16. $x^2 + 12x + 36$ **(A)** **17.** $x^2 + 8x + 16$ **(A)** **18.** $x^2 - 14x + 49$ **(A)**

19. $x^2 - 100$ **(A)** **20.** $16x^2 - 1$ **(A)** **21.** $4x^2 - 25$ **(A)**

22. $4x^2 + 12x + 9$ **(A)** **23.** $9x^2 - 30x + 25$ **(A)** **24.** $x^3 + 3x^2 + 3x + 2$ **(M)**

25. $x^3 - x^2 - 10x + 12$ **(M)** **26.** $2x^3 - 9x^2 + 4x + 15$ **(M)** **27.** Length YZ = $\sqrt{2x^2 + 2}$ **(M)**

28. $A = 2x^2 + 3x - 18$ **(M)** **29.** $10x + 25$ **(M)** **30.** $x^3 + 6x^2 + 12x + 8$ **(M)**

Unit 11.1 Activity 11C: Basic factorising (page 104)

Many of these problems practise basic skills.

1. $4(a + 3)$ **2.** $7(x - 2)$ **3.** $5(y + 5)$ **4.** $3(3x + 5y)$

5. $x(y - 3)$ **6.** $y(y^2 + 5)$ **7.** $a(a + b)$ **8.** $x(x - 7)$

9. $2(3x + 4y)$ **10.** $a(b - a)$ **11.** $3p(p - 2)$ **12.** $2(2x + 7y)$

13. $3x(x^2 + 2x - 1)$ **14.** $4(2x + 3y - 5z)$ **15.** $5(3x + 2y - 5)$

16. $2xy^2(4xy + 6y^2 - 5)$ **(A)** **17.** $4(2x - 6y + 5)$ **(A)** **18.** $ab(3a + 5a^2b + 1)$ **(A)**

19. $2(2x^2 + 4x - 5)$ **(A)** **20.** $y(2 - 6y + y^2)$ **(A)** **21.** $3pq(p - 3q)$ **(A)**

22. $p^2(p^2 - 3)$ **(A)** **23.** $2a^2(2 - 9a)$ **(A)** **24.** $a^2b^2(b + a)$ **(A)**

25. $(x + 3)(2x + 5)$ **(A)** **26.** $(x - 2)(6x - 5)$ **(A)** **27.** $(a + x)(b + c)$ **(M)**

28. $(x + 2y)(3 + a)$ **(M)** **29.** $(x + 3y)(5 - p)$ **(M)** **30.** $(a - b)(x + y)$ **(M)**

Unit 11.1 Activity 11D: Factorising quadratic expressions (page 107)

1. $(x + 4)(x + 2)$ **(A)** **2.** $(x + 5)^2$ **(A)** **3.** $(x - 4)^2$ **(A)**

4. $(x + 5)(x - 3)$ **(A)** **5.** $(x + 8)(x - 3)$ **(A)** **6.** $(x - 7)(x + 2)$ **(A)**

7. $(x + 5)(x + 4)$ **(A)** **8.** $(x - 5)^2$ **(A)** **9.** $(x + 5)(x + 2)$ **(A)**

10. $(x - 3)(x - 1)$ **(A)** **11.** $(x - 7)(x - 4)$ **(A)** **12.** $(x + 5)(x - 4)$ **(A)**

13. $(x + 7)(x - 3)$ **(A)** **14.** $(x + 6)(x + 3)$ **(A)** **15.** $(x - 6)^2$ **(A)**

16. $(x - 3)(x + 3)$ **(A)** **17.** $(x - 7)(x + 7)$ **(A)** **18.** $(x + 8)(x - 8)$ **(A)**

19. $(5x-4)(5x+4)$ **(A)** **20.** $(2x+9)(2x-9)$ **(A)** **21.** $5(x-2)(x+2)$ **(A)**

22. $3(x-3)(x+7)$ **(A)** **23.** $2(x+1)(x-2)$ **(A)** **24.** $4(x-3)(x-1)$ **(A)**

25. $(2x-1)(x+3)$ **(M)** **26.** $(2x+5)(x+2)$ **(M)** **27.** $(3x-1)(x-2)$ **(M)**

28. $(5x+2)(x-3)$ **(M)** **29.** $(3x+7)(x-1)$ **(M)** **30.** $(3x-4)(2x+3)$ **(M)**

Unit 11.1 Activity 11E: Simplifying rational expressions by factorising (page 108)

1. 9 **(M)** **2.** 7 **(M)** **3.** $x+3$ **(M)** **4.** 3 **(M)** **5.** x **(M)**

6. $x+6$ **(M)** **7.** $x+4$ **(M)** **8.** $\frac{x+4}{x-2}$ **(M)** **9.** $\frac{x-8}{x+2}$ **(M)** **10.** $\frac{x+4}{x-2}$ **(M)**

11. $\frac{x+2}{x+3}$ **(M)** **12.** $\frac{x+4}{2(x+5)}$ **(M)**

Unit 11.1 Activity 11F: Multiple choice – factorising quadratic expressions (page 108)

1. A **2.** B **3.** D **4.** C **5.** D **6.** C **7.** A

8. B **9.** A **10.** B **11.** D **12.** C **13.** A **14.** B

Unit 11.1 Activity 12A: Solving quadratic equations (page 113)

1. **a.** $-2, -3$ **(A)** **b.** $2, 4$ **(A)** **c.** $-1, 3$ **(A)** **d.** $5, -2$ **(A)**

e. $0, 4$ **(A)** **f.** $0, -5$ **(A)** **g.** $\frac{3}{2}, -6$ **(A)** **h.** $-\frac{1}{2}, 1$ **(A)**

i. $-\frac{3}{5}, -\frac{5}{2}$ **(A)** **j.** $\frac{3}{4}, \frac{9}{2}$ **(A)** **k.** $\frac{4}{3}, 5$ **l.** $\frac{2}{3}, -\frac{2}{3}$ **(A)**

2. **a.** $0, -3$ **(M)** **b.** $0, 5$ **(M)** **c.** $-3, 2$ **(M)** **d.** $-1, -4$ **(M)**

e. $8, -5$ **(M)** **f.** 3 **(M)** **g.** $2, 6$ **(M)** **h.** $-5, 2$ **(M)**

i. ± 2 **(M)**

3. **a.** $5, -3$ **(M)** **b.** $-6, 3$ **(M)** **c.** $1, 3$ **(M)** **d.** $0, 6$ **(M)** **e.** $0, \frac{5}{2}$ **(M)**

f. ± 7 **(M)** **g.** $-2, 4$ **(M)** **h.** ± 2 **(M)** **i.** $\frac{3}{2}, -2$ **(M)**

4. **a.** ± 8 **(M)** **b.** $-3, 9$ **(M)** **c.** ± 12 **(M)** **d.** ± 5 **(M)**

e. $\frac{-9}{2}, 4$ **(M)** **f.** $\frac{-19}{6}, 2$ **(M)**

Unit 11.1 Activity 12B: Solving word problems with quadratic equations (page 115)

1. a. Sam's age. **b.** Sam = 14, Sam's sister = 11, Sam's brother = 16. **(M)**

2. Day 3 and Day 9. **(M)** **3.** 12 m by 32 m **(M)** **4.** 2 m square **(M)**

5. 5 **(M)** **6.** 8 cm **(M)** **7.** 4, 14 or –4, –14 **(M)**

8. 16 cm square **(E)** **9.** (18 and 20) or (–10 and –36) **(E)** **10.** 8, 12, 18 points **(E)**

Unit 11.1 Activity 12C: Multiple choice – Solving quadratic equations (page 116)

1. C **2.** D **3.** B **4.** A **5.** D **6.** B **7.** C **8.** C

Unit 11.1 Activity 13A: Quadratic graphs (page 121)

1. a.

x	–2	–1	0	1	2
x^2	4	1	0	1	4
$y = x^2 - 2$	2	–1	–2	–1	2

b.

x	0	1	2	3	4
$x - 2$	–2	–1	0	1	2
$y = (x - 2)^2$	4	1	0	1	4

c.

x	–5	–4	–3	–2	–1
$x + 3$	–2	–1	0	1	2
$y = (x + 3)^2$	4	1	0	1	4

d.

x	–2	–1	0	1	2
x^2	4	1	0	1	4
$y = 2 - x^2$	–2	1	2	1	–2

e.

x	–2	–1	0	1	2
x^2	4	1	0	1	4
$y = x^2 + 3$	7	4	3	4	7

f.

x	–2	–1	0	1	2
x^2	4	1	0	1	4
$y = 2x^2$	8	2	0	2	8

2. a. $y = x^2 + 3$ **b.** $y = (x + 3)^2$ **c.** $y = 2x^2$ **d.** $y = 2 - x^2$

e. $y = x^2 - 2$ **f.** $y = (x - 2)^2$

3. a.

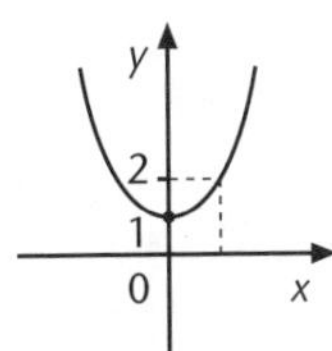

(A)

b.

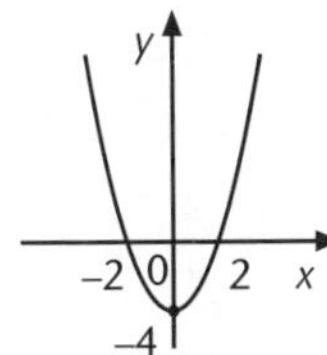

(A)

c.

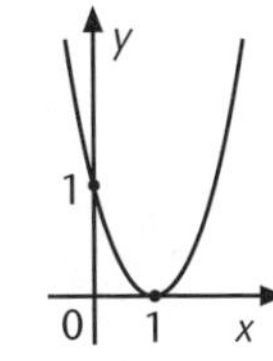

(A)

d.

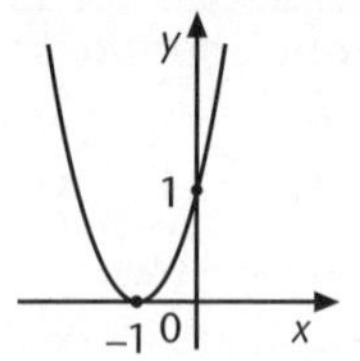

(A)

e.

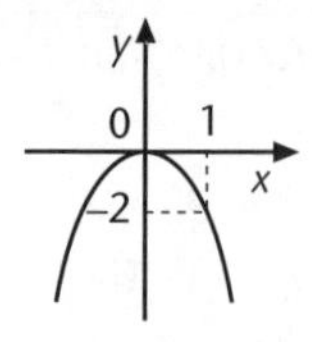

(M)

f.

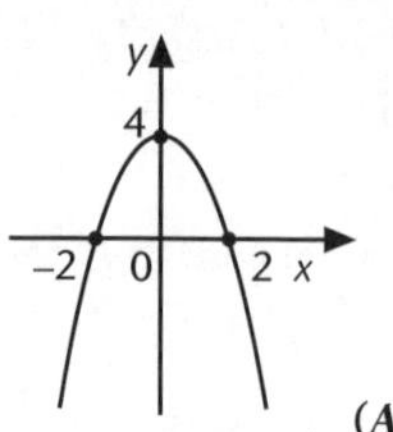

(A)

4. a.

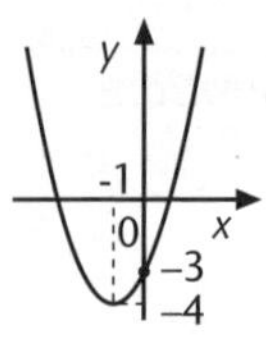

(M)

b.

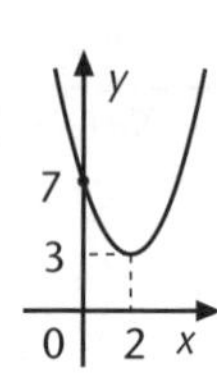

(M)

c.

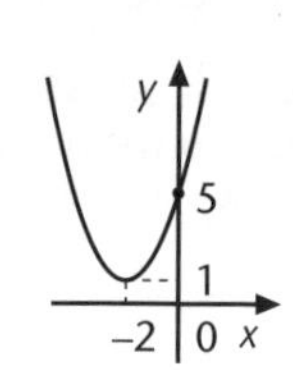

(M)

d.

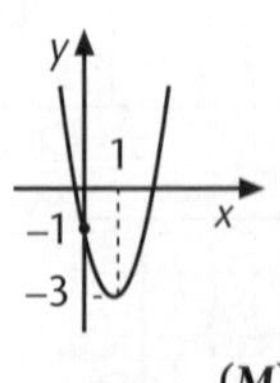

(M)

e.

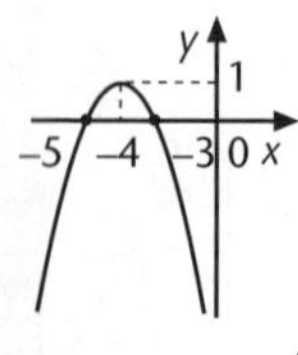

(M)

f.

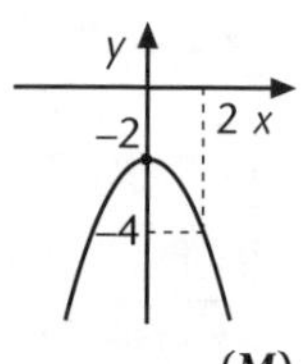

(M)

Unit 11.1 Activity 13B: Quadratic graphs (page 125)

1. a.

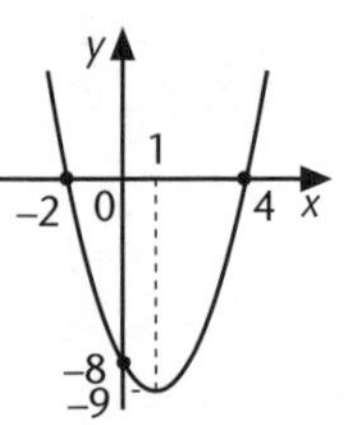

(A)

b.

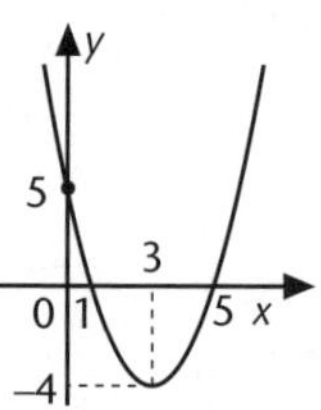

(A)

c.

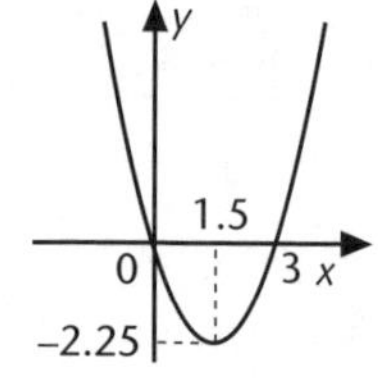

(A)

d.

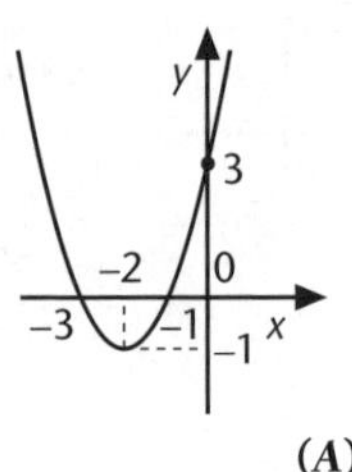

(A)

e.

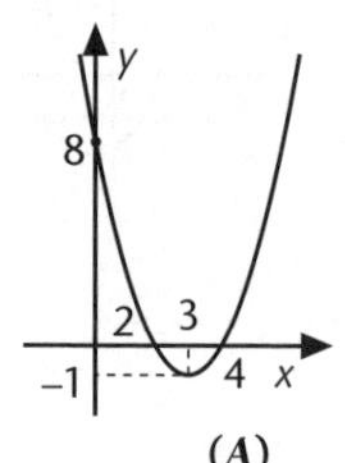

(A)

f.

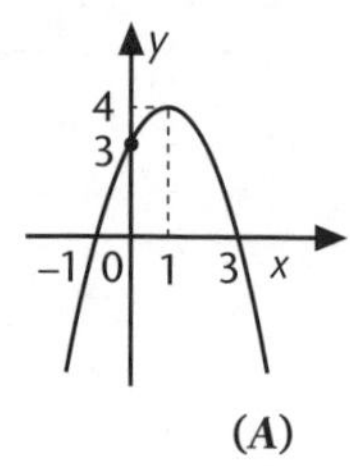

(A)

g.

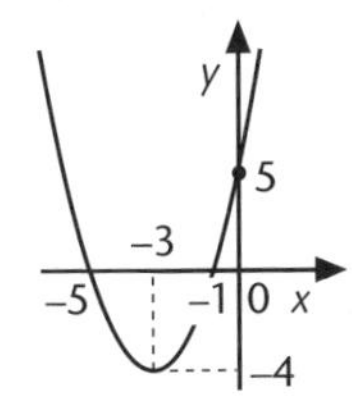

(**A**)

h.

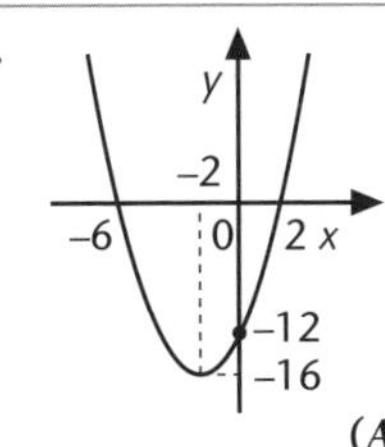

(**A**)

i.

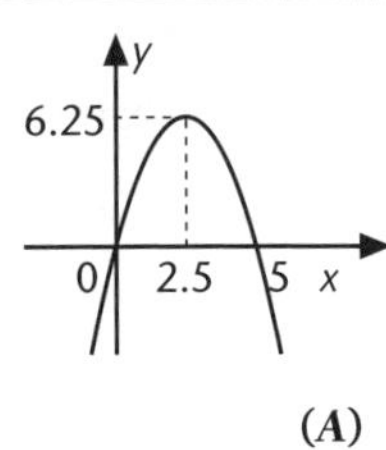

(**A**)

2. **a.** **i.** $(-4, 0)$ (**A**) **ii.** $(2, 0)$ (**A**) **iii.** $(-1, 0)$ (**A**) **iv.** $(-1, -9)$ (**A**)

b. -9 **c.** $y = (x + 1)^2 - 9$ (**M**)

3. **a.** $y = (x - 15)(x - 55)$ *or* $y = (x - 35)^2 - 400$ (**E**)

b. $y = 3(x - 5)(x - 9)$ *or* $y = 3(x - 7)^2 - 12$ (**E**)

c. $y = 2(3 - x)(x - 15)$ *or* $y = -2(x - 3)(x - 15)$ *or* $y = -2(x - 9)^2 + 72$ (**E**)

d. $y = \frac{-1}{2}x^2 + 800$ *or* $y = \frac{-1}{2}(x + 40)(x - 40)$ (**E**)

e. $y = -(x - 2)^2 + 9$ *or* $y = -(x + 1)(x - 5)$ (**E**)

f. $y = \frac{1}{3}(x - 9)^2 + 3$ (**E**)

Unit 11.1 Activity 13C: Using quadratic graphs to solve problems (page 128)

1. **a.**

x	0	1	2	3	4	5	6	7	8	9	10	11	12
A	0	22	40	54	64	70	72	70	64	54	40	22	0

b.

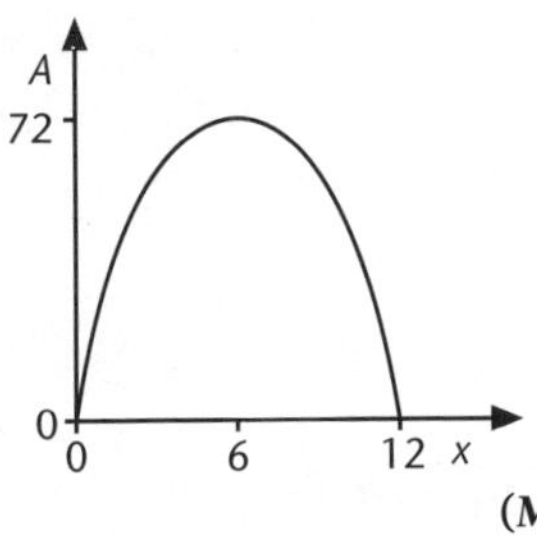

(**M**)

c. 72 m^2 (**M**)

d. Cannot have a negative length. (**A**)

e. 3.6 m or 8.4 m (**M**)

2. a.

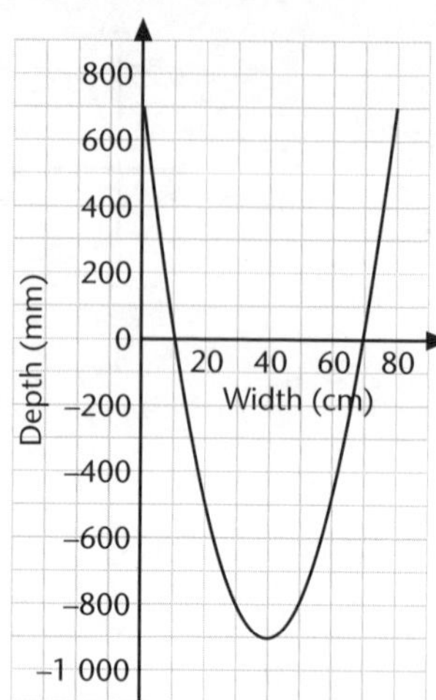

(*A*)

b. i. 40 cm **(*A*)**

ii. 900 mm **(*A*)**

c. i. 60 cm **(*A*)**

ii. Distance between intercepts on x-axis. **(*A*)**

d. 20 cm and 60 cm **(*A*)**

e. For $10 \le W < 70$ (a depth of +700 cm at the pole is inapplicable). **(E)**

f. Cross-section of the drain may not exactly 'fit' the parabola and the sides and bottom of the drain may be uneven. **(E)**

3. a.

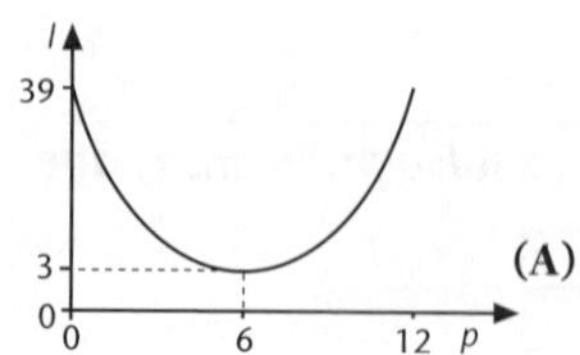

(A)

b. 3 m **(*A*)**

c. i. 12 m **(*A*)**

ii. Symmetry and distance from first pole to the lowest point is 6 m. **(*A*)**

4. a. $y = \frac{1}{8}(x-2)^2 + 1.5$ *or* $y = \frac{1}{8}x^2 + 1.5$

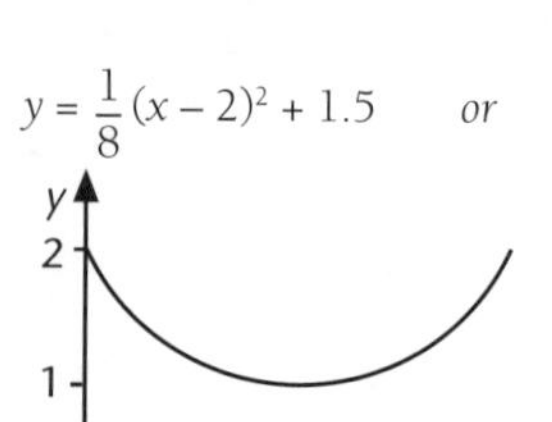

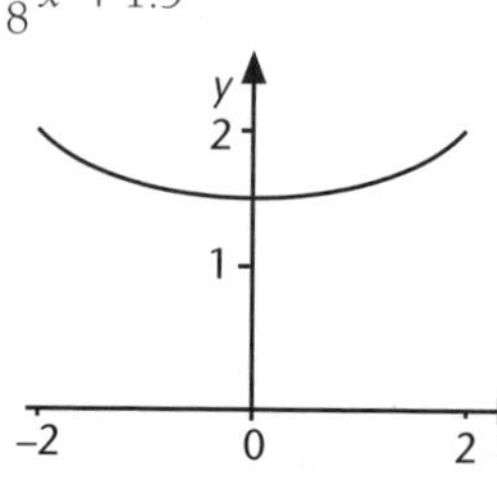

(E)

b. No, height of string is 1.68 m. **(E)**

5. a. $h = 50d + 10$ **(M)**

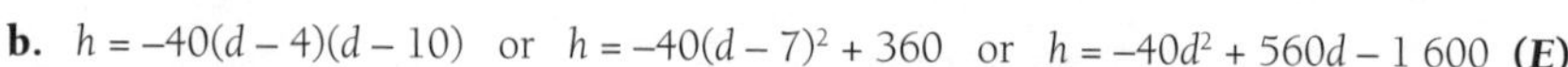

b. $h = -40(d-4)(d-10)$ or $h = -40(d-7)^2 + 360$ or $h = -40d^2 + 560d - 1\,600$ **(E)**

6. a.

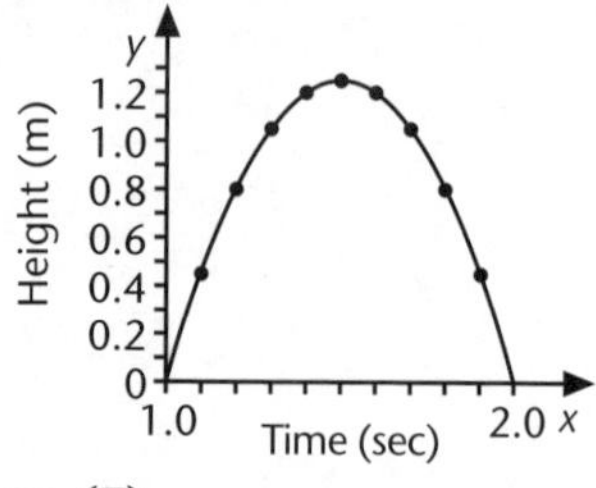

b. 1.25 m **(*A*)**

c. $y = 5(x-1)(2-x)$ or $y = -5x^2 + 15x - 10$ **(E)**

d. 0.9375 m **(*A*)**

7. 7.5 cm **(E)**

Unit 11.1 Activity 14A: Simultaneous equations (page 135)

1. **a.** $x = 14, y = 19$ **(M)** **b.** $x = 11, y = 15$ **(M)** **c.** $x = 7, y = 17$ **(M)**
d. $x = 12, y = 27$ **(M)**

2. **a.** $x = 4, y = 3$ **(M)** **b.** $x = 5, y = 4$ **(M)** **c.** $x = 3, y = 5$ **(M)**
d. $x = 7, y = -3$ **(M)** **e.** $x = 2\frac{1}{2}, y = 1$ **(M)** **f.** $x = 1, y = -5$ **(M)**
g. $x = 4, y = 5$ **(M)** **h.** $x = 5, y = 0$ **(M)**

3. **a.** $x = 3, y = 4$ **(M)** **b.** $x = 2, y = 7$ **(M)** **c.** $x = 5, y = 5$ **(M)**
d. $x = -3, y = 2$ **(M)** **e.** $x = 8, y = 3$ **(M)** **f.** $x = -\frac{1}{2}, y = 2$ **(M)**
g. $x = 2, y = -1$ **(M)** **h.** $x = -1, y = 10$ **(M)**

4. **a.** $x = 2, y = 0$ **(M)** **b.** $x = 0, y = -1$ **(M)** **c.** $x = 3, y = -1$ **(M)**
d. $x = 4, y = 5.5$ **(M)**

5.

a.

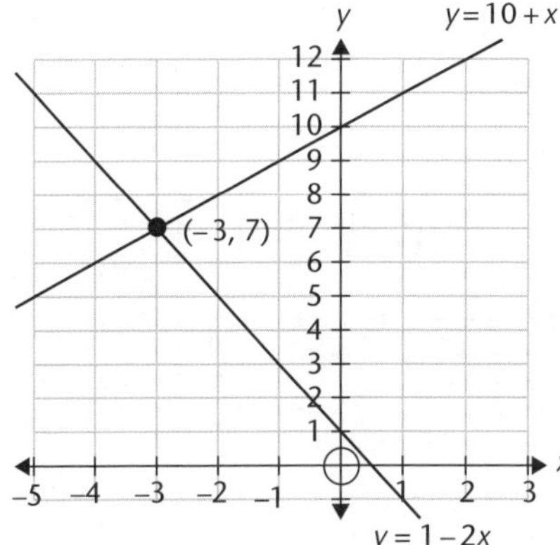

b.

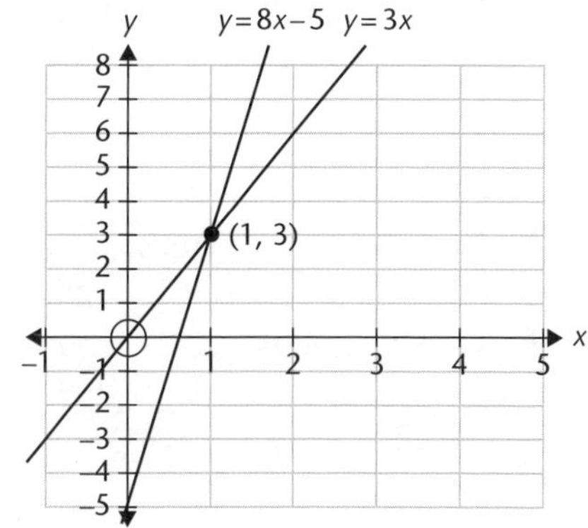

c.

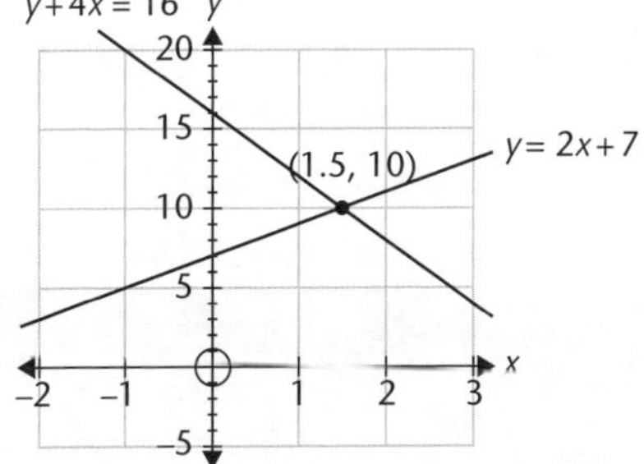

d.

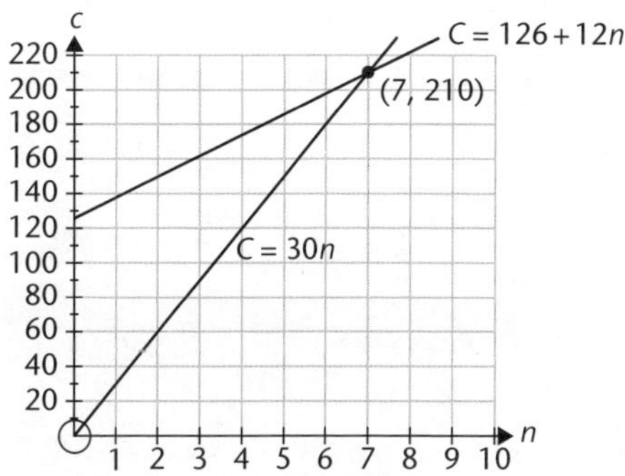

Unit 11.1 Activity 14B: Word problems with simultaneous equations (page 136)

1. K4.40 **(M)** **2.** 3.5 hours **(M)** **3.** Artemius 32, Brown 18. **(M)**

4. 6 beans, 9 aibika. **(M)**

5. a. $7s + 3h = 50$ and $3s + 2h = 24$ where s = Kcost socks, h = Kcost handkerchiefs **(E)**

b. Socks cost K5.60 a pair. **(M)**

6. Flower garden is 10 m by 13 m; vegetable garden is 12 m by 28 m. **(E)**

Unit 11.1 Activity 15A: Inequations (page 138)

1. a. i. $x \leq 2$ **ii.** –5 –4 –3 –2 –1 0 1 2 3 4 5 x **iii.** $(-\infty, 2)$

b. i. $x \leq -2$ **ii.** –5 –4 –3 –2 –1 0 1 2 3 4 5 x **iii.** $(-\infty, -2)$

c. i. $x \geq -4$ **ii.** –5 –4 –3 –2 –1 0 1 2 3 4 5 x **iii.** $(-4, \infty)$

d. i. $x \geq \frac{3}{2}$ **ii.** –5 –4 –3 –2 –1 0 1 2 3 4 5 x **iii.** $\left(\frac{3}{2}, \infty\right)$

e. i. $x > -7.5$ **ii.** –10 –9 –8 –7 –6 –5 –4 –3 –2 –1 0 1 2 3 x **iii.** $(-7.5, \infty)$

f. i. $x > \frac{7}{2}$ **ii.** –2 –1 0 1 2 3 4 5 6 7 8 x **iii.** $\left(\frac{7}{2}, \infty\right)$

2. a. $x > \frac{-5}{3}$ **b.** $x \leq 3$ **c.** $x \leq 21$ **d.** $x \geq 2$

e. $x \geq \frac{9}{2}$ **f.** $x < -11$ **g.** $x \leq 14$ **h.** $x \geq \frac{5}{2}$

3. a. $a > \frac{-3}{2}$ **b.** $a > \frac{4}{3}$ **c.** $a \leq \frac{1}{3}$ **d.** $a \geq 1$

e. $a > \frac{5}{2}$ **f.** $a \leq \frac{-1}{2}$ **g.** $a < 2$ **h.** $a \leq \frac{9}{5}$

4. a. $b > -11$ **b.** $b \leq -13$ **c.** $b < 12$

d. $b \leq \frac{13}{7}$ **e.** $b \geq 1$ **f.** $b \leq \frac{11}{5}$

Unit 11.1 Activity 15B: Regions (page 141)

1. a.

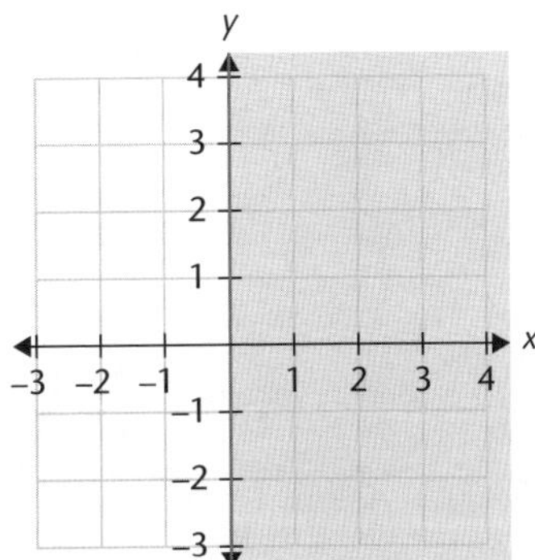

b.

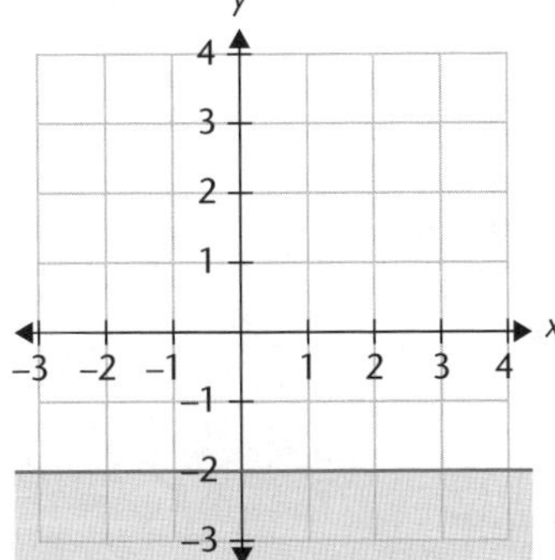

c.

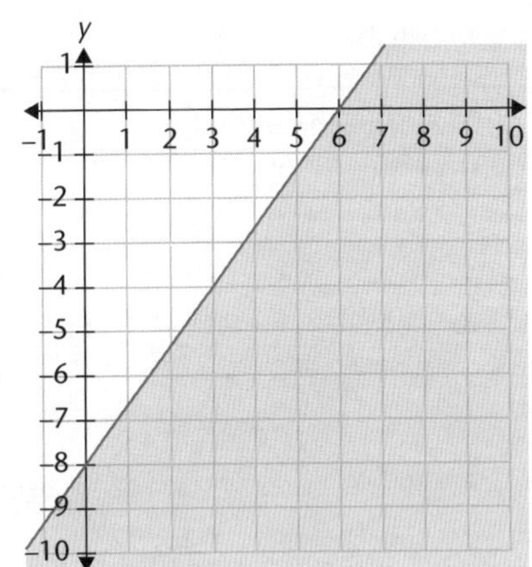

d.

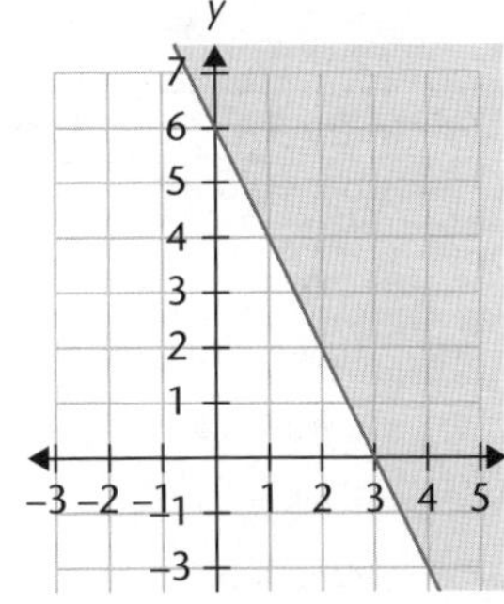

e.

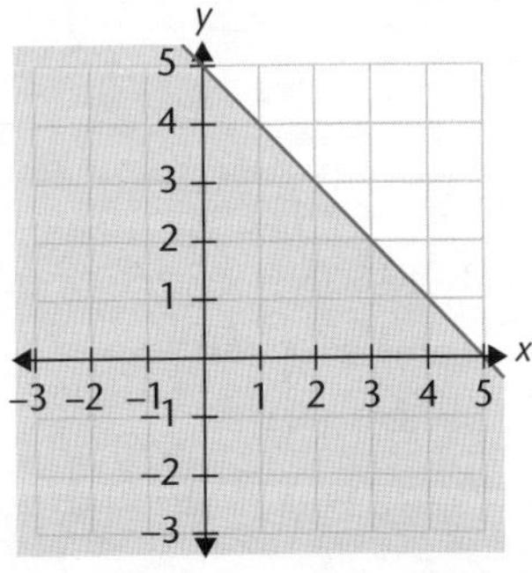

f.

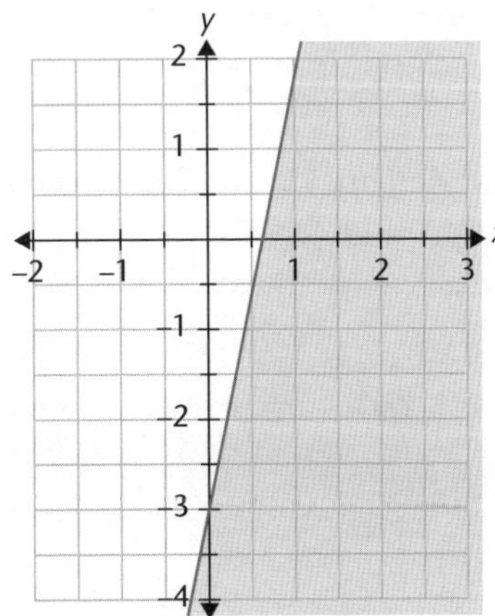

g.

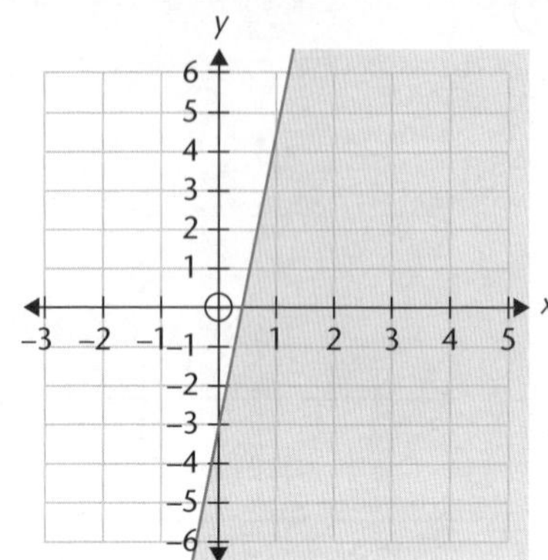

2. a. $x > -1$ b. $y \leq 0$ c. $x + y < 4$
d. $y \leq -2x + 4$ e. $y \leq \frac{2}{3}x + 2$ f. $y \leq 3x$

Unit 11.2 Activity 1A: Calculations with percentages (page 146)

1. a. 12 b. 18 c. 202.8 d. 288 e. 0.63
f. 7.2 g. 12 h. K217 i. 27 j. K120
2. a. 96 b. 70.8 c. 464 d. K96.60 e. K18.85 f. K141.75
3. a. 72 b. 102 c. 180 d. K66.88 e. K11.97 f. K61.21
4. a. 75% b. 32% c. 40% d. 64% e. 55% f. 11%
g. 1% h. 20%
5. a. 25% increase. b. 50% increase. c. 30% increase. d. 12.5% decrease.
e. 25% decrease. f. 15% decrease. g. 16% increase. h. 12.5% decrease.
6. a. 78% **(A)** b. 13 338 **(A)** c. 13% **(A)**
7. 75% **(A)** 8. 7.3% **(A)**
9. K221.25 **(A)** 10. 23.4 km **(A)**

Unit 11.2 Activity 1B: Finding original quantities (page 147)

1. a. K50 b. K15 c. K120 d. K18.45
e. K24.50 f. K596 g. K52.40 h. K108.24 **(M)**
2. K120 **(M)** 3. K84 **(M)** 4. 35 m **(M)** 5. K95 **(M)**
6. K253 000 **(M)** 7. a. Loss since $1.40 \times 0.65 = 0.91$ b. 9% **(M)**

Unit 11.2 Activity 2A: Calculating wages (page 150)

1. **a.** K31.50 **b.** K45.00 **c.** K105.75 **d.** K339.75
2. K928.00
3.

Total Hours Worked	Hourly Rate (K)	Regular Wages (K)	Overtime Wages (K)
30	6.50	195	0
50	12.00	480	240
40	10.00	0	600.00
25	8.00	200	0
35	8.75	306.25	0

4. K425
5. B
6. K450.00
7. Miller 1
8. **a.** K340.00 **b.** K100.00
9. **a.** K97.50 **b.** K317.50
10. K2544.00
11. K3947.41
12. 6.25%
13. 6.5%

Unit 11.2 Activity 2B: Calculating salaries (page 152)

1. **a.** K3066.67 **b.** K1415.38
2. K32 000
3. **a.** K48 190.00 **b.** K926.73
4. **a.** K16.75 **b.** K423.08
   c. K246.15 **d.** K1900.00
5. **a.** 3.125% **b.** K38.46 **c.** K28.19
6. **a.** 6% **b.** K22 048.00
7. K32 006.00
8. **a.** K19843.20 **b.** K22 295.82 **c.** K137.53
9. K752
10. **a.** K1240 **b.** K93.00

Unit 11.2 Activity 2C: Calculating piece work and casual work (page 154)

1. K618.75
2. K50.00
3. K7500.00
4. K129.95
5. **a.** the first family **b.** the first family **c.** K463.60
6. K261.00
7. K96.00
8. K190.20
9. K6355.75

Unit 11.2 Activity 3A: Calculating taxable income, superannuation and allowances (page 159)

1.

Employee	Taxation (K)	Tax as a percentage of taxable income
Lore	10.11	3.1%
Mary	232.26	42.1%
Laki	49.20	7.9%
Hiari	336.00	42%

2. **a.** K297.50
   b. K17.85, K279.65
   c. K118.02
   d. K0.00
   e. K3068.52
3. **a.** K20.76 **b.** K755.66
4. **a.**

Employee	Taxable Income (K)	Tax/fortnight (K)	Tax as a % of income
Flora	284	1.78	0.63%
Martin	568	64.76	11.4%

   b. No, the more you earn the higher the rate of tax.
5. **a.** K751.50 **b.** K45.09 **c.** K61.91
   d. 8.76% **e.** K53.76 **f.** 8.25%
6. **a. i.** K36.91 **ii.** K117.18
   b. K154.09 **c.** 19.91% **d.** K67.19
7. K80.59 **8. a.** K12.34 **b.** K295.66
9. **a.** K311.64 **b.** K430.36 **c.** K66.2.46

Unit 11.2 Activity 3B: Calculating tax payable (page 164)

1. **a.** K0.00 **b.** K23 290 **c.** K10 717.50 **d.** K37 870.00 **e.** K2637.80
2. **a.** D **b.** I **c.** D **d.** D **e.** D **f.** I **g.** I **h.** I **i.** I **j.** I
3. **a.** K45 300 **b.** K11 225.00
4. **a.** K66 510 **b.** K20 188.50 **c.** K19 438.50
5. **a.** K10 661 **b.** K36 054.00 **c.** K35 304.00
6. **a.** K62 309 **b.** K17 178.15 **c.** K16 128.15

Unit 11.2 Activity 3C: Calculating GST (page 167)

1. **a.** K4.20 **b.** K3.63 **c.** K235.00
2. **a.** K4.45 **b.** K36.36 **c.** K185.91
3. **a.** K113 **b.** K144.30 **c.** K1587.30 **d.** K31.30
4. **a.** K2.51 **b.** K3.87 **c.** K46.87 **d.** K8.16
5. **a.** K997.92 **b.** K45.82 **c.** K53.97 **d.** K439.95
6. **a.** K51.41 **b.** K34.27 **c.** K3.12 **d.** K2.03 **e.** K20.25
7. **a.** sausages K3.41, steak K7.66, chicken K4.87 **b.** K136.89 **c.** 28.6%

Unit 11.2 Activity 4A: Calculations using foreign exchange (page 170)

1.

PNG Kina	Currency	Equivalent amount
300.00	USD	USD131.73
50 900.00	EUR	EUR15635.97
15.50	IDR	IDR59 028.70
234 800.00	AUD	AUD98 770.97
86.75	SGD	SGD162.30

2.

Amount	Currency	Equivalent amount in Kina
45.50	USD	103.62
568.95	EUR	1852.10
10000	IDR	2.60
5650.00	AUD	13 431.41
450.00	SGD	240.53

3. K240.53 **4.** K65 is cheaper by K5.84

5. K844.46 **6.** K435.50

7. **a.** K2.183 **b.** K74.95

8. **a.** K50.00 **b.** NZD1277.60 **c.** K513.74

9. **a.** Large amounts eg 100 000 are often exchanged.

b. If accuracy to two decimal places is required then amounts greater than 999 need more than 5 decimal places if the exchange rate is less than 1.

c. AUD9999

10. **a.** Yes, use the exchange rates in Table 1 to find equivalent values of 1 kina.

b. 1.0438 **c.** USD104.38 **d.** 1.14615 **e.** GBP872.49

Unit 11.2 Activity 5A: Budgeting (page 176)

1. **a.**

Income		Expenses	
	Amount (K)		Amount (K)
Cow sales Pig sales Chicken sales Coconut sales	3 000 1 600 2 500 7 500	Chicken feed Fuel Workers' pay Other expenses	450 500 3 000 1 000
	14 600		**4 950**

b. K9 650 **c.** K31 633.33 **d.** K10 725

e. **i.** K250 900 **ii.** K379 600 **iii.** K128 700

2. **a.**

Expenses			
Item	Unit price (K)	Number	Amount (K)
Sausages	20	40	800
Tomato sauce	9	2	18
Bread	3.70	20	74
Butter	6.90	2	13.80
Cordial	9	2	18
Cooking oil	5.95	1	5.95
Onion	4	1	4
			933.33

b. K933.75

3.

Income	Amount (K)	Expense	Amount (K)
Wage	350	Food	150
Commission	1 200	Toiletries	50
		PMV fare	20
		Power	50
		Water Bill	20
		Clothes	50
		Rent	150
		Entertainment	50
	1 550		**540**

a. K1 550 **b.** K540 **c.** K1010

4.

Expense	Amount (K)
Admission fee	
- student	700
- Adult	21
Transport (both ways)	1 200
Lunch pack	1 430
Drinks	429
	3 780

a. K27.00 **b.** K3 891

5. a.

Expenses	Amount (K)
Accommodation	1 650
Meal	4 240
Transport	2 000
	6 890

b. K65 **c.** K70

Unit 11.2 Activity 6A: Interest and loans (page 181)

1. 15% **2.** 9.42%

3. 12% **4.** 8.84%

5. 9.1% **6.** 16%; 18%

7. a. K41 000 **b.** K82 000

8. 14.1% **9.** K830.00

10. K1043.00 **11.** K252

12. a. K1380 **b.** K380 **c.** K31.67 **d.** 82.3%

13. a. K5390.10 **b.** K2390.10 **c.** K79.67 **d.** 69%

14. a. i. K277 329.00 **ii.** K153 329 **iii.** 4.95%

b. i.

Year	7.6% interest charged (K)	Balance after interest (K)	Repayment (K)	Amount owing at the end of the year (K)
0				124000
1	9424	133424	11221.87	122202.13
2	9287.42	131489.55	11221.87	120267.68
3	9140.34	129408.02	11221.87	118186.15
4	8982.15	127168.30	11221.87	115946.43

ii. It is decreasing.

Unit 11.2 Activity 6B: Calculating interest (page 185)

1. 26%

2. K4.55

3. a. K2.27 **b.** K250.74

4. K480.00

5. K200; 20%

6. K157.20

7. a. K360 **b.** K248.89

8. 5.97%

9. 10.05%

10. a. 13 **b.** K120 **c.** 8%

Unit 11.3 Activity 1A: Statistical variables (page 191)

1. Discrete – a, d, f. Continuous – b, c, e. **2.** Univariate a, c; bivariate b, d.

3. 'Bad' b (closed one-answer type question); c (avoid 'cause' questions); 'Good' a, d (open questions).

4. **a.** Students deliberately not telling the truth; pupils not remembering exactly.

b. Students not remembering exactly since they have not kept a record; time-consuming to ask each student as they leave the tuck shop how much they spent.

c. Time-consuming gathering the information; difficult to know where to collect the information since many dogs are not registered.

d. People, when interviewed, may have just had a good/bad lesson affecting their feeling at the time; numbers of students taking a subject does not necessarily indicate popularity – some subjects are compulsory and others are not.

5. **a.** Students at Goroka Secondary School and Bena Bena Secondary School.

b. Number of cigarettes smoked (univariate).

c. Answers will vary. Fitness, money spent on cigarettes, age at which smoking started.

d. Students not giving accurate information about the number of cigarettes smoked per day – improve by asking the student being interviewed to keep a record of the number of cigarettes smoked each day for a week, for example.

6. **a.** Students at Mora Mora Technical School.

b. Mass of a school bag. Grade level at school. Bivariate.

c. Answers will vary. For example:

- What time of the day the bag is measured. Standardise the time, eg 8.30 am.
- Daily variation in bag weight (take average over a week).
- Differences between senior and junior bag weights (sample proportionally so numbers are fairly represented).

d. Provincial Government (eg install lockers), Health Department.

Unit 11.3 Activity 1B: Averages (page 197)

1. **a.** **i.** 3 **(A)** **ii.** 3 **(A)** **iii.** 3.4 **(A)**

b. **i.** 4.5 **(A)** **ii.** 4, 6 **(A)** **iii.** 4.6 **(A)**

c. **i.** 14 **(A)** **ii.** None. **(A)** **iii.** 13.8 **(A)**

d. **i.** 13.5 **(A)** **ii.** None **(A)** **iii.** 13.9 **(A)**

e. **i.** 105 **(A)** **ii.** 105 **(A)** **iii.** 105 **(A)**

f. **i.** 203.5 **(A)** **ii.** 203 **(A)** **iii.** 204 **(A)**

2. **a.** 2.44 **(A)** **b.** 2 **(A)**

3. a. 2.75 (A) b. 2.5 (A)

4. a.

Distance	Tally	f	m	f.m
32 < 34	𝍸	5	33	165
34 < 36	II	2	35	70
36 < 38	𝍸 𝍸	10	37	370
38 < 40	𝍸	5	39	195
40 < 42	𝍸 𝍸 I	11	41	451
42 < 44	𝍸 II	7	43	301
		40		1 552

(A)

b. $40 \le x < 42$ (A) c. 38.8 m (A)

d. $\frac{18}{40}$ or 45% (A)

Unit 11.3 Activity 1C: Measures of spread (page 201)

1. a. i. 11 (A) ii. 9 (A) iii. 3 (A) iv. 6 (A)
   b. i. 2.6 (A) ii. 18.3 (A) iii. 17.35 (A) iv. 0.95 (A)
   c. i. 48 (A) ii. 134 (A) iii. 108 (A) iv. 26 (A)
   d. i. 3.51 (A) ii. 25.1 (A) iii. 23.47 (A) iv. 1.63 (A)

2. a. 7.6 seconds (A) b. 20.3 seconds (A)
   c. 17.6 seconds (A) d. 2.7 seconds (A)

3. a. 222 (A) b. 89 (A)

4. a. $\bar{x} = 5$, s = 3.29

   b. i. $\bar{x} = 15$, s = 3.29

   ii. Mean is increased by 10 and standard deviation remains the same.

   c. i. $\bar{x} = 50$, s = 32.9

   ii. Both the mean and the standard deviation are multiplied by 10.

5. a. $\bar{x} = 11, s = 2$ b. $\bar{x} = 30, s = 10$ c. $\bar{x} = 26, s = 2$ d. $\bar{x} = 120, s = 40$

   e. $\bar{x} = 4, s = 2$ f. $\bar{x} = 3, s = 1$ g. $\bar{x} = 6, s = 2$

6. a. 4 b. 1.7 c. 2.45

7. a. 28 b. 8 c. 10.33

8. a. 6 b. 2 c. 2.45

9. a. 3 b. 2.45

Unit 11.3 Activity 2A: Bar graphs (page 206)

1.

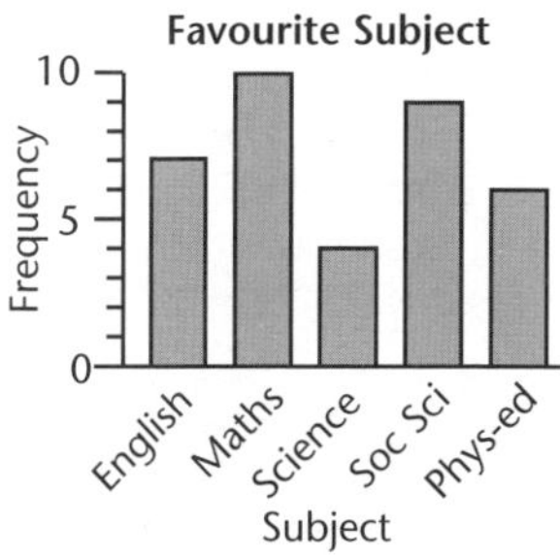

(*A*)

2.

Score	Frequency
0	1
1	4
2	3
3	5
4	2
5	4
6	3
7	1
8	3
9	2
10	2

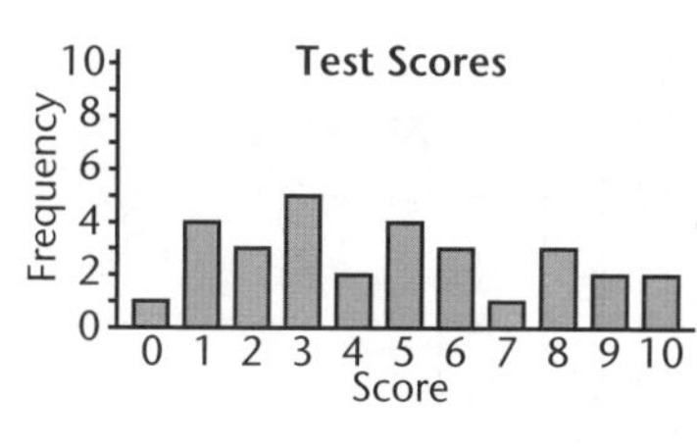

(*A*)

3.

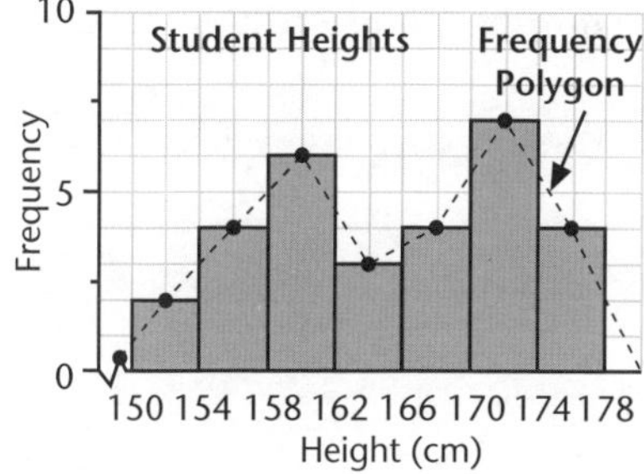

(*A*)

4.

Height (cm)	f
150–	2
155–	2
160–	6
165–	7
170–	6
175–	4
180–185	3
Total	30

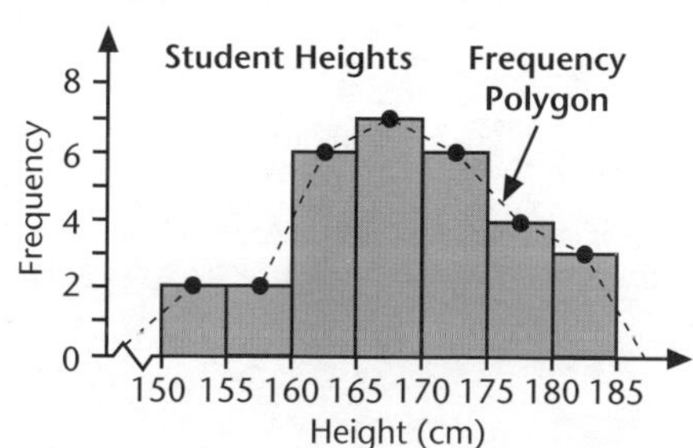

(*A*)

5. a.

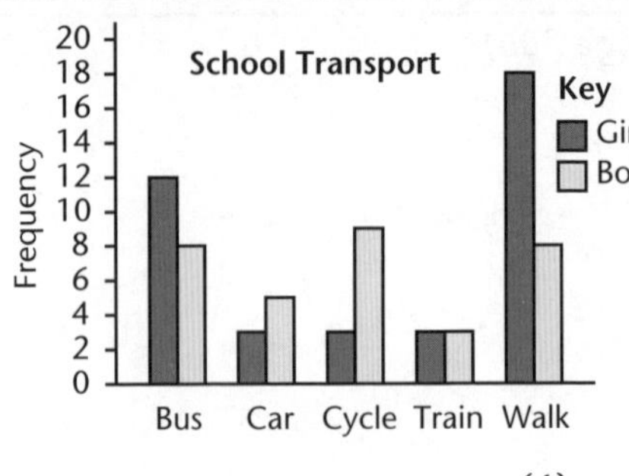

(A)

b. Girls outnumber boys when travelling by bus or walking. Equal numbers of boys and girls travel by train. More boys than girls cycle or travel by car. This could be because boys get licences earlier than girls on average, and some boys drive to school. ***(M)***

6. a.

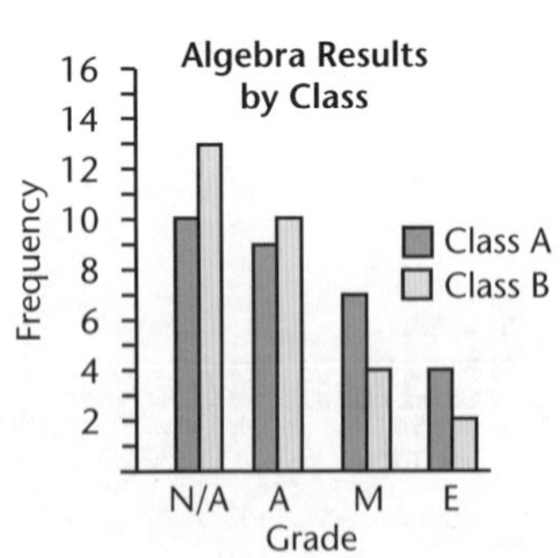

(A)

b. More of Class B gained 'Satisfactory achievement' or 'Low achievement' than Class A. More of Class A got 'High achievement' or 'Very high achievement'. Overall Class A did better, since fewer students did not achieve, and more students reached a higher standard. ***(M)***

Unit 11.3 Activity 2B: Cumulative frequency graphs (page 210)

1. a. i. 182 cm

ii. 189 cm

iii. 180 cm

iv. 191 cm

v. 172 cm

b. i. 182 cm

ii. 180 cm

iii. 78%

iv. 60%

2. a.

Maximum daily temperature (°C)	Cumulative frequency
< 15	0
< 20	1
< 25	6
< 30	20
< 35	25
< 40	28
< 45	30

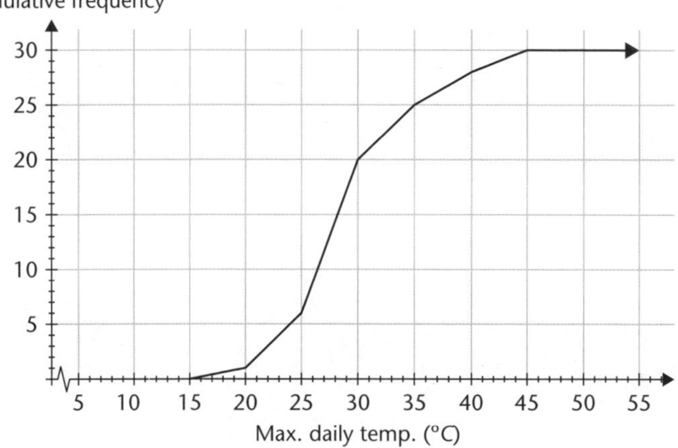

b. i. 28°C **ii.** 32°C

iii. 22°C **iv.** 7.5°C

c. i. On half of the days the maximum daily temperature(MDT) was 28°C or more.

ii. 80% of the days had a MDT of 32°C or less.

iii. 10% of the days had a MDT of 22°C or less.

iv. the middle 50% of the MDT's were spread over 7.5°C

3. a. 600 **b.** 162

c.

Speed (km/hr)	% cumulative frequency
< 50	0
< 60	1.3
< 70	4.3
< 80	18.3
< 90	37.7
< 100	73
< 110	98
< 120	99.7
< 130	100

d.

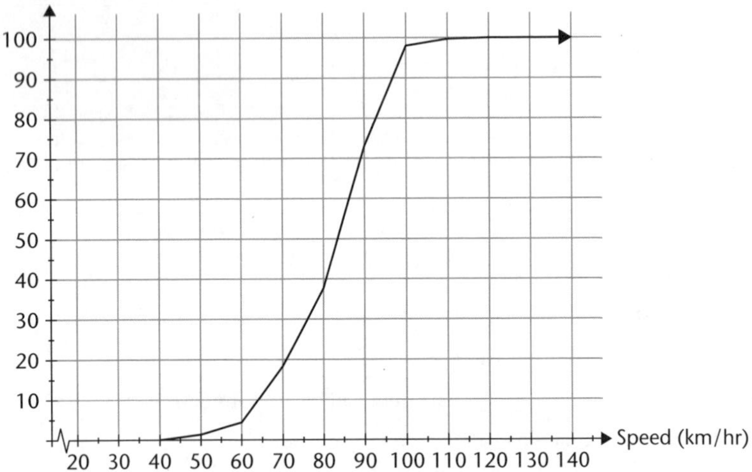

e. i. 83 km/hr; half of the cars are travelling at 83km/hr or more.

ii. 74 km/hr; 25% of the cars are travelling at 74 km/hr or less.

iii. 91 km/hr; 75% of the cars are travelling at 91 km/hr or more.

iv. 97 km/hr; 90% of the cars are travelling at 97km/hr or less.

v. 64 km/hr; 10% of the cars are travelling at 64 km/hr or less.

vi. 17 km/hr; the speed of the middle 50% of the cars is spread over 17 km/hr.

4. a.

Swim time (secs)	% cumulative frequency
< 50	0
< 55	2.5
< 60	20
< 65	62.5
< 70	90
< 75	95
< 80	100

b.

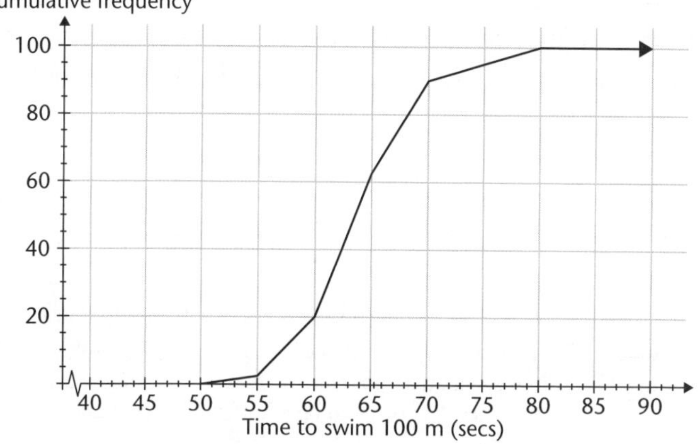

c. i. 63 secs; half of the swimmers had a time of 63 seconds or less

ii. 60.5 secs; 25% of the swimmers had a time of 62.5 secs or more

iii. 67.5 secs; 75% of the swimmers swam a time of 67.5 secs or less.

iv. 70 secs; 90% of the swimmers swam a time of 70 seconds or less.

v. 57.5%; the fastest 10% of the swimmers swam a time of 57.5 secs. or less.

vi. 7 secs; the middle 50% of the swimmers had swim times spread over 7 secs.

Unit 11.3 Activity 2C: Further data displays (page 215)

1. a. Length of leaves

5	1 1 3
4	1 1 2 2 2 3 4
3	0 1 3 4 6 7 8 8
2	8 9

(A)

b. i. 39.5 **(A)** **ii.** 42.5 **(A)** **iii.** 33.5 **(A)**

c. **(A)**

20 30 40 50 60

2. a.

7	0 1 2 4
6	1 3 8 8 9
5	0 1 1 2 2 3 4 5
4	4 6 7 7 8 8 9
3	1 2 2 3 5 6

(A)

b. i. 51 **(A)** **ii.** 63 **(A)** **iii.** 46 **(A)**

c. **(A)**

30 40 50 60 70 80

3. a. i. 78.5 **(A)** **ii.** 88 **(A)** **iii.** 64.5 **(A)** **iv.** 54 **(A)** **v.** 23.5 **(A)**

b.

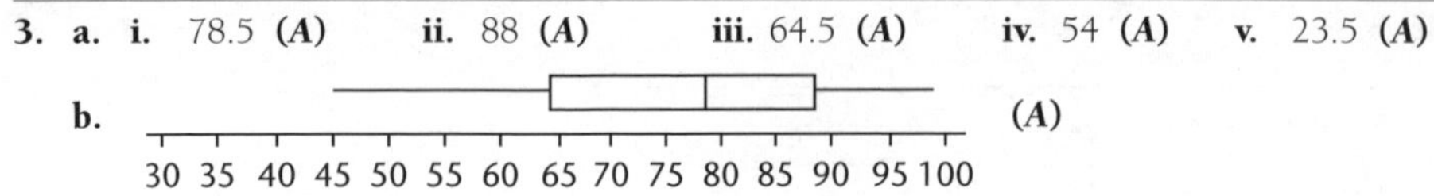

(A)

4. a.

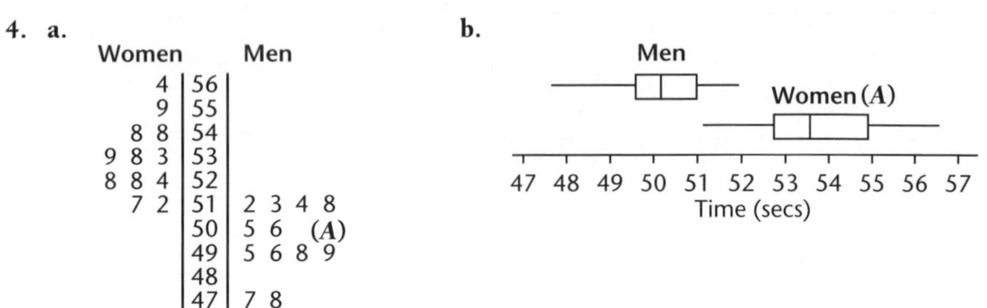

Women		Men
4	56	
9	55	
8 8	54	
9 8 3	53	
8 8 4	52	
7 2	51	2 3 4 8
	50	5 6 **(A)**
	49	5 6 8 9
	48	
	47	7 8

b.

c. Moving the women's times down 4 seconds (or moving the men's times up 4 seconds) would have both sets of data in a similar position on the box-and-whisker graph.

A 4-second start for the women appears to be a very fair 'handicap'. **(M)**

5. a.

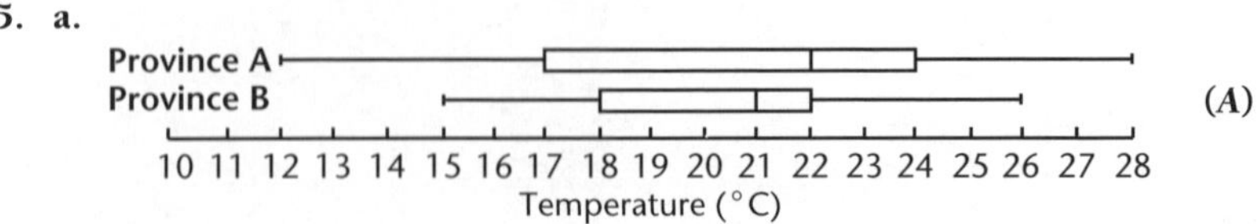

(A)

b. Province A has a median equal to the UQ of Province B, hence 50% of Province A's temperatures are higher than 75% of Province B's. Also, Province A has a higher maximum, UQ and higher median temperature than Province B's equivalent temperatures. Thus, Province A appears to be a hotter Province than B. **(M)**

Unit 11.3 Activity 2D: Time series graphs (page 218)

1. a. Estimated volume of cured vanilla bean exports, 1997–2006

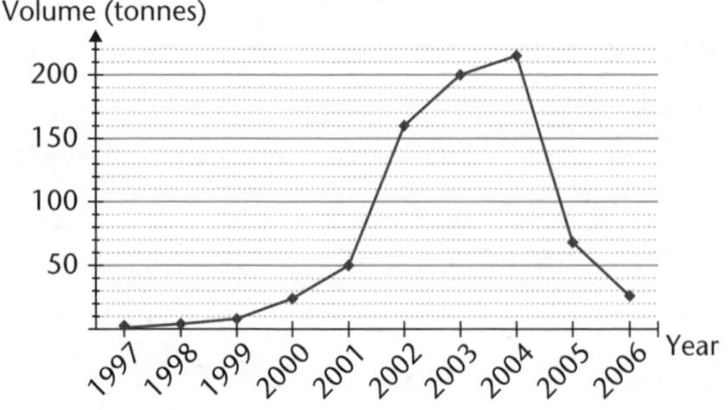

b. The vanilla bean exports showed an increase from 1997 to 2001, then there was a large increase in the year 2002, from 50 to 160 tonnes. The exports continued to increase in 2003 and 2004 then there was a large decrease in 2005, from 215 down to 68 tonnes. The decrease continued into 2006 with only 26 tonnes exported.

2. a. i. Yes ii. No iii. Yes, a downward trend

b. i. 1985 ii. 1997

c. There is a general downward trend over the years 1980 to 1997 although large fluctuations in production occur in this time; the highest production (950 tonnes) is in 1985 and the lowest in 1987(50 tonnes). After 1997 there appears to be a steady upward trend.

d. Approx. 450 tonnes.

3. a. i. Fruit and vegetable imports from Australia showed an increasing trend for the years 1983 to 1997. For the years 1997 to 2000 there was a dramatic decline in imports(750 tonnes to 350 tonnes) and then imports were steady for two years before declining slightly in 2003.

ii. Fruit and vegetable imports from New Zealand showed random fluctuations over the years 1983 to 2003 and a slight downward trend.

b. The imports from Australia and New Zealand show the same pattern over the years 2000 to 2003; steady for the first three years and then a decline in 2003. Australia imported approximately 500 tonnes more of fruit and vegetables.

c. Approximately 2500 tonnes and 2000 tonnes.

4. a. The coffee production shows a yearly seasonal trend with high production in June to August and the lowest production in February. Production was higher in 1983 than in 1982.

b. The food sales show a yearly seasonal trend with high sales in June to August and the lowest sales in February. Food sales were similar in both 1982 and 1983.

c. The patterns follow the same pattern over each year.

5. a. There appears to be an upward trend in the price of sweet potato in Port Moresby for 1998 to 2001, no increase in 2003 then a decline in the price(from 130t to 105 t) in 2003. The price increased again in 2004 and 2005 to 130t.

b. The price of sweet potato was steady for 1998 to 2000 and then there was a large increase in the price (from 35t to 110t) over the next two years. In 2003 the price decreased again to 55t and remained about that price until 2005.

c. The price of sweet potato appears to follow the same pattern over the years 1998 to 2005. However, the price was always less in Madang than in Port Moresby and the fluctuations were more pronounced.

6. a. The data shows seasonal fluctuations with a seven-day repeat of the pattern. The highs generally occur on Wednesdays and the lows on Sundays. There also appears to be an upward trend in the data.

b.

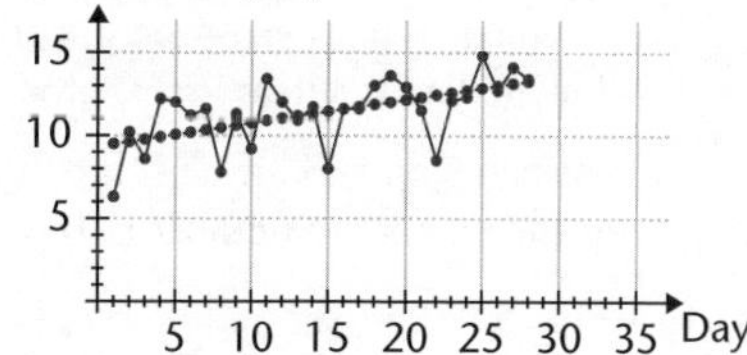

c. i. 17 km

ii. 18 km

iii. 19 km

d. Saturday's prediction. Saturday's values are always closer to the prediction line than Sundays and Wednesdays.

Unit 11.3 Activity 3A: Interpreting graphs and statistics (page 224)

1. a. That for every 1000 children under 5, in 2009 in PNG, 68 will die.

b. India

c. South Africa

d. i. PNG, Indonesia, India, Turkey and South Africa.

ii. PNG, India, South Africa

e. 25.3% decrease

2. a.

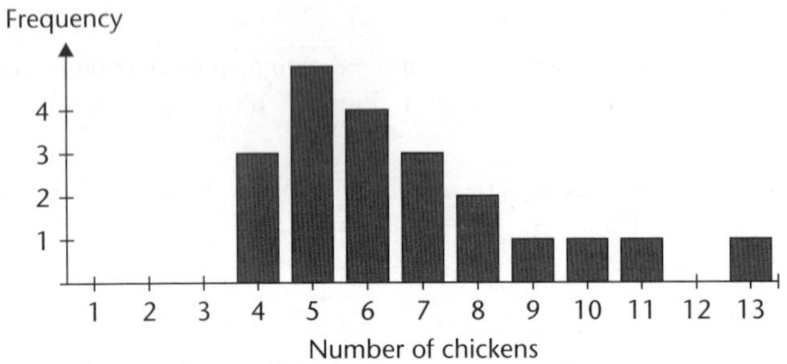

b. Median = 6, upper quartile = 8, lower quartile = 5

c.

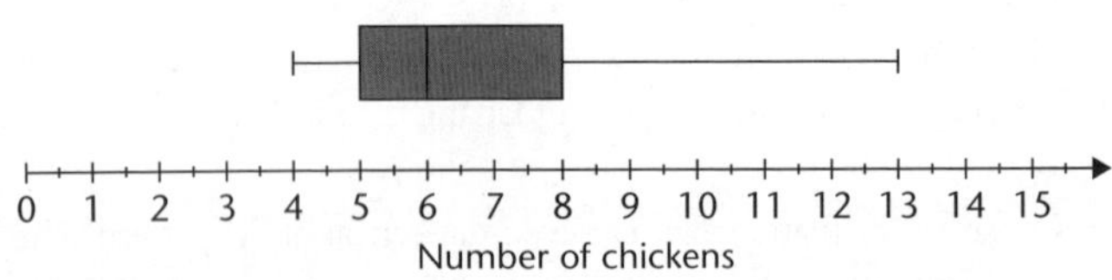

d. The data is positively skewed with a median of 6 chickens per household. (The mean is 6.7 chickens, suggesting that the data is positively skewed, and the mode is 5.) The number of chickens per household above the median is spread from 6 to 13 whereas the number of chickens per household below the median is spread from 4 to 6.

The middle 50% of the households have between 5 and 8 chickens. the minimum number of chickens in a household is 4 and the maximum number is 13.

3. **a.** In 1975 the production of Palm oil was about 17 000 tonnes and in 2005 it was about 330000 tonnes; the production in 2005 was about 20 times the production in 1975.

b. The production of coffee has about doubled over the thirty years from 1975 to 2005.

c. The production of Copra has declined from about 90 000 tonnes in 1975 to 20 000 tonnes in 2005.

4. Both the graph and the boxplot show that the data is negatively skewed; the spread of the number of avocados per tree below the median is from 192 to 221 whereas the spread of the number of avocados per tree above the median is from 221 to 235. The median number of avocados per tree was 221 with 25% of the trees producing 211 or fewer avocados and 25% producing 226 or more avocados. The minimum number of avocados produced on a tree was 192 and the maximum number was 235.

5. The distributions of scores in 2009 and 2010 were symmetric about a median score of 18, however the scores in 2009 were more variable. The range for 2009 was 28 compared with a range of 19 for 2010.

The scores for 2011 were positively skewed with a median of 19. The range of scores in 2011 was 21 so the scores were not as variable as 2009 but slightly more variable than 2010.

Overall the scores in 2011 were higher; the median was higher than the other two years and 2011 had the highest minimum score(11) and the highest maximum score (32).

6. **a.** It means that 65% of the males in PNG can read and write.

b. PNG, Malaysia, South Africa and the Philippines.

c. India, Afganistan and Egypt.

d. Afganistan

e.

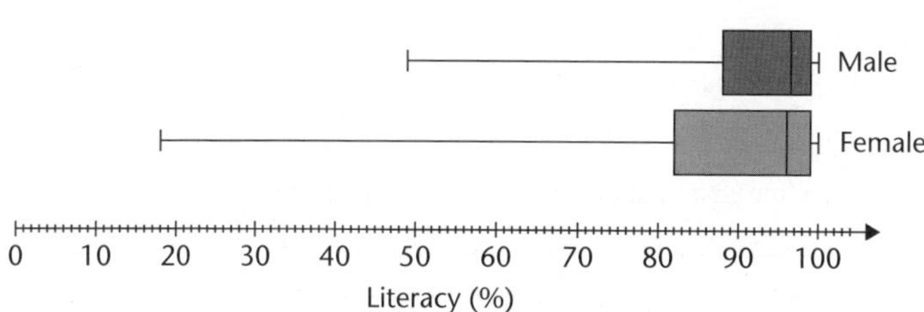

f. The distribution of the data for both males and females is negatively skewed indicating a much larger spread of literacy rates below the median literacy rate. The spread of the data for females is larger with a range of 82 compared to a range of 51 for males. The median literacy rate for males was slightly higher than that for the females; 97% compared to 96%. For females the middle 50% of the countries had a literacy rate that was more variable; an IQR of 17 compared to an IQR of 11 for males.

Overall it can be said (based on this sample of 14 countries) that female literacy is slightly lower than male literacy and that the female literacy is more variable.

Unit 11.4 Activity 1A: Angles (page 229)

1. **a.** $a = 80°$ (∠s on line) **b.** $b = 92°$ (∠s on line)

c. $c = 25°$ (∠s on line)

d. $d = 76°$ (∠s on line), $e = 12°$ (∠s on line) or (∠s at point)

e. $f = 155°$ (∠s on line), $g = 95°$ (∠s on line) or (∠s at point)

f. $h = 148°$ (∠s at point)

g. $i = 50°$ (∠s on line), $j = 130°$ (vert opp ∠s)

h. $k = 37°$ (vert opp ∠s), $l = 143°$ (∠s on line)

i. $m = 70°$ (∠s on line)

2. 60° **(A)**

3. **a.** $u = 25°$ (∠s on line) **b.** $v = 50°$ (∠s on line) **c.** $w = 45°$ (∠s on line)

d. $x = 72°$ (∠s at point) **e.** $y = 50°$ (∠s on line) **f.** $z = 55°$ (vert opp ∠s)

4. **a.** $s + t + u = 90°$ **b.** $s + t + u = 180°$ **c.** $2s + u + t = 270°$

Unit 11.4 Activity 1B: Angles and parallel lines (page 233)

1. **a.** $i = 78°$ (vert opp ∠s), $j = 78°$ (alt ∠s, // lines) or (corr ∠s, // lines)

b. $k = 80°$ (vert opp ∠s), $l = 80°$ (corr ∠s, // lines)

c. $m = 115°$ (corr ∠s, // lines), $n = 65°$ (∠s on line)

d. $p = 48°$ (alt ∠s, // lines), $q = 50°$ (∠s on line), $r = 128°$ (coint ∠s, // lines)

e. $m = 128°$ (coint ∠s, // lines), $n = 52°$ (∠s on line) or (alt ∠s, // lines)

f. $p = 61°$ (∠s on line), $q = 66°$ (alt ∠s, // lines)

$r = 127°$ (coint ∠s, // lines) or (alt ∠s, // lines)

2. 45° **(A)**

3. **a.** $x = 40°$ (∠s on a line). Not parallel. (corr ∠s not equal) **(M)**

b. $y = 80°$ (vert opp ∠s). Parallel. (coint ∠s add to 180°) **(M)**

c. $z = 115°$ (∠s on a line). Not parallel. (alt ∠s not equal) **(M)**

4. **a.** $x = 54°$, $y = 112°$

b. EF//GH (corr ∠s =), AB//CD (alt ∠s =) **(M)**

5. **a.** $x = 70°$ (alt ∠s, // lines and ∠s at a point) **(M)**

b. $a = 76°$ (alt ∠s, // lines after drawing a parallel line through vertex of angle a) **(M)**

6. ∠CBQ = a (vert opp ∠s), ∠CBQ = b (corr ∠s, // lines) ∴ $a = b$

7. ∠BAD = x (alt ∠s, // lines), $x = y$ (corr ∠s, // lines)

8. $\angle QST = 180° - x$ (coint $\angle$s, // lines)

$\angle QST + \angle STR = 180° - x + x$

$= 180°$

$\therefore$ SQ//TR (coint $\angle$s add to 180°) **(E)**

9. $\angle CDG = x$ (corr $\angle$s, // lines)

$\angle CDG = y$ (given $x = y$)

$\angle CDG = \angle CEF$

$\therefore$ BG//EF (corr $\angle$s =) **(E)**

10. a. $s + t = 180°$ ($\angle$s on line)

$t + u = 180°$ (coint $\angle$s, // lines)

$s + 2t + u = 360°$ [adding above two equations] **(M)**

b. $s + t = 180°$ (vert opp and coint $\angle$s, // lines)

$s + u + 90 = 180°$ ($\angle$s on line)

$2s + t + u + 90 = 360°$ [adding above two equations]

$2s + t + u = 270°$ [subtracting 90°] **(M)**

c. Using alternate angles and vertically opposite angles, $t = s + u$.

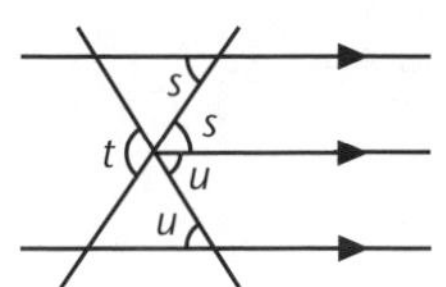

(E)

Unit 11.4 Activity 2A: Angles and triangles (page 237)

1. a. $a = 92°$ ($\angle$ sum Δ)

b. $b = 46°$ ($\angle$ sum isos Δ)

c. $c = 74°$ (ext $\angle$ Δ)

d. $d = 70°$ ($\angle$ sum isos Δ), $e = 125°$ (ext $\angle$ Δ)

e. $f = 81°$ ($\angle$ sum Δ), $g = 45°$ (ext $\angle$ Δ) **(A)**

f. $h = 84°$ (corr $\angle$s, // lines), $i = 22°$ (ext $\angle$ Δ)

g. $d = 70°$ ($\angle$ sum isos Δ), $e = 110°$ (ext $\angle$ Δ)

h. $f = 71°$ ($\angle$ sum Δ), $g = 51°$ (ext $\angle$ Δ)

i. $h = 40°$ (alt $\angle$s, // lines), $i = 148°$ (ext $\angle$ Δ)

2. $\angle$ sum Δ

QT

40°

alt $\angle$s = **(M)**

3. x

x

base $\angle$s isos Δ **(M)**

4. Let $\angle BFC = c$

$\therefore \angle DFE = c$ (vert opp $\angle$s)

$x = a + c$ (ext $\angle$ Δ)

$y = b + c$ (ext $\angle$ Δ)

$\therefore x = y$ **(E)**

(Other possible solutions exist.)

Unit 11.4 Activity 2B: Angles and quadrilaterals (page 240)

1. 120° **(A)**

2. **a.** u = 28° (∠ sum Δ), v = 17° (∠BAE = 45°, AE axis of symmetry) **(A)**,

w = 28° (alt ∠s, // lines) **(A)**

b. a = 118° (BG // CF (rhombus) so ∠ABG = 62° (coint ∠s, // lines) then coint ∠s, // lines) **(M)**

b = 35° (∠FCD = 70° (rhombus) so ∠DCE is half of 70°) **(M)**

c = 360 – 62 – 35 × 2 = 228° (∠s at a point) **(M)**

c. a = 104° (angle sum quad, ∠FIG = 76° vert opp ∠s) **(M)**

b = 21° (angle sum arrowhead, since ∠FHG = 256° ∠s at a point) **(M)**

d. u = 360 – 75 – 80 = 205° (int ∠s parallelogram plus ∠s at pt) **(M)**

v = 360 – 105 – 100 = 155° (int ∠s parallelogram plus ∠s at pt) **(M)**

3. 360°, 360°, 180°, 180°

4. Interior angles of quadrilateral are a, b, c and d (vert opp ∠s) so

$a + b + c + d$ = 360° (sum int ∠s Δ) **(M)**

5. ∠BEA = ∠EAB (base ∠s isos Δ)

∠EAB = ∠FCB (opp ∠s, parallelogram =)

∠AEB = ∠FBC (alt ∠s, // lines)

∴ ∠ FCB = ∠ FBC

∴ ΔBCF is isosceles because it has two equal angles. **(E)** (*Other possible solutions exist.*)

Unit 11.4 Activity 2C: Angles and polygons (page 243)

1. **a.** g = 66° (sum int ∠s quad), h = 114° (∠s on line)

b. j = 102° (∠s on line), k = 132° (∠ sum pentagon = 3 × 180 = 540°) **(A)**

c. l = 108° (∠s on line), m = 112° (∠s on line), n = 82° (int ∠s pentagon) **(A)**

d. p = 250° (∠ sum octagon = 6 × 180 = 1 080°) **(A)**

e. u = 225° (∠ sum decagon = 8 × 180 = 1 440°, 6 × 90 + 4u = 1 440°) **(A)**

f. v = 135° (∠ sum hexagon = 4 × 180 = 720°) **(A)**

2. 140° (each ext ∠ is 40°) **(A)**

3. 150° **(A)**

4. 12 sides (each ext ∠ = 30°) **(A)**

5. **a.** 15 **(A)** **b.** 156° **(A)** **c.** 2 340° **(A)**

6. 5 040° **(A)**

7. a. 90° **(A)**

 b. Where these paving stones meet, the angles are 90°, 135° and 135°.

 90 + 135 + 135 = 360° so no 'gaps'. **(M)**

8. a. 45° **(A)** b. 135° **(M)**

 c. 135 is not a factor of 360 so corners cannot meet without 'gaps'. **(M)**

Unit 11.4 Activity 2D: Scale factor and enlargements (page 246)

1. a. $\frac{4}{5}$ b. $\frac{33}{28}$ c. $\frac{43}{76}$

2. a. i. $\frac{1}{2}$ ii. 1.5m

 b. i. $\frac{2}{3}$ ii. 32 cm

 c. i. $\frac{7}{4}$ ii. 8.75 cm

 d. i. $\frac{1}{2}$ ii. 2 cm

3. a. No b. Yes c. No d. No

4. Show your answer to your teacher.

5. Show your answer to your teacher.

6. Show your answer to your teacher.

7. 10 000

8. Show your answer to your teacher.

Unit 11.4 Activity 2E: Triangles and quadrilaterals (page 251)

1. a. 4.5 b. 9

2. a. B and C b. B and C c. A and B d. A and B

3. Show your answers to your teacher.

4. $x = 40$, $y = 45$

5. $18\frac{2}{3}$, 10.5

6. 36 cm

7. a. $x = 6\frac{2}{3}$ cm, $y = 6.75$ cm b. $a = 12.5$ cm, $b = 9.6$ cm

 c. p = 4 cm, q = 7 cm d. x = 6 cm, y = 6 cm

Unit 11.4 Activity 2F: Congruent figures (page 254)

1. b. and h.; c. and e.
2. a and j; d and e; h and i.
3. 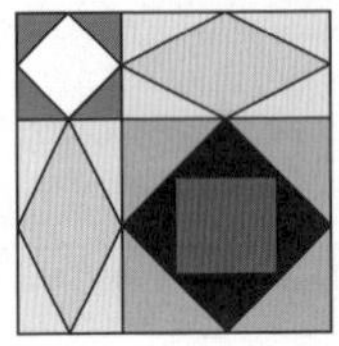

Unit 11.4 Activity 2G: Congruent triangles (page 257)

1. b.
2. a. and c.; b. and d.; e. and f.
3. Show your answers to your teacher.

Unit 11.4 Activity 3A: Perpendiculars and bisectors (page 265)

1. 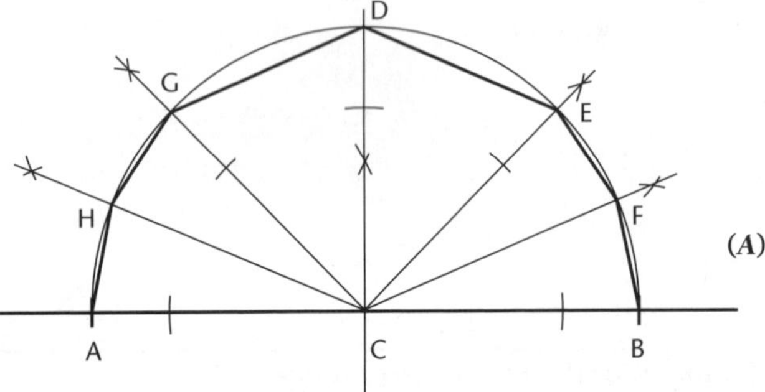

2. 3.

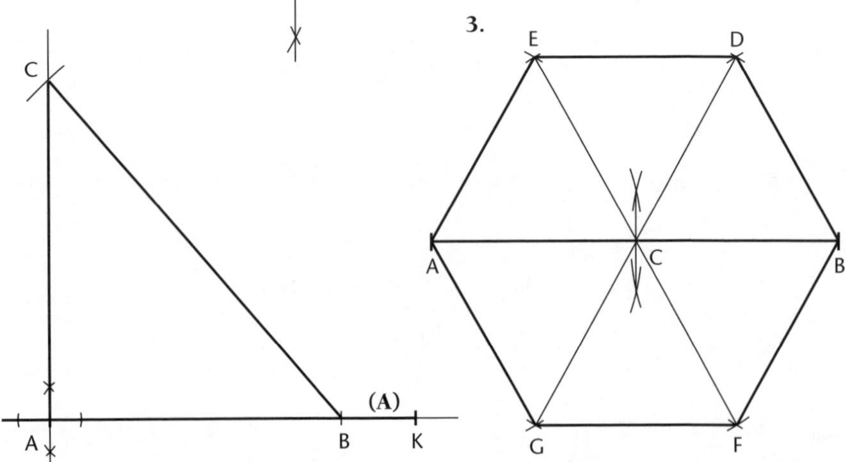

Shape is a regular hexagon. **(A)**

4.

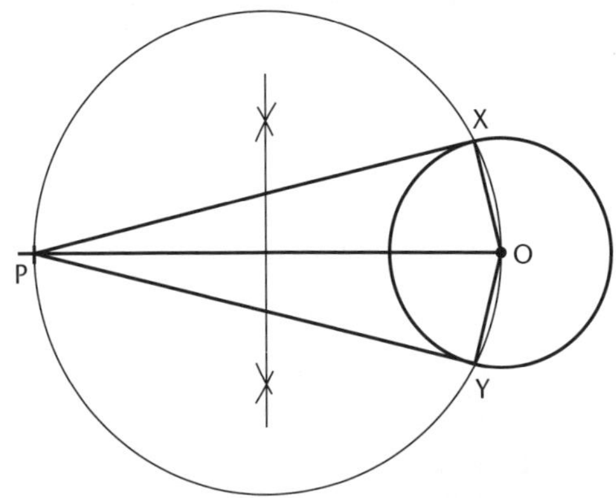

Shape is a kite. **(A)**

5.

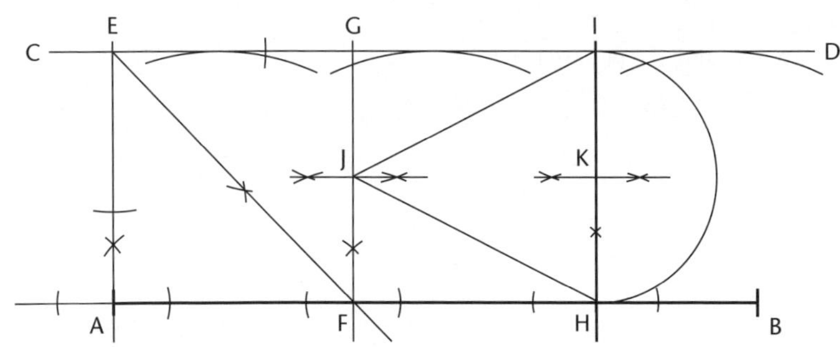

(A)

Unit 11.4 Activity 3B: Three-dimensional shapes (page 270)

1. i.

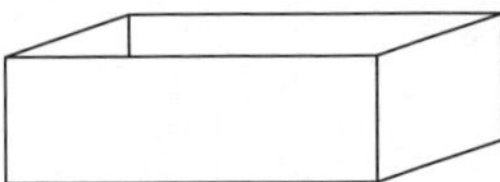

ii.

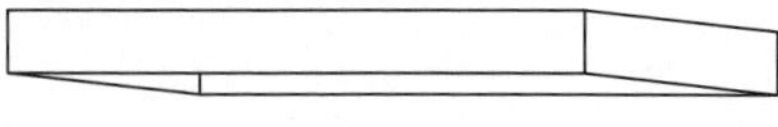

2.

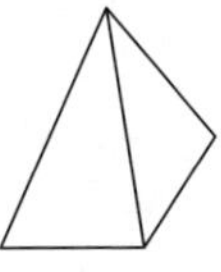

3.

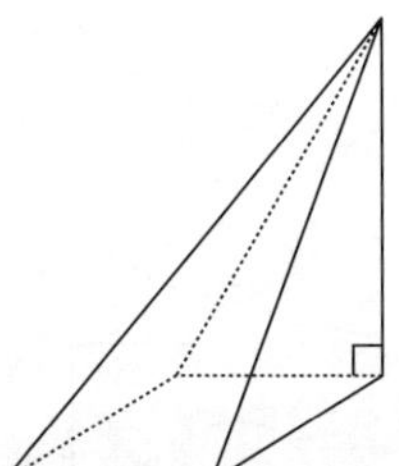

4.

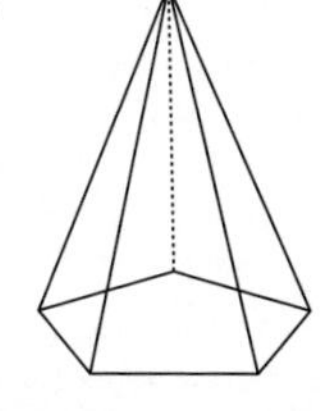

5. Show your construction to your teacher.

6.

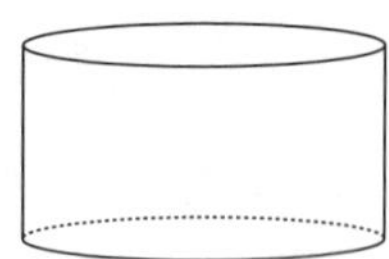

7. i.

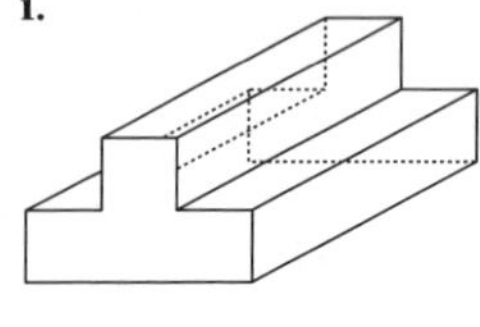

ii.

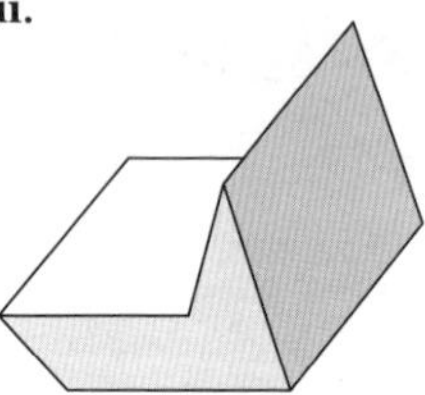

8. Show your construction to your teacher.

9. i.

ii.

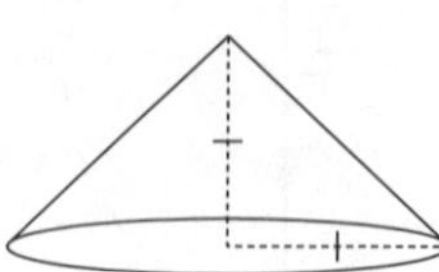

10.

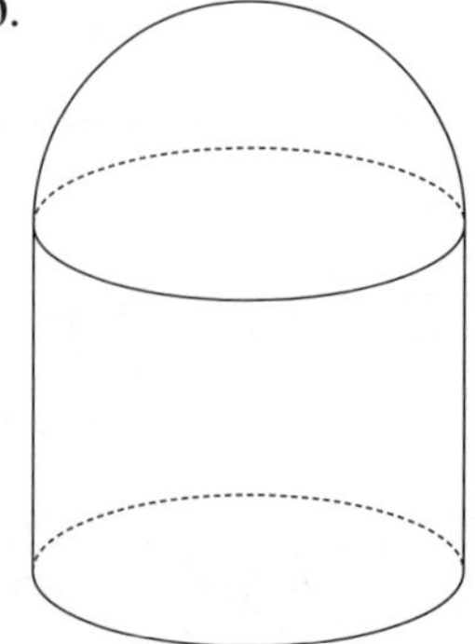

11. a.

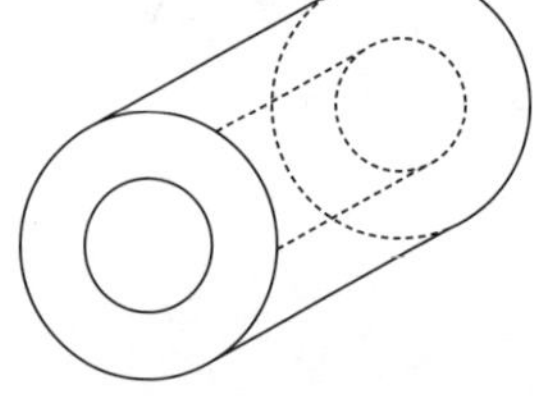

b.

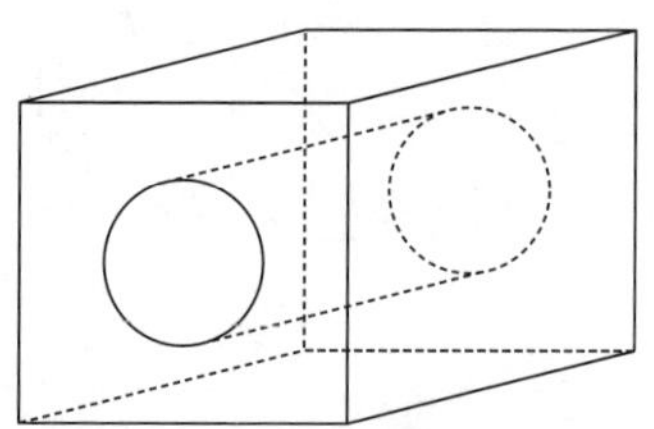

Unit 11.4 Activity 3C: Truncated solids (page 273)

1. Vertical and horizontal cuts all produce rectangular shapes. Cutting-off one corner will produce a triangular surface and an oblique cut will produce a trapezoidal surface.

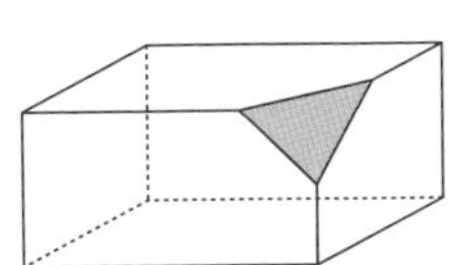
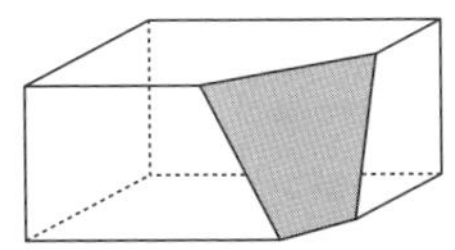

2. All cuts through a sphere produce a circular surface; the largest circle is produced by a cut through the centre of the sphere.

3.

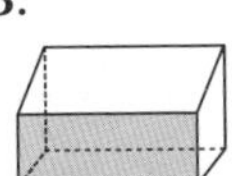

Vertial cut parallel to the back.

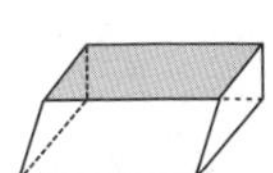

Horizontal cut

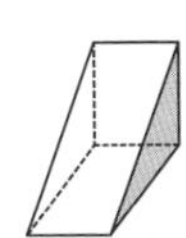

Vertical cut parallel to the side.

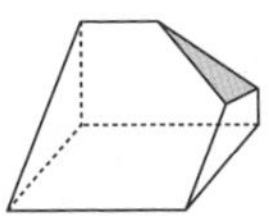

Oblique cut removing one of the vertices.

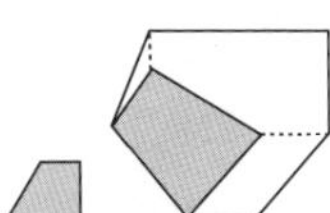

Oblique cut removing two of the vertices.

4.

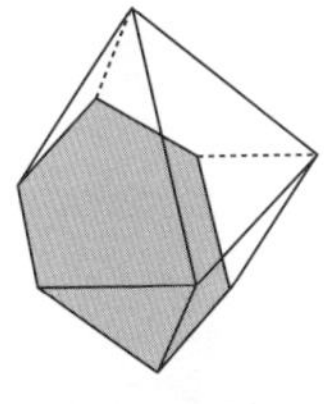

hexagon

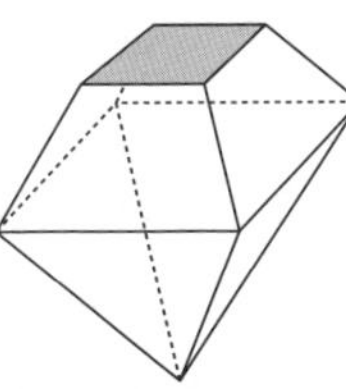

square

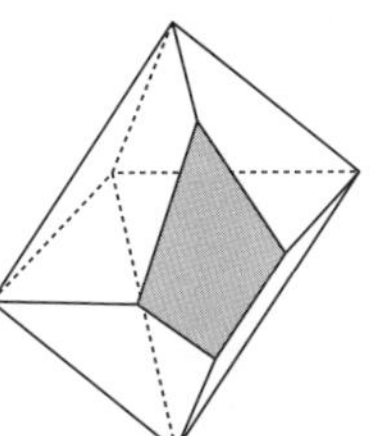

kite

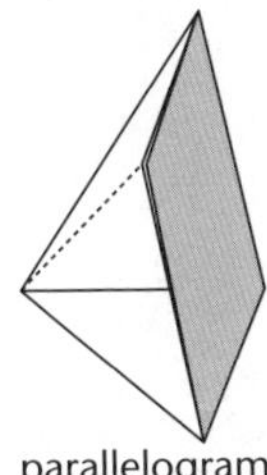

parallelogram

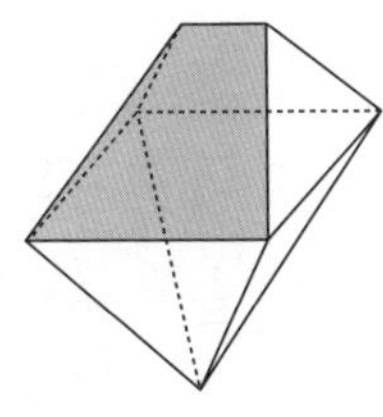

trapezium

Unit 11.4 Activity 4A: Circle geometry (page 278)

1. **a.** $a = 35°$ (∠s same arc)
   b. $b = 92°$ (∠ at centre)
   c. $c = 21°$ (∠ at centre)
   d. $d = 90°$ (∠ in semi-circle), $e = 70°$ (base ∠ isos Δ), $f = 110°$ (ext ∠ Δ)
   e. $g = 90°$ (∠ in semi), $h = 25°$ (∠ in semi), $i = 65°$ (base ∠ isos Δ), $j = 130°$ (∠ sum Δ)
   f. $k = 32°$ (base ∠ isos Δ), $l = 64°$ (∠ on line), $m = 90°$ (∠ in semi), $n = 50°$ (∠ sum Δ)

2. **a.** $a = 90°$ ($\angle$ in semi-circle), $b = 28°$ ($\angle$s same arc), $c = 62°$ ($\angle$ sum Δ) **(M)**

b. $d = 26°$ (base $\angle$ isos Δ), $e = 52°$ ($\angle$ at centre) **(M)**

c. $f = 56°$ ($\angle$s same arc), $g = 34°$ ($\angle$ in semi-circle, sum $\angle$s Δ) **(M)**

d. $r = 80°$ ($\angle$ at centre), $s = 50°$ (base $\angle$ isos Δ) **(M)**

e. $t = 42°$ (alt $\angle$s // lines), $u = 42°$ ($\angle$s same arc), $v = 96°$ ($\angle$ sum Δ, vert opp $\angle$s) **(M)**

f. $w = 40°$ ($\angle$s same arc), $x = 90°$ ($\angle$ in semi), $y = 130°$ (ext $\angle$ Δ),
$z = 25°$ (base $\angle$ isos Δ) **(M)**

3.
AO = OD (radii)
$\angle$AOD = 52° ($\angle$s in isos Δ)
$\angle$BOC = 52° (symmetry)
$\angle$DOC = 76° ($\angle$s on line) **(M)**

4.
$\angle$OQS = 35° (base $\angle$ isos Δ OQS)
$\angle$QOS = 110° ($\angle$ sum Δ)
$\angle$QPS = 55° ($\angle$ at centre) **(M)**

5. **a.** $\angle$OBA = 63° (base $\angle$ isos Δ), $\angle$OBC = 27° ($\angle$ in semi) **(M)**

b. $\angle$AOD = 126° ($\angle$s on line), $\angle$ODA = 27° (base $\angle$ isos Δ) **(M)**

c. Alternate angles $\angle$OBC and $\angle$ODA are equal so AD is parallel to BC. **(M)**

6.
$x = 2a$ ($\angle$ at centre)
$x = 2b$ ($\angle$ at centre)
$\therefore 2a = 2b$
$\therefore a = b$ **(E)**

7.
$\angle$BAO $= a$ (base $\angle$ isos Δ)
$\angle$AOD $= 2a$ (ext $\angle$ Δ)
$\angle$BCO $= b$ (base $\angle$ isos Δ)
$\angle$DOC $= 2b$ (ext $\angle$ Δ)
$\angle$AOC = $\angle$AOD + $\angle$DOC
$= 2a + 2b$
$= 2(a + b)$
$= 2\angle$ABC **(E)**

8.
$x + y = 180°$ ($\angle$s on on line)
$\angle$ACO $= \frac{1}{2}x$ ($\angle$ at centre)
$\angle$BCO $= \frac{1}{2}y$ ($\angle$ at centre)
$\angle$ACB = $\angle$ACO + $\angle$OCB
$= \frac{1}{2}x + \frac{1}{2}y$
$= \frac{1}{2}(x + y)$
$= \frac{1}{2} \times 180°$
$= 90°$ **(E)**

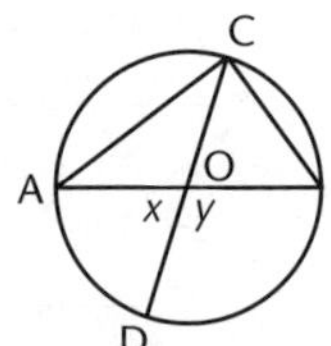

Unit 11.4 Activity 4B: Cyclic quadrilaterals (page 282)

No grading given for one-step problems.

1. **a.** $i = 80°$ (opp ∠s cyc quad), $e = 84°$ (ext ∠ cyc quad)

b. $j = 58°$ (rad ⊥ tan), $k = 122°$ (opp ∠s cyc quad)

c. $m = 65°$ (rad ⊥ tan), $n = 130°$ (∠ sum isos Δ)

d. $t = 75°$ (∠s same arc), $u = 61°$ (ext ∠ cyc quad)

e. $v = 48°$ (rad ⊥ tan), $w = 144°$ (∠ sum isos Δ), $x = 72°$ (∠ at centre),
$y = 108°$ (opp ∠s cyc quad)

f. $a = 90°$ (rad ⊥ tan), $b = 48°$ (∠s Δ), $c = 42°$ (symmetry), $d = 48°$ (symmetry)

2. **a.** ∠ADC = 40° (alt ∠s // lines), $x = 50°$ (rad ⊥ tan) **(M)**
$w = 80°$ (∠ at centre), ∠ODB = 50° (base ∠ isos Δ), z = 40° (rad ⊥ tan) **(M)**

b. ∠OAC = 40° (symmetry), ∠OCA = 90° (rad ⊥ tan), $a = 50°$ (∠ sum Δ) **(M)**
∠COB = 100° (symmetry), $b = 50°$ (∠ at centre) **(M)**

c. ∠OBT = ∠OAT = 90° (rad ⊥ tan), $c = 130°$ (∠ sum quad) **(M)**
∠BCA = 65° (∠ at centre), $d = 30°$ (∠BCO + d = 65°) **(M)**

d. ∠ADC = 80° (∠ at centre), $e = 80°$ (ext ∠ cyclic quad) **(M)**

e. ∠SQR = 55° (∠s on same arc), $f = 35°$ (∠ in semi) **(M)**
∠PSQ = 55° (alt ∠s // lines), g = 35° (angles on line) **(M)**

f. ∠TWU = 48° (alt ∠s // lines), $h = 42°$ (∠ in semi (TWV)) **(M)**
∠WUV = 90° (corr ∠s and rad ⊥ tan), $i = 48°$ (∠ sum Δ) **(M)**

3. Reflex angle DOB = 216° (∠ at centre)
a = 144 (∠s at point) **(M)**
∠DAB = 72° (opp ∠s cyclic quad)
∠OAB = 60° (Δ OAB is equilateral)
b = 12° **(M)**

4. **a.** A, B, D, E **(M)**

b. A, B, C, D and A, B, D, E **(M)**

c. B, C, E, F and A, B, D, F **(M)**

d. A, B, C, D **(M)**

e. A, B, C, D **(M)**

f. A, B, F, E or B, C, D, E **(M)**

5. **a.** and **b.**

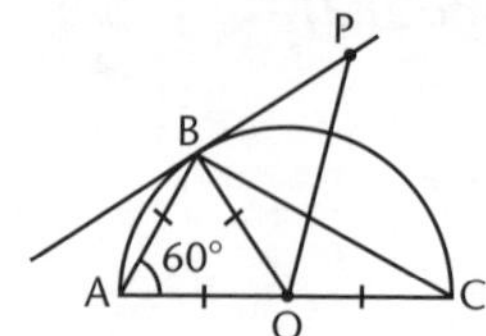

c. **i.** Right-angled. **ii.** Isosceles.

iii. Right-angled. iv. Equilateral.

d. No, because ∠PBO = 90° no matter what position P is. Hence the other two angles of ΔOBP are acute. **(*M*)**

6. ∠RCB = ∠BQR (∠s same arc)

∠BQR = ∠PAB (ext ∠ cyc quad)

∠RCB = ∠PAB

PA is parallel to CR (alt ∠s =) **(*E*)**

7. ∠OAC = 90° – x (rad ⊥ tan)

∠OCA = 90° – x (OAC isos Δ)

∠AOC = 180° – 2(90° – x) (∠s Δ)

= 180° – 180° + $2x$

= $2x$

∠AOC = $2y$ (∠ at centre)

∴ $2x = 2y$

$x = y$ **(*E*)**

8. Obtuse ∠BOD = 2C (∠ at centre)

Reflex ∠BOD = 2A (∠ at centre)

2A + 2C = 360° (∠s at point)

A + C = 180° **(*E*)** [dividing by 2]

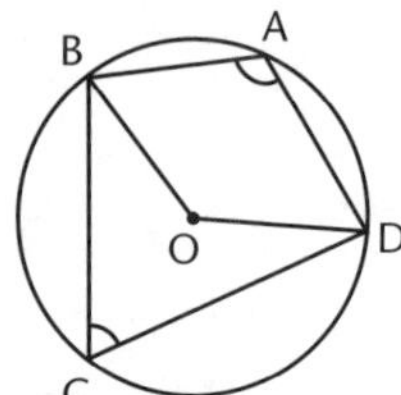

Unit 11.5 Activity 1A: Pythagoras' Theorem (page 287)

1. **a.** 4.47 **(*A*)** **b.** 8.94 m **(*A*)** **c.** 7.62 cm **(*A*)**

d. 9.90 **(*A*)** **e.** 6.93 m **(*A*)** **f.** 30.74 **(*A***

g. 8.31 cm **(*A*)** **h.** 17 mm **(*A*)** **i.** 7.55 cm **(*A*)**

2. 7.86 m **(*A*)**

3. 139.6 cm **(*A*)**

4. 13.27 mm **(*A*)**

5. **a.** Right-angled. **b.** Not right-angled. **c.** Right-angled.

Unit 11.5 Activity 1B: Solving problems using Pythagoras' Theorem (page 289)

1. 240 m **(M)**

2. 28.84 m **(A)**

3. a. When the walls are at right angles, the lengths of the walls and the diagonal will obey Pythagoras' Theorem. **(M)**

b. 9.44 m **(A)**

4. 7.16 m **(M)**

5. 82 cm **(M)**

6. Yes – maximum possible length is 1.08 m. **(M)**

7. 38.47 m **(A)**

8. 9.17 m **(M)**

9. 66.4 m **(A)**

10. a. 5.3 cm **(M)** **b.** 3.6 cm **(M)**

Unit 11.5 Activity 2A: Trigonometric ratios (page 292)

1. a. 13 **b.** 12 **c.** 12 **d.** $\frac{12}{13}$ **e.** $\frac{5}{12}$ **f.** $\frac{12}{13}$ **g.** $\frac{5}{13}$

2. a. 10 **b.** 8 **c.** 6 **d.** $\frac{8}{6}$ **e.** $\frac{6}{10}$ **f.** $\frac{8}{10}$ **g.** $\frac{8}{10}$

Unit 11.5 Activity 2B: Using your calculator (page 294)

1. a. 0.3420 **b.** 0.9848 **c.** 2.7475 **d.** 0.7880 **e.** 1.1106 **f.** 0.4384
g. 0.8320 **h.** 4.8716 **i.** 0.2147

2. a. 40.0° **b.** 40.0° **c.** 60.0° **d.** 50.0° **e.** 38.4° **f.** 26.9°
g. 28.9° **h.** 51.0° **i.** 6.7°

Unit 11.5 Activity 2C: Finding unknown sides (page 295)

1. a. 5 **(A)** **b.** 7.2 **(A)** **c.** 6.7 **(A)** **d.** 14 **(A)** **e.** 14.3 **(A)**
f. 13.5 **(A)** **g.** 14.2 **(A)** **h.** 7.4 **(A)** **i.** 16.2 **(A)**

2. a. 13.7 **(M)** **b.** 4.4 **(M)** **c.** 14.8 **(M)** **d.** 41.5 **(M)** **e.** 6.57 **(M)**
f. 12.7 **(M)** **g.** 19.2 **(M)** **h.** 13.8 **(M)** **i.** 52.9 **(M)**

Unit 11.5 Activity 2D: Finding unknown angles (page 297)

1. a. 34.8° **(A)** **b.** 53.1° **(A)** **c.** 29.9° **(A)** **d.** 56.3° **(A)** **e.** 36.9° **(A)**
f. 53.1° **(A)** **g.** 38.9° **(A)** **h.** 51.3° **(A)** **i.** 60.3° **(A)**

2. a. $s = 58.0°$, $t = 32.0°$ **(A)** **b.** $u = 74.5°$, $v = 15.5°$ **(A)**
c. $w = 52.3°$, $x = 37.7°$ **(A)** **d.** $y = 19.0°$, $z = 71.0°$ **(A)**
e. $a = 35.7°$, $b = 54.3°$ **(A)** **f.** $c = 55.4°$, $d = 34.6°$ **(A)**

Unit 11.5 Activity 2E: Problem solving using trigonometry (page 299)

1. 56.4° ***(A)***

2. 1.33 m ***(A)***

3. 508 m ***(M)***

4. **a.** 176 m ***(M)*** **b.** 19.4° ***(M)***

5. 2.16 m ***(A)***

6. 119 m ***(M)***

7. **a.** 10.4 m ***(M)***

b.
- Measurement error.
- Mistake in calculation when working out height.
- Equipment used not suitable for accurate measurement. ***(M)***

8. **a.** 5.0 m ***(M)*** **b.** 9.05 m ***(M)***

9. 49.3° ***(E)*** **10.** 139.5° ***(E)***

Unit 11.5 Activity 3A: 3-D trigonometry (page 304)

1. 1.27 m ***(M)***

2. **a.** ∠AHE **b.** ∠CEG **c.** ∠HBD **d.** ∠DFC

3. **a.** 6.7 ***(M)*** **b.** ∠CDF = 26.6° ***(M)***

4. **a.** 20 ***(M)*** **b.** 21.54 ***(M)*** **c.** ∠CEG = 21.8° ***(M)***

5. **a.** 56.3° ***(M)*** **b.** 16.97 m or 1 697 cm ***(M)***

6. 5.7 cm ***(E)***

7. 6.9 cm ***(M)***

8. 29.8° ***(E)***

Unit 11.5 Activity 3B: Further 3-D trigonometry (page 307)

1. **a.** ∠AHE or ∠BGF **b.** ∠ADE or ∠BCF **c.** ∠DEH or ∠CFG **d.** ∠ACB or ∠EGF

2. 56.3° ***(M)***

3. 90° ***(M)***

4. ∠BCA = 32.0° ***(M)***

5. ∠EXH = 71.6° ***(M)***

6. 52.7° ***(E)***

7. 15 m (other alternative was 17.46 m) ***(E)***

8. **a.** 20.83 m (to nearest cm) ***(E)*** **b.** 15.89 m along roof/wall line from end at T. ***(E)***

Unit 11.5 Activity 4A: The cosine rule and applications (page 310)

1. a. 5.0 **(A)** **b.** 60° **(A)**

2. a. 12.4 **(A)** **b.** 51.0° **(A)**

3. a. 6.0 **(A)** **b.** 60.5° **(A)**

4. a. 6.5 **(A)** **b.** 132.5° **(A)**

5. a. 78.54° **(A)** **b.** 1.9 m **(A)**

6. a. 328 m **(A)** **b.** 79° **(A)**

Unit 11.5 Activity 4B: The sine and area rules and applications (page 312)

1. a. 7.3 **b.** 9.7 **c.** 25 **(A)**

2. a. 6.0 **b.** 8.7 **(A)**

3. a. 22.4 **b.** 14.4 **c.** 233.2 **(A)**

4. a. 187.86 km **b.** E48.8°N **(A)**

5. 162 cm **(A)**

6. a. 3.5 cm **b.** 7.3 cm^2 **c.** 374 cm^3 **(A)**

7. a. \$1 114 **b.** 1 282.6 m^2 **(M)**

8. a. 5.27 km **b.** 5.44 km **(M)**

9. a. W46.2°N **b.** 15.07 km **(M)**

10. $\frac{2}{\sqrt{3}}$ **(M)**

11. a. 53.1° **b.** 199.2 m **c.** 2 220 m^2 **(M)**

12. 5.2 km **(M)**

Unit 11.5 Activity 5A: Bearings (page 318)

1. a. 325° **b.** 117° **c.** 112.5° **d.** 252°

2. a. 222° **b.** 299° **c.** 068° **d.** 128°

3. a. 075° **b.** 110° **c.** 170° **d.** 255°

e. 290° **f.** 350°

4. a.

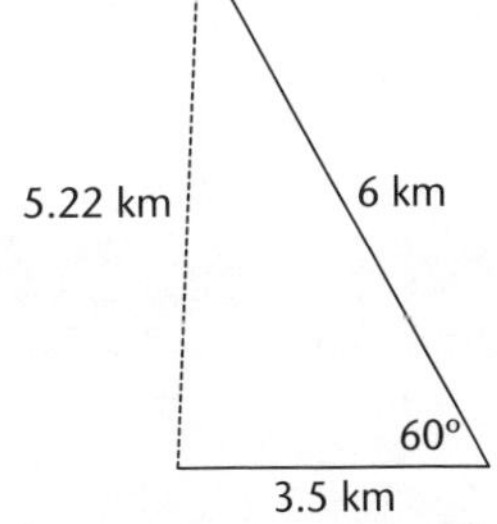

b.

town C
11.03 km
35°
30°
town A
20 km
town B

c.

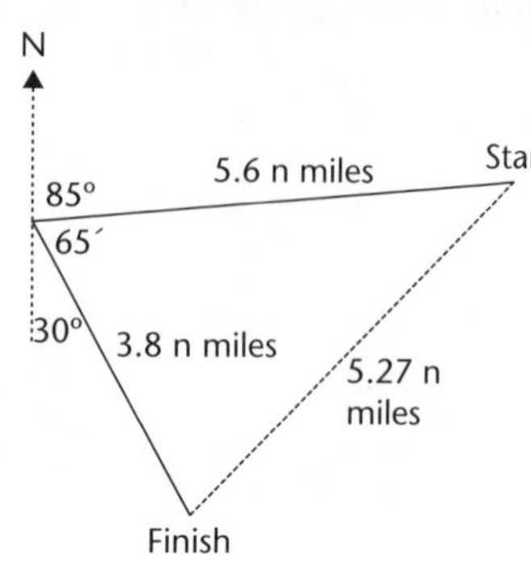

d. Bearing 170°

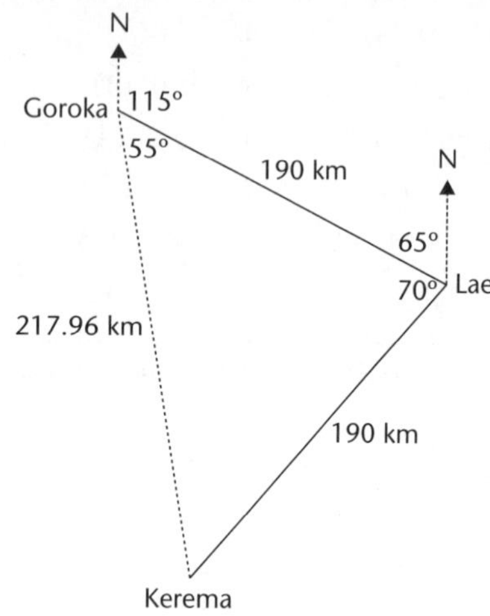

5. From town P the village Q is 23 kilometres on a bearing of 325° and the mine is 18 km on a bearing of 035°

6. a.

Resting point

N

5 km

40°

320°

N

35°

4 km

Starting point

b. 7.166 km

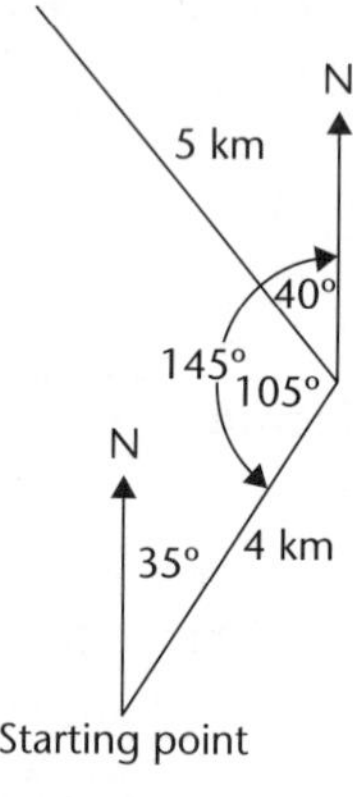

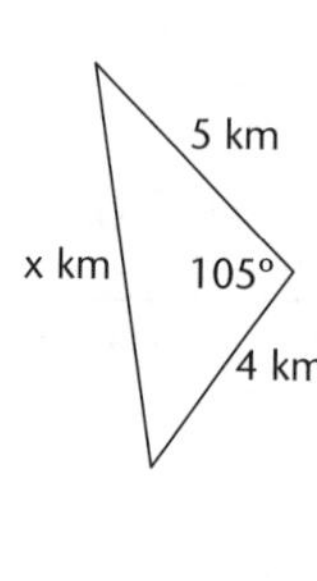

c. 7.2 km **d.** 173°

Unit 11.5 Activity 5B: Contour maps (page 322)

1. **a.** CD **b.** AB **c.** GH **d.** FE

2. **a.**

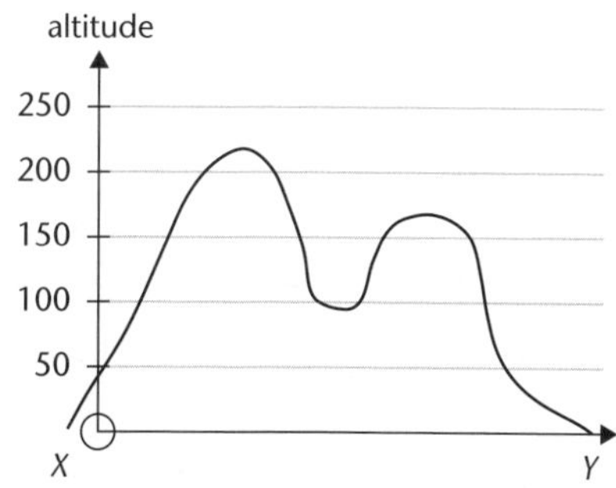

b. **i.** $\frac{150}{2000} = 0.075$ **ii.** $\frac{50}{800} = 0.0625$

c. **i.** 2005.6 m **ii.** 801.6 m

3. **a.** **i.**

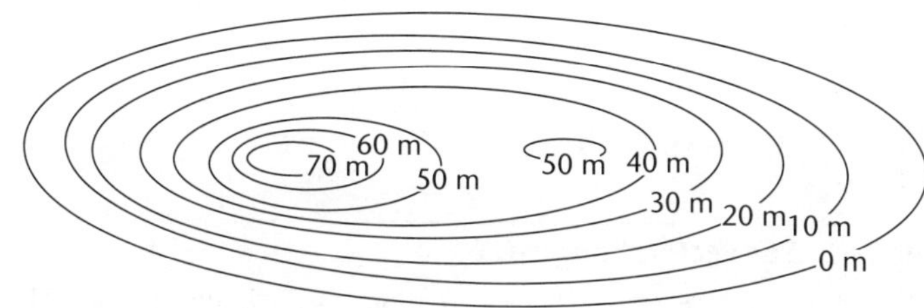

ii.

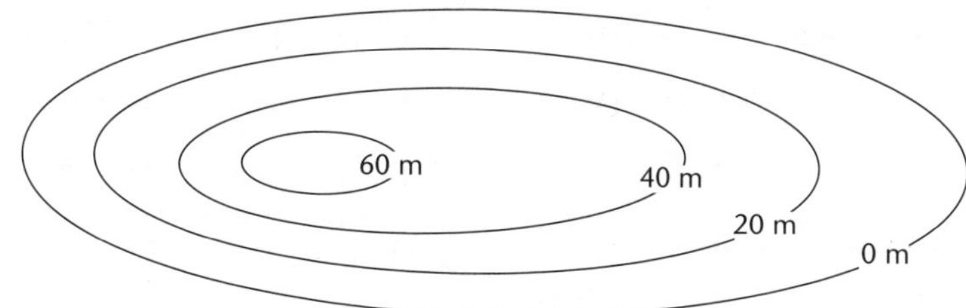

b. There is more detail in the contour map with the 10 metre contour interval.

c. i. 0.12 **ii.** $\frac{4}{75} \approx 0.05$

4. a.

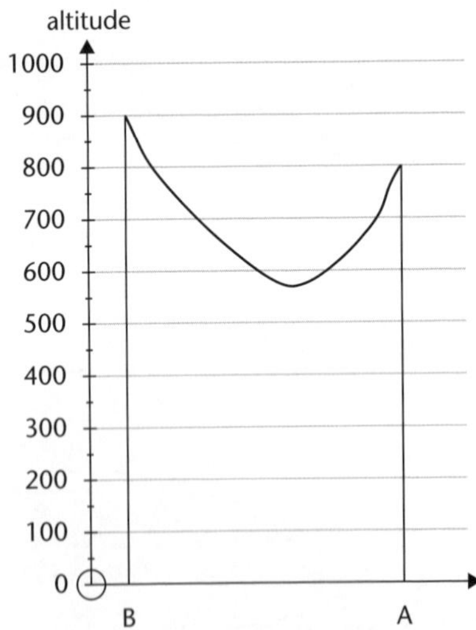

b. i. $\frac{10}{33}$ **ii.** 345 m

Unit 11.5 Activity 5C: Vectors (page 328)

1. a. $|\underset{\sim}{a}| = 5.4$, direction N68.2°E **(M)**

b. $|\underset{\sim}{b}| = 5$, direction W **(M)**

c. $|\underset{\sim}{c}| = 4.2$, direction NW **(M)**

d. $|\underset{\sim}{d}| = 5$, direction N **(M)**

e. $|\underset{\sim}{e}| = 4.5$, direction S26.6°E **(M)**

f. $|\underset{\sim}{f}| = 5.1$, direction S78.7°E **(M)**

g. $|\underset{\sim}{g}| = 5.1$, direction S78.7°W **(M)**

h. $|\underset{\sim}{h}| = 5.4$, direction N68.2°W **(M)**

i. $|\underset{\sim}{i}| = 3$, direction E **(M)**

j. $|\underset{\sim}{j}| = 6.1$, direction N9.5°W **(M)**

k. $|\underset{\sim}{k}| = 4$, direction S **(M)**

2.

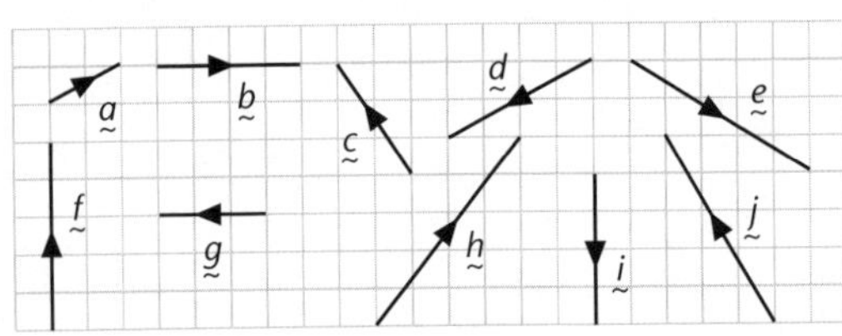

a. $|\underset{\sim}{a}| = 2.2$, direction N63.4°E **(M)**

b. $|\underset{\sim}{b}| = 4$, direction E **(M)**

c. $|\underset{\sim}{c}| = 3.6$, direction N33.7°W **(M)**

d. $|\underset{\sim}{d}| = 4.5$, direction S63.4°W **(M)**

e. $|\underset{\sim}{e}| = 5.8$, direction S59.0°E **(M)**

f. $|\underset{\sim}{f}| = 5$, direction N **(M)**

g. $|\underset{\sim}{g}| = 3$, direction W **(M)**

h. $|\underset{\sim}{h}| = 6.4$, direction N38.7°E **(M)**

i. $|\underset{\sim}{i}| = 4$, direction S **(M)**

j. $|\underset{\sim}{j}| = 5.8$, direction N31.0°W **(M)**

3. a. $2\underset{\sim}{a}$ is 8 units long in direction East.

b. $-\underset{\sim}{b}$ is 3 units long in direction South.

c. $\underset{\sim}{a} + \underset{\sim}{b}$ is 5 units long in direction N53.1°E. **(M)**

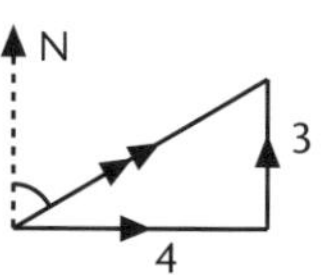

4. a. $\begin{pmatrix} 8.19 \\ 5.74 \end{pmatrix}$ **(M)** **b.** $\begin{pmatrix} 11.74 \\ -22.07 \end{pmatrix}$ **(M)** **c.** $\begin{pmatrix} -334.57 \\ -371.57 \end{pmatrix}$ **(M)**

Unit 11.5 Activity 5D: Problem solving using vectors (page 329)

1. a. 939.7 units **(M)** **b.** 342.0 units **(M)** **2.** 346.4 km **(M)**

3. a.

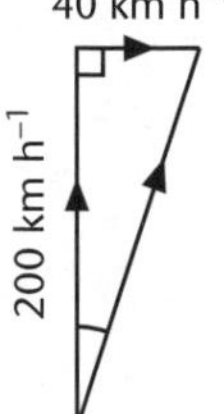

b. 204 km h^{-1} **(M)** **c.** 011° **(M)**

4. 1 931.9 units **(M)** **5.** 17.3 km **(M)**

6.

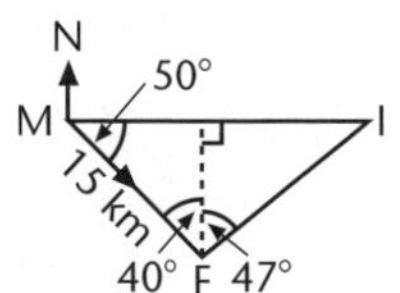

Second journey took 40.4 minutes.

The island is 22.0 km from the mooring. **(E)**

Glossary/Index

12-hour clock (70): Instrument for recording the passing of time in hours, minutes and seconds from midnight or noon. Morning times use am and afternoon times use pm.

24-hour clock (70): Instrument for recording the passing of time in hours, minutes and seconds from midnight, eg 1430 is 30 minutes after 1400 hours, ie half-past two in the afternoon.

acute angle (227): An angle between 0° and 90°.

acute-angled triangle (237): A *triangle* in which all angles are *acute*.

adjacent side (291): The side of a triangle, not the hypotenuse, which is next to a particular angle.

algebra (91–116): The use of numbers and *variables* to solve problems.

algebraic expression (91, 92): A collection of *algebraic* (and maybe *numerical*) *terms*, linked by + and –.

algebraic fractions (95): Fractions containing *variables*. Also called *rational expressions*.

algebraic term (91): An algebraic term consists of a combination of a *coefficient*, *variable*, and *exponent*. Addition and subtraction **(93)**; multiplication and division **(94)**.

allied angles (232): See *co-interior angles*.

alternate angles (231): The angles between parallel lines and on different sides of the *transversal*.

altitude (262): A line from the *vertex* of a triangle which meets the opposite side at *right angles*.

analysis (187): The process of deriving information from *data* in a *statistical investigation*.

angle (∠ or **^**) **(263):** A turning, usually measured in degrees. Construction of angles **(264)**. Types of angle **(227)**. Angle rules **(227–228)**.

angle between a line and plane (302): The *angle* between a line and its *projection* on the plane.

angle between two planes (305): The largest *angle* between two *planes*.

angle bisector (263): A line which divides an *angle* into two equal parts.

angle of depression (299): Angle from the horizontal *down* to an object.

angle of elevation (299): Angle from the horizontal *up* to an object.

approximate answer (32): A less accurate answer which is close in value to the actual answer.

arc (275): Part of the circumference of a circle. In constructions **(260–265)**.

area rule (311): A relationship used to find the area of any triangle given the length of two sides and the angle between them.

arithmetic mean (194): See *mean*.

arrowhead (239): A *quadrilateral* with two pairs of equal adjacent sides and an interior *reflex angle* between one pair.

average (194): A measure of the central or middle value of a *data* set, eg the *median*, *mean* and *mode*. Comparing averages **(197)**.

axis of symmetry (117): A line about which an object may be folded so that one half sits exactly on the other half. Also called a *mirror line*.

back-to-back stem-and-leaf plots (214): *Stem-and-leaf plots* which share the stem in order to compare two sets of *univariate data*.

bar graph (204): A data display using bars to represent the frequency of the data. Also called a *column graph*.

base (47): The number or letter being raised to a power, eg in 5^4, the base is 5.

base (236): The unequal side of an *isosceles triangle*.

base angles (236): The equal angles on the *base* of an *isosceles triangle*.

base changing rule (62): Used to change logarithms from one base to another.

basic unit (67): The fundamental units (eg metre) from which other units (eg centimetre and millimetre) are derived.

bearing (316): An *angle* measured clockwise from North, expressed using three digits, used for describing direction.

BEMA (or **BEDMAS) (5, 92):** A mnemonic used to indicate the order in which the operations are done.

biased (190): Unrepresentative.

bivariate data (187): Data which have two variables.

box-and-whisker plots (214): Data displays showing five key features of the data distribution.

brackets (**( )** or **[ ]**) **(92, 99):** Punctuation used to establish the order of basic operations.

calculators, use of with: indices **(54)**.

cancelling (16): Simplification of a fraction by dividing the numerator and denominator by the same factor.

capacity (69): Amount of liquid a container holds, measured in *litres*, *millilitres*, etc.

Celsius (69): See *degrees Celsius*.

census (188): Investigation of an entire *population*.

centigrade (69): See *degrees Celsius*.

centimetre (**cm**) **(69):** A smaller unit of length than the metre. 100 cm = 1 m.

centre (275): The middle of a *circle*.

chord (276): A straight line joining any two points on the circumference of a circle.

circle (275–280): A set of points equidistant from a fixed point (its *centre*). Angles in a circle rules **(276–277)**.

circumference (275): The *perimeter* of a *circle*.

class interval (192): Range of values for a group in a table of *grouped data*.

cluster sampling (189): A sampling technique which collects data from a representative subgroup (cluster) of the *population*.

coefficient (91, 104): A number which multiplies the *variable*.

cointerior angles (232): The angles between parallel lines and on the same side of the *transversal*. Also called *allied angles*.

column graph (204): See *bar graph*.

comparing fractions (16): Ordering fractions by size.

compass points (315): Method of describing direction based on the directions North, South, East and West.

components (327): Two vectors which add to give the original vector.

composite bar graph (205): Statistical graph used for comparing data sets.

concertinaing (205): The broken line on the horizontal axis of a statistical graph to indicate that there is nothing plotted between zero and the first value.

conclusions (189): Opinions arrived at and justified by statistical analysis.

concurrent (263): Concurrent lines pass through the same point.

concyclic points (281): Points which all lie on the circumference of the same circle.

constant (91, 326): A fixed amount which does not vary.

construction (259): An accurate drawing, using compass, protractor etc.

continuous data (187, 205): Data obtained by measuring (no gaps between possible values).

conversion (68): Changing from one unit to another in the metric system.

corresponding angles (230): The angles on the same side of the *transversal* and on the corresponding side of each parallel line.

cosine (cos) (291): A *trigonometric ratio*. Relative to a particular angle it is the ratio $\frac{\text{adjacent side}}{\text{hypotenuse}}$.

cosine rule (309): A relationship between three sides and one angle of a triangle.

counting numbers (5): The numbers 1, 2, 3, 4, . . . Also called the *natural numbers* (N).

cube numbers (12): The cube of a number such as 5 is $5^3 = 5 \times 5 \times 5 = 125$.

cube root (13): A number which when cubed gives the number under the $\sqrt[3]{\ }$ sign.

cyclic quadrilateral (280–281): A quadrilateral with all four corners on the *circumference* of a circle.

data (187): Pieces of information; collected by observation, interviews or questionnaires.

decagon (235): A 10-sided *polygon*.

decimal (27): A number which has a *decimal point*. Percentages and decimals **(143)**.

decimal places (31): The number of *digits* after the decimal point.

decimal point (27): A full stop used to separate the whole part from the fractional part in a *decimal*.

degrees (227): Unit used to measure angles. 1 degree (1°) is $\frac{1}{360}$ revolution

degrees Celsius (°C) (69): A unit for *temperature*. Water freezes at 0°C and boils at 100°C. Also called *degrees centigrade*.

denominator (15, 27): The integer below the horizontal line in a *fraction*.

diagonal (235): A line segment joining two vertices of a polygon and which is not a side.

dial (72): A circular *scale* used in measurement.

diameter (275): A straight line passing through the *centre* of a circle and joining two points on the *circumference*.

difference of two squares (102): *Quadratic expression* of the form $a^2 - b^2$. Factorising **(106)**.

digits (28): The ten symbols, 0, 1, 2, 3, 4, 5, 6, 7, 8, 9, used in numbers.

direction (325, 326): Line or course of a *vector's* movement.

discrete data (187): Data obtained by counting (possible values can be listed).

distribution (192): The spread of data values between their lowest and highest values.

divisors (68): Numbers used to divide by.

dodecagon (235): A 12-sided *polygon*.

elimination method (132): A method for solving simultaneous equations where the two equations are added to or subtracted from each other, in order to create an equation in only one variable.

equilateral triangle (236): A *triangle* in which all three sides are of equal length.

equivalent fraction (15): A different way of representing the same *fraction*.

estimate (32): An approximate value. A number *rounded* to fewer *significant figures* than it originally had.

expanding (99, 105): Removing *brackets* from an expression using the rule $a(b + c) = ab + ac$.

exponent (47, 91): A superscript (raised) number or variable used to indicate repeated multiplication of a factor. Also called a *power* or *index*

exponential function (59): A function of the form $y = a^x$, whose inverse is the logarithmic function $y = \log_a x$.

exterior angles (235, 241): The outside angles of a polygon formed by extending the sides of a *polygon*.

factor (103): Any *natural number* (or *algebraic expression*) which divides into another number or expression without a remainder.

factorising (103): Writing an *expression* as a product of its *factors*.

FOIL (101): A mnemonic to indicate the order in which to expand a pair of brackets.

fraction (15): A part of a whole, often represented by an *integer* divided by another non-zero *integer*. Addition and subtraction **(18, 19)**. Multiplication and division **(20–22)**. Algebraic fractions **(95)**.

frequency (192): The number of times a score occurs.

frequency table (192): A table which lists data values and their frequencies. (Also called a *frequency distribution table* or a *frequency distribution*.)

full turn (227): A 360° angle.

gram (g) (69): The base unit of *mass* in the *metric system*. 1 000 g = 1 kg.

graph (189): Visual display of information, often involving axes and a grid.

grouped data (192): *Data* values that are put into groups or classes.

GST (Goods and Services Tax) (155, 165): A tax of 10% added to all goods and services. Finding the GST in a price **(166)**.

hexagon (235): A 6-sided *polygon*.

highest common factor (**HCF**) **(103):** The largest *factor* shared by two or more numbers or expressions.

histogram (205): Display of continuous data with bars which touch. The area of each bar is *proportional* to the *frequency* of the interval on which it stands.

horizontal component (325): The amount a *vector* shifts parallel to the *x*-axis.

horizontal line (259): Line parallel to the horizon.

hypotenuse (285, 291): The side opposite the right angle (the longest side) of a right-angled *triangle*.

hypothesis (187): An idea or claim to be tested.

improper fraction (17): A fraction where the *size* of the *numerator* is larger than the *size* of the *denominator*.

incentre (263): The centre of an incircle where the angle bisectors meet.

index, indices (47): An alternative expression for exponent. Working with indices **(47–56)**.

inequality (75): One of the four *relations*, $<$, $>$, $\leq$ or $\geq$.

integers (**I**) **(5, 7–9):** The set of *whole numbers* and their opposites: …, –3, –2, –1, 0, 1, 2, 3, …

interior angles (235, 241): The angles inside a *polygon* and between adjacent sides.

interpret (221): Make sense of. Understand the implications of.

interquartile range (198): The difference between the *upper* and *lower quartiles* of a data set.

inverse function (60): The function which reverses the effect of another function.

inverted (118): Turned upside down.

irrational number (2, 39): A number which cannot be written as the quotient of two integers.

irrational numbers (5): Numbers which are not *rational*, ie which cannot be written as a *fraction*.

isosceles trapezium (239): A *trapezium* in which the non-parallel sides are equal in length.

isosceles triangle (236): A *triangle* which has two equal sides and two equal *base angles*.

kilogram (**kg**) **(69):** A larger unit for *mass* than the *gram*. 1 kg = 1 000 g.

kilometre (km) (69): A longer unit of length than the metre. 1 km = 1 000 m.

kina (K) (69): A unit for money. K1 = 100 toea.

kite (239): A *quadrilateral* with two pairs of equal adjacent sides.

leaf (213): See *stem-and-leaf plots*.

length (69): The distance from one end of an object to the other. The base unit is the *metre*.

like terms (93): Terms that have the same *variable*(s) and *exponents*.

limits of accuracy (74): Range of values within which the true value of a measurement falls.

line segment (260): Part of a line (with two end points). Also called an interval.

litre (L) (69): A unit for the *capacity* (or *volume*) of an object. Related to volume by the rule 1 L = 1 000 cm^3.

logarithm (log) (59–66): The inverse of $y = a^x$ is $y = \log_a x$, a function with domain ($x > 0$).

lottery method (189): Method of selecting a *random sample*, eg by selecting names from a hat.

lower limit (75): The smallest value possible for a particular measurement.

lower quartile (198): The *median* of the lower half of values in a set of data.

magnitude (325, 326): Size. The length of a *vector*.

major arc (275): Arc of length more than a semi-circle.

map (81): A representation of a geographical area, drawn to scale often with grid lines for reference.

mass (69): The amount of material in an object. Often referred to as weight.

matrix (plural **matrices**) **(325):** An array of numbers in rows and columns enclosed in *brackets*.

maximum (118): The *y* value for a function that has a peak at its turning point.

mean (189, 195): Central value in data, often called the *average*. The mean is found by adding all the scores and then dividing by the number of scores.

measurement (67): The process of assigning a numerical value to properties of an object, such as mass.

measures of central tendency (194): See *average*.

measures of spread (194): A measure of spread describes how data are distributed about a central value, eg the *range* and the *interquartile range*.

median (260): A line joining a *vertex* of a triangle to the midpoint of the opposite side. The central (or middle) value found by *ranking* data and selecting the middle number **(189, 194)**.

mediator (239, 280): See *perpendicular bisector*.

memory (6): Function of calculator used for storing the answer to a calculation.

metre (m) (69): The base unit for *length* in the metric system.

metric system (67): A base 10 system of measurement.

milligram (mg) (69): A unit of *mass* smaller than the *gram*. 1 000 mg = 1 g.

millilitre (mL) (69): A unit of *capacity* (or *volume*) smaller than the *litre*. 1 000 mL = 1 L.

millimetre (mm) (69): A unit of *length* smaller than the centimetre and metre. 1 000 mm = 1 m, 10 mm = 1 cm.

minimum (118): The *y* value for a function that has a trough at its turning point.

minor arc (275): *Arc* of length less than a semi-circle.

mixed number (17): An *integer* and *fraction* written together.

modal class (196): The class in a grouped frequency table which has the highest frequency.

mode (189, 194): The most frequently occurring number in a set of data.

money (69): Currency used to exchange goods and services. The basic unit is the *kina*.

multipliers (68): Numbers used to multiply by.

natural numbers (**N**) **(5):** The set of counting numbers: 1, 2, 3, . . .

Normal table (appendix): A table giving the areas under a standardised Normal probability curve.

***n*th root (13):** The inverse of raising a number to the *n*th power. Raising the *n*th root (of a number) to the *n*th power results in the original number.

numerator (15, 27): The integer above the horizontal line in a *fraction*.

numerical expression (91): Numbers combined using operations such as +, –, × etc.

obtuse angle (227): An angle between 90° and 180°.

obtuse-angled triangle (237): A triangle which has one obtuse angle.

octagon (235): An 8-sided *polygon*.

operation (92): A process which changes one or more elements to other elements, eg addition.

opposite (326): The opposite of a vector is a vector with the same *magnitude* but opposite direction.

opposite side (291): The side of the triangle which is directly opposite a particular angle.

ordered stem-and-leaf plot (213): *Stem-and-leaf plot* in which leaves are arranged in order of size.

original quantity (147): Amount *before* an increase or decrease.

outliers (197): Data values that do not fit a general pattern, eg values that are extra large or small.

parabola (117): The curve which is the graph of a *quadratic equation*.

parallel lines (230, 264): Two lines are parallel if they are always the same distance apart. Construction of a line parallel to another line **(264)**.

parallelogram (239): A *quadrilateral* in which the opposite sides are parallel to each other.

pentagon (235): A polygon with 5 sides.

percentage (%) **(143):** A percentage means 'out of every 100'.

perfect square (102): A *square number* such as 25 or an algebraic expression such as $(a + b)^2$, formed by multiplying a factor by itself.

perpendicular bisector (260): A line that intersects a given line at right angles and cuts the given line into two equal lengths. Also called *mediator*.

perpendicular line (259): A line at *right angles* to a given line. Construction of the perpendicular at a point on a line **(261)**. Construction of a perpendicular to a line from a point **(262)**.

pie graph (203): Circular statistical graph in which sectors show data distribution. Also called pie chart.

place holder (29): A zero used to fill a position in a number.

plane (302, 305): A flat (2-D) surface.

polygon (235): A closed figure with many straight sides.

population (188): The entire group of interest in a *statistical investigation*.

power (47): An alternative expression for exponent.

prefix (67): Part words, such as kilo, which precede other words, such as metre, to change their meaning, eg a kilometre is 1 000 metres. The prefixes milli and centi are also commonly used in the metric system.

product (103): The result of multiplying two or more terms together.

pronumeral (91): A letter or symbol representing an unknown quantity or number.

proper fraction (16): A fraction where the *numerator* is smaller than the denominator.

proportion (189): A comparison of one quantity with another (often a part with a whole). Sample proportions **(197, 203)**.

proportional (204): Of equal proportions; in the same ratio.

Pythagoras' theorem (285–290, 306): A relationship ($a^2 + b^2 = c^2$) involving the three sides of a *right-angled triangle*. 3D trigonometry and Pythagoras **(301)**.

quadratic equation (111, 113): An equation which can be expressed in the form $ax^2 + bx + c = 0$.

quadratic expressions (104): *Expressions* of the form $ax^2 + bx^2 + c$. Factorising **(102)**.

quadratic function (117): A function with equation of the form $y = ax^2 + bx + c$.

quadrilateral (235, 238–239): A 4-sided *polygon*. Types of quadrilateral **(239)**.

qualitative data (187): Data described by names or words.

quantitative data (187): Numerical data collected by measuring or counting.

quartiles (189): See *lower quartile* and *upper quartile*.

radius (plural radii) (275): A straight line from the *centre* to a point on the *circumference* of a *circle* or *sphere*.

random sample (189): A sample in which every member of the *population* had an equal chance of being selected.

range (189, 198): The difference between the highest and lowest scores in a set of data.

ranked (194): Ranked data are listed in order of size, eg from lowest value to highest value.

ratio (24): A comparison of two or more like quantities (measured using the same unit). Sharing quantities in a given ratio **(25)**.

rational expressions (107): See *algebraic fractions*.

rational numbers (5): Numbers which can be written as a *fraction* (an integer over an integer).

rational powers (52): An index which is a rational number.

raw data (188, 192): Initial, unsorted *data*.

real numbers (R) (5): The set of all possible numbers on the number line.

rearranging (112): Using the *basic operations* to write an expression in another way, or to make another *variable* the subject.

reciprocal (97): The multiplicative inverse. The reciprocal of x is $\frac{1}{x}$. The reciprocal of $\frac{a}{b}$ is $\frac{b}{a}$.

reciprocal powers (55): Used for solving equations with indices.

rectangle (239): A *quadrilateral* in which all the *interior angles* are right angles.

recurring decimal (28): A decimal with the same number (or sequence of numbers) repeated infinitely. Also called a repeating decimal.

reflex angle (227): An angle between 180° and 360°.

regular (235, 242): A regular *polygon* has the lengths of all sides equal and the sizes of all *interior angles* equal. A regular polyhedron has all faces made up of the same regular polygons.

repeating decimal (28): See *recurring decimal*.

representative (189): A representative (*unbiased*) sample has features similar to the population from which it was selected.

resolving (327): Resolving a *vector* means to find its *components*.

resultant (326): The vector which results after vectors have been added or subtracted together.

rhombus (239): A *parallelogram* in which all the sides are equal.

right angle (∟) (227): An angle of 90°.

right-angled triangle (237, 285): A *triangle* which has a *right angle*.

rounding (31, 32): The writing of a number with fewer significant figures.

sample (188): Part of a *population*.

scalar (325): A quantity that has only magnitude and no direction. Also called a *constant*.

scale (72): A line of values at set intervals marked on a measuring device.

scalene triangle (236): A *triangle* which has no equal sides and no equal angles.

sector (276): A fraction of a whole *circle* formed by two radii and the arc in between.

segment (276): The region of a *circle* enclosed by a chord and an *arc*.

semi-circle (275): A shape enclosed by a diameter of a circle and half the *circumference* of the circle.

sign (+ or –) (7): The positive or negative nature of a number.

significant figure (sf) (30, 74): Any digit in a number which is not a zero at the beginning of a decimal or a zero at the end of a whole number. The number of significant figures shows how accurate a number is.

simplest form (16, 79): *Simplified* as far as possible.

simplify (79, 99): To write an expression in a briefer form (by combining like terms, etc). Simplifying fractions **(16)**. Simplifying algebraic fractions **(95, 107)**.

simultaneous equations (131): Two or more equations which are true at the same time.

sine (sin) (291): A *trigonometric ratio*. Relative to a particular angle it is the ratio $\frac{\text{opposite side}}{\text{hypotenuse}}$.

sine rule (311): The relationship between one side of a triangle and its opposite angle, and a second side and its opposite angle.

size (7): The distance (positive) of a number from 0 on the number line.

sketch (259): A freehand drawing showing the main features of an object.

SOH CAH TOA (292): Mnemonic relating the *trigonometric ratios* to the sides of a *right-angled triangle*.

solving: Finding the value of the *variable* which makes an *equation* true. Simultaneous equations **(131)**.

spread (198): A description of the *distribution* of data values between their highest and lowest values.

square (239): A *quadrilateral* in which all four sides are equal and meet at *right angles*.

square centimetre (cm^2): Unit of measurement of *area* in the *metric system*. $10\ 000\ cm^2 = 1\ m^2$.

square kilometre (km^2): Unit of measurement of *area* in the *metric system*. $1\ km^2 = 1\ 000\ 000\ m^2$.

square metre (m^2): Base unit of measurement of *area* in the *metric system*.

square millimetre (mm^2): Unit of measurement of *area* in the *metric system*. $100\ mm^2 = 1\ cm^2$.

square numbers (11): The square of a number is the number multiplied by itself, eg $4^2 = 4 \times 4 = 16$.

square root ($\sqrt{\ }$) (11): A number which, when *squared*, gives the original number. The symbol $\sqrt{\ }$ indicates the positive square root only.

standard form (34, 36): A number between 1 and 10, multiplied by a power of 10. Used to write and use large or small numbers. Standard form and calculators **(35)**.

statistical graphs (203): Data displays. See *graph*.

statistical investigation (187): A sequence of activities involving data collection and analysis in order to reach conclusions about a question or hypothesis.

statistics (187): The collection, display and analysis of data.

stem (213): See *stem-and-leaf plots*.

stem-and-leaf plots (213): Data display which splits data into a *stem* (the first significant figure(s) of a data value) and a *leaf* (last significant figure of a data value).

straight angle (227): An angle on a straight line, ie 180° or a half turn.

stratified sampling (189): *Proportional* sampling method.

substitution (112): Replacing a *variable* with a number or with another equivalent expression.

substitution method (132): A method for solving simultaneous equations. One unknown is the subject of one equation and is then substituted in the other equation.

subtend (276): Lies under (holds up).

supplementary (281): Adding to 180°.

systematic sampling (189): Sample taken in a systematic manner, eg every 10th name on a list.

table (189, 221): Information arranged in columns and rows.

tally chart (192): A simple recording system for the number of times an outcome occurs.

tangent (tan) (291): A trigonometric ratio. Relative to a particular angle it is the ratio $\frac{\text{opposite side}}{\text{adjacent side}}$.

tangent to a circle (275): A straight line that touches a circle at one point on the circumference. Tangent to a circle rules **(279–280)**.

temperature (69): How hot or how cold an object is. The unit used is *degrees Celsius* or *centigrade*.

term (91, 99): See *algebraic term*.

time (69): How long an event takes to occur. The basic unit is the second (s).

toea (t) (69): A unit of *money*. 100t = K1.

tonne (69): A unit of *mass*. 1 000 kg = 1 tonne.

transformed (119): Changed in position or size.

translation (119): A *transformation* which changes the position of an object by sliding.

transversal (230): A line which crosses a pair of lines.

trapezium (239): A *quadrilateral* that has one pair of parallel sides.

triangle (235–237): A 3-sided *polygon*.

triangles: Convention for labelling of **(309)**.

trigonometric ratios (291, 292): Relationships between side lengths and angle sizes in a *right-angled triangle*. The three basic trigonometric ratios are *sine, cosine* and *tangent*.

turning point (117): A point on a graph where it reaches a *maximum* or *minimum*. See also *vertex*.

unbiased data (189): Data which are *representative* of the *population* from which they are sampled.

ungrouped data (192, 195): Data for which each possible value is listed individually.

univariate data (187): Data with only one variable.

unlike terms (93): Terms which do not have the same *variable*(s) and *exponents*.

upper limit (75): The greatest possible value for a particular measurement.

upper quartile (198): The *median* of the upper half of values in a set of data.

variable (91): A letter or symbol representing an unknown number. Statistical variables **(187, 188)**.

vector (325–330): A quantity with both *magnitude* and *direction* which is represented by an arrowed line. Equal and opposite vectors **(326)**.

vector diagram (325): A diagram showing vectors.

vertex (plural **vertices**) **(117):** The *turning point* of a parabola. The junction where two sides of a polygon *intersect* **(235)**.

vertical component (325): The amount a *vector* shifts parallel to the y-axis.

vertical line (259): Line at *right angles* to a *horizontal line*.

vertically opposite angles (228): Equal angles that are directly opposite each other when two lines intersect.

whole numbers (**W**) **(5):** The set of numbers 0, 1, 2, 3, 4, 5, 6, …

word problem (21, 114, 135, 148): A description of a mathematical problem in words.

ZFS (113): A mnemonic for the strategy required to solve quadratic equations (zero, factorise, solve).